THE

THEORY OF SOUND

BY

JOHN WILLIAM STRUTT, BARON RAYLEIGH, Sc.D., F.R.S.

HONORARY FELLOW OF TRINITY COLLEGE, CAMBRIDGE

WITH A HISTORICAL INTRODUCTION BY

ROBERT BRUCE LINDSAY

HAZARD PROFESSOR OF PHYSICS IN BROWN UNIVERSITY

IN TWO VOLUMES

VOLUME I

SECOND EDITION REVISED AND ENLARGED

NEW YORK

DOVER PUBLICATIONS

Published in Canada by General Publishing Com-
pany, Ltd., 30 Lesmill Road, Don Mills, Toronto,
Ontario.
Published in the United Kingdom by Constable
and Company, Ltd., 10 Orange Street, London
WC 2.

This Dover edition, first published in 1945, is an
unabridged republication of the second revised and
enlarged edition of 1894. It is reprinted through
special arrangement with The Macmillan Company,
publishers of the original edition, and contains a
new Historical Introduction by Robert Bruce
Lindsay.

Standard Book Number: 486-60292-3
Library of Congress Catalog Card Number: 45-8804

Manufactured in the United States of America
Dover Publications, Inc.
180 Varick Street
New York, N. Y. 10014

HISTORICAL INTRODUCTION
BY ROBERT BRUCE LINDSAY

THE current reprinting of Lord Rayleigh's "The Theory of Sound," first published in 1877, stimulates an inquiry into the reason why such a treatise still retains a position of importance in the literature of its field, when most scientific treatises of sixty-five years ago now possess for the most part historical interest only, and have long since been superseded by twentieth century standard works. It has seemed appropriate on this occasion to review briefly the historical development of the subject of acoustics in which this situation has occurred, and to pay some tribute to the character and contributions of the author of a book which continues to show such vitality. It is hoped that the following introductory comments will enhance the pleasure of those who continue to turn to Rayleigh's treatise for enlightenment and guidance in acoustics.

I. BIOGRAPHICAL SKETCH OF JOHN WILLIAM STRUTT, THIRD BARON RAYLEIGH (1842-1919)

The author of "The Theory of Sound" occupies an unusual position in the history of British physics if only because, while there are numerous examples of men raised to the peerage as a reward for outstanding scientific work, it is rare to find a peer by inheritance devoting himself to science. Lord Rayleigh was born John William Strutt, the eldest son of the second Baron Rayleigh of Terling Place, Witham in the county of Essex. His immediate ancestors were country gentlefolk with little or no interest in scientific pursuits, though one of his grandmothers was descended from a brother of Robert Boyle. In his boyhood Rayleigh exhibited no unusual precocity but apparently displayed the average boy's interest in the world about him. His schooling was rather scattered, short stays at Eton and Harrow being terminated by ill-health. He finally spent the four years preceding college at a small boarding school kept by a Rev. Mr. Warner in Highstead, Torquay, where he showed no interest in classics but began to develop decided competence in mathematics.

In 1861 at the age of nearly 20, young Rayleigh went up to Cambridge and entered Trinity College. Here he became a pupil of

E. J. Routh, the famous "coach" in applied mathematics. It was under the guidance of Routh that he acquired the grasp of mathematics which stood him in such good stead in his later research. The system has often been criticized, but it ground the methods of advanced mathematical analysis essential for the physical scientist so thoroughly into the candidate that they became a natural part of his very being. It was not rigorous mathematics in the pure sense, but it was vigorous mathematics, which served to cultivate a keen appreciation of the particular method best suited for the solution of any particular problem. Rayleigh also stated in after life that he had profited greatly from the Cambridge lectures of Sir George G. Stokes, who though Lucasian professor of mathematics was greatly interested in experimental physics and performed many experimental demonstrations for his classes. In the Mathematical Tripos of 1865, Rayleigh came out as Senior Wrangler and also became first Smith's Prizeman. By this time he had clearly decided on a scientific career, though the propriety of this was considered by some members of his family rather doubtful in view of the social obligations inherent in his ultimate succession to his father's title and position. Rayleigh seems to have felt that such obligations should not be allowed to interfere with his scientific work. In 1866 he was elected Fellow of Trinity College, thus further emphasizing his scholarly leanings. Curiously enough he replaced the usual grand tour of the continent with a trip to the United States, then in the throes of reconstruction after the Civil War.

In 1868 immediately after his return from America Rayleigh purchased an outfit of experimental equipment. There was at that time no university physical laboratory, though certain professors possessed apparatus for their own experimental purposes and for demonstrations. Students received little or no direct encouragement to embark on experimental investigations for themselves. This may seem strange when one recalls that Cambridge had been for long the home of Newton. Moreover, long before Rayleigh's undergraduate days the immortal experiments of Young, Davy and Faraday, to mention only a few, had already shed undying lustre on British science. But this research had been carried on, by and large, outside the universities, which thus remained quite out of the current of real scientific progress in physics well past the first half of the nineteenth century. It was not until 1871 that Cambridge University established a professorship of experimental physics; in 1873 the Cavendish Laboratory was erected through the munificence of the Chancellor of the University, the eighth Duke of Devonshire. James Clerk Maxwell was elected the first Cavendish professor and served from 1871 to his untimely death in 1879. For the first time practical instruction from a distinguished physicist was provided at Cambridge.

To return to Rayleigh: it is interesting to observe that his first experimental investigations were on electricity and concerned the action of alternating currents on a galvanometer. The results were presented in a paper (his first) to the Norwich meeting of the British Association for the Advancement of Science in 1868. But he was soon thereafter deeply immersed in other things, including color vision and the pitch of resonators. The latter was his first work in acoustics and was apparently stimulated by his reading Helmholtz's famous work "On the Sensations of Tone" (1863). There was much correspondence about this and kindred matters with Maxwell, who was always eager to help along a youthful colleague. Rayleigh's experimental work was carried out at Terling in a rather crudely improvised laboratory. Later when the estate became his home by inheritance, more elaborate arrangements were made.

In 1871 Rayleigh married Evelyn Balfour, the sister of Arthur James Balfour, who was destined to gain much celebrity as a scholar, philosopher and statesman. He had become acquainted with Balfour as a fellow student at Cambridge. Shortly after his marriage a serious attack of rheumatic fever threatened for a time to cut short his career and left him much weakened in health. An excursion to Egypt was undertaken as a recuperative measure, and it was on a house boat trip up the Nile late in 1872 that the "Theory of Sound" had its genesis, the first part having been written with no access to a large library. The preparation of the treatise eventually extended over many years, and the two well-known volumes did not make their appearance from the press until 1877. In the meantime Rayleigh had succeeded to his father's title and had settled down at Terling. Changes were made to enable him to embark on more elaborate laboratory work, including experiments in acoustics and optics. It was during the period from 1871 to 1879 that he gave much attention to the diffraction of light and made copies of diffraction gratings. These investigations led to the introduction of the present standard definition of resolving power, a quantity of the utmost importance in specifying the performance of any optical instrument.

The premature death of Clerk Maxwell in 1879 left the Cavendish professorship vacant. Pressed by many scientific friends to stand for the post, Rayleigh finally consented, being partly influenced in his decision by the loss of income from his estate due to the agricultural depression of the late 70's. It does not appear that he ever contemplated retaining the professorship for an indefinite period, and indeed he ultimately limited his tenure to five years. The pedagogical duties of the Cavendish professor were not onerous: he was required to be in residence for eighteen weeks during the academic year and to deliver at least forty lectures in the course of this period. Rayleigh, however, had no desire to interpret the job as a sinecure. He embarked vigorously on a program of developing elementary laboratory instruction in a really elaborate way. It is difficult to appreciate today what a task such a program involved sixty years ago. Collegiate instruction in practical physics was almost a new thing, and there was little to go on save the teacher's imagination. Under Rayleigh's direction his demonstrators Glazebrook and Shaw, both of whom later became men of note, the former in applied physics and the latter in meteorology, developed laboratory courses for large classes in heat, electricity and magnetism, properties of matter, optics and acoustics. This was pioneer work of high order and had a beneficial influence on the teaching of physics throughout England and ultimately elsewhere.

Rayleigh was impressed at this time with the desirability of cooperative research on a problem of importance and selected for this purpose the redetermination of the standard electrical units. In particular he wished to undertake a new evaluation of the relation between the ohm, the practical unit of electrical resistance, and the electromagnetic unit of resistance. The first precision work on this problem had been carried out in 1863-64 under the auspices of the British Association with Maxwell in charge. Later work by others had disclosed considerable discrepancies. Rayleigh and his collaborators devoted three years of labor to a repetition of the original experiments with greater attention to sources of error. It is a tribute to Rayleigh's great experimental care that his final results have not been appreciably altered by more modern work. He appeared to possess the uncanny power to make the simplest of equipment produce the utmost in precision.

In December 1884 Rayleigh returned to Terling, which he made his scientific headquarters for the remainder of his life. It was close enough to London to permit frequent visits to the metropolis for the performance of official duties in connection with government or the various professional societies in which he played a prominent role. But he clearly enjoyed having his laboratory in his own home. Probably many contemporaries in the peerage, as well as the tenants on the estate, thought him a trifle queer, but he went his way with typical British imperturbability. The laboratory could hardly be considered elaborate even when judged by contemporary standards. Rayleigh had a hatred of superfluous elegance and always stressed the desirability of simplicity in all research apparatus. Some of this feeling was undoubtedly inspired by his constitutional aversion to unnecessary expenditure; there was also a profound philosophical implication in the method which may be of value to the present day investigator, even when surrounded by highly intricate and sophisticated apparatus.

The life of a scientist working at his desk or in his laboratory has little to offer in the way of the dramatic, at least to the man in the street. It is inevitable that mankind in the large should find more emotional satisfaction in the contemplation of man's relations with his fellow creatures than in his relations with the physical environment. For the most part, too, scientific investigations involve a train of reasoning unfamiliar and intricate to the general run of people. Occasionally, however, a scientific discovery will be made which involves a relatively simple and clear cut situation, while at the same time it solves a puzzle originally as baffling as any detective story mystery. This was the case with the most dramatic popular episode in Rayleigh's career, namely the discovery of the rare gas argon in the atmosphere.

Already in his address to the Mathematics and Physics Section of the British Association at the Southampton meeting in 1882, Rayleigh had called attention to the desirability of a more precise determination of the densities of the so-called permanent gases, oxygen, hydrogen and nitrogen. The importance of this lay in its bearing on the problem of the atomic weights of the elements and hence the whole foundation of chemistry. This job Rayleigh now set for himself and devoted to it a good part of his own time and that of a skilled assistant for the better part of ten years, culminating in the famous joint announcement with Sir William Ramsay of the isolation of argon in 1895. The story is too well known for detailed repetition here. It furnishes a classic example of the importance of following up a small experimental discrepancy lying outside the limit of reasonable experimental error, in this case the difference between the density of nitrogen prepared from nitrogen compounds and nitrogen obtained by removing the oxygen of the air. It seems easy to say now that the larger value of the latter points directly to the existence in the air of a small amount of a gas heavier than nitrogen. But in 1895 this was not so simple and neither was the task of isolating the new gas. It is not too much to say that the subsequent discovery of all the other rare gases of the atmosphere was directly due to Rayleigh's patient, ingenious and methodical investigation.

From 1887 to 1905, Lord Rayleigh served as Professor of Natural Philosophy at the Royal Institution of Great Britain as successor to John Tyndall, who in turn had succeeded Faraday. Unlike his predecessors Rayleigh spent comparatively little time in the laboratory of the Institution, confining his activity to the annual course of public lectures. These continued the tradition established by Faraday and Tyndall in covering the whole gamut of topics of physical interest with a profusion of experimental demonstrations. Sir Arthur Schuster says of Rayleigh in this connection: "Though not by nature a ready speaker, his lectures were effective." At any rate the auditor could always be confident that the speaker thoroughly understood what he was talking about.

In 1896 Rayleigh was appointed Scientific Adviser to Trinity House, a very ancient organization, dating back to Henry VIII, and having as its duties the erection and maintenance of such coastal installations as lighthouses, buoys and the like. For the next fifteen years he served faithfully and made numerous inspection trips. Some of his later work in optics and acoustics was suggested by problems arising in connection with the tests of lights and fog-signals. In spite of his devotion to his laboratory research, Rayleigh gave willingly of his time and energy to the deliberations of scientific committees of government and the various societies to which he belonged. Thus he was one of the leaders of the movement which led to the establishment of the National Physical Laboratory (the British counterpart of the National Bureau of Standards in Washington), and presided over the Executive Committee of the Laboratory until shortly before his death. He also served as President of the Advisory committee on Aeronautics from its inception in 1909 (at the instance of Prime Minister Asquith) until the time of his death. The activities of this committee were particularly important during the first world war from 1914-1918.

Among Lord Rayleigh's other public positions there is space only to mention his presidency of the Royal Society from 1905-1908 and his service as Chancellor of Cambridge University from 1908 until his death. Honors came to him in heaping measure, notable among them the Order of Merit, of which he was one of the first recipients in 1902, and the Nobel Prize in Physics in 1904.

Unlike most scientific men, Rayleigh was able to continue his work until his death, though he survived to the ripe old age of 76. He died on June 30, 1919, with three papers still unpublished. It is interesting that the last of these was one on acoustics: he never got over his interest in sound.

The opinion of his contemporaries and successors places Rayleigh in that great group of nineteenth century physicists that have made British science famous all over the world, the group whose other members were Kelvin, Maxwell and Stokes. His position in the history of science is a great one. It is good to recall that he was above all a modest man and it is impossible to accept as otherwise than sincere the remarks he made when he received the Order of Merit: "the only merit of which he personally was conscious was that of having pleased himself by his studies, and any results that may have been due to his researches were owing to the fact that it had been a pleasure to him to become a physicist."

II. HISTORICAL DEVELOPMENT OF ACOUSTICS TO THE TIME OF RAYLEIGH

Introduction. Sound plays in our daily lives a part scarcely less important than motion and light, and the sense of hearing though by no means esteemed so precious as the sense of sight and the ability to locomote is yet so prized that the production of efficient hearing aids for the deaf is fast becoming a major industry. Life is full of sounds and we want to hear the pleasant and vital ones, while shunning the unpleasant and dangerous variety. All told we are becoming steadily more sound conscious, as the relatively enormous growth of the telephone, radio, phonographic recording and talking motion picture industries sufficiently attests.

In view of its importance, it might be supposed that the science of sound, technically known as acoustics, would loom as a substantial item in the history of the development of physical ideas. Strangely enough, in the standard histories this is by no means the case: the history of acoustics has been largely a neglected subject. A possible reason for this has been advanced by Whewell in his "History of the Inductive Sciences." The basic theory of the origin, propagation and reception of sound was proposed at a very early stage in the development of human thought in substantially the form which we accept today: the ancient Greeks, according to the most reasonable interpretation of the records, evidently were aware that sound somehow arises from the motion of the parts of bodies, that it is transmitted by the air through some undefined motion of the latter and in this way ultimately striking the ear produces the sensation of hearing. Vague as these ideas were they were yet clarity itself compared with the ancient views on the motion of solid bodies as well as on light and heat. The latter branches of physics had to go through a long course of development in which theory succeeded ,heory until the present stage was reached. As Whewell emphasizes, in acoustics the basic theory was laid down early and all that was needed was its implementation by the necessary analysis and its application to new problems as they arose. On the theoretical side the history of acoustics thus tends to be merged in the larger development of mechanics as a whole.

It has seemed eminently worth while, however, in connection with a re-issue of the greatest single work ever published in acoustics to take advantage of the occasion to review the history of those parts of mechanics and other branches of physics which have had a definite bearing on acoustical theory. In a small measure this may serve to supplement D. C. Miller's interesting "Anecdotal History of the Science of Sound" (1935), which is devoted mainly to a resumé of the experimental phenomena.

The problems of acoustics as already indicated are most conveniently divided into three main groups, viz: 1) the production of sound, 2) the propagation of sound, and 3) the reception of sound. We shall find it advantageous to organize the following historical outline accordingly.

The Production of Sound. The fact that when a solid body is struck a sound is produced must have been observed from the very earliest times. The additional fact that under certain circumstances the sounds produced are particularly agreeable to the ear furnished the basis for the creation of music, which also originated long before the beginning of recorded history. But music was an art for centuries before its nature began to be examined in a scientific manner. It is usually assumed that the first Greek philosopher to study the origin of musical sounds was Pythagoras in the 6th century B.C. He is supposed to have discovered that of two stretched strings fastened at the ends the higher note is emitted by the shorter one, and that indeed if one has twice the length of the other, the shorter will emit a note an octave above the other. By this time the notion of pitch had, of course, been developed, but its association with the frequency of the vibrations of the sounding body was probably not understood, and it does not appear that this concept emerged until the time of Galileo Galilei (1564-1642), the founder of modern physics. At the very end of the "First Day" of Galileo's "Discourses Concerning Two New Sciences," first published in 1638, the reader will find a remarkable discussion of the vibrations of bodies. Beginning with the well known observations on the isochronism of the simple pendulum and the dependence of the frequency of vibration on the length of the suspension, Galileo goes on to describe the phenomenon of sympathetic vibrations or resonance by which the vibrations of one body can produce similar vibrations in another distant body. He reviews the common notions about the relation of the pitch of a vibrating string to its length and then expresses the opinion that the physical meaning of the relation is to be found in the number of vibrations per unit time. He says he was led to this point of view by an experiment in which he scraped a brass plate with an iron chisel and found that when a pure note of definite pitch was emitted the chisel cut the plate in a number of fine lines. When the pitch was high the lines were close together, while when the pitch was lower they were farther apart. Galileo was actually able to tune two spinet strings with two of these scraping tones; when the musical interval between the string notes was judged by the ear to be a fifth, the number of lines produced in the corresponding scrapings in the same total time interval bore precisely the ratio 3:2. The presumption is that if the octave had been tuned the ratio would have been 2:1, etc. It seems plain from a careful reading of Galileo's writings that he had a clear understanding of the dependence of the frequency of a stretched string on the length, tension and density. There was, of course, no question then of a dynamical discussion of the actual motion of the string: the theory of mechanics had not advanced far enough for that. But Galileo did make an interesting comparison between the vibrations of strings and pendulums in the endeavor to understand the reason why sounds of certain frequencies, i.e., those whose frequencies are in the ratio of two small integers, appear to the ear to combine pleasantly whereas others not possessing this property sound discordant. He observed that a set of pendulums of different lengths, set oscillating about a common axis and viewed in the original plane of their equilibrium positions present to the eye a pleasing pattern if the frequencies are simply commensurable, whereas they form a complicated jumble otherwise. This is a kinematic observation of great ingenuity and illustrates the fondness of the great Italian genius for analogy in physical description.

Credit is usually given to the Franciscan friar, Marin Mersenne (1588-1648) for the first correct published account of the vibrations of strings. This occurred in his "Harmonicorum Liber" published in Paris in 1636, two years before the appearance of Galileo's famous treatise on mechanics. However, it is now clear that Galileo's actual discovery antedated that of Mersenne. The latter did add one very important point: he actually measured the frequency of vibration of a long string and from this inferred the frequency of a shorter one of the same density and tension which gave a musical note. This was apparently the first direct determination of the frequency of a musical sound.

Though later experimenters like Robert Hooke (1635-1703) tried to connect frequency of vibration with pitch by allowing a cog wheel to run against a piece of cardboard, the most thorough-going pioneer studies of this matter were made by Joseph Sauveur (1653-1716), who incidentally first suggested the name *acoustics* for the science of sound. He employed an ingenious use of the beats between the sounds from two organ pipes which were adjudged by the ear to be a semi tone apart, i.e., having frequencies in the ratio 15/16. By experiment he found that when sounded together the pipes gave 6 beats a second. By treating this number as the difference between the frequencies of the pipes the conclusion was that these latter numbers were 90 and 96 respectively. Sauveur also worked with strings and calculated (1700) by a somewhat dubious method the frequency of a given stretched string from the measured sag of the central point. It was reserved to the English methematician Brook Taylor (1685-1731), the celebrated author of Taylor's Theorem on infinite series, to be the first to work out a strictly dynamical solution of the problem of the vibrating string. This was published in 1713 and was based on an assumed curve for the shape of the string of such a character that every point would reach the rectilinear position in the same time. From the equation of this curve and the Newtonian equation of motion he was able to derive a formula for the frequency of vibration agreeing with the experimental law of Galileo and Mersenne. Though only a special case, Taylor's treatment paved the way for the more elaborate mathematical techniques of Daniel Bernoulli (1700-1782), D'Alembert (1717-1783) and Euler (1707-1783), involving the introduction of partial derivatives and the representation of the equation of motion in the modern fashion.

In the meantime it had already been observed, notably by Wallis (1616-1703) in England as well as by Sauveur in France, that a stretched string can vibrate in parts with certain points, which Sauveur called *nodes*, at which no motion ever takes place, whereas very violent motion takes place at intermediate points called *loops*. It was soon recognized that such vibrations correspond to higher frequencies than that associated with the simple vibration of the string as a whole without nodes, and indeed that the frequencies are integral multiples of the frequency of the simple vibration. The corresponding emitted sounds were called by Sauveur the *harmonic* tones, while the sound associated with the simple vibration was named the *fundamental*. The notation thus introduced (about 1700) has survived to the present

day. Sauveur noted the additional important fact that a vibrating
string could produce the sounds corresponding to several of its harmonics at the same time. The dynamical explanation of this vibration was provided by Daniel Bernoulli in a celebrated memoir
published by the Berlin Academy in 1755. Here he showed that it is
possible for a string to vibrate in such a way that a multitude of simple harmonic oscillations are present at the same time and that each
contributes independently to the resultant vibration, the displacement
at any point of the string at any instant being the algebraic sum of
the displacements for each simple harmonic node. This is the famous
principle of the *coexistence of small oscillations*, also referred to as
the *superposition principle*. It has proved of the utmost importance
in the development of the theory of oscillations, though curiously
enough its validity was at first strenuously doubted by D'Alembert
and Euler, who saw at once that it led to the possibility of expressing
any arbitrary function, e.g., the initial shape of a vibrating string, in
terms of an infinite series of sines and cosines. The state of mathematics in the middle of the 18th century hardly permitted so bold a
result. However, in 1822 Fourier (1768-1830) in his "Analytical
Theory of Heat" did not hesitate to develop his celebrated theorem on
this type of expansion with consequences of the greatest value for
the advancement of acoustics.

The problem of the vibrating string was fully solved in elegant
analytical fashion by J. L. Lagrange (1736-1813) in an extensive
memoir of the Turin Academy in 1759. Here he supposed the string
made up of a finite number of equally spaced identical mass particles
and studied the motion of this system, establishing the existence of a
number of independent frequencies equal to the number of particles.
When he passed to the limit and allowed the number of particles to
become infinitely great and the mass of each correspondingly small,
these frequencies were found to be precisely the harmonic frequencies
of the stretched string. The method of Lagrange was adopted by
Rayleigh in his "Theory of Sound" and is indeed standard practise
to-day, though most elementary books now develop the differential
equation of motion of the string treated as a continuous medium by
the method first set forth by D'Alembert in a memoir of the Berlin
Academy of 1750. This differential equation we now call the wave
equation, though the savants of the middle 18th century did not stress
this interpretation.

In the memoir of Lagrange just referred to there is also a treatment
of the sounds produced by organ pipes and musical wind instruments
in general. The basic experimental facts were already known and

Lagrange was able to predict theoretically the approximate harmonic frequencies of closed and open pipes. The boundary conditions gave some trouble, as indeed they do to this day; in any case the problem impinges rather closely on the propagation of sound and as such is better treated in the next section.

The extension of the methods described in the preceding paragraphs to the vibrations of extended solid bodies like bars and plates naturally demanded a knowledge of the relation between the deformability of a solid body and the deforming force. Fortunately this problem had already been solved by Hooke, who in 1660 discovered and in 1676 announced in the form of the anagram $C E I I I N O S S S T T U V$ the law "ut tensio sic vis" connecting the stress and strain for bodies undergoing *elastic* deformation. This law of course forms the basis for the whole mathematical theory of elasticity including elastic vibrations giving rise to sound. Its application to the vibrations of bars supported and clamped in various ways appears to have been made first by Euler in 1744 and Daniel Bernoulli in 1751, though it must be emphasized that dates of publication of memoirs do not always reflect accurately the time of discovery. The method used involved the variation of the expression for the work done in bending the bar. It is essentially that employed by Rayleigh in his treatise and leads of course to the well known equation of the fourth order in the space derivatives.

The corresponding analytical solution of the vibrations of a solid elastic plate came much later, though much experimental information was obtained in the latter part of the 18th century by the German E. F. F. Chladni (1756-1824), one of the greatest experimental acousticians. In 1787 he published his celebrated treatise "Entdeckungen über die Theorie des Klanges" in which he described his method of using sand sprinkled on vibrating plates to show the nodal lines. His figures were very beautiful and in a general way could be accounted for by considerations similar to those relating to vibrating strings. The exact forms, however, defied analysis for many years, even after the publication of Chladni's classic work "Die Akustik" in 1802. Napoleon provided for the Institute of France a prize of 3000 francs to be awarded for a satisfactory mathematical theory of the vibrations of plates. The prize was awarded in 1815 to Mlle. Sophie Germain, who gave the correct fourth order differential equation. Her choice of boundary conditions proved, however, to be incorrect. It was not until 1850 that Kirchhoff (1824-1887) gave a more accurate theory. The problem still provides considerable interest for workers even at the present time, both along theoretical and experimental lines.

In the meantime the analogous problem of the vibrations of a flexible membrane, important for the understanding of the sounds emitted by drum heads, was solved first by S. D. Poisson (1781-1840), though he did not complete the case of the circular membrane. This was done by Clebsch (1833-1872) in 1862. It is significant that most of the theoretical work on vibration problems during the 19th century was done by persons who called themselves mathematicians. This was natural though perhaps somewhat unfortunate, since the choice of conditions did not always reflect actual experimentally attainable situations. Rayleigh's own work did much to rectify this condition, and nowadays the experimental and theoretical acousticians work hand in hand. The importance of this is evident in the design of such modern instruments as loud speakers and quartz crystal vibrators.

A more complete description of the historical development of sound producers would, of course, necessarily pay much attention to musical instruments. Unfortunately this development lay rather aside from the scientific progress in acoustics, a situation which has persisted in large measure even to recent times. There are signs, however, that the designers of new musical instruments are paying more attention to acoustical principles than previously was the case, and that the theory of acoustics will have a greater influence on music in the future than it has had in the past.

We have now brought our brief sketch of the production of sound up to the time of Rayleigh. We shall therefore proceed with the equally important problem of the propagation of sound.

The Propagation of Sound. From the earliest recorded observations there has been rather general agreement that sound is conveyed from one point in space to another through some activity of the air. Aristotle, indeed, emphasizes that there is actual motion of the air involved, but as was often the case with his notions on physics his expressions are rather vague. Since in the transmission of sound the air certainly does not appear to move, it is not surprising that other philosophers denied Aristotle's view. Thus even during the Galilean period the French philosopher Gassendi (1592-1655), in his revival of the atomic theory, attributed the propagation of sound to the emission of a stream of fine, invisible particles from the sounding body which, after moving through the air, are able to affect the ear. Otto von Guericke (1602-1686) expressed great doubt that sound is conveyed by a motion of the air, observing that sound is transmitted better when the air is still than when there is a wind. Moreover he had tried around the middle of the 17th century the experiment of ringing a bell in a jar which was evacuated by means of his air pump, and claimed that he could still hear the sound. As a matter of fact, the first to try the bell-in-vacuum experiment was apparently the Jesuit Athanasius Kircher (1602-1680). He described it in his book "Musurgia Universalis", published in 1650, and concluded that air is not necessary for the transmission of sound. Undoubtedly there was not sufficient care to avoid transmission through the walls of the vessel. In 1660 Robert Boyle (1627-1691) in England repeated the experiment with a much improved air pump and more careful arrangements, and finally observed the now well known decrease in the intensity of the sound as the air is pumped out. He definitely concluded that the air is a medium for acoustic transmission, though presumably not the only one.

If air is the principal medium for the transmission of sound, the next question is: how rapidly does the propagation take place? As early as 1635 Gassendi while in Paris made measurements of the velocity of sound in air, using fire arms and assuming the passage of light as effectively infinite. His value was 1473 Paris feet per second. (The Paris foot is equivalent approximately to 32.48 cm.) Later by more careful measurements Mersenne showed this figure to be too high, obtaining 1380 Paris feet per second or about 450 meters/sec. Gassendi did note one matter of importance, namely that the velocity is independent of the pitch of the sound, thus discrediting the view of Aristotle, who had taught that high notes are transmitted faster than low notes. On the other hand Gassendi made the mistake of believing that the wind has no effect on the measured velocity of sound. In 1656 the Italian Borelli (1608-1679) and his colleague Viviani (1622-1703) made a more careful measurement and obtained 1077 Paris feet per second or 350 meters/sec. It is clear that all these values suffer from the lack of reference to the temperature, humidity and wind velocity conditions. It was not until 1740 that the Italian Branconi showed definitely by some experiments performed at Bologna that the velocity of sound in air increases with the temperature. Probably the first open air measurement of the velocity of sound that can be reckoned at all precise in the modern sense was carried out under the direction of the Academy of Sciences of Paris in 1738, using cannon fire. When reduced to 0°C the result was 332 meters/sec. Careful repetitions during the rest of the 18th century and the first half of the 19th century gave results differing by only a few meters per second from this value. The best modern value is 331.36 + 0.08 meters per second in still air under standard conditions of temperature and pressure (0°C and 76 cm of Hg. pressure).

This value is taken from D. C. Miller's "Sound Waves: Their Shape and Speed" (1937).

In 1808 the French physicist J. B. Biot (1774-1862) made the first experiments on the velocity of sound in solid media, using for this purpose an iron water pipe in Paris nearly 1000 meters long. By comparing the times of arrival of sound through the metal and the air respectively it was established that the velocity of the compressional wave in the solid metal is many times greater than that through the air. As a matter of fact Chladni, whose work on the vibrations of solids has been mentioned earlier in this sketch, had already measured the velocity of elastic waves in rods in connection with his study of the vibration of solids, with results in general agreement with those of Biot.

J. D. Colladon and the mathematician J. C. F. Sturm (1803-1855) in the year 1826 investigated the transmission of sound through water in Lake Geneva, in Switzerland, using a sound and flash arrangement. The velocity was found to be 1435 meters/sec. at 8°C.

To return to the propagation of sound through air, though it had very early been compared with the motion of ripples on the surface of water, the first theoretical attempt to theorize seriously about a *wave* theory of sound was made by Isaac Newton (1642-1727), who in the second book of his Principia (1687) (Propositions 47, 48 and 49) compares the propagation of sound to pulses produced when a vibrating body moves the adjacent portions of the surrounding medium and these in turn move those next adjacent to themselves and so on. Newton here made some rather specific and arbitrary assumptions, among them the hypothesis that when a pulse is propagated through a fluid the particles of the fluid always move in simple harmonic motion, or, as he puts it "are always accelerated or retarded according to the law of the oscillating pendulum". He indeed affects to prove this as a theorem, but inspection fails to reveal any demonstration save that if it is true for one particle it will be true for all. He then assumes that the elastic medium under consideration is subject to the pressure produced by a homogeneous medium of height h and density equal to the density of the medium under consideration. Newton further imagines a pendulum whose length between the point of suspension and center of oscillation is h. It is then proved that in one period of the pendulum the pulse will travel a distance of $2\pi h$. But since the period of the pendulum is $2\pi \sqrt{h/g}$, it follows that the velocity of the pulse is $\sqrt{gh}$, and since for a homogeneous fluid of density ρ the pressure p produced at the bottom of a column of height h is $p = \rho h g$, it follows that the pulse velocity is $\sqrt{p/\rho}$.

This demonstration was severely criticized by Lagrange in his Turin memoir of 1759 (already mentioned) as well as in the later one of 1760, and indeed one must admit the conditions laid down by Newton are highly specialized: an elastic wave need not be harmonic, nor should the velocity depend on this assumption. Lagrange gave a more rigorous general derivation, the outcome of which, however, must have surprised him, for it led to precisely Newton's result. When the relevant data for air at 60°F are substituted into Newton's formula, the velocity proves to be about 945 feet/second. At the time of his deduction this was not in bad order of magnitude agreement with the observed velocity of sound in air under the conditions cited. However, the more accurate measurements consistently turned out higher, and Newton was himself dissatisfied; hence, in the second edition of the Principia (1713) he revised his theory to try to bring it into better agreement with the best experimental value of the time, viz., 1142 feet/second. His explanation was so obviously *ad hoc* that it should have failed to carry conviction. However, no further serious question about the matter appears to have been raised until 1816 when Laplace suggested that in the previous determinations an error had been made in using the isothermal volume elasticity of the air, i.e. the pressure itself, thereby assuming that the elastic motions of the air particles take place at constant temperature. In view of the rapidity of the motions, however, it seemed to him more reasonable to suppose that the compressions and rarefactions follow the adiabatic law in which the changes in temperature lead to a higher value of the elasticity, namely, the product of the pressure by the ratio γ of the two specific heats of the air. At the time of Laplace's first investigation rather crude experiments had indeed indicated the existence of two specific heats of a gas, but the values were not known with much precision. Laplace used some data of the experimentalists, LaRoche and Berard, giving $\gamma = 1.5$ and leading to a value of the velocity of sound at 6°C equal to 345.9 meters/sec. The best experimental value obtained up to that time by members of the Academy was 337.18 meters/sec. for this temperature. Laplace did not consider this discrepancy serious. He returned to the problem later and included a chapter on the velocity of sound in air in his famous "Méchanique Céleste" (1825). By that time Clément and Désormes had performed their well-known experiment on the determination of γ (1819) and had obtained the value of 1.35 leading to 332.9 meters/sec. for the velocity. Some years later the more accurate value of $\gamma = 1.41$ led to complete agreement with the measured velocity. The theory of

Laplace is so well established that it is now common practice to determine γ for various gases by precision measurements of the velocity of sound.

As has already been remarked the first treatment of the partial differential equation of wave motion came with D'Alembert in 1750 in connection with the vibration of strings. The rest of the 18th century saw numerous attempts to theorize about waves in continuous media, such as waves on the surface of water and the like. These had value in connection with acoustics only to the extent that they rendered the use of the wave equation familiar to workers in sound. By the end of the 18th century the general treatment of the solution of the wave equation for sound in tubes, for example, subject to the boundary conditions at the ends, had been pretty well established, and the predicted harmonic frequencies checked with experiment with reasonable accuracy. Of course there were discrepancies leading to end corrections and so forth, which were never fully cleared up until the time of Rayleigh. It was not until 1868 that A. Kundt (1839-1894) developed his simple but effective method of dust figures for studying experimentally the propagation of sound in tubes and in particular measuring sound velocity from standing wave patterns.

In the meantime the more difficult problem of the propagation of a compressional wave in a three dimensional fluid medium had been attacked by Poisson in a celebrated memoir of 1820. The method was essentially that adopted by Rayleigh in Chap. XIV of "Theory of Sound". Three years before in a similar memoir, 100 pages long, Poisson had given the most elaborate theory up to that time of the propagation of sound in tubes, including the theory of stationary air waves for tubes of finite length, both open and closed. He even considered the possibility of an end correction in the case of an open tube to take care of the fact that the condensation cannot be considered precisely zero at the open end. It remained, however, for Hermann von Helmholtz (1821-1894) in 1860 to give a more thorough treatment of this question. The special case of an abrupt change in cross-section was also studied by Poisson along with the reflection and transmission of sound at normal incidence on the boundary of two different fluids. Much modern work of practical significance was anticipated in this great study of Poisson.

The more difficult problem of the reflection and refraction of a plane sound wave incident *obliquely* on the boundary of two different fluids was solved by the self-taught Nottingham genius George Green (1793-1841) in 1838. This served to emphasize both the similarities and differences between the reflection and refraction of sound and light. It should be recalled that sound waves in fluids, being strictly compressional, are longitudinal, whereas light waves are transverse. Hence light waves can be polarized, and sound waves in fluids can not. Of course elastic waves in an extended solid can be both longitudinal and transverse, more accurately irrotational and solenoidal. This was realized by Poisson in his study of isotropic elastic media of 1829. The direct significance of this for acoustics is, of course, not great, but it had an important bearing on the elastic solid theory of light, which was actively pursued during the middle decades of the 19th century. The connection with modern geophysics (seismological waves) is obvious.

The Reception of Sound. In the historical development of acoustics up to very recent times the only sound receiver of interest has been the human ear and the reception of sound has been largely the study of the acoustical behavior of this organ. In this connection it is interesting to observe that no completely acceptable theory of audition has ever been proposed, and how we hear still remains a puzzling problem in modern psychophysics.

After the relation between pitch and frequency had been established it became an interesting task to determine the frequency limits of audibility. F. Savart (1791-1841) using fans and toothed wheels (1830) placed the minimum audible frequency at 8 vibrations per second and the upper limit at 24,000 vibrations per second. The later investigators Seebeck (1770-1831), Biot (1774-1862), K. R. Koenig (1832-1901) and Hermann von Helmholtz obtained values for the lower limit ranging from 16 to 32 vibrations per second. In such matters there are bound to be individual differences. These play an even greater role in the upper limit of audibility, which not only can vary many thousand vibrations per second from person to person, but for each individual usually decreases with age. The most elaborate studies on audibility during the 19th century were made by Koenig, who devoted a lifetime to the production of precision sources of sound of controlled frequency, such as tuning forks, rods, strings and pipes. The electrically driven fork also originated with him.

The closely related problem of the minimum sound amplitude or intensity necessary for audibility was apparently first studied by Toepler (1836-1912) and Boltzmann (1844-1906) in 1870. The more recent work dates from Rayleigh.

In 1843, Georg S. Ohm, the author of the famous law of electric currents, put forward a law of audition according to which all musical tones arise from simple harmonic vibrations of definite frequency, and the particular quality of actual musical sounds is due to combinations of simple tones of commensurable frequencies. He held, moreover, that the ear is able to analyze any complex note into the set of simple tones in terms of which it may be expanded mathematically by means of Fourier's theorem. This law has stimulated a host of researches in physiological acoustics. The greatest of these in the pre-Rayleigh period were undoubtedly those of Helmholtz, whose treatise "Die Lehre von den Tonempfindungen als Physiologische Grundlage für die Theorie der Musik", published in 1862, ranks as one of the great masterpieces of acoustics. Here he gave the first elaborate theory of the mechanism of the ear, the so-called resonance theory, and was able to justify theoretically the law of Ohm. In the course of his investigations he invented the resonator, now so well known by his name and employed in modern acoustics for many applications. Helmholtz developed the theory of summation and difference tones and in general laid the ground work for all subsequent research in the field of audition. One of the greatest physicists of the 19th century, he touched no field that he did not enrich with his experimental and theoretical genius.

Since the reception of sound by the ear in enclosed spaces like rooms and auditoriums is a common experience, it is proper that some attention should be paid here to what has come to be called architectural acoustics. The first discussion of the problem of improving hearing in rooms was limited to purely geometrical considerations, such as the installation of sounding boards and other reflectors. A Boston physician, J. B. Upham, in 1853 wrote several papers indicating a much clearer grasp of the more important matter involved, namely the reverberation or multiple reflection of the sound from all the surfaces of the room. He also showed how the reverberation time could be reduced by the installation of fabric curtains and upholstered furnishings. In 1856 Joseph Henry, the celebrated American physicist, who became the first secretary of the Smithsonian Institution, made a study of auditorium acoustics which reflects a thorough understanding of all the factors involved, though his suggestions were all of a qualitative character. In spite of this the subject was completely neglected by architects, and attempts were often made to correct gross acoustical defects by such inadequate, if not absurd, devices as stringing wires, etc. The real quantitative foundation of architectural acoustics dates from W. C. Sabine (1868-1919) in 1900.

Special devices for the amplification of sound received by the ear go back a long way. Horns for the production of sounds are of great antiquity. It is uncertain just when the suggestion arose that they might be used to improve the *reception* of sound. At all events the Jesuit Athanasius Kircher, already mentioned in this sketch, in 1650 designed a parabolic horn as a hearing aid as well as a speaking trumpet, and evidently realized the importance of the flare in the amplification. Robert Hooke suggested the possibility of a device to magnify the sounds of the body, but it seems to have been reserved for the French physician René Laënnec (1781-1826) actually to invent and employ the stethoscope for clinical purposes (1819). Sir Charles Wheatstone (1802-1875) in 1827 developed a similar instrument which he termed a microphone, a name now applied to an exclusively electrical device for the reception of sound. Koenig also invented a new type of stethoscope. The theoretical and experimental improvement of instruments like horns and other sound receivers of similar type has been and still is an important feature of modern acoustics.

All through the historical development of physics there has been a tendency to reduce the observation of physical phenomena and particularly experimental measurements to something which can be *seen*. Practically all physical instruments involve this principle and employ a pointer or a spot of light moving on a scale. It was therefore inevitable that attempts would be made to study sound phenomena visually, and this of course was especially necessary for the investigation of sounds whose frequencies lie outside the range of audibility of the ear. One of the first moves in this direction was the observation by John LeConte (1818-1891) that musical sounds can produce jumping in a gas flame if the pressure is properly adjusted (1858). The sensitive flame, as it later came to be called, was developed to a high pitch of excellence by John Tyndall (1820-1893), who used it for the detection of high frequency sounds and the study of the reflection, refraction and diffraction of sound waves. It still provides a very effective lecture demonstration but for practical purposes has been superseded in recent times by various types of electrical microphones.

In the endeavor to make visible the form of a sound wave Koenig about 1860 invented the manometric flame device which consists of a box through which gas flows to a burner. One side of the box is a flexible membrane. When sound waves impinge on the membrane the changes in pressure produce corresponding fluctuations in the flame which can be made visible by reflecting the light of the flame from a rapidly rotating mirror. Another attempt to visualize sound waves was made by Leon Scott in 1857 in his "phonautograph" in which a flexible diaphragm at the throat of a receiving horn was attached to a stylus which in turn touched a smoked rotating drum surface and traced out a curve corresponding to the incident sound. This was the precursor of the phonograph. An equally ambitious attempt to record sound was made by Eli Whitney Blake (1836-1895), the first Hazard Professor of Physics in Brown University, who in 1878 made a microphone by attaching a small metallic mirror to a vibrating disc at the back of a telephone mouthpiece. By reflecting a beam of light from the mirror Blake succeeded in photographing the sounds of human speech. Such studies were much advanced by D. C. Miller, (1866-1941) who invented a similar instrument in the "phonodeik" and made very elaborate photographs of sound wave forms.

III. RAYLEIGH'S CONTRIBUTIONS TO ACOUSTICS AND THEIR SIGNIFICANCE FOR MODERN DEVELOPMENTS

The results of Rayleigh's work in acoustics are embodied in his treatise "The Theory of Sound" and in 128 published articles, the first of which appeared in 1870 (his fourth paper) and the last in 1919—this was his last published paper and appeared in print after his death. Except for the years 1895, 1896 and 1906, there was not a year from 1870 to 1919 in which an article having a definite connection with acoustics did not appear. This record of devotion to a single department of thought is undoubtedly unique in the annals of science and becomes all the more remarkable when we recall that this activity was accompanied by unchecked attention to a host of other problems extending over the whole field of physics, leading to a total of nearly 450 publications.

Lord Rayleigh appeared on the acoustical scene when the time was precisely ripe for a synthesis of experimental phenomena and rather highly developed theory, much of which was, however, too idealized for practical application. On the other hand much of the experimental work had been discussed in rather empirical fashion with little attempt at a dynamical explanation. Rayleigh's interest in acoustics appears to have been started through the advice of Professor W. F. Donkin, Savilian Professor of Astronomy at Oxford, that he ought to learn to read German. Rayleigh followed the suggestion and the first scientific work he read was Helmholtz's treatise "Lehre von der Tonempfindung". Certain references here to the properties of acoustic resonators attracted his attention and led to his first elaborate research, reported on in a long paper on the theory of resonance in the *Philosophical Transactions of the Royal Society* in 1870. This article furnishes a clear indication of the method of thinking about problems that remained characteristic of all Rayleigh's later work. He endeavored to develop the mathematical theory of the subject in a form related as closely as possible to experimentally realizable situations, and then followed up the results by the attempt at direct experimental verification. There was no pretense of an over-elaborate method of measurement, but the precision was fully sufficient in view of the inevitable limitations of the theory of aerial vibrations. In this paper Rayleigh first introduced the useful concept of the *acoustic conductivity* of an orifice. It has remained a standard acoustical quantity ever since, even if rather difficult to to estimate theoretically for all sorts of openings.

It was evidently not long after the publication of his researches on resonance that Rayleigh conceived the desirability of writing a treatise on acoustics. His reasons for the step are amply set forth in the preface to the first edition of "The Theory of Sound" and need no repetition here. In preparation for his task he studied in detail the general theory of vibrations of a dynamical system about a state of equilibrium and uncovered a number of general results of great interest. These were presented in the *Proceedings of the London Mathematical Society* in 1873 and include such theorems as that the increase in the mass of any part of a vibrating system can never lead to a decrease in any period of the motion. Here he also introduced his famous dissipation function for a system subject to damping forces proportional to the component velocities and finally proved a very general reciprocity theorem of which the one generally known by the name of Helmholtz is a special case. This theorem has been of the greatest importance in comparing the efficiency of acoustical devices as emitters and receivers of radiation energy. As before, a characteristic feature of these articles is the skillful combination of theory and observation or experiment. Rarely does one find a mass of analysis without illustrations from experience, and Rayleigh was always very keen to follow up supposed experimental exceptions to theoretically deduced laws. Usually his uncanny insight into the important things led him to the correct explanation of apparent difficulties.

In 1877, the year of the publication of "The Theory of Sound", Rayleigh inaugurated the custom of publishing collections of miscellaneous acoustical phenomena which he had himself observed. These were continued at intervals for the rest of his life, being published for the most part in the *Philosophical Magazine*. Among the earlier subjects investigated were the perception of the direction of a sound source, the diffracting effect of the head on spoken and received sound, the end correction of an organ pipe, sensitive flames, Aeolian tones, acoustical shadows, etc.

"The Theory of Sound" was published in June, 1877. Though, as his son remarks "the sale was not wholly unprofitable", it was hardly a best seller. Those interested in the general field realized its importance, but the possible fundamental significance of the work for future applications of sound to a host of practical problems could scarcely be properly estimated at that time. Helmholtz, it is true, reviewed the volumes in *Nature* and compared the treatment to the famous unfinished "Treatise on Natural Philosophy" of Thomson and Tait. He pressed, indeed, for a third volume on physiological acoustics and the maintenance of acoustic vibrations. Rather wisely, it seems, Rayleigh refrained from this and contented himself with enriching the literature of acoustics for the following forty years with a succession of attractive papers on a wide variety of topics, many if not most of which were a direct outgrowth of the treatment inaugurated in his treatise.

While it would be gratuitous in the extreme to present a detailed analysis of the contents of "The Theory of Sound" to the reader who has the book before him, it is difficult to refrain from emphasizing briefly some of the features which have made the treatise such a mine of information for all workers in acoustics from Rayleigh's day to the present time.

Though written in the rather informal style which characterized practically all of Rayleigh's published work, the book reflects clearly a great deal of careful planning with respect to its logical structure. The author was evidently impressed by the importance of the subdivision of the subject into the two principal sections: the production and propagation of sound. Hence the whole of the first half (the first volume in the original edition), with the exception of an introductory chapter, is devoted to the vibrations of dynamical systems, naturally with special emphasis on those giving rise to acoustically interesting radiation. In contrast to the usual continental European method of writing a treatise, Rayleigh's treatment opens with the simplest possible case, namely the oscillations of a system of one degree of freedom, and each element of the theory is accompanied by a definite experimental illustration.

The simple case is followed by two chapters on the general theory of vibrations of a system of n degrees of freedom, largely a development of his 1873 paper mentioned just above. It was here he emphasized the value of the method of obtaining an approximation to the lowest frequency of vibration of a complicated system in which the direct solution of the differential equations is impracticable. This

procedure, which makes use of the expressions for the maximum
potential and kinetic energies, was later generalized by Ritz and is
now usually known as the Rayleigh-Ritz method. It has proved of
value in handling not only all sorts of involved vibration problems
but also problems in quantum mechanics. Applications to acoustics
occur frequently throughout "The Theory of Sound", particularly
with reference to non-uniform strings, bars, membranes and plates.
Throughout his treatise Rayleigh displays great fondness for the
use of energy considerations and uses the energy method (virtual
work) freely for setting up the differential equations of motion of
different types of vibrating systems. It is scarcely an exaggeration to
say that there is no vibrating system likely to be encountered in
practice which cannot be tackled successfully by the methods set
forth in the first ten chapters of Rayleigh's treatise. Even the worker
in the field of non-linear systems, a department of increasing practical importance in modern vibration theory, will find useful basic
hints in Rayleigh. The reader should, indeed, be cautioned not to
consider "The Theory of Sound" as a mere reference book. One who
goes to it in this frame of mind is apt to be disappointed. It is a
rather closely knit work in which the author, having developed certain methods, feels free to refer the reader to them again and again.
Hence reading Rayleigh is a real process of discovery, not always
easy but constantly challenging and illuminating. One rather trivial
mathematical detail may properly be mentioned at this point. Rayleigh's mathematical notation is standard in nearly all respects from
the standpoint of present-day fashions, but he never uses the round ∂
to denote the distinction between partial and ordinary derivatives.
Presumably he felt that the reader with a suitable grasp of the physical meaning of the mathematical processes would have no difficulty
in distinguishing the one type of derivative from the other.

The last thirteen chapters of "The Theory of Sound" are devoted
primarily to acoustic radiation through fluid media. This is by far
the more difficult part of the subject matter of acoustics and has
remained so to the present time. Since there is no such thing as a
perfect fluid the exact hydrodynamical equations describing with
precision the motion of a compressional disturbance in a fluid
medium like air or water must necessarily be extremely complicated.
It has therefore proved desirable to approximate, and it is just here
that the judgment of the physicist plays a significant role. Rayleigh
possessed the power of assessing a problem from the point of view
of the best possible approximation to lead to a physically useful
result. This is particularly well illustrated by his studies of the

diffraction and scattering of sound by obstacles, which is by no means so easy to study theoretically as in the analogous case of optics, largely because of the approximate character of the equations and the relatively large wave length of audible sound. Another illustration is the acoustic radiation into a surrounding fluid medium from a vibrating sphere or plate. The whole modern study of high frequency acoustic beams is based on this work. Still another example is provided by the effect of viscosity and heat conduction on the propagation of sound. Here the fundamental theory had already been worked out by other men like Stokes and Kirchhoff, but Rayleigh seemed able to seize on the useful applications to the transmission of sound through narrow tubes and the interstices of fabrics. He was aware that these effects are inadequate to account for the actually observed absorption of sound in three dimensional fluid media like the atmosphere or the sea. Progress in the solution of this problem at the present time is actually being made along the lines of a hint thrown out by Rayleigh in a paper on the cooling of air by radiation and conduction published in 1899.

A second revised and enlarged edition of "The Theory of Sound" was brought out by the author between the years 1894 and 1896, embodying the results of his investigations in the seventeen years which had elapsed since the first appearance of the book. No further revisions or reprintings were made until after Rayleigh's death. This would seem to reflect a rather stagnant state of acoustics during the first two decades of the twentieth century. Compared with the activity of university physical laboratories in other fields this must be considered the truth: academically, acoustics became, by and large, an uninteresting subject. In the meantime, however, the development of certain technological fields such as telephony, both with and without wires, as well as acoustic signalling under water and architectural acoustics, made it imperative for engineers to gain a better understanding of the theory of acoustics. The large industrial concerns began to make use of the subject in their research and development laboratories, and the whole field received a stimulus such as it probably never could have gained from the side of academic workers. We may say that acoustics was rediscovered and along with it Rayleigh's book. Reprintings were called for in rapid succession in 1926 and 1929 at about the time of the founding of the Acoustical Society of America (1928). At the same time numerous books began to appear whose main purpose was largely to interpret Rayleigh's work to the new workers in the subject, and to apply the methods of his treatise to a multitude of new and practical problems.

It would be absurd to maintain that the whole of acoustics is to be found within the covers of "The Theory of Sound". Rayleigh himself in some 60 papers published between 1900 and his death advanced the subject mightily and called attention to many problems which have turned out to be of great significance in recent applications. Among these foreshadowings of the future must be reckoned his use of electric circuit analogies in connection with the forced vibrations of acoustical resonators and other systems. This procedure has developed to such an extent that the modern acoustical engineer, using electrical equipment for most of his practical work, invariably insists on expressing all acoustical systems in terms of their electrical analogues. Other striking anticipations by Rayleigh of modern acoustical considerations concern the use of conical horns for the production and reception of sound in signalling, the acoustic shadow of a sphere (of particular significance in the diffraction effect of a microphone), the pressure of acoustic radiation (used in the measurement of sound intensity, especially in supersonics), the binaural effect in sound perception, the possible regime of sound waves of finite amplitude (explosion waves and those associated with gun fire) and the selective transmission of waves through stratified media (acoustical filtration). The list could easily be extended, but this will suffice to suggest to the contemporary worker in acoustics his debt to Rayleigh's foresight.

No one can foresee the future of the science of acoustics as, on the one hand, it reaches out into new realms of application in the engineering fields of the recording and reproduction of sound, the creation of more comfortable environments for the hearing of sound and the development of adequate hearing aids for the deaf, and, on the other hand, joins forces with pure physics and chemistry in the endeavor to learn more about the solid, liquid and gaseous states of matter, particularly through the agency of supersonics. It is safe to predict, however, that for a long time to come Lord Rayleigh's "The Theory of Sound" will be a *vade mecum* for both the pure and applied acoustician.

IV. BRIEF BIBLIOGRAPHY

This bibliography is not intended to be complete, being suggestive rather than exhaustive.

1. *Lord Rayleigh and His Work*

GLAZEBROOK, R. T., *The Rayleigh Period* (in *A History of the Cavendish Laboratory* [London, 1910]).

SCHUSTER, SIR ARTHUR, *Obituary Notice of Lord Rayleigh* (Proc. Roy. Soc. 98-A i, 1921).

STRUTT, ROBERT JOHN, 4th Baron Rayleigh, *Life of John William Strutt, 3d Baron Rayleigh.* (London, 1924).

STRUTT, JOHN WILLIAM, 3d Baron Rayleigh, *Scientific Papers* (Six volumes, Cambridge, 1899-1920). (Volume Six contains a complete list of Rayleigh's collected papers classified according to subject. Those relating to sound are found on pages 681-688.)

STRUTT, JOHN WILLIAM, 3d Baron Rayleigh, *The Theory of Sound* (London, Vol. 1, 1877. Vol. 2, 1878. Second edition, London, Vol. 1, 1894. Vol. 2, 1896).

THOMSON, SIR J. J. and GLAZEBROOK, R. T., *Obituary Notices of Lord Rayleigh* (Nature *103*, 365, 1919).

2. History of Acoustics

BERNOULLI, DANIEL, *Reflexions et Eclaircissemens sur les Nouvelles Vibrations des Cordes Exposees dans les Memoires de l'Academie, de 1747 et 1748.* (Royal Academy of Berlin, 1750 p. 147 ff.)

BOYLE, ROBERT, *New Experiments Physio-Mechanical Touching the Spring of the Air.* (Second Edition, Oxford, 1662, p. 105 ff.)

CAJORI, FLORIAN (Editor)—*Sir Isaac Newton's Mathematical Principles of Natural Philosophy and his System of the World.* (A revision of Motte's translation.) (Berkeley, California, 1934.)

CHLADNI, E. F. F. *Die Akustik* (Leipzig, 1802).

D'ALEMBERT, *Recherches sur le Courbe que Forme une Corde Tendue Mise en Vibration* (Royal Academy of Berlin, 1747, p. 214 ff).

FLETCHER, HARVEY, *Speech and Hearing* (New York, 1929).

GALILEI, GALILEO, *Dialogues concerning Two New Sciences* (Translated from the Italian and Latin by Henry Crew and Alfonso de Salvio, Evanston and Chicago, 1939).

HELMHOLTZ, HERMANN VON, *Sensations of Tone* (English translation of *Die Lehre von den Tonempfindungen* by A. J. Ellis, London, 1895).

KNUDSEN, V. O., *Architectural Acoustics*, (New York, 1932). (Chapter I contains material of historical interest.)

LAGRANGE, J. L., *Recherches sur la Nature et la Propagation du Son* (Miscellanea Taurinensia, t. I, 1759). (Vol. 1, p. 39 of *Oeuvres de Lagrange*, Paris, 1867).

LAPLACE, P. S. DE, *Traité de Méchanique Céleste* (Paris, 1825). (Vol. 5, p. 133 ff: "De la vitesse du son . . .".)

LOVE, A. E. H., *A Treatise on the Mathematical Theory of Elasticity* (4th edition, New York, 1944). (The historical introduction contains much material of importance to acoustics.)

MAGIE, W. F., *A Source Book in Physics* (New York, 1935).

MERSENNE, MARIN, *Harmonicorum Liber* (Paris, 1636).

MILLER, D. C., *Anecdotal History of the Science of Sound* (New York, 1935). (This possesses a good bibliography.)

MILLER, D. C. *Sound Waves: Their Shape and Speed.* (New York, 1937.)

POISSON, S. D. *Sur le mouvement des fluides elastiques dans les tuyaux cylindriques, et sur la theorie des instrumens a vent.* (Mémoires de l'Academie Royale des Sciences de l'Institut de France, Annee 1817, Tome II, Paris, 1819, p. 305.)

SAUVEUR, JOSEPH, *Systéme General des Intervalles des Sons* (L'Academie Royale des Sciences, Paris 1701, p. 297 ff. The word "acoustics" was first introduced in this memoir).

TYNDALL, JOHN, *Sound* (New York, 1867).

VIOLLE, J. *Sur la Vitesse de Propagation du Son* (Congrès International de Physique, Vol. 1, Paris 1900, p. 228).

WEBER, E. H. and W., *Wellenlehre* (Leipzig, 1825).

WHEWELL, WILLIAM, *History of the Inductive Sciences* (Two volumes, 3d edition, New York, 1874). (The part on acoustics begins on p. 23 of Vol. II.)

WOLF, A., *A History of Science, Technology and Philosophy in the 16th and 17th Centuries.* (London, 1935.)

WOLF, A., *A History of Science, Technology and Philosophy in the 18th Century* (New York, 1939.)

3. Recent Developments in Acoustics

The reader interested in recent developments is referred to the *Journal of the Acoustical Society of America*, Vol. 1 of which appeared in 1929. The issue of Vol. 16 began in July, 1944.

Another very useful review of current work in acoustics is to be found in *Reports on Progress in Physics*, published annually by the Physical Society of London. Vol. 1 appeared in 1934. The articles on Sound by E. G. Richardson are inclusive and illuminating.

PREFACE.

IN the work, of which the present volume is an instalment, my endeavour has been to lay before the reader a connected exposition of the theory of sound, which should include the more important of the advances made in modern times by Mathematicians and Physicists. The importance of the object which I have had in view will not, I think, be disputed by those competent to judge. At the present time many of the most valuable contributions to science are to be found only in scattered periodicals and transactions of societies, published in various parts of the world and in several languages, and are often practically inaccessible to those who do not happen to live in the neighbourhood of large public libraries. In such a state of things the mechanical impediments to study entail an amount of unremunerative labour and consequent hindrance to the advancement of science which it would be difficult to overestimate.

Since the well-known Article on Sound in the *Encyclopædia Metropolitana*, by Sir John Herschel (1845), no complete work has been published in which the subject is treated mathematically. By the premature death of Prof. Donkin the scientific world was deprived of one whose mathematical attainments in combination with a practical knowledge of music qualified him in a special manner to write on Sound. The first part of his *Acoustics* (1870), though little more than a fragment, is sufficient to shew that my labours would have been unnecessary had Prof. Donkin lived to complete his work.

In the choice of topics to be dealt with in a work on Sound, I have for the most part followed the example of my predecessors. To a great extent the theory of Sound, as commonly understood, covers the same ground as the theory of Vibrations in general; but, unless some limitation were admitted, the consideration of such subjects as the Tides, not to speak of Optics, would have to be included. As a general rule we shall confine ourselves to those classes of vibrations for which our ears afford a ready made and wonderfully sensitive instrument of investigation. Without ears we should hardly care much more about vibrations than without eyes we should care about light.

The present volume includes chapters on the vibrations of systems in general, in which, I hope, will be recognised some novelty of treatment and results, followed by a more detailed consideration of special systems, such as stretched strings, bars, membranes, and plates. The second volume, of which a considerable portion is already written, will commence with aërial vibrations.

My best thanks are due to Mr H. M. Taylor of Trinity College, Cambridge, who has been good enough to read the proofs. By his kind assistance several errors and obscurities have been eliminated, and the volume generally has been rendered less imperfect than it would otherwise have been.

Any corrections, or suggestions for improvements, with which my readers may favour me will be highly appreciated.

TERLING PLACE, WITHAM,
April, 1877.

IN this second edition all corrections of importance are noted, and new matter appears either as fresh sections, e.g. § 32 *a*, or enclosed in square brackets [   ]. Two new chapters X A, X B are interpolated, devoted to *Curved Plates or Shells*, and to *Electrical Vibrations*. Much of the additional matter relates to the more difficult parts of the subject and will be passed over by the reader on a first perusal.

In the mathematical investigations I have usually employed such methods as present themselves naturally to a physicist. The pure mathematician will complain, and (it must be confessed) sometimes with justice, of deficient rigour. But to this question here are two sides. For, however important it may be to maintain a uniformly high standard in pure mathematics, the physicist may occasionally do well to rest content with arguments which are fairly satisfactory and conclusive from his point of view. To his mind, exercised in a different order of ideas, the more severe procedure of the pure mathematician may appear not more but less demonstrative. And further, in many cases of difficulty to insist upon the highest standard would mean the exclusion of the subject altogether in view of the space that would be required.

In the first edition much stress was laid upon the establishment of general theorems by means of Lagrange's method, and I am more than ever impressed with the advantages of this procedure. It not unfrequently happens that a theorem can be thus demonstrated in all its generality with less mathematical apparatus than is required for dealing with particular cases by special methods.

During the revision of the proof-sheets I have again had the very great advantage of the cooperation of Mr H. M. Taylor, until he was unfortunately compelled to desist. To him and to several other friends my thanks are due for valuable suggestions.

July, 1894.

EDITORIAL NOTE FOR THE 1929 RE-ISSUE

In this re-issue, a few pencilled corrections and references in the Author's own copy have been made use of. Otherwise no change has been made.

EDITORIAL NOTE FOR THE PRESENT 1945 RE-ISSUE

In this re-issue, a Historical Introduction by Robert Bruce Lindsay had been added, and both volumes are bound as one. The text remains the same as the 1929 re-issue.

CONTENTS.

CHAPTER I.

CHAPTER II.

CHAPTER VI.

CHAPTER VII.

CHAPTER VIII.

CHAPTER IX.

CHAPTER X.

CHAPTER X A.

CHAPTER X B.

THE

THEORY OF SOUND

CHAPTER I.

1. THE sensation of sound is a thing *sui generis*, not comparable with any of our other sensations. No one can express the relation between a sound and a colour or a smell. Directly or indirectly, all questions connected with this subject must come for decision to the ear, as the organ of hearing; and from it there can be no appeal. But we are not therefore to infer that all acoustical investigations are conducted with the unassisted ear. When once we have discovered the physical phenomena which constitute the foundation of sound, our explorations are in great measure transferred to another field lying within the dominion of the principles of Mechanics. Important laws are in this way arrived at, to which the sensations of the ear cannot but conform.

2. Very cursory observation often suffices to shew that sounding bodies are in a state of vibration, and that the phenomena of sound and vibration are closely connected. When a vibrating bell or string is touched by the finger, the sound ceases at the same moment that the vibration is damped. But, in order to affect the sense of hearing, it is not enough to have a vibrating instrument; there must also be an uninterrupted communication between the instrument and the ear. A bell rung *in vacuo*, with proper precautions to prevent the communication of motion, remains inaudible. In the air of the atmosphere, however, sounds have a universal vehicle, capable of conveying them without break from the most variously constituted sources to the recesses of the ear.

3. The passage of sound is not instantaneous. When a gun is fired at a distance, a very perceptible interval separates the report from the flash. This represents the time occupied by sound in travelling from the gun to the observer, the retardation of the flash due to the finite velocity of light being altogether negligible. The first accurate experiments were made by some members of the French Academy, in 1738. Cannons were fired, and the retardation of the reports at different distances observed. The principal precaution necessary is to reverse alternately the direction along which the sound travels, in order to eliminate the influence of the motion of the air in mass. Down the wind, for instance, sound travels relatively to the earth faster than its proper rate, for the velocity of the wind is added to that proper to the propagation of sound in still air. For still dry air at a temperature of 0°C., the French observers found a velocity of 337 metres per second. Observations of the same character were made by Arago and others in 1822; by the Dutch physicists Moll, van Beek and Kuytenbrouwer at Amsterdam; by Bravais and Martins between the top of the Faulhorn and a station below; and by others. The general result has been to give a somewhat lower value for the velocity of sound—about 332 metres per second. The effect of alteration of temperature and pressure on the propagation of sound will be best considered in connection with the mechanical theory.

4. It is a direct consequence of observation, that within wide limits, the velocity of sound is independent, or at least very nearly independent, of its intensity, and also of its pitch. Were this otherwise, a quick piece of music would be heard at a little distance hopelessly confused and discordant. But when the disturbances are very violent and abrupt, so that the alterations of density concerned are comparable with the whole density of the air, the simplicity of this law may be departed from.

5. An elaborate series of experiments on the propagation of sound in long tubes (water-pipes) has been made by Regnault[1]. He adopted an automatic arrangement similar in principle to that used for measuring the speed of projectiles. At the moment when a pistol is fired at one end of the tube a wire conveying an electric current is ruptured by the shock. This causes the withdrawal of a tracing point which was previously marking a line on a revolving drum. At the further end of the pipe is a stretched membrane so arranged that when on the arrival of the sound it yields to the

[1] *Mémoires de l'Académie de France*, t. XXXVII.

impulse, the circuit, which was ruptured during the passage of the sound, is recompleted. At the same moment the tracing point falls back on the drum. The blank space left unmarked corresponds to the time occupied by the sound in making the journey, and, when the motion of the drum is known, gives the means of determining it. The length of the journey between the first wire and the membrane is found by direct measurement. In these experiments the velocity of sound appeared to be not quite independent of the diameter of the pipe, which varied from $0^{m}.108$ to $1^{m}.100$. The discrepancy is perhaps due to friction, whose influence would be greater in smaller pipes.

6. Although, in practice, air is usually the vehicle of sound, other gases, liquids and solids are equally capable of conveying it. In most cases, however, the means of making a direct measurement of the velocity of sound are wanting, and we are not yet in a position to consider the indirect methods. But in the case of water the same difficulty does not occur. In the year 1826, Colladon and Sturm investigated the propagation of sound in the Lake of Geneva. The striking of a bell at one station was simultaneous with a flash of gunpowder. The observer at a second station measured the interval between the flash and the arrival of the sound, applying his ear to a tube carried beneath the surface. At a temperature of 8°C., the velocity of sound in water was thus found to be 1435 metres per second.

7. The conveyance of sound by solids may be illustrated by a pretty experiment due to Wheatstone. One end of a metallic wire is connected with the sound-board of a pianoforte, and the other taken through the partitions or floors into another part of the building, where naturally nothing would be audible. If a resonance-board (such as a violin) be now placed in contact with the wire, a tune played on the piano is easily heard, and the sound seems to emanate from the resonance-board. [Mechanical telephones upon this principle have been introduced into practical use for the conveyance of speech.]

8. In an open space the intensity of sound falls off with great rapidity as the distance from the source increases. The same amount of motion has to do duty over surfaces ever increasing as the squares of the distance. Anything that confines the sound will tend to diminish the falling off of intensity. Thus over the flat surface of still water, a sound carries further than over broken ground; the corner between a smooth pavement and a vertical wall is still better; but the most effective of all is a tube-like enclosure, which prevents spreading altogether. The use of speaking tubes to facilitate communication between the different parts of a building is well known. If it were not for certain effects (frictional and other) due to the sides of the tube, sound might be thus conveyed with little loss to very great distances.

9. Before proceeding further we must consider a distinction, which is of great importance, though not free from difficulty. Sounds may be classed as musical and unmusical; the former for convenience may be called *notes* and the latter *noises*. The extreme cases will raise no dispute; every one recognises the difference between the note of a pianoforte and the creaking of a shoe. But it is not so easy to draw the line of separation. In the first place few notes are free from all unmusical accompaniment. With organ pipes especially, the hissing of the wind as it escapes at the mouth may be heard beside the proper note of the pipe. And, secondly, many noises so far partake of a musical character as to have a definite pitch. This is more easily recognised in a sequence, giving, for example, the common chord, than by continued attention to an individual instance. The experiment may be made by drawing corks from bottles, previously tuned by pouring water into them, or by throwing down on a table sticks of wood of suitable dimensions. But, although noises are sometimes not entirely unmusical, and notes are usually not quite free from noise, there is no difficulty in recognising which of the two is the simpler phenomenon. There is a certain smoothness and continuity about the musical note. Moreover by sounding together a variety of notes—for example, by striking simultaneously a number of consecutive keys on a pianoforte—we obtain an approximation to a noise; while no combination of noises could ever blend into a musical note.

10. We are thus led to give our attention, in the first instance, mainly to musical sounds. These arrange themselves naturally in a certain order according to *pitch*—a quality which all can appreciate to some extent. Trained ears can recognise an enormous number of gradations—more than a thousand, probably, within the compass of the human voice. These gradations of pitch are not, like the degrees of a thermometric scale, without special mutual relations. Taking any given note as a starting point,

musicians can single out certain others, which bear a definite relation to the first, and are known as its octave, fifth, &c. The corresponding differences of pitch are called *intervals*, and are spoken of as always the same for the same relationship. Thus, wherever they may occur in the scale, a note and its octave are separated by *the interval of the octave*. It will be our object later to explain, so far as it can be done, the origin and nature of the consonant intervals, but we must now turn to consider the physical aspect of the question.

Since sounds are produced by vibrations, it is natural to suppose that the simpler sounds, viz. musical notes, correspond to *periodic* vibrations, that is to say, vibrations which after a certain interval of time, called the *period*, repeat themselves with perfect regularity. And this, with a limitation presently to be noticed, is true.

11. Many contrivances may be proposed to illustrate the generation of a musical note. One of the simplest is a revolving wheel whose milled edge is pressed against a card. Each projection as it strikes the card gives a slight tap, whose regular recurrence, as the wheel turns, produces a note of definite pitch, *rising in the scale, as the velocity of rotation increases.* But the most appropriate instrument for the fundamental experiments on notes is undoubtedly the Siren, invented by Cagniard de la Tour. It consists essentially of a stiff disc, capable of revolving about its centre, and pierced with one or more sets of holes, arranged at equal intervals round the circumference of circles concentric with the disc. A windpipe in connection with bellows is presented perpendicularly to the disc, its open end being opposite to one of the circles, which contains a set of holes. When the bellows are worked, the stream of air escapes freely, if a hole is opposite to the end of the pipe; but otherwise it is obstructed. As the disc turns, a succession of puffs of air escape through it, until, when the velocity is sufficient, they blend into a note, whose pitch rises continually with the rapidity of the puffs. We shall have occasion later to describe more elaborate forms of the Siren, but for our immediate purpose the present simple arrangement will suffice.

12. One of the most important facts in the whole science is exemplified by the Siren—namely, that the pitch of a note depends upon the period of its vibration. The size and shape of the holes, the force of the wind, and other elements of the problem may be varied; but if the number of puffs in a given time, such as one second, remains unchanged, so also does the pitch. We may even dispense with wind altogether, and produce a note by allowing the corner of a card to tap against the edges of the holes, as they revolve; the pitch will still be the same. Observation of other sources of sound, such as vibrating solids, leads to the same conclusion, though the difficulties are often such as to render necessary rather refined experimental methods.

But in saying that pitch depends upon period, there lurks an ambiguity, which deserves attentive consideration, as it will lead us to a point of great importance. If a variable quantity be periodic in any time τ, it is also periodic in the times 2τ, 3τ, &c. Conversely, a recurrence within a given period τ, does not exclude a more rapid recurrence within periods which are the aliquot parts of τ. It would appear accordingly that a vibration really recurring in the time $\frac{1}{2}\tau$ (for example) may be regarded as having the period τ, and therefore by the law just laid down as producing a note of the pitch defined by τ. The force of this consideration cannot be entirely evaded by defining as the period the *least* time required to bring about a repetition. In the first place, the necessity of such a restriction is in itself almost sufficient to shew that we have not got to the root of the matter; for although a right to the period τ may be denied to a vibration repeating itself rigorously within a time $\frac{1}{2}\tau$, yet it must be allowed to a vibration that may differ indefinitely little therefrom. In the Siren experiment, suppose that in one of the circles of holes containing an even number, every alternate hole is displaced along the arc of the circle by the same amount. The displacement may be made so small that no change can be detected in the resulting note; but the periodic time on which the pitch depends has been doubled. And secondly it is evident from the nature of periodicity, that the superposition on a vibration of period τ, of others having periods $\frac{1}{2}\tau$, $\frac{1}{3}\tau$...&c., does not disturb the period τ, while yet it cannot be supposed that the addition of the new elements has left the quality of the sound unchanged. Moreover, since the pitch is not affected by their presence, how do we know that elements of the shorter periods were not there from the beginning?

13. These considerations lead us to expect remarkable relations between the notes whose periods are as the reciprocals of the natural numbers. Nothing can be easier than to investigate the question by means of the Siren. Imagine two circles of holes, the inner containing any convenient number, and the outer twice as many. Then at whatever speed the disc may turn, the period of the vibration engendered by blowing the first set will necessarily be the double of that belonging to the second. On making the experiment the two notes are found to stand to each other in the relation of octaves; and we conclude that *in passing from any note to its octave, the frequency of vibration is doubled.* A similar method of experimenting shews, that to the ratio of periods 3 : 1 corresponds the interval known to musicians as the *twelfth*, made up of an octave and a fifth; to the ratio of 4 : 1, the double octave; and to the ratio 5 : 1, the interval made up of two octaves and a *major third*. In order to obtain the intervals of the fifth and third themselves, the ratios must be made 3 : 2 and 5 : 4 respectively.

14. From these experiments it appears that if two notes stand to one another in a fixed relation, then, no matter at what part of the scale they may be situated, their periods are in a certain constant ratio characteristic of the relation. The same may be said of their *frequencies*[1], or the number of vibrations which they execute in a given time. The ratio 2 : 1 is thus characteristic of the octave interval. If we wish to combine two intervals,—for instance, starting from a given note, to take a step of an octave and then another of a fifth in the same direction, the corresponding ratios must be compounded :

$$\frac{2}{1} \times \frac{3}{2} = \frac{3}{1}.$$

The twelfth part of an octave is represented by the ratio $\sqrt[12]{2} : 1$, for this is the step which repeated twelve times leads to an octave above the starting point. If we wish to have a measure of intervals in the proper sense, we must take not the characteristic ratio itself, but the logarithm of that ratio. Then, and then only, will the measure of a compound interval be the *sum* of the measures of the components.

---

[1] A single word to denote the number of vibrations executed in the unit of time is indispensable : I know no better than ' frequency,' which was used in this sense by Young. The same word is employed by Prof. Everett in his excellent edition of Deschanel's *Natural Philosophy.*

15. From the intervals of the octave, fifth, and third considered above, others known to musicians may be derived. The difference of an octave and a fifth is called a *fourth*, and has the ratio $2 \div \frac{3}{2} = \frac{4}{3}$. This process of subtracting an interval from the octave is called *inverting* it. By inverting the major third we obtain the minor sixth. Again, by subtraction of a major third from a fifth we obtain the minor third; and from this by inversion the major sixth. The following table exhibits side by side the names of the intervals and the corresponding ratios of frequencies:

Octave .... 2 : 1
Fifth .... 3 : 2
Fourth .... 4 : 3
Major Third .... 5 : 4
Minor Sixth .... 8 : 5
Minor Third .... 6 : 5
Major Sixth .... 5 : 3

These are all the consonant intervals comprised within the limits of the octave. It will be remarked that the corresponding ratios are all expressed by means of *small* whole numbers, and that this is more particularly the case for the more consonant intervals.

The notes whose frequencies are multiples of that of a given one, are called its *harmonics*, and the whole series constitutes a *harmonic scale*. As is well known to violinists, they may all be obtained from the same string by touching it lightly with the finger at certain points, while the bow is drawn.

The establishment of the connection between musical intervals and definite ratios of frequency—a fundamental point in Acoustics—is due to Mersenne (1636). It was indeed known to the Greeks in what ratios the lengths of strings must be changed in order to obtain the octave and fifth; but Mersenne demonstrated the law connecting the length of a string with the period of its vibration, and made the first determination of the actual rate of vibration of a known musical note.

16. On any note taken as a key-note, or *tonic*, a *diatonic* scale may be founded, whose derivation we now proceed to explain. If the key-note, whatever may be its absolute pitch, be called Do, the fifth above or dominant is Sol, and the fifth below

or subdominant is Fa. The common chord on any note is produced by combining it with its major third, and fifth, giving the
ratios of frequency $1 : \frac{5}{4} : \frac{3}{2}$ or $4 : 5 : 6$. Now if we take the
common chord on the tonic, on the dominant, and on the sub-
dominant, and transpose them when necessary into the octave
lying immediately above the tonic, we obtain notes whose frequencies arranged in order of magnitude are:

Do	Re	Mi	Fa	Sol	La	Si	Do
1,	$\frac{9}{8}$,	$\frac{5}{4}$,	$\frac{4}{3}$,	$\frac{3}{2}$,	$\frac{5}{3}$,	$\frac{15}{8}$,	2.

Here the common chord on Do is Do—Mi—Sol, with the
ratios $1 : \frac{5}{4} : \frac{3}{2}$; the chord on Sol is Sol—Si—Re, with the ratios
$\frac{3}{2} : \frac{15}{8} : 2 \times \frac{9}{8} = 1 : \frac{5}{4} : \frac{3}{2}$; and the chord on Fa is Fa—La—Do,
still with the same ratios. The scale is completed by repeating
these notes above and below at intervals of octaves.

If we take as our Do, or key-note, the lower c of a tenor voice,
the diatonic scale will be

c   d   e   f   g   a   b   c'.

Usage differs slightly as to the mode of distinguishing the
different octaves; in what follows I adopt the notation of Helmholtz. The octave below the one just referred to is written with
capital letters—C, D, &c.; the next below that with a suffix—
C,, D,, &c.; and the one beyond that with a double suffix—C,,, &c.
On the other side accents denote elevation by an octave—c', c'',
&c. The notes of the four strings of a violin are written in this
notation, g—d'—a'—e''. The middle c of the pianoforte is c'.
[In French notation c' is denoted by ut₃.]

17. With respect to an absolute standard of pitch there has
been no uniform practice. At the Stuttgard conference in 1834,
c' = 264 complete vibrations per second was recommended. This
corresponds to a' = 440. The French pitch makes a' = 435. In
Handel's time the pitch was much lower. If c' were taken at 256
or 2^8, all the c's would have frequencies represented by powers
of 2. This pitch is usually adopted by physicists and acoustical
instrument makers, and has the advantage of simplicity.

The determination *ab initio* of the frequency of a given note is
an operation requiring some care. The simplest method in prin-

ciple is by means of the Siren, which is driven at such a rate as to give a note in unison with the given one. The number of turns effected by the disc in one second is given by a counting apparatus, which can be thrown in and out of gear at the beginning and end of a measured interval of time. This multiplied by the number of effective holes gives the required frequency. The consideration of other methods admitting of greater accuracy must be deferred.

18. So long as we keep to the diatonic scale of c, the notes above written are all that are required in a musical composition. But it is frequently desired to change the key-note. Under these circumstances a singer with a good natural ear, accustomed to perform without accompaniment, takes an entirely fresh departure, constructing a new diatonic scale on the new key-note. In this way, after a few changes of key, the original scale will be quite departed from, and an immense variety of notes be used. On an instrument with fixed notes like the piano and organ such a multiplication is impracticable, and some compromise is necessary in order to allow the same note to perform different functions. This is not the place to discuss the question at any length; we will therefore take as an illustration the simplest, as well as the commonest case—modulation into the key of the dominant.

By definition, the diatonic scale of c consists of the common chords founded on c, g and f. In like manner the scale of g consists of the chords founded on g, d and c. The chords of c and g are then common to the two scales; but the third and fifth of d introduce new notes. The third of d written f♯ has a frequency $\frac{9}{8} \times \frac{5}{4} = \frac{45}{32}$, and is far removed from any note in the scale of c. But the fifth of d, with a frequency $\frac{9}{8} \times \frac{3}{2} = \frac{27}{16}$, differs but little from a, whose frequency is $\frac{5}{3}$. In ordinary keyed instruments the interval between the two, represented by $\frac{81}{80}$, and called a *comma*, is neglected, and the two notes by a suitable compromise or *temperament* are identified.

19. Various systems of temperament have been used; the simplest and that now most generally used, or at least aimed at, is the *equal* temperament. On referring to the table of frequencies for the diatonic scale, it will be seen that the intervals from Do to Re, from Re to Mi, from Fa to Sol, from Sol to La, and from La to Si, are nearly the same, being represented by $\frac{9}{8}$ or $\frac{10}{9}$; while the intervals from Mi to Fa and from Si to Do, represented by $\frac{16}{15}$, are about half as much. The equal temperament treats these approximate relations as exact, dividing the octave into twelve equal parts called mean semitones. From these twelve notes the diatonic scale belonging to any key may be selected according to the following rule. Taking the key-note as the first, fill up the series with the third, fifth, sixth, eighth, tenth, twelfth and thirteenth notes, counting upwards. In this way all difficulties of modulation are avoided, as the twelve notes serve as well for one key as for another. But this advantage is obtained at a sacrifice of true intonation. The equal temperament third, being the third part of an octave, is represented by the ratio $\sqrt[3]{2}:1$, or approximately 1·2599, while the true third is 1·25. The tempered third is thus higher than the true by the interval 126 : 125. The ratio of the tempered fifth may be obtained from the consideration that seven semitones make a fifth, while twelve go to an octave. The ratio is therefore $2^{\frac{7}{12}}:1$, which $= 1·4983$. The tempered fifth is thus too low in the ratio 1·4983 : 1·5, or approximately 881 : 882. This error is insignificant; and even the error of the third is not of much consequence in quick music on instruments like the pianoforte. But when the notes are *held*, as in the harmonium and organ, the consonance of chords is materially impaired.

20. The following Table, giving the twelve notes of the chromatic scale according to the system of equal temperament, will be convenient for reference[1]. The standard employed is $a' = 440$; in

	$C_{,,}$	$C_{,}$	C	c	c′	c″	c‴	c‴′
C	16·35	32·70	65·41	130·8	261·7	523·3	1046·6	2093·2
C♯	17·32	34·65	69·30	138·6	277·2	554·4	1108·8	2217·7
D	18·35	36·71	73·42	146·8	293·7	587·4	1174·8	2349·6
D♯	19·44	38·89	77·79	155·6	311·2	622·3	1244·6	2489·3
E	20·60	41·20	82·41	164·8	329·7	659·3	1318·6	2637·3
F	21·82	43·65	87·31	174·6	349·2	698·5	1397·0	2794·0
F♯	23·12	46·25	92·50	185·0	370·0	740·0	1480·0	2960·1
G	24·50	49·00	98·00	196·0	392·0	784·0	1568·0	3136·0
G♯	25·95	51·91	103·8	207·6	415·3	830·6	1661·2	3322·5
A	27·50	55·00	110·0	220·0	440·0	880·0	1760·0	3520·0
A♯	29·13	58·27	116·5	233·1	466·2	932·3	1864·6	3729·2
B	30·86	61·73	123·5	246·9	493·9	987·7	1975·5	3951·0

[1] Zamminer, *Die Musik und die musikalischen Instrumente.* Giessen, 1855.

order to adapt the Table to any other absolute pitch, it is only necessary to multiply throughout by the proper constant.

The ratios of the intervals of the equal temperament scale are given below (Zamminer):—

Note.	Frequency.		Note.	Frequency.
c	$= 1 \cdot 00000$		f♯	$2^{\frac{6}{12}} = 1 \cdot 41421$
c♯	$2^{\frac{1}{12}} = 1 \cdot 05946$		g	$2^{\frac{7}{12}} = 1 \cdot 49831$
d	$2^{\frac{2}{12}} = 1 \cdot 12246$		g♯	$2^{\frac{8}{12}} = 1 \cdot 58740$
d♯	$2^{\frac{3}{12}} = 1 \cdot 18921$		a	$2^{\frac{9}{12}} = 1 \cdot 68179$
e	$2^{\frac{4}{12}} = 1 \cdot 25992$		a♯	$2^{\frac{10}{12}} = 1 \cdot 78180$
f	$2^{\frac{5}{12}} = 1 \cdot 33484$		b	$2^{\frac{11}{12}} = 1 \cdot 88775$

$$c' = 2 \cdot 000$$

21. Returning now for a moment to the physical aspect of the question, we will assume, what we shall afterwards prove to be true within wide limits,—that, when two or more sources of sound agitate the air simultaneously, the resulting disturbance at any point in the external air, or in the ear-passage, is the simple sum (in the extended geometrical sense) of what would be caused by each source acting separately. Let us consider the disturbance due to a simultaneous sounding of a note and any or all of its harmonics. By definition, the complex whole forms a note having the same period (and therefore pitch) as its gravest element. We have at present no criterion by which the two can be distinguished, or the presence of the higher harmonics recognised. And yet—in the case, at any rate, where the component sounds have an independent origin—it is usually not difficult to detect them by the ear, so as to effect an analysis of the mixture. This is as much as to say that a strictly periodic vibration may give rise to a sensation which is not simple, but susceptible of further analysis. In point of fact, it has long been known to musicians that under certain circumstances the harmonics of a note may be heard along with it, even when the note is due to a single source, such as a vibrating string; but the significance of the fact was not understood. Since attention has been drawn to the subject, it has been proved (mainly by the labours of Ohm and Helmholtz) that almost all musical notes are highly compound, consisting in fact of the notes of a harmonic scale, from which in particular cases one or more members may be missing. The reason of the uncertainty and difficulty of the analysis will be touched upon presently.

22. That kind of note which the ear cannot further resolve is called by Helmholtz in German a '*ton*.' Tyndall and other recent writers on Acoustics have adopted 'tone' as an English equivalent, —a practice which will be followed in the present work. The thing is so important, that a convenient word is almost a matter of necessity. *Notes* then are in general made up of *tones*, the pitch of the note being that of the gravest tone which it contains.

23. In strictness the quality of pitch must be attached in the first instance to simple tones only ; otherwise the difficulty of discontinuity before referred to presents itself. The slightest change in the nature of a note may lower its pitch by a whole octave, as was exemplified in the case of the Siren. We should now rather say that the effect of the slight displacement of the alternate holes in that experiment was to introduce a new feeble tone an octave lower than any previously present. This is sufficient to alter the period of the whole, but the great mass of the sound remains very nearly as before.

In most musical notes, however, the fundamental or gravest tone is present in sufficient intensity to impress its character on the whole. The effect of the harmonic overtones is then to modify the *quality* or *character*[1] of the note, independently of pitch. That such a distinction exists is well known. The notes of a violin, tuning fork, or of the human voice with its different vowel sounds, &c., may all have the same pitch and yet differ independently of loudness ; and though a part of this difference is due to accompanying noises, which are extraneous to their nature as notes, still there is a part which is not thus to be accounted for. Musical notes may thus be classified as variable in three ways: First, *pitch*. This we have already sufficiently considered. Secondly, *character*, depending on the proportions in which the harmonic overtones are combined with the fundamental : and thirdly, *loudness*. This has to be taken last, because the ear is not capable of comparing (with any precision) the loudness of two notes which differ much in pitch or character. We shall indeed in a future chapter give a mechanical measure of the intensity of sound, including in one system all gradations of pitch ; but this is nothing to the point.

---

[1] German, 'Klangfarbe'—French, 'timbre.' The word 'character' is used in this sense by Everett.

We are here concerned with the intensity of the sensation of sound, not with a measure of its physical cause. The difference of loudness is, however, at once recognised as one of more or less ; so that we have hardly any choice but to regard it as dependent *cœteris paribus* on the magnitude of the vibrations concerned.

24. We have seen that a musical note, as such, is due to a vibration which is necessarily periodic; but the converse, it is evident, cannot be true without limitation. A periodic repetition of a noise at intervals of a second—for instance, the ticking of a clock—would not result in a musical note, be the repetition ever so perfect. In such a case we may say that the fundamental tone lies outside the limits of hearing, and although some of the harmonic overtones would fall within them, these would not give rise to a musical note or even to a chord, but to a noisy mass of sound like that produced by striking simultaneously the twelve notes of the chromatic scale. The experiment may be made with the Siren by distributing the holes quite irregularly round the circumference of a circle, and turning the disc with a moderate velocity. By the construction of the instrument, everything recurs after each complete revolution.

25. The principal remaining difficulty in the theory of notes and tones, is to explain why notes are sometimes analysed by the ear into tones, and sometimes not. If a note is really complex, why is not the fact immediately and certainly perceived, and the components disentangled ? The feebleness of the harmonic overtones is not the reason, for, as we shall see at a later stage of our inquiry, they are often of surprising loudness, and play a prominent part in music. On the other hand, if a note is sometimes perceived as a whole, why does not this happen always ? These questions have been carefully considered by Helmholtz[1], with a tolerably satisfactory result. The difficulty, such as it is, is not peculiar to Acoustics, but may be paralleled in the cognate science of Physiological Optics.

The knowledge of external things which we derive from the indications of our senses, is for the most part the result of inference. When an object is before us, certain nerves in our retinæ are excited, and certain sensations are produced, which we are accustomed to associate with the object, and we forthwith infer its presence. In the case of an unknown object the process is much

---

[1] *Tonempfindungen*, 3rd edition, p. 98.

the same. We interpret the sensations to which we are subject so
as to form a pretty good idea of their exciting cause. From the
slightly different perspective views received by the two eyes we
infer, often by a highly elaborate process, the actual relief and
distance of the object, to which we might otherwise have had no
clue. These inferences are made with extreme rapidity and quite
unconsciously. The whole life of each one of us is a continued
lesson in interpreting the signs presented to us, and in drawing
conclusions as to the actualities outside. Only so far as we succeed
in doing this, are our sensations of any use to us in the ordinary
affairs of life. This being so, it is no wonder that the study of our
sensations themselves falls into the background, and that subjective
phenomena, as they are called, become exceedingly difficult of
observation. As an instance of this, it is sufficient to mention the
' blind spot ' on the retina, which might *a priori* have been
expected to manifest itself as a conspicuous phenomenon, though
as a fact probably not one person in a hundred million would find
it out for themselves. The application of these remarks to the
question in hand is tolerably obvious. In the daily use of our ears
our object is to disentangle from the whole mass of sound that
may reach us, the parts coming from sources which may interest
us at the moment. When we listen to the conversation of a friend,
we fix our attention on the sound proceeding from him and
endeavour to grasp that as a whole, while we ignore, as far as
possible, any other sounds, regarding them as an interruption.
There are usually sufficient indications to assist us in making this
partial analysis. When a man speaks, the whole sound of his
voice rises and falls together, and we have no difficulty in recognising its unity. It would be no advantage, but on the contrary
a great source of confusion, if we were to carry the analysis further,
and resolve the whole mass of sound present into its component
tones. Although, as regards sensation, a resolution into tones
might be expected, the necessities of our position and the practice
of our lives lead us to stop the analysis at the point, beyond
which it would cease to be of service in deciphering our sensations, considered as signs of external objects[1].

But it may sometimes happen that however much we may
wish to form a judgment, the materials for doing so are absolutely

[1] Most probably the power of attending to the important and ignoring the
unimportant part of our sensations is to a great extent inherited—to how great an
extent we shall perhaps never know.

wanting.   When a note and its octave are sounding close together
and with perfect uniformity, there is nothing in our sensations to
enable us to distinguish, whether the notes have a double or a
single origin.   In the mixture stop of the organ, the pressing down
of each key admits the wind to a group of pipes, giving a note and
its first three or four harmonics.   The pipes of each group always
sound together, and the result is usually perceived as a single
note, although it does not proceed from a single source.

26.   The resolution of a note into its component tones is a
matter of very different difficulty with different individuals.   A
considerable effort of attention is required, particularly at first;
and, until a habit has been formed, some external aid in the shape
of a suggestion of what is to be listened for, is very desirable.

The difficulty is altogether very similar to that of learning to
draw.   From the machinery of vision it might have been expected
that nothing would be easier than to make, on a plane surface, a
representation of surrounding solid objects; but experience shews
that much practice is generally required.

We shall return to the question of the analysis of notes at a
later stage, after we have treated of the vibrations of strings, with
the aid of which it is best elucidated; but a very instructive
experiment, due originally to Ohm and improved by Helmholtz,
may be given here.   Helmholtz[1] took two bottles of the shape
represented in the figure, one about twice as large as the other.
These were blown by streams of air directed
across the mouth and issuing from gutta-percha
tubes, whose ends had been softened and pressed
flat, so as to reduce the bore to the form of a
narrow slit, the tubes being in connection with
the same bellows.   By pouring in water when
the note is too low and by partially obstructing
the mouth when the note is too high, the bottles
may be made to give notes with the exact
interval of an octave, such as b and b'.   The
larger bottle, blown alone, gives a somewhat muffled sound similar
in character to the vowel U; but, when both bottles are blown,
the character of the resulting sound is sharper, resembling rather
the vowel O.   For a short time after the notes had been heard
separately Helmholtz was able to distinguish them in the mixture ;

F I G . I.

[1]   *Tonempfindungen*, p. 109.

but as the memory of their separate impressions faded, the higher note seemed by degrees to amalgamate with the lower, which at the same time became louder and acquired a sharper character. This blending of the two notes may take place even when the high note is the louder.

27. Seeing now that notes are usually compound, and that only a particular sort called tones are incapable of further analysis, we are led to inquire what is the physical characteristic of tones, to which they owe their peculiarity? What sort of periodic vibration is it, which produces a simple tone ? According to what mathematical function of the time does the pressure vary in the passage of the ear ? No question in Acoustics can be more important.

The simplest periodic functions with which mathematicians are acquainted are the circular functions, expressed by a sine or cosine ; indeed there are no others at all approaching them in simplicity. They may be of any period, and admitting of no other variation (except magnitude), seem well adapted to produce simple tones. Moreover it has been proved by Fourier, that the most general single-valued periodic function can be resolved into a series of circular functions, having periods which are submultiples of that of the given function. Again, it is a consequence of the general theory of vibration that the particular type, now suggested as corresponding to a simple tone, is the only one capable of preserving its integrity among the vicissitudes which it may have to undergo. Any other kind is liable to a sort of physical analysis, one part being differently affected from another. If the analysis within the ear proceeded on a different principle from that effected according to the laws of dead matter outside the ear, the consequence would be that a sound originally simple might become compound on its way to the observer. There is no reason to suppose that anything of this sort actually happens. When it is added that according to all the ideas we can form on the subject, the analysis within the ear must take place by means of a physical machinery, subject to the same laws as prevail outside, it will be seen that a strong case has been made out for regarding tones as due to vibrations expressed by circular functions. We are not however left entirely to the guidance of general considerations like these. In the chapter on the vibration of strings, we shall see that in many cases theory informs us beforehand of the nature of the vibration executed by a string, and in particular whether any specified simple vibration is a component or not.　Here we have a decisive test.　It is found by experiment that, whenever according to theory any simple vibration is present, the corresponding tone can be heard, but, whenever the simple vibration is absent, then the tone cannot be heard.　We are therefore justified in asserting that simple tones and vibrations of a circular type are indissolubly connected.　This law was discovered by Ohm.

CHAPTER II.

28. THE vibrations expressed by a circular function of the time and variously designated as *simple, pendulous,* or *harmonic,* are so important in Acoustics that we cannot do better than devote a chapter to their consideration, before entering on the dynamical part of our subject. The quantity, whose variation constitutes the 'vibration,' may be the displacement of a particle measured in a given direction, the pressure at a fixed point in a fluid medium, and so on. In any case denoting it by u, we have

$$u = a \cos\left(\frac{2\pi t}{\tau} - \epsilon\right) \dots\dots\dots\dots\dots (1),$$

in which a denotes the *amplitude*, or extreme value of u; τ is the *periodic time*, or *period*, after the lapse of which the values of u recur; and ϵ determines the *phase* of the vibration at the moment from which t is measured.

Any number of harmonic vibrations *of the same period* affecting a variable quantity, compound into another of the same type, whose elements are determined as follows:

$$u = \Sigma a \cos\left(\frac{2\pi t}{\tau} - \epsilon\right)$$

$$= \cos\frac{2\pi t}{\tau} \Sigma a \cos \epsilon + \sin\frac{2\pi t}{\tau} \Sigma a \sin \epsilon$$

$$= r \cos\left(\frac{2\pi t}{\tau} - \theta\right) \dots\dots\dots\dots\dots (2),$$

if

$$r = \{(\Sigma a \cos \epsilon)^2 + (\Sigma a \sin \epsilon)^2\}^{\frac{1}{2}} \dots\dots\dots\dots (3),$$

and

$$\tan \theta = \Sigma a \sin \epsilon \div \Sigma a \cos \epsilon \dots\dots\dots\dots\dots (4).$$

For example, let there be two components,

$$u = a \cos \left(\frac{2\pi t}{\tau} - \epsilon \right) + a' \cos \left(\frac{2\pi t}{\tau} - \epsilon' \right);$$

then

$$r = \{a^2 + a'^2 + 2aa' \cos (\epsilon - \epsilon')\}^{\frac{1}{2}} \dots\dots\dots\dots\dots(5),$$

$$\tan \theta = \frac{a \sin \epsilon + a' \sin \epsilon'}{a \cos \epsilon + a' \cos \epsilon'} \dots\dots\dots\dots\dots\dots\dots (6).$$

Particular cases may be noted. If the phases of the two components agree,

$$u = (a + a') \cos \left(\frac{2\pi t}{\tau} - \epsilon \right).$$

If the phases differ by half a period,

$$u = (a - a') \cos \left(\frac{2\pi t}{\tau} - \epsilon \right),$$

so that if $a' = a$, u vanishes. In this case the vibrations are often said to *interfere*, but the expression is rather misleading. Two sounds may very properly be said to interfere, when they together cause silence; but the mere superposition of two vibrations (whether rest is the consequence, or not) cannot properly be so called. At least if this be interference, it is difficult to say what non-interference can be. It will appear in the course of this work that when vibrations exceed a certain intensity they no longer compound by mere addition; *this* mutual action might more properly be called interference, but it is a phenomenon of a totally different nature from that with which we are now dealing.

Again, if the phases differ by a quarter or by three-quarters of a period, $\cos (\epsilon - \epsilon') = 0$, and

$$r = \{a^2 + a'^2\}^{\frac{1}{2}}.$$

Harmonic vibrations of given period may be represented by lines drawn from a pole, the lengths of the lines being proportional to the amplitudes, and the inclinations to the phases of the vibrations. The resultant of any number of harmonic vibrations is then represented by the geometrical resultant of the corresponding lines. For example, if they are disposed symmetrically round the pole, the resultant of the lines, or vibrations, is zero.

29. If we measure off along an axis of x distances proportional to the time, and take u for an ordinate, we obtain the harmonic curve, or curve of sines,

$$u = a \cos\left(\frac{2\pi x}{\lambda} - \epsilon\right),$$

where λ, called the wave-length, is written in place of τ, both quantities denoting the range of the independent variable corresponding to a complete recurrence of the function. The harmonic curve is thus the locus of a point subject at once to a uniform motion, and to a harmonic vibration in a perpendicular direction. In the next chapter we shall see that the vibration of a tuning fork is simple harmonic; so that if an excited tuning fork be moved with uniform velocity parallel to the line of its handle, a tracing point attached to the end of one of its prongs describes a harmonic curve, which may be obtained in a permanent form by allowing the tracing point to bear gently on a piece of smoked paper. In Fig. 2 the continuous lines are two harmonic curves of the same wave-length and amplitude, but of different

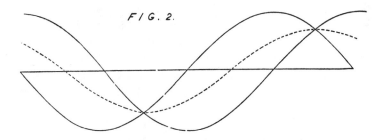

F I G . 2.

phases; the dotted curve represents half their resultant, being the locus of points midway between those in which the two curves are met by any ordinate.

30. If two harmonic vibrations of different periods coexist,

$$u = a \cos\left(\frac{2\pi t}{\tau} - \epsilon\right) + a' \cos\left(\frac{2\pi t}{\tau'} - \epsilon'\right).$$

The resultant cannot here be represented as a simple harmonic motion with other elements. If τ and τ' be incommensurable, the value of u never recurs; but, if τ and τ' be in the ratio of two whole numbers, u recurs after the lapse of a time equal to the least common multiple of τ and τ'; but the vibration is not simple harmonic. For example, when a note and its fifth are sounding together, the vibration recurs after a time equal to twice the period of the graver.

One case of the composition of harmonic vibrations of different periods is worth special discussion, namely, when the difference of the periods is small. If we fix our attention on the course of things during an interval of time including merely a few periods, we see that the two vibrations are nearly the same as if their periods were absolutely equal, in which case they would, as we know, be equivalent to another simple harmonic vibration of the same period. For a few periods then the resultant motion is approximately simple harmonic, but the same harmonic will not continue to represent it for long. The vibration having the shorter period continually gains on its fellow, thereby altering the difference of phase on which the elements of the resultant depend. For simplicity of statement let us suppose that the two components have equal amplitudes, frequencies represented by m and n, where $m - n$ is small, and that when first observed their phases agree. At this moment their effects conspire, and the resultant has an amplitude double of that of the components. But after a time $1 \div 2 (m - n)$ the vibration m will have gained half a period relatively to the other; and the two, being now in complete disagreement, neutralize each other. After a further interval of time equal to that above named, m will have gained altogether a whole vibration, and complete accordance is once more re-established. The resultant motion is therefore approximately simple harmonic, with an amplitude not constant, but varying from zero to twice that of the components, the frequency of these alterations being $m - n$. If two tuning forks with frequencies 500 and 501 be equally excited, there is every second a rise and fall of sound corresponding to the coincidence or opposition of their vibrations. This phenomenon is called beats. We do not here fully discuss the question how the ear behaves in the presence of vibrations having nearly equal frequencies, but it is obvious that if the motion in the neighbourhood of the ear almost cease for a considerable fraction of a second, the sound must appear to fall. For reasons that will afterwards appear, beats are best heard when the interfering sounds are simple tones. Consecutive notes of the stopped diapason of the organ shew the phenomenon very well, at least in the lower parts of the scale. A permanent interference of two notes may be obtained by mounting two stopped organ pipes of similar construction and identical pitch side by side on the same wind chest. The vibrations of the two pipes adjust themselves to complete opposition, so that at a. little distance nothing can be heard, except the hissing of the wind If by a rigid wall between the two pipes one sound could be cut off, the other would be instantly restored. Or the balance, on which silence depends, may be upset by connecting the ear with a tube, whose other end lies close to the mouth of one of the pipes.

By means of beats two notes may be tuned to unison with great exactness. The object is to make the beats as slow as possible, since the number of beats in a second is equal to the difference of the frequencies of the notes. Under favourable circumstances beats so slow as one in 30 seconds may be recognised, and would indicate that the higher note gains only two vibrations a *minute* on the lower. Or it might be desired merely to ascertain the difference of the frequencies of two notes nearly in unison, in which case nothing more is necessary than to count the number of beats. It will be remembered that the difference of frequencies does not determine the *interval* between the two notes; that depends on the *ratio* of frequencies. Thus the rapidity of the beats given by two notes nearly in unison is doubled, when both are taken an exact octave higher.

Analytically

$$u = a \cos (2\pi mt - \epsilon) + a' \cos (2\pi nt - \epsilon'),$$

where $m - n$ is small.

Now $\cos (2\pi nt - \epsilon')$ may be written

$$\cos \{2\pi mt - 2\pi (m - n) t - \epsilon'\},$$

and we have

$$u = r \cos (2\pi mt - \theta) \dots\dots\dots\dots (1),$$

where

$$r^2 = a^2 + a'^2 + 2aa' \cos \{2\pi (m - n) t + \epsilon' - \epsilon\}\dots\dots(2),$$

$$\tan \theta = \frac{a \sin \epsilon + a' \sin \{2\pi (m - n) t + \epsilon'\}}{a \cos \epsilon + a' \cos \{2\pi (m - n) t + \epsilon'\}}\dots\dots(3).$$

The resultant vibration may thus be considered as harmonic with elements r and θ, which are not constant but slowly varying functions of the time, having the frequency $m - n$. The amplitude r is at its maximum when

$$\cos \{2\pi (m - n) t + \epsilon' - \epsilon\} = + 1,$$

and at its minimum when

$$\cos \{2\pi (m - n) t + \epsilon' - \epsilon\} = - 1,$$

the corresponding values being $a + a'$ and $a - a'$ respectively.

31. Another case of great importance is the composition of vibrations corresponding to a tone and its harmonics. It is known that the most general single-valued finite periodic function can be expressed by a series of simple harmonics—

$$u = a_0 + \Sigma_{n=1}^{n=\infty} a_n \cos\left(\frac{2\pi n t}{\tau} - \epsilon_n\right) \dots\dots\dots(1),$$

a theorem usually quoted as Fourier's. Analytical proofs will be found in Todhunter's *Integral Calculus* and Thomson and Tait's *Natural Philosophy;* and a line of argument almost if not quite amounting to a demonstration will be given later in this work. A few remarks are all that will be required here.

Fourier's theorem is not obvious. A vague notion is not uncommon that the infinitude of arbitrary constants in the series of necessity endows it with the capacity of representing an arbitrary periodic function. That this is an error will be apparent, when it is observed that the same argument would apply equally, if one term of the series were omitted; in which case the expansion would not in general be possible.

Another point worth notice is that simple harmonics are not the only functions, in a series of which it is possible to expand one arbitrarily given. Instead of the simple elementary term

$$\cos\left(\frac{2\pi n t}{\tau} - \epsilon_n\right),$$

we might take

$$\cos\left(\frac{2\pi n t}{\tau} - \epsilon_n\right) + \frac{1}{2}\cos\left(\frac{4\pi n t}{\tau} - \epsilon_n\right),$$

formed by adding a similar one in the same phase of half the amplitude and period. It is evident that these terms would serve as well as the others; for

$$\cos\left(\frac{2\pi n t}{\tau} - \epsilon_n\right) = \left\{\cos\left(\frac{2\pi n t}{\tau} - \epsilon_n\right) + \frac{1}{2}\cos\left(\frac{4\pi n t}{\tau} - \epsilon_n\right)\right\}$$

$$- \frac{1}{2}\left\{\cos\left(\frac{4\pi n t}{\tau} - \epsilon_n\right) + \frac{1}{2}\cos\left(\frac{8\pi n t}{\tau} - \epsilon_n\right)\right\}$$

$$+ \frac{1}{4}\left\{\cos\left(\frac{8\pi n t}{\tau} - \epsilon_n\right) + \frac{1}{2}\cos\left(\frac{16\pi n t}{\tau} - \epsilon_n\right)\right\}$$

$$- \dots\dots ad\ infin.,$$

so that each term in Fourier's series, and therefore the sum of the series, can be expressed by means of the double elementary terms now suggested. This is mentioned here, because students, not being acquainted with other expansions, may imagine that simple harmonic functions are by nature the only ones qualified to be the elements in the development of a periodic function. The reason of the preeminent importance of Fourier's series in Acoustics is the mechanical one referred to in the preceding chapter, and to be explained more fully hereafter, namely, that, in general, simple harmonic vibrations are the only kind that are propagated through a vibrating system without suffering decomposition.

32. As in other cases of a similar character, e.g. Taylor's theorem, if the possibility of the expansion be known, the coefficients may be determined by a comparatively simple process. We may write (1) of § 31

$$u = A_0 + \Sigma_{n=1}^{n=\infty} A_n \cos \frac{2n\pi t}{\tau} + \Sigma_{n=1}^{n=\infty} B_n \sin \frac{2n\pi t}{\tau} \ \ldots \ldots (1).$$

Multiplying by $\cos (2n\pi t/\tau)$ or $\sin (2n\pi t/\tau)$, and integrating over a complete period from $t = 0$ to $t = \tau$, we find

$$\left. \begin{aligned} A_n &= \frac{2}{\tau} \int_0^\tau u \cos \frac{2n\pi t}{\tau} \, dt \\ B_n &= \frac{2}{\tau} \int_0^\tau u \sin \frac{2n\pi t}{\tau} \, dt \end{aligned} \right\} \ \ldots \ldots \ldots \ldots \ (2).$$

An immediate integration gives

$$A_0 = \frac{1}{\tau} \int_0^\tau u \, dt \ldots \ldots \ldots \ldots \ldots \ldots \ldots \ldots \ldots (3),$$

indicating that A_0 is the *mean* value of u throughout the period.

The degree of convergency in the expansion of u depends in general on the continuity of the function and its derivatives. The series formed by successive differentiations of (1) converge less and less rapidly, but still remain convergent, and arithmetical representatives of the differential coefficients of u, so long as these latter are everywhere finite. Thus (Thomson and Tait, § 77), if all the derivatives up to the mth inclusive be free from infinite values, the series for u is more convergent than one with

$$1, \ \frac{1}{2^m}, \ \frac{1}{3^m}, \ \frac{1}{4^m}, \ldots \ldots \&c.,$$

for coefficients.

32 a. The general explanation of the beats heard when two pure tones nearly in unison are sounded simultaneously has been discussed in § 30. But the occurrence of beats is not confined to the case of approximate unison, at least when we have to deal with compound notes. Suppose for example that the interval is an octave. The graver note then usually includes a tone coincident in pitch with the fundamental tone of the higher note. If the interval be disturbed, the previously coincident tones separate from one another, and give rise to beats of the same frequency as if they existed alone. There is usually no difficulty in observing these beats; but if one or both of the component tones concerned be very faint, the aid of a resonator may be invoked.

In general we may consider that each consonant interval is characterized by the coincidence of certain component tones, and if the interval be disturbed the previously coincident tones give rise to beats. Of course it may happen in any particular case that the tones which would coincide in pitch are absent from one or other of the notes. The disturbance of the interval would then, according to the above theory, not be attended by beats. In practice faint beats are usually heard; but the discussion of this phenomenon, as to which authorities are not entirely agreed, must be postponed.

33. Another class of compounded vibrations, interesting from the facility with which they lend themselves to optical observation, occur when two harmonic vibrations affecting the same particle are executed *in perpendicular directions*, more especially when the periods are not only commensurable, but in the ratio of two *small* whole numbers. The motion is then completely periodic, with a period not many times greater than those of the components, and the curve described is re-entrant. If u and v be the co-ordinates, we may take

$$u = a \cos (2\pi nt - \epsilon), \quad v = b \cos 2\pi n't \dots\dots\dots\dots(1).$$

First let us suppose that the periods are equal, so that $n' = n$; the elimination of t gives for the equation of the curve described,

$$\frac{u^2}{a^2} + \frac{v^2}{b^2} - \frac{2uv}{ab} \cos \epsilon - \sin^2 \epsilon = 0 \dots\dots\dots\dots(2),$$

representing in general an ellipse, whose position and dimensions depend upon the amplitudes of the original vibrations and upon the difference of their phases. If the phases differ by a quarter period, cos $\epsilon = 0$, and the equation becomes

$$\frac{u^2}{a^2} + \frac{v^2}{b^2} = 1.$$

In this case the axes of the ellipse coincide with those of co-ordinates. If further the two components have equal amplitudes, the locus degenerates into the circle

$$u^2 + v^2 = a^2,$$

which is described with uniform velocity. This shews how a uniform circular motion may be analysed into two rectilinear harmonic motions, whose directions are perpendicular.

If the phases of the components agree, $\epsilon = 0$, and the ellipse degenerates into the coincident straight lines

$$\left(\frac{u}{a} - \frac{v}{b}\right)^2 = 0 \, ;$$

or if the difference of phase amount to half a period, into

$$\left(\frac{u}{a} + \frac{v}{b}\right)^2 = 0.$$

When the unison of the two vibrations is exact, the elliptic path remains perfectly steady, but in practice it will almost always happen that there is a slight difference between the periods. The consequence is that though a fixed ellipse represents the curve described with sufficient accuracy for a few periods, the ellipse itself gradually changes in correspondence with the alteration in the magnitude of ϵ. It becomes therefore a matter of interest to consider the system of ellipses represented by (2), supposing a and b constants, but ϵ variable.

Since the extreme values of u and v are $\pm a$, $\pm b$ respectively, the ellipse is in all cases inscribed in the rectangle whose sides are $2a$, $2b$. Starting with the phases in agreement, or $\epsilon = 0$, we have the ellipse coincident with the diagonal $\frac{u}{a} - \frac{v}{b} = 0$. As ϵ increases from 0 to $\frac{1}{2}\pi$, the ellipse opens out until its equation becomes

$$\frac{u^2}{a^2} + \frac{v^2}{b^2} = 1.$$

From this point it closes up again, ultimately coinciding with the other diagonal $\frac{u}{a} + \frac{v}{b} = 0$, corresponding to the increase of ϵ from $\frac{1}{2}\pi$ to π. After this, as ϵ ranges from π to 2π, the ellipse retraces

its course until it again coincides with the first diagonal. The
sequence of changes is exhibited in Fig. 3.

FIG.3.

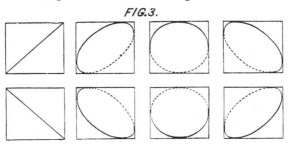

The ellipse, having already four given tangents, is completely
determined by its point of contact P (Fig. 4) with the line $v = b$.

FIG. 4.

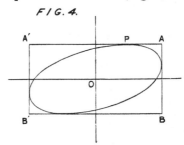

In order to connect this with ϵ, it is sufficient to observe that
when $v = b$, $\cos 2\pi nt = 1$; and therefore $u = a \cos \epsilon$. Now if the
elliptic paths be the result of the superposition of two harmonic
vibrations of nearly coincident pitch, ϵ varies uniformly with the
time, so that P itself executes a harmonic vibration along AA'
with a frequency equal to the difference of the two given frequencies.

34. Lissajous[1] has shewn that this system of ellipses may be
regarded as the different aspects of one and the same ellipse
described on the surface of a transparent cylinder. In Fig. 5

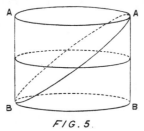

FIG. 5.

[1] *Annales de Chimie* (3) LI. 147, 1857.

$AA'B'B$ represents the cylinder, of which AB' is a plane section. Seen from an infinite distance in the direction of the common tangent at A to the plane sections, the cylinder is projected into a rectangle, and the ellipse into its diagonal. Suppose now that the cylinder turns upon its axis, carrying the plane section with it. Its own projection remains a constant rectangle in which the pro-

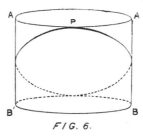

F I G. 6.

jection of the ellipse is inscribed. Fig. 6 represents the position of the cylinder after a rotation through a right angle. It appears therefore that by turning the cylinder round we obtain in succession all the ellipses corresponding to the paths described by a point subject to two harmonic vibrations of equal period and fixed amplitudes. Moreover if the cylinder be turned continuously with uniform velocity, which insures a harmonic motion for P, we obtain a complete representation of the varying orbit described by the point when the periods of the two components differ slightly, each complete revolution answering to a gain or loss of a single vibration[1]. The revolutions of the cylinder are thus synchronous with the beats which would result from the composition of the two vibrations, if they were to act in the same direction.

35. Vibrations of the kind here considered are very easily realized experimentally. A heavy pendulum-bob, hung from a fixed point by a long wire or string, describes ellipses under the action of gravity, which may in particular cases, according to the circumstances of projection, pass into straight lines or circles. But in order to see the orbits to the best advantage, it is necessary that they should be described so quickly that the impression on the retina made by the moving point at any part of its course has not time to fade materially, before the point comes round again to renew its action. This condition is fulfilled by the vibration of a silvered bead (giving by reflection a luminous point), which is

---

[1] By a vibration will always be meant in this work a *complete* cycle of changes.

attached to a straight metallic wire (such as a knitting-needle),
firmly clamped in a vice at the lower end. When the system is set
into vibration, the luminous point describes ellipses, which appear
as fine lines of light. These ellipses would gradually contract in
dimensions under the influence of friction until they subsided
into a stationary bright point, without undergoing any other
change, were it not that in all probability, owing to some want
of symmetry, the wire has slightly differing periods according to
the plane in which the vibration is executed. Under these circumstances the orbit is seen to undergo the cycle of changes
already explained.

36. So far we have supposed the periods of the component
vibrations to be equal, or nearly equal; the next case in order of
simplicity is when one is the double of the other. We have

$$u = a \cos(4n\pi t - \epsilon), \quad v = b \cos 2n\pi t.$$

The locus resulting from the elimination of t may be written

$$\frac{u}{a} = \cos \epsilon \left(2\frac{v^2}{b^2} - 1 \right) + 2 \sin \epsilon \frac{v}{b} \sqrt{1 - \frac{v^2}{b^2}} \dots\dots\dots(1),$$

which for all values of ϵ represents a curve inscribed in the rectangle $2a$, $2b$. If $\epsilon = 0$, or π, we have

$$v^2 = \frac{b^2}{2}\left(1 \pm \frac{u}{a} \right),$$

FIG. 7.

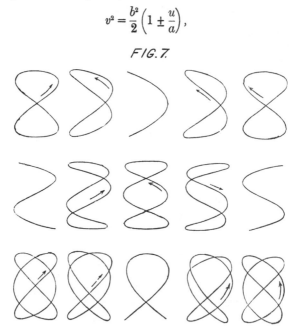

representing parabolas. Fig. 7 shews the various curves for the intervals of the octave, twelfth, and fifth.

To all these systems Lissajous' method of representation by the transparent cylinder is applicable, and when the relative phase is altered, whether from the different circumstances of projection in different cases, or continuously owing to a slight deviation from exactness in the ratio of the periods, the cylinder will appear to turn, so as to present to the eye different aspects of the same line traced on its surface.

37. There is no difficulty in arranging a vibrating system so that the motion of a point shall consist of two harmonic vibrations in perpendicular planes, with their periods in any assigned ratio. The simplest is that known as Blackburn's pendulum. A wire ACB is fastened at A and B, two fixed points at the same level. The bob P is attached to its middle point by another wire CP. For vibrations in the plane of the diagram, the point of suspension is practically C, provided that the wires are sufficiently stretched; but for a motion perpendicular to this plane, the bob turns about D, carrying the wire ACB with it. The periods of vibration in

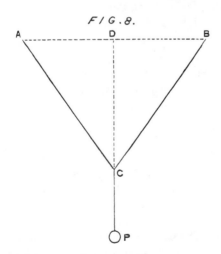

F I G . 8.

the principal planes are in the ratio of the square roots of CP and DP. Thus if $DC = 3CP$, the bob describes the figures of the octave. To obtain the sequence of curves corresponding to approximate unison, ACB must be so nearly tight, that CD is relatively small.

38. Another contrivance called the kaleidophone was originally invented by Wheatstone. A straight thin bar of steel carrying a bead at its upper end is fastened in a vice, as explained in a previous paragraph. If the section of the bar is square, or circular, the period of vibration is independent of the plane in which it is performed. But let us suppose that the section is a rectangle with unequal sides. The stiffness of the bar—the force with which it resists bending—is then greater in the plane of greater thickness, and the vibrations in this plane have the shorter period. By a suitable adjustment of the thicknesses, the two periods of vibration may be brought into any required ratio, and the corresponding curve exhibited.

The defect in this arrangement is that the same bar will give only one set of figures. In order to overcome this objection the following modification has been devised. A slip of steel is taken whose rectangular section is very elongated, so that as regards bending in one plane the stiffness is so great as to amount practically to rigidity. The bar is divided into two parts, and the broken ends reunited, the two pieces being turned on one another through a right angle, so that the plane, which contains the small thickness of one, contains the great thickness of the other. When the compound rod is clamped in a vice at a point below the junction, the period of the vibration in one direction, depending almost entirely on the length of the upper piece, is nearly constant; but that in the second direction may be controlled by varying the point at which the lower piece is clamped.

39. In this arrangement the luminous point itself executes the vibrations which are to be observed; but in Lissajous' form of the experiment, the point of light remains really fixed, while its *image* is thrown into apparent motion by means of successive reflection from two vibrating mirrors. A small hole in an opaque screen placed close to the flame of a lamp gives a point of light, which is observed after reflection in the mirrors by means of a small telescope. The mirrors, usually of polished steel, are attached to the prongs of stout tuning forks, and the whole is so disposed that when the forks are thrown into vibration the luminous point appears to describe harmonic motions in perpendicular directions, owing to the angular motions of the reflecting surfaces. The amplitudes and periods of these harmonic motions depend upon those of the corresponding forks, and may be made such as to give with enhanced brilliancy any of the figures possible with the kaleidophone. By a similar arrangement it is possible to project the figures on a screen. In either case they gradually contract as the vibrations of the forks die away.

40. The principles of this chapter have received an important application in the investigation of rectilinear periodic motions. When a point, for instance a particle of a sounding string, is vibrating with such a period as to give a note within the limits of hearing, its motion is much too rapid to be followed by the eye; so that, if it be required to know the character of the vibration, some indirect method must be adopted. The simplest, theoretically, is to compound the vibration under examination with a uniform motion of translation in a perpendicular direction, as when a tuning-fork draws a harmonic curve on smoked paper. Instead of moving the vibrating body itself, we may make use of a revolving mirror, which provides us with an *image* in motion. In this way we obtain a representation of the function characteristic of the vibration, with the abscissa proportional to time.

But it often happens that the application of this method would be difficult or inconvenient. In such cases we may substitute for the uniform motion a harmonic vibration of suitable period in the same direction. To fix our ideas, let us suppose that the point, whose motion we wish to investigate, vibrates vertically with a period τ, and let us examine the result of combining with this a horizontal harmonic motion, whose period is some multiple of τ, say, $n\tau$. Take a rectangular piece of paper, and with axes parallel to its edges draw the curve representing the vertical motion (by setting off abscissæ proportional to the time) on such a scale that the paper just contains n repetitions or waves, and then bend the paper round so as to form a cylinder, with a re-entrant curve running round it. A point describing this curve in such a manner that it revolves uniformly about the axis of the cylinder will appear from a distance to combine the given vertical motion of period τ, with a horizontal harmonic motion of period $n\tau$. Conversely therefore, in order to obtain the representative curve of the vertical vibrations, the cylinder containing the apparent path must be imagined to be divided along a generating line, and developed into a plane. There is less difficulty in conceiving the cylinder and the situation of the curve upon it, when the adjustment of the periods is not quite exact, for then the cylinder appears to turn, and the contrary motions serve to distinguish those parts of the curve which lie on its nearer and further face.

41. The auxiliary harmonic motion is generally obtained optically, by means of an instrument called a vibration-microscope invented by Lissajous. One prong of a large tuning-fork carries a lens, whose axis is perpendicular to the direction of vibration; and which may be used either by itself, or as the object-glass of a compound microscope formed by the addition of an eye-piece independently supported. In either case a stationary point is thrown into apparent harmonic motion along a line parallel to that of the fork's vibration.

The vibration-microscope may be applied to test the rigour and universality of the law connecting *pitch* and *period*. Thus it will be found that any point of a vibrating body which gives a pure musical note will appear to describe a re-entrant curve, when examined with a vibration-microscope whose note is in strict unison with its own. By the same means the ratios of frequencies characteristic of the consonant intervals may be verified; though for this latter purpose a more thoroughly acoustical method, to be described in a future chapter, may be preferred.

42. Another method of examining the motion of a vibrating body depends upon the use of intermittent illumination[1]. Suppose, for example, that by means of suitable apparatus a series of electric sparks are obtained at regular intervals τ. A vibrating body, whose period is also τ, examined by the light of the sparks must appear at rest, because it can be seen only in one position. If, however, the period of the vibration differ from τ ever so little, the illuminated position varies, and the body will appear to vibrate slowly with a frequency which is the difference of that of the spark and that of the body. The type of vibration can then be observed with facility.

The series of sparks can be obtained from an induction-coil, whose primary circuit is periodically broken by a vibrating fork, or by some other interrupter of sufficient regularity. But a better result is afforded by sunlight rendered intermittent with the aid of a fork, whose prongs carry two small plates of metal, parallel to the plane of vibration and close together. In each plate is a slit

---

[1] Plateau, *Bull. de l'Acad. roy. de Belgique*, t. III, p. 364, 1836.

parallel to the prongs of the fork, and so placed as to afford a free passage through the plates when the fork is at rest, or passing through the middle point of its vibrations. On the opening so formed, a beam of sunlight is concentrated by means of a burning-glass, and the object under examination is placed in the cone of rays diverging on the further side[1]. When the fork is made to vibrate by an electro-magnetic arrangement, the illumination is cut off except when the fork is passing through its position of equilibrium, or nearly so. The flashes of light obtained by this method are not so instantaneous as electric sparks (especially when a jar is connected with the secondary wire of the coil), but in my experience the regularity is more perfect. Care should be taken to cut off extraneous light as far as possible, and the effect is then very striking.

A similar result may be arrived at by looking at the vibrating body through a series of holes arranged in a circle on a revolving disc. Several series of holes may be provided on the same disc, but the observation is not satisfactory without some provision for securing uniform rotation.

Except with respect to the sharpness of definition, the result is the same when the period of the light is any multiple of that of the vibrating body. This point must be attended to when the revolving wheel is used to determine an unknown frequency.

When the frequency of intermittence is an exact multiple of that of the vibration, the object is seen without apparent motion, but generally in more than one position. This condition of things is sometimes advantageous.

Similar effects arise when the frequencies of the vibrations and of the flashes are in the ratio of two small whole numbers. If, for example, the number of vibrations in a given time be half as great again as the number of flashes, the body will appear stationary, and in general double.

42 a. We have seen (§ 28) that the resultant of two isoperiodic vibrations of equal amplitude is wholly dependent upon their phase relation, and it is of interest to inquire what we are to expect from the composition of a large number (n) of equal vibrations of amplitude unity, of the same period, and of phases accidentally determined. The intensity of the resultant, represented by the square of the amplitude § 245, will of course depend upon the

---

[1] Töpler, *Phil. Mag.* Jan. 1867.

precise manner in which the phases are distributed, and may vary from n^2 to zero. But is there a definite intensity which becomes more and more probable when n is increased without limit?

The nature of the question here raised is well illustrated by the special case in which the possible phases are restricted to two *opposite* phases. We may then conveniently discard the idea of phase, and regard the amplitudes as at random *positive* or *negative*. If all the signs be the same, the intensity is n^2; if, on the other hand, there be as many positive as negative, the result is zero. But although the intensity may range from 0 to n^2, the smaller values are more probable than the greater.

The simplest part of the problem relates to what is called in the theory of probabilities the "expectation" of intensity, that is, the mean intensity to be expected after a great number of trials, in each of which the phases are taken at random. The chance that all the vibrations are positive is $(\frac{1}{2})^n$, and thus the expectation of intensity corresponding to this contingency is $(\frac{1}{2})^n . n^2$. In like manner the expectation corresponding to the number of positive vibrations being $(n-1)$ is

$$(\tfrac{1}{2})^n n (n-2)^2,$$

and so on. The whole expectation of intensity is thus

$$\frac{1}{2^n}\left\{1 . n^2 + n(n-2)^2 + \frac{n(n-1)}{1 . 2}(n-4)^2 \right.$$
$$\left. + \frac{n(n-1)(n-2)}{1 . 2 . 3}(n-6)^2 + ...\right\} ...... (1).$$

Now the sum of the $(n+1)$ terms of this series is simply n, as may be proved by comparison of coefficients of x^2 in the equivalent forms

$$(e^x + e^{-x})^n = 2^n(1 + \tfrac{1}{2}x^2 + ...)^n$$
$$= e^{nx} + n e^{(n-2)x} + \frac{n(n-1)}{1 . 2} e^{(n-4)x} + ...$$

The expectation of intensity is therefore n, and this whether n be great or small.

The same conclusion holds good when the phases are unrestricted. From (3) § 28, if $a_1 = a_2 = ... = 1$,

$$r^2 = (\cos \epsilon_1 + \cos \epsilon_2 + ...)^2 + (\sin \epsilon_1 + \sin \epsilon_2 + ...)^2$$
$$= n + 2\Sigma \cos(\epsilon_2 - \epsilon_1) ....(2),$$

where under the sign of summation are to be included the cosines of the $\frac{1}{2}n(n-1)$ differences of phase. When the phases are accidental, the sum is as likely to be positive as negative, and thus the mean value of r^2 is n.

The reader must be on his guard here against a fallacy which has misled some eminent authors. We have not proved that when n is large there is any tendency for a single combination to give an intensity equal to n, but the quite different proposition that in a large number of trials, in each of which the phases are distributed at random, the *mean* intensity will tend more and more to the value n. It is true that even in a single combination there is no reason why any of the cosines in (2) should be positive rather than negative. From this we may infer that when n is increased the sum of the terms tends to vanish in comparison with the number of terms; but, the number of the terms being of the order n^2, we can infer nothing as to the value of the sum of the series in comparison with n.

So far there is no difficulty; but a complete investigation of this subject involves an estimate of the relative probabilities of resultants lying within assigned limits of magnitude. For example, we ought to be able to say what is the probability that the intensity due to a large number (n) of equal components is less than $\frac{1}{2}n$. This problem may conveniently be considered here, though it is naturally beyond the reach of elementary methods. We will commence by taking it under the restriction that the phases are of two opposite kinds only.

Adopting the statistical method of statement, let us suppose that there are an immense number N of independent combinations, each consisting of n unit vibrations, positive or negative, and combined accidentally. When N is sufficiently large, the statistics become regular; and the number of combinations in which the resultant amplitude is found equal to x may be denoted by $N \cdot f(n, x)$, where f is a definite function of n and x. Now suppose that each of the N combinations receives another random contribution of ± 1, and inquire how many of them will subsequently possess a resultant x. It is clear that those only can do so which originally had amplitudes $x - 1$, or $x + 1$. *Half* of the former, and *half* of the latter number will acquire the amplitude x, so that the number required is
$$\tfrac{1}{2}Nf(n, x-1) + \tfrac{1}{2}Nf(n, x+1).$$
But this must be identical with the number corresponding to $n + 1$ and x, so that
$$f(n + 1, x) = \tfrac{1}{2}f(n, x - 1) + \tfrac{1}{2}f(n, x + 1) \ldots\ldots\ldots(3).$$

This equation of differences holds good for all integral values of x and for all positive integral values of n. If $f(n, x)$ be given for one value of n, the equation suffices to determine $f(n, x)$ for all higher integral values of n. For the present purpose the initial value of n is zero. In that case we know that $f(x) = 0$ for all values of x other than zero, and that when $x = 0$, $f(0, 0) = 1$.

The problem proposed in the above form is perfectly definite; but for our immediate object it suffices to limit ourselves to the supposition that n is great, regarding $f(n, x)$ as a continuous function of continuous variables n and x, much as in the analogous problem of §§ 120, 121, 122.

Writing (3) in the form

$$f(n+1, x) - f(n, x) = \tfrac{1}{2}f(n, x-1) + \tfrac{1}{2}f(n, x+1) - f(n, x)...(4),$$

we see that the left-hand member may then be identified with df/dn, and the right-hand member with $\tfrac{1}{2}d^2f/dx^2$, so that under these circumstances the differential equation to which (3) reduces is of the well-known form

$$\frac{df}{dn} = \frac{1}{2}\frac{d^2f}{dx^2} \dots\dots\dots\dots\dots\dots(5).$$

The analogy with the conduction of heat is indeed very close; and the methods developed by Fourier for the solution of problems in the latter subject are at once applicable. The special condition here is that initially, that is when $n = 0$, f must vanish for all values of x other than zero. As may be verified by differentiation, the special solution of (5) is then

$$f(n, x) = \frac{A}{\sqrt{n}}\, e^{-x^2/2n} \dots\dots\dots\dots\dots(6),$$

in which A is an arbitrary constant to be determined from the consideration that the whole number of combinations is N. Thus, if dx be large in comparison with unity, the number of combinations which have amplitudes between x and $x + dx$ is

$$\frac{AN}{\sqrt{n}}\, e^{-x^2/2n}\, dx\,;$$

while

$$\frac{AN}{\sqrt{n}} \int_{-\infty}^{+\infty} e^{-x^2/2n}\, dx = N,$$

so that in virtue of the known equality

$$\int_{-\infty}^{+\infty} e^{-z^2}\, dz = \sqrt{\pi},$$

$$A \, . \, \sqrt{2\pi} = 1.$$

The final result for the number of combinations which have amplitudes between x and $x + dx$ is accordingly

$$\frac{N}{\sqrt{(2\pi n)}} e^{-x^2/2n} \, dx \quad \dots\dots\dots\dots\dots \quad (7).$$

The *mean* intensity is expressed by

$$\frac{1}{\sqrt{(2\pi n)}} \int_{-\infty}^{+\infty} x^2 e^{-x^2/2n} \, dx = n,$$

as before.

We will now pass on to the more important problem in which the phases of the n unit vibrations are distributed at random over the entire period. In each combination the resultant amplitude is denoted by r and the phase (referred to a given epoch) by θ; and rectangular coordinates are taken so that

$$x = r \cos \theta, \quad y = r \sin \theta.$$

Thus any point (x, y) in the plane of reference represents a vibration of amplitude r and phase θ, and the whole system of N vibrations is represented by a distribution of points, whose density it is our object to determine. Since no particular phase can be singled out for distinction, we know beforehand that the density of distribution will be independent of θ.

Of the infinite number N of points we suppose that

$$Nf(n, x, y) \, dx \, dy$$

are to be found within the infinitesimal area $dx \, dy$, and we will inquire as before how this number would be changed by the addition to the n component vibrations of one more unit vibration of accidental phase. Any vibration which after the addition is represented by the point x, y must before have corresponded to the point

$$x' = x - \cos \phi, \quad y' = y - \sin \phi,$$

where ϕ represents the phase of the additional unit vibration. And, if for the moment ϕ be regarded as given, to the area $dx \, dy$ corresponds an equal area $dx' \, dy'$. Again, all values of ϕ being equally probable, the factor necessary under this head is $d\phi/2\pi$. Accordingly the whole number to be found in $dx \, dy$ after the superposition of the additional unit is

$$N dx \, dy \int_{0}^{2\pi} f(n, x', y') \, d\phi/2\pi \, ;$$

and this is to be equated to

$$N dx \, dy \, f(n + 1, x, y) \, ;$$

so that

$$f(n + 1, x, y) = \int_{0}^{2\pi} f(n, x', y') \, d\phi/2\pi \quad \dots\dots\dots \quad (8).$$

The value of $f(n, x', y')$ is obtained by introduction of the values of x', y' and expansion:

$$f(x', y') = f(x, y) - \frac{df}{dx}\cos\theta - \frac{df}{dy}\sin\theta + \frac{1}{2}\frac{d^2f}{dx^2}\cos^2\theta$$

$$+ \frac{d^2f}{dxdy}\cos\theta\sin\theta + \frac{1}{2}\frac{d^2f}{dy^2}\sin^2\theta + \ldots,$$

so that

$$\int_0^{2\pi} f(n, x', y')\,d\phi/2\pi = f(n, x, y) + \frac{1}{4}\frac{d^2f}{dx^2} + \frac{1}{4}\frac{d^2f}{dy^2} + \ldots\ldots$$

Also, n being very great,

$$f(n+1, x, y) - f(n, x, y) = df/dn ;$$

and (8) reduces to

$$\frac{df}{dn} = \frac{1}{4}\left(\frac{d^2f}{dx^2} + \frac{d^2f}{dy^2}\right)\ldots\ldots\ldots\ldots\ldots\ldots\ldots(9),$$

the usual equation for the conduction of heat in two dimensions.

In addition to (9), f has to satisfy the special condition of evanescence when $n = 0$ for all points other than the origin. The appropriate solution is necessarily symmetrical round the origin, and takes the form

$$f(n, x, y) = An^{-1}e^{-(x^2+y^2)/n}\ldots\ldots\ldots\ldots(10),$$

as may be verified by differentiation. The constant A is to be determined by the condition that the whole number is N. Thus

$$N = NAn^{-1}\iint e^{-(x^2+y^2)/n}\,dxdy = NA2\pi n^{-1}\int_0^\infty e^{-r^2/n}r\,dr = \pi AN ;$$

and the number of vibrations within the area $dxdy$ becomes

$$\frac{N}{\pi n}e^{-(x^2+y^2)/n}\,dx\,dy\ldots\ldots\ldots\ldots\ldots(11).$$

If we wish to find the number of vibrations which have amplitudes between r and $r + dr$, we must introduce polar coordinates and integrate with respect to θ. The required number is thus

$$2Nn^{-1}e^{-r^2/n}r\,dr\ldots\ldots\ldots\ldots\ldots\ldots(12)[1].$$

The result may also be expressed by saying that the *probability* of a resultant amplitude between r and $r + dr$ when a large number n of unit vibrations are compounded at random is

$$2n^{-1}e^{-r^2/n}r\,dr\ldots\ldots\ldots\ldots\ldots(13).$$

[1] *Phil. Mag.* Aug. 1880.

The mean intensity is given by

$$2n^{-1} \int_0^\infty e^{-r^2/n} r^3 \, dr = n,$$

as was to be expected.

The probability of a resultant amplitude less than r is

$$2n^{-1} \int_0^r e^{-r^2/n} r \, dr = 1 - e^{-r^2/n} \dots\dots\dots\dots (14),$$

or, which is the same thing, the probability of a resultant amplitude *greater* than r is

$$e^{-r^2/n} \dots\dots\dots\dots\dots\dots\dots\dots\dots (15).$$

The following table gives the probabilities of *intensities* less than the fractions of n named in the first column. For example, the probability of intensity less than n is ·6321.

·05	·0488	·80	·5506
·10	·0952	1·00	·6321
·20	·1813	1·50	·7768
·40	·3296	2·00	·8647
·60	·4512	3·00	·9502

It will be seen that, however great n may be, there is a reasonable chance of considerable relative fluctuations of intensity in different combinations.

If the amplitude of each component be α, instead of unity, as we have hitherto supposed for brevity, the probability of a resultant amplitude between r and $r + dr$ is

$$\frac{2}{n\alpha^2} e^{-r^2/n\alpha^2} r \, dr \dots\dots\dots\dots\dots (16).$$

The result is thus a function of n and α only through $n\alpha^2$, and would be unchanged if for example the amplitude became $\frac{1}{2}\alpha$ and the number $4n$. From this it follows that the law is not altered, even if the components have different amplitudes, provided always that the whole number of each kind is very great; so that if there be n components of amplitude α, n' of amplitude β, and so on, the probability of a resultant between r and $r + dr$ is

$$\frac{2}{n\alpha^2 + n'\beta^2 + \dots} e^{-\frac{r^2}{n\alpha^2 + n'\beta^2 + \dots}} r \, dr \dots\dots (17).$$

That this is the case may perhaps be made more clear by the consideration of a particular case. Let us suppose in the first place that $n + 4n'$ unit vibrations are compounded at random.

The appropriate law is given at once by (13) on substitution of $n + 4n'$ for n, that is

$$2\,(n + 4n')^{-1}\, e^{-r^2/(n+4n')}\, r\,dr \dots\dots\dots\dots(18).$$

Now the combination of $n + 4n'$ unit vibrations may be regarded as arrived at by combining a random combination of n unit vibrations with a second random combination of $4n'$ units, and the second random combination is the same as if due to a random combination of n' vibrations each of amplitude 2. Thus (18) applies equally well to a random combination of $(n + n')$ vibrations, n of which are of amplitude unity and n' of amplitude 2.

Although the result has no application to the theory of vibrations, it may be worth notice that a similar method applies to the composition *in three dimensions* of unit vectors, whose directions are accidental. The equation analogous to (8) gives in place of (9)

$$\frac{df}{dn} = \frac{1}{6}\left(\frac{d^2 f}{dx^2} + \frac{d^2 f}{dy^2} + \frac{d^2 f}{dz^2}\right).$$

The appropriate solution, analogous to (13), is

$$3\,\sqrt{\left(\frac{6}{\pi n^3}\right)}\, e^{-r^2/\frac{2}{3}n}\, r^2 dr \dots\dots\dots\dots(18),$$

expressing the probability of a resultant amplitude lying between r and $r + dr$.

Here again the mean value of r^2, to be expected in a great number of independent combinations, is n.

CHAPTER III.

43. THE material systems, with whose vibrations Acoustics is concerned, are usually of considerable complication, and are susceptible of very various modes of vibration, any or all of which may coexist at any particular moment. Indeed in some of the most important musical instruments, as strings and organ-pipes, the number of independent modes is theoretically infinite, and the consideration of several of them is essential to the most practical questions relating to the nature of the consonant chords. Cases, however, often present themselves, in which one mode is of paramount importance; and even if this were not so, it would still be proper to commence the consideration of the general problem with the simplest case—that of one degree of freedom. It need not be supposed that the mode treated of is the only one possible, because so long as vibrations of other modes do not occur their possibility under other circumstances is of no moment.

44. The condition of a system possessing one degree of freedom is defined by the value of a single co-ordinate u, whose origin may be taken to correspond to the position of equilibrium. The kinetic and potential energies of the system for any given position are proportional respectively to $\dot{u}^2$ and u^2:—

$$T = \tfrac{1}{2} m\dot{u}^2, \quad V = \tfrac{1}{2} \mu u^2 \dots\dots\dots\dots\dots\dots(1),$$

where m and μ are in general functions of u. But if we limit ourselves to the consideration of positions *in the immediate neighbourhood of that corresponding to equilibrium*, u is a small quantity, and m and μ are sensibly constant. On this understanding we now proceed. If there be no forces, either resulting from internal friction or viscosity, or impressed on the system from without, the whole energy remains constant. Thus

$$T + V = \text{constant}.$$

Substituting for T and V their values, and differentiating with respect to the time, we obtain the equation of motion

$$m\ddot{u} + \mu u = 0 \dots\dots\dots\dots\dots\dots\dots(2)$$

of which the complete integral is

$$u = a \cos(nt - \alpha) \dots\dots\dots\dots\dots\dots(3),$$

where $n^2 = \mu \div m$, representing a *harmonic* vibration. It will be seen that the period alone is determined by the nature of the system itself; the amplitude and phase depend on collateral circumstances. If the differential equation were exact, that is to say, if T were strictly proportional to $\dot{u}^2$, and V to u^2, then, without any restriction, the vibrations of the system about its configuration of equilibrium would be accurately harmonic. But in the majority of cases the proportionality is only approximate, depending on an assumption that the displacement u is always small—how small depends on the nature of the particular system and the degree of approximation required; and then of course we must be careful not to push the application of the integral beyond its proper limits.

But, although not to be stated without a limitation, the principle that the vibrations of a system about a configuration of equilibrium have a period depending on the structure of the system and not on the particular circumstances of the vibration, is of supreme importance, whether regarded from the theoretical or the practical side. If the pitch and the loudness of the note given by a musical instrument were not within wide limits independent, the art of the performer on many instruments, such as the violin and pianoforte, would be revolutionized.

The periodic time

$$\tau = \frac{2\pi}{n} = 2\pi\sqrt{\frac{m}{\mu}} \dots\dots\dots\dots\dots\dots(4),$$

so that an increase in m, or a decrease in μ, protracts the duration of a vibration. By a generalization of the language employed in the case of a material particle urged towards a position of equilibrium by a spring, m may be called the inertia of the system, and

μ the force of the equivalent spring. Thus an augmentation of mass, or a relaxation of spring, increases the periodic time. By means of this principle we may sometimes obtain limits for the value of a period, which cannot, or cannot easily, be calculated exactly.

45. The absence of all forces of a frictional character is an ideal case, never realized but only approximated to in practice. The original energy of a vibration is always dissipated sooner or later by conversion into heat. But there is another source of loss, which though not, properly speaking, dissipative, yet produces results of much the same nature. Consider the case of a tuning-fork vibrating *in vacuo*. The internal friction will in time stop the motion, and the original energy will be transformed into heat. But now suppose that the fork is transferred to an open space. In strictness the fork and the air surrounding it constitute a single system, whose parts cannot be treated separately. In attempting, however, the exact solution of so complicated a problem, we should generally be stopped by mathematical difficulties, and in any case an approximate solution would be desirable. The effect of the air during a few periods is quite insignificant, and becomes important only by accumulation. We are thus led to consider its effect as a *disturbance* of the motion which would take place *in vacuo*. The disturbing force is periodic (to the same approximation that the vibrations are so), and may be divided into two parts, one proportional to the acceleration, and the other to the velocity. The former produces the same effect as an alteration in the mass of the fork, and we have nothing more to do with it at present. The latter is a force arithmetically proportional to the velocity, and always acts in opposition to the motion, and therefore produces effects of the same character as those due to friction. In many similar cases the loss of motion by communication may be treated under the same head as that due to dissipation proper, and is represented in the differential equation with a degree of approximation sufficient for acoustical purposes by a term proportional to the velocity. Thus

$$\ddot{u} + \kappa\dot{u} + n^2 u = 0 \dots\dots\dots\dots\dots\dots\dots (1)$$

is the equation of vibration for a system with one degree of freedom subject to frictional forces. The solution is

$$u = A e^{-\frac{1}{2}\kappa t} \cos\left\{\sqrt{n^2 - \tfrac{1}{4}\kappa^2} \cdot t - \alpha\right\} \dots\dots\dots\dots (2).$$

If the friction be so great that $\frac{1}{4}\kappa^2 > n^2$, the solution changes its form, and no longer corresponds to an oscillatory motion; but in all acoustical applications κ is a small quantity. Under these circumstances (2) may be regarded as expressing a harmonic vibration, whose amplitude is not constant, but diminishes in geometrical progression, when considered after equal intervals of time. The difference of the logarithms of successive extreme excursions is nearly constant, and is called the Logarithmic Decrement. It is expressed by $\frac{1}{4}\kappa\tau$, if τ be the periodic time.

The frequency, depending on $n^2 - \frac{1}{4}\kappa^2$, involves only the second power of κ; so that to the first order of approximation *the friction has no effect on the period*,—a principle of very general application.

The vibration here considered is called the *free* vibration. It is that executed by the system, when disturbed from equilibrium, and then *left to itself.*

46. We must now turn our attention to another problem, not less important,—the behaviour of the system, when subjected to an external force varying as a harmonic function of the time. In order to save repetition, we may take at once the more general case including friction. If there be no friction, we have only to put in our results $\kappa = 0$. The differential equation is

$$\ddot{u} + \kappa\dot{u} + n^2 u = E \cos pt \dots\dots\dots\dots\dots (1).$$

Assume
$$u = a \cos (pt - \epsilon) \dots\dots\dots\dots\dots\dots (2),$$

and substitute:

$$a (n^2 - p^2) \cos (pt - \epsilon) - \kappa pa \sin (pt - \epsilon)$$
$$= E \cos \epsilon \cos (pt - \epsilon) - E \sin \epsilon \sin (pt - \epsilon);$$

whence, on equating coefficients of $\cos (pt - \epsilon)$, $\sin (pt - \epsilon)$,

$$\left. \begin{array}{l} a (n^2 - p^2) = E \cos \epsilon \\ a \cdot p\kappa = E \sin \epsilon \end{array} \right\} \dots\dots\dots\dots\dots (3),$$

so that the solution may be written

$$u = \frac{E \sin \epsilon}{p\kappa} \cos (pt - \epsilon) \dots\dots\dots\dots\dots (4),$$

where
$$\tan \epsilon = \frac{p\kappa}{n^2 - p^2} \dots\dots\dots\dots\dots\dots (5).$$

This is called a *forced* vibration; it is the response of the system to a force imposed upon it from without, and is maintained by the continued operation of that force. The amplitude is proportional to E—the magnitude of the force, and the period is the same as that of the force.

Let us now suppose E given, and trace the effect on a given system of a variation in the period of the force. The effects produced in different cases are not strictly similar; because the frequency of the vibrations produced is always the same as that of the force, and therefore variable in the comparison which we are about to institute. We may, however, compare the energy of the system in different cases at the moment of passing through the position of equilibrium. It is necessary thus to specify the moment at which the energy is to be computed in each case, because the total energy is not invariable throughout the vibration. During one part of the period the system receives energy from the impressed force, and during the remainder of the period yields it back again.

From (4), if $u = 0$,
$$\text{energy} \propto \dot{u}^2 \propto \sin^2 \epsilon,$$
and is therefore a maximum, when $\sin \epsilon = 1$, or, from (5), $p = n$. If the maximum kinetic energy be denoted by T_0, we have
$$T = T_0 \sin^2 \epsilon \dots\dots\dots\dots\dots\dots\dots(6).$$

The kinetic energy of the motion is therefore the greatest possible, when the period of the force is that in which the system would vibrate freely under the influence of its own elasticity (or other internal forces), *without friction*. The vibration is then by (4) and (5),
$$u = \frac{E}{n\kappa} \sin nt \,;$$

and, if κ be small, its amplitude is very great. Its phase is a quarter of a period behind that of the force.

The case, where $p = n$, may also be treated independently. Since the period of the actual vibration is the same as that natural to the system.
$$\ddot{u} + n^2 u = 0,$$
so that the differential equation (1) reduces to
$$\kappa \dot{u} = E \cos pt,$$
whence by integration
$$u = \frac{E}{\kappa} \int \cos pt \, dt = \frac{E}{p\kappa} \sin pt,$$
as before.

If p be less than n, the retardation of phase relatively to the force lies between zero and a quarter period, and when p is greater than n, between a quarter period and a half period.

In the case of a system devoid of friction, the solution is

$$u = \frac{E}{n^2 - p^2} \cos pt \dots\dots\dots\dots\dots (7).$$

When p is smaller than n, the phase of the vibration agrees with that of the force, but when p is the greater, the sign of the vibration is changed. The change of phase from complete agreement to complete disagreement, which is gradual when friction acts, here takes place abruptly as p passes through the value n. At the same time the expression for the amplitude becomes infinite. Of course this only means that, in the case of equal periods, friction *must* be taken into account, however small it may be, and however insignificant its result when p and n are not approximately equal. The limitation as to the magnitude of the vibration, to which we are all along subject, must also be borne in mind.

That the excursion should be at its maximum in one direction while the generating force is at its maximum in the opposite direction, as happens, for example, in the canal theory of the tides, is sometimes considered a paradox. Any difficulty that may be felt will be removed by considering the extreme case, in which the "spring" vanishes, so that the natural period is infinitely long. In fact we need only consider the force acting on the bob of a common pendulum swinging freely, in which case the excursion on one side is greatest when the action of gravity is at its maximum in the opposite direction. When on the other hand the inertia of the system is very small, we have the other extreme case in which the so-called equilibrium theory becomes applicable, the force and excursion being in the same phase.

When the period of the force is longer than the natural period, the effect of an increasing friction is to introduce a retardation in the phase of the displacement varying from zero up to a quarter period. If, however, the period of the natural vibration be the longer, the original retardation of half a period is diminished by something short of a quarter period; or the effect of friction is to *accelerate* the phase of the displacement estimated from that corresponding to the absence of friction. In either case the influence of friction is to cause an approximation to the state of things that would prevail if friction were paramount.

If a force of nearly equal period with the free vibrations vary slowly to a maximum and then slowly decrease, the displacement does not reach its maximum until after the force has begun to diminish. Under the operation of the force at its maximum, the vibration continues to increase until a certain limit is approached, and this increase continues for a time even although the force, having passed its maximum, begins to diminish. On this principle the retardation of spring tides behind the days of new and full moon has been explained[1].

47. From the linearity of the equations it follows that the motion resulting from the simultaneous action of any number of forces is the simple sum of the motions due to the forces taken separately. Each force causes the vibration proper to itself, without regard to the presence or absence of any others. The peculiarities of a force are thus in a manner transmitted into the motion of the system. For example, if the force be periodic in time τ, so will be the resulting vibration. Each harmonic element of the force will call forth a corresponding harmonic vibration in the system. But since the retardation of phase ϵ, and the ratio of amplitudes $a : E$, is not the same for the different components, the resulting vibration, though periodic in the same time, is different in *character* from the force. It may happen, for instance, that one of the components is isochronous, or nearly so, with the free vibration, in which case it will manifest itself in the motion out of all proportion to its original importance. As another example we may consider the case of a system acted on by two forces of nearly equal period. The resulting vibration, being compounded of two nearly in unison, is intermittent, according to the principles explained in the last chapter.

To the motions, which are the immediate effects of the impressed forces, must always be added the term expressing free vibrations, if it be desired to obtain the most general solution. Thus in the case of one impressed force,

$$u = \frac{E \sin \epsilon}{p\kappa} \cos (pt - \epsilon) + A e^{-\frac{1}{2}\kappa t} \cos \{\sqrt{n^2 - \tfrac{1}{4}\kappa^2} . t - \alpha\} \dots\dots (1),$$

where A and α are arbitrary.

48. The distinction between *forced* and *free* vibrations is very important, and must be clearly understood. The period of the

[1] Airy's *Tides and Waves*, Art. 328.

former is determined solely by the force which is supposed to act on the system from without; while that of the latter depends only on the constitution of the system itself. Another point of difference is that so long as the external influence continues to operate, a forced vibration is permanent, being represented strictly by a harmonic function; but a free vibration gradually dies away, becoming negligible after a time. Suppose, for example, that the system is at rest when the force $E \cos pt$ begins to operate. Such finite values must be given to the constants A and α in (1) of § 47, that both u and $\dot{u}$ are initially zero. At first then there is a free vibration not less important than its rival, but after a time friction reduces it to insignificance, and the forced vibration is left in complete possession of the field. This condition of things will continue so long as the force operates. When the force is removed, there is, of course, no discontinuity in the values of u or $\dot{u}$, but the forced vibration is at once converted into a free vibration, and the period of the force is exchanged for that natural to the system.

During the coexistence of the two vibrations in the earlier part of the motion, the curious phenomenon of beats may occur, in case the two periods differ but slightly. For, n and p being nearly equal, and κ small, the initial conditions are approximately satisfied by

$$u = a \cos (pt - \epsilon) - a e^{-\frac{1}{2}\kappa t} \cos \left\{ \sqrt{n^2 - \tfrac{1}{4}\kappa^2} \, . \, t - \epsilon \right\}.$$

There is thus a rise and fall in the motion, so long as $e^{-\frac{1}{2}\kappa t}$ remains sensible. This intermittence is very conspicuous in the earlier stages of the motion of forks driven by electro-magnetism (§ 63), [and may be utilized when it is desired to adjust n and p to equality. The initial beats are to be made slower and slower, until they cease to be perceptible. The vibration then swells continuously to a maximum.]

49. Vibrating systems of one degree of freedom may vary in two ways according to the values of the constants n and κ. The distinction of pitch is sufficiently intelligible; but it is worth while to examine more closely the consequences of a greater or less degree of damping. The most obvious is the more or less rapid extinction of a free vibration. The effect in this direction may be measured by the number of vibrations which must elapse before the amplitude is reduced in a given ratio. Initially the amplitude may be taken as unity; after a time t, let it be θ. Thus $\theta = e^{-\frac{1}{2}\kappa t}$.

If $t = x\tau$, we have $x = -\dfrac{2}{\kappa\tau} \log \theta$. In a system subject to only a moderate degree of damping, we may take approximately,

$$\tau = 2\pi \div n\,;$$

so that
$$x = -\frac{n}{\kappa\pi} \log \theta \dots\dots\dots\dots\dots\dots\dots(1).$$

This gives the number of vibrations which are performed, before the amplitude falls to θ.

The influence of damping is also powerfully felt in a forced vibration, when there is a near approach to isochronism. In the case of an exact equality between p and n, it is the damping alone which prevents the motion becoming infinite. We might easily anticipate that when the damping is small, a comparatively slight deviation from perfect isochronism would cause a large falling off in the magnitude of the vibration, but that with a larger damping the same precision of adjustment would not be required. From the equations

$$T = T_0 \sin^2 \epsilon, \quad \tan \epsilon = \frac{\kappa p}{n^2 - p^2},$$

we get
$$\frac{n^2 - p^2}{\kappa p} = \sqrt{\frac{T_0 - T}{T}} \dots\dots\dots\dots\dots\dots(2)\,;$$

so that if κ be small, p must be very nearly equal to n, in order to produce a motion not greatly less than the maximum.

The two principal effects of damping may be compared by eliminating κ between (1) and (2). The result is

$$\frac{\log \theta}{x} = \pi \left(\frac{p}{n} - \frac{n}{p}\right) \sqrt{\frac{T}{T_0 - T}} \dots\dots\dots\dots(3),$$

where the sign of the square root must be so chosen as to make the right-hand side negative.

If, when a system vibrates freely, the amplitude be reduced in the ratio θ after x vibrations; then, when it is acted on by a force (p), the energy of the resulting motion will be less than in the case of perfect isochronism in the ratio $T : T_0$. It is a matter of indifference whether the forced or the free vibration be the higher; all depends on the *interval*.

In most cases of interest the interval is small; and then, putting $p = n + \delta n$, the formula may be written,

$$\frac{\log \theta}{x} = \frac{2\pi\delta n}{n} \sqrt{\frac{T}{T_0 - T}} \dots\dots\dots\dots\dots(4).$$

The following table calculated from these formulæ has been given by Helmholtz[1]:

Interval corresponding to a reduction of the resonance to one-tenth. $T : T_0 = 1 : 10.$	Number of vibrations after which the intensity of a free vibration is reduced to one-tenth. $\theta^2 = \tfrac{1}{10}.$
$\frac{1}{8}$ tone.	38·00
$\frac{1}{4}$ tone.	19·00
$\frac{1}{2}$ tone.	9·50
$\frac{3}{4}$ tone.	6·33
Whole tone.	4·75
$\frac{5}{4}$ tone.	3·80
$\frac{6}{4}$ tone = minor third.	3·17
$\frac{7}{4}$ tone.	2·71
Two whole tones = major third.	2·37

Formula (4) shews that, when δn is small, it varies *cœteris paribus* as $\dfrac{1}{x}$.

50. From observations of forced vibrations due to known forces, the natural period and damping of a system may be determined. The formulæ are

$$u = \frac{E \sin \epsilon}{p\kappa} \cos (pt - \epsilon),$$

where

$$\tan \epsilon = \frac{p\kappa}{n^2 - p^2}.$$

On the equilibrium theory we should have

$$u = \frac{E}{n^2} \cos pt.$$

The ratio of the actual amplitude to this is

$$\frac{E \sin \epsilon}{p\kappa} : \frac{E}{n^2} = \frac{n^2 \sin \epsilon}{p\kappa}.$$

If the equilibrium theory be known, the comparison of amplitudes tells us the value of $\dfrac{n^2 \sin \epsilon}{p\kappa}$, say

$$\frac{n^2 \sin \epsilon}{p\kappa} = a,$$

[1] *Tonempfindungen*, 3rd edition, p. 221.

and ϵ is also known, whence

$$n^2 = p^2 \div \left(1 - \frac{\cos \epsilon}{a}\right), \text{ and } \kappa = \frac{p \sin \epsilon}{a - \cos \epsilon}\dots\dots\dots(1).$$

51. As has been already stated, the distinction of forced and free vibrations is important; but it may be remarked that most of the forced vibrations which we shall have to consider as affecting a system, take their origin ultimately in the motion of a second system, which influences the first, and is influenced by it. A vibration may thus have to be reckoned as forced in its relation to a system whose limits are fixed arbitrarily, even when that system has a share in determining the period of the force which acts upon it. On a wider view of the matter embracing both the systems, the vibration in question will be recognized as free. An example may make this clearer. A tuning-fork vibrating in air is part of a compound system including the air and itself, and in respect of this compound system the vibration is free. But although the fork is influenced by the reaction of the air, yet the amount of such influence is small. For practical purposes it is convenient to consider the motion of the fork as given, and that of the air as forced. No error will be committed if the *actual* motion of the fork (as influenced by its surroundings) be taken as the basis of calculation. But the peculiar advantage of this mode of conception is manifested in the case of an approximate solution being required. It may then suffice to substitute for the actual motion, what would be the motion of the fork in the absence of air, and afterwards introduce a correction, if necessary.

52. Illustrations of the principles of this chapter may be drawn from all parts of Acoustics. We will give here a few applications which deserve an early place on account of their simplicity or importance.

A string or wire ACB is stretched between two fixed points A and B, and at its centre carries a mass M, which is supposed to be so considerable as to render the mass of the string itself negligible. When M is pulled aside from its position of equilibrium, and then let go, it executes along the line CM vibrations, which are the subject of inquiry. $AC = CB = a$. $CM = x$. The tension of the string in the position of equilibrium depends on the amount of the stretching to which it has been subjected. In any other position the tension is greater; but we limit ourselves to the case of vibrations so small that the additional stretching is a negligible fraction of the whole. On this condition the tension may be treated as constant. We denote it by T.

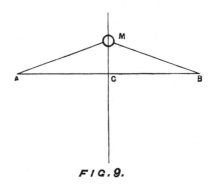

FIG. 9.

Thus, kinetic energy $= \frac{1}{2}M\dot{x}^2$,

and

potential energy $= 2T\{\sqrt{a^2 + x^2} - a\} = T\dfrac{x^2}{a}$ approximately.

The equation of motion (which may be derived also independently) is therefore

$$M\ddot{x} + 2T\frac{x}{a} = 0 \dots\dots\dots\dots\dots\dots(1),$$

from which we infer that the mass M executes harmonic vibrations, whose period

$$\tau = 2\pi \div \sqrt{\frac{2T}{aM}} \dots\dots\dots\dots\dots\dots (2).$$

The amplitude and phase depend of course on the initial circumstances, being arbitrary so far as the differential equation is concerned.

Equation (2) expresses the manner in which τ varies with each of the independent quantities T, M, a: results which may all be obtained by consideration of the *dimensions* (in the technical sense) of the quantities involved. The argument from dimensions is so often of importance in Acoustics that it may be well to consider this first instance at length.

In the first place we must assure ourselves that of all the quantities on which τ may depend, the only ones involving a reference to the three fundamental units—of length, time, and mass—are a, M, and T. Let the solution of the problem be written

$$\tau = f(a, M, T) \dots\dots\dots\dots\dots\dots(3).$$

This equation must retain its form unchanged, whatever may be the fundamental units by means of which the four quantities are numerically expressed, as is evident, when it is considered that in deriving it no assumptions would be made as to the magnitudes of those units. Now of all the quantities on which f depends, T is the only one involving time; and since its dimensions are (Mass) (Length) (Time)$^{-2}$, it follows that when a and M are constant, $\tau \propto T^{-\frac{1}{2}}$; otherwise a change in the unit of time would necessarily disturb the equation (3). This being admitted, it is easy to see that in order that (3) may be independent of the unit of length, we must have $\tau \propto T^{-\frac{1}{2}} \cdot a^{\frac{1}{2}}$, when M is constant; and finally, in order to secure independence of the unit of mass,

$$\tau \propto T^{-\frac{1}{2}} \cdot M^{\frac{1}{2}} \cdot a^{\frac{1}{2}}.$$

To determine these indices we might proceed thus :—assume

$$\tau \propto T^x \cdot M^y \cdot a^z;$$

then by considering the dimensions in time, space, and mass, we obtain respectively

$$1 = -2x, \quad 0 = x + z, \quad 0 = x + y,$$

whence as above $\quad x = -\frac{1}{2}, \quad y = \frac{1}{2}, \quad z = \frac{1}{2}.$

There must be no mistake as to what this argument does and does not prove. We have *assumed* that there is a definite periodic time depending on no other quantities, having dimensions in space, time, and mass, than those above mentioned. For example, we have not proved that τ is independent of the amplitude of vibration. That, so far as it is true at all, is a consequence of the linearity of the approximate differential equation.

From the necessity of a complete enumeration of all the quantities on which the required result may depend, the method of dimensions is somewhat dangerous; but when used with proper care it is unquestionably of great power and value.

53. The solution of the present problem might be made the foundation of a method for the absolute measurement of pitch. The principal impediment to accuracy would probably be the difficulty of making M sufficiently large in relation to the mass of the wire, without at the same time lowering the note too much in the musical scale.

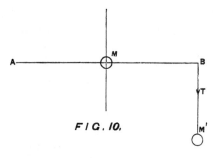

F I G . 10.

The wire may be stretched by a weight M' attached to its further end beyond a bridge or pulley at B. The periodic time would be calculated from

$$\tau = 2\pi . \sqrt{\frac{aM}{2gM'}} \dots\dots\dots\dots\dots\dots(1).$$

The ratio of $M' : M$ is given by the balance. If a be measured in feet, and $g = 32\cdot2$, the periodic time is expressed in seconds.

54.   In an ordinary musical string the weight, instead of being concentrated in the centre, is uniformly distributed over its length. Nevertheless the present problem gives some idea of the nature of the gravest vibration of such a string.   Let us compare the two cases more closely, supposing the amplitudes of vibration the same at the middle point.

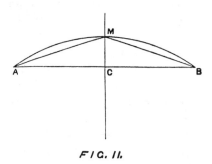

F I G . II.

When the uniform string is straight, at the moment of passing through the position of equilibrium, its different parts are animated with a variable velocity, increasing from either end towards the centre. If we attribute to the whole mass the velocity of the centre, it is evident that the kinetic energy will be considerably over-estimated. Again, at the moment of maximum excursion, the uniform string is more stretched than its substitute, which follows the straight courses AM, MB, and accordingly the potential energy is diminished by the substitution. The concentration of the mass at the middle point at once increases the kinetic energy when $x = 0$, and decreases the potential energy when $\dot{x} = 0$, and therefore, according to the principle explained in § 44, prolongs the periodic time. For a string then the period is less than that calculated from the formula of the last section, on the supposition that M denotes the mass of the string. It will afterwards appear that in order to obtain a correct result we should have to take instead of M only $(4/\pi^2)M$. Of the factor $4/\pi^2$ by far the more important part, viz. $\frac{1}{2}$, is due to the difference of the kinetic energies.

55. As another example of a system possessing practically but one degree of freedom, let us consider the vibration of a spring, one end of which is clamped in a vice or otherwise held fast, while the other carries a heavy mass.

In strictness, this system like the last has an infinite number of independent modes of vibration; but, when the mass of the spring is relatively small, that vibration which is nearly independent of its inertia becomes so much the most important that the others may be ignored. Pushing this idea to its limit, we may regard the spring merely as the origin of a force urging the attached mass towards the position of equilibrium, and, if a certain point be not exceeded, in simple proportion to the displacement. The result is a harmonic vibration, with a period dependent on the stiffness of the spring and the mass of the load.

FIG 12.

56. In consequence of the oscillation of the centre of inertia, there is a constant tendency towards the communication of motion to the supports, to resist which adequately the latter must be very firm and massive. In order to obviate this inconvenience, two precisely similar springs and loads may be mounted on the same framework in a symmetrical manner. If the two loads perform vibrations of equal amplitude in such a manner that the motions are always opposite, or, as it may otherwise be expressed, with a phase-difference of half a period, the centre of inertia of the whole system remains at rest, and there is no tendency to set the framework into vibration. We shall see in a future chapter that this peculiar relation of phases will quickly establish itself, whatever may be the original disturbance. In fact, any part of the motion which does not conform to the condition of leaving the centre of inertia unmoved is soon extinguished by damping, unless indeed the supports of the system are more than usually firm.

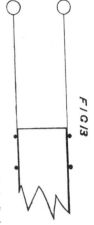

FIG 13

57. As in our first example we found a rough illustration of the fundamental vibration of a musical string, so here with the spring and attached load we may compare a uniform slip, or bar, of elastic material, one end of which is securely fastened, such for instance as the *tongue* of a *reed* instrument. It is true of course that the mass is not concentrated at one end, but distributed over the whole length; yet on account of the smallness of the motion near the point of support, the inertia of that part of the bar is of but little account. We infer that the fundamental vibration of a uniform rod cannot be very different in character from that which we have been considering. Of course for purposes requiring precise calculation, the two systems are sufficiently distinct; but where the object is to form clear ideas, precision may often be advantageously exchanged for simplicity.

In the same spirit we may regard the combination of two springs and loads shewn in Fig. 13 as a representation of a tuning-fork. The instrument, which has been much improved of late years, is indispensable to the acoustical investigator. On a large scale and for rough purposes it may be made by welding a cross piece on the middle of a bar of steel, so as to form a T, and then bending the bar into the shape of a horse-shoe. On the handle a screw should be cut. But for the better class of tuning-forks it is preferable to shape the whole out of one piece of steel.

A division running from one end down the middle of a bar is first made, the two parts opened out to form the prongs of the fork, and the whole worked by the hammer and file into the required shape. The two prongs must be exactly symmetrical with respect to a plane passing through the axis of the handle, in order that during the vibration the centre of inertia may remain unmoved, — unmoved, that is, in the direction in which the prongs vibrate.

The tuning is effected thus. To make the note higher, the equivalent inertia of the system must be reduced. This is done by filing away the ends of the prongs, either diminishing their thickness, or actually shortening them. On the other hand, to lower the pitch, the substance of the prongs near the bend may be reduced, the effect of which is to diminish the force of the spring, leaving the inertia practically unchanged ; or the inertia may be increased (a method which would be preferable for temporary purposes) by loading the ends of the prongs with wax, or other material. Large forks are sometimes provided with moveable weights, which slide along the prongs, and can be fixed in any position by screws. As these approach the ends (where the velocity is greatest) the equivalent inertia of the system increases. In this way a considerable range of pitch may be obtained from one fork. The number of vibrations per second for any position of the weights may be marked on the prongs.

The relation between the pitch and the size of tuning-forks is remarkably simple. In a future chapter it will be proved that, provided the material remains the same and the shape constant, the period of vibration varies directly as the linear dimension. Thus, if the linear dimensions of a tuning-fork be doubled, its note falls an octave.

58. The note of a tuning-fork is a nearly pure tone. Immediately after a fork is struck, high tones may indeed be heard, corresponding to modes of vibration, whose nature will be subsequently considered ; but these rapidly die away, and even while they exist, they do not blend with the proper tone of the fork, partly on account of their very high pitch, and partly because they do not belong to its harmonic scale. In the forks examined by Helmholtz the first of these overtones had a frequency from 5·8 to 6·6 times that of the proper tone.

Tuning-forks are now generally supplied with resonance cases, whose effect is greatly to augment the volume and purity of the sound, according to principles to be hereafter developed. In order to excite them, a violin or cello bow, well supplied with rosin, is drawn across the prongs in the direction of vibration. The sound so produced will last a minute or more.

59. As standards of pitch tuning-forks are invaluable. The pitch of organ-pipes rapidly varies with the temperature and with the pressure of the wind; that of strings with the tension, which can never be retained constant for long; but a tuning-fork kept clean and not subjected to violent changes of temperature or magnetization, preserves its pitch with great fidelity.

[But it must not be supposed that the vibrations of a fork are altogether independent of temperature. According to the observations of McLeod and Clarke[1] the frequency falls by ·00011 of its value for each degree Cent. of elevation.]

By means of beats a standard tuning-fork may be copied with very great precision. The number of beats heard in a second is the difference of the frequencies of the two tones which produce them; so that if the beats can be made so slow as to occupy half a minute each, the frequencies differ by only 1-30th of a vibration. Still greater precision might be obtained by Lissajous' optical method.

Very slow beats being difficult of observation, in consequence of the uncertainty whether a falling off in the sound is due to interference or to the gradual dying away of the vibrations, Scheibler adopted a somewhat modified plan. He took a fork slightly different in pitch from the standard—whether higher or lower is not material, but we will say, lower,—and counted the number of beats, when they were sounded together. About four beats a second is the most suitable, and these may be counted for perhaps a minute. The fork to be adjusted is then made slightly higher than the auxiliary fork, and tuned to give with it precisely the same number of beats, as did the standard. In this way a copy as exact as possible is secured. To facilitate the counting of the beats Scheibler employed pendulums, whose periods of vibration could be adjusted.

[The question between slow and quick beats depends upon the circumstances of the case. It seems to be sometimes supposed that quick beats have the advantage as admitting of greater

[1] *Phil. Trans.* 1880, p. 12.

relative accuracy of counting. But a little consideration shews
that in a comparison of frequencies we are concerned not with the
relative, but with the *absolute* accuracy of the counting. If we
miscount the beats in a minute by one, it makes just the same
error in the result, whether the whole number of beats be 60 or
240.

When the sounds are pure tones and are well maintained, it is
advisable to use beats much slower than four per second. By
choosing a suitable position it is often possible to make the
intensities at the ear equal; and then the phase of silence,
corresponding to antagonism of equal and opposite vibrations, is
extremely well marked. Taking advantage of this we may determine slow beats with very great accuracy by observing the time
which elapses between recurrences of silence. In favourable cases
the whole number of beats in the period of observation may be
fixed to within one-tenth or one-twentieth of a single beat, a
degree of accuracy which is out of the question when the beats
are quick. In this way beats of periods exceeding 30 seconds may
be utilised with excellent effect [1].]

60. The method of beats was also employed by Scheibler to
determine the absolute pitch of his standards. Two forks were
tuned to an octave, and a number of others prepared to bridge
over the interval by steps so small that each fork gave with its
immediate neighbours in the series a number of beats that could
be easily counted. The difference of frequency corresponding to
each step was observed with all possible accuracy. Their sum,
being the difference of frequencies for the interval of the octave,
was equal to the frequency of that fork which formed the starting
point at the bottom of the series. The pitch of the other forks
could be deduced.

If consecutive forks give four beats per second, 65 in all will
be required to bridge over the interval from c' (256) to c'' (512).
On this account the method is laborious; but it is probably the
most accurate for the original determination of pitch, as it is
liable to no errors but such as care and repetition will eliminate.
It may be observed that the essential thing is the measurement
of the *difference* of frequencies for two notes, whose *ratio* of
frequencies is independently known. If we could be sure of its
accuracy, the interval of the fifth, fourth, or even major third, might

---

[1] Acoustical Observations, *Phil. Mag.* May, 1882, p. 342.

be substituted for the octave, with the advantage of reducing the number of the necessary interpolations. It is probable that with the aid of optical methods this course might be successfully adopted, as the corresponding Lissajous' figures are easily recognised, and their steadiness is a very severe test of the accuracy with which the ratio is attained.

[It is essential to the success of this method that the pitch of each of the numerous sounds employed should be definite, and in particular that the vibrations of any fork should take place at the same rate whether that fork be sounding in conjunction with its neighbour above or with its neighbour below. There is no reason to doubt that this condition is sufficiently satisfied in the case of independent tuning-forks; but an attempt to replace forks by a set of reeds, mounted side by side on a common wind-chest, has led to error, owing to a disturbance of pitch by mutual inter-action [1].]

The frequency of large tuning-forks may be determined by allowing them to trace a harmonic curve on smoked paper, which may conveniently be mounted on the circumference of a revolving drum. The number of waves executed in a second of time gives the frequency.

In many cases the use of intermittent illumination described in § 42 gives a convenient method of determining an unknown frequency.

61. A series of forks ranging at small intervals over an octave is very useful for the determination of the frequency of any musical note, and is called Scheibler's Tonometer. It may also be used for tuning a note to any desired pitch. In either case the frequency of the note is determined by the number of beats which it gives with the forks, which lie nearest to it (on each side) in pitch.

For tuning pianofortes or organs, a set of twelve forks may be used giving the notes of the chromatic scale on the equal temperament, or any desired system. The corresponding notes are adjusted to unison, and the others tuned by octaves. It is better, however, to prepare the forks so as to give four vibrations per second less than is above proposed. Each note is then tuned a little higher than the corresponding fork, until they give when sounded together exactly four beats in the second. It will be

---

[1] *Nature*, xvii. pp. 12, 26; 1877.

observed that the addition (or subtraction) of a constant number to the frequencies is not the same thing as a mere displacement of the scale in absolute pitch.

In the ordinary practice of tuners a' is taken from a fork, and the other notes determined by estimation of fifths. It will be remembered that twelve true fifths are slightly in excess of seven octaves, so that on the equal temperament system each fifth is a little flat. The tuner proceeds upwards from a' by successive fifths, coming down an octave after about every alternate step, in order to remain in nearly the same part of the scale. Twelve fifths should bring him back to a. If this be not the case, the work must be readjusted, until all the twelve fifths are too flat by, as nearly as can be judged, the same small amount. The inevitable error is then impartially distributed, and rendered as little sensible as possible. The octaves, of course, are all tuned true. The following numbers indicate the order in which the notes may be taken :

$a\sharp$	b	c'	$c'\sharp$	d'	$d'\sharp$	e'	f'	$f'\sharp$	g'	$g'\sharp$	a'	$a'\sharp$	b'	c''	$c''\sharp$	d''	$d''\sharp$	e''
13	5	16	8	19	11	3	14	6	17	9	1	12	4	15	7	18	10	2

In practice the equal temperament is only approximately attained; but this is perhaps not of much consequence, considering that the system aimed at is itself by no means perfection.

Violins and other instruments of that class are tuned by true fifths from a'.

62. In illustration of *forced* vibration let us consider the case of a pendulum whose point of support is subject to a small horizontal harmonic motion. Q is the bob attached by a fine wire to a moveable point P. $OP = x_0$. $PQ = l$, and x is the horizontal co-ordinate of Q. Since the vibrations are supposed small, the vertical motion may be neglected, and the tension of the wire equated to the weight of Q. Hence for the horizontal motion $\ddot{x} + \kappa\dot{x} + \dfrac{g}{l}(x - x_0) = 0$.

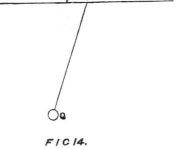

F I C 14.

Now $x_0 \propto \cos pt$; so that putting $g \div l = n^2$, our equation takes the form already treated of, viz.

$$\ddot{x} + \kappa\dot{x} + n^2 x = E \cos pt.$$

If p be equal to n, the motion is limited only by the friction. The assumed horizontal harmonic motion for P may be realized by means of a second pendulum of massive construction, which carries P with it in its motion. An efficient arrangement is shewn in the figure. A, B are iron rings screwed into a beam, or other firm

F I G I5.

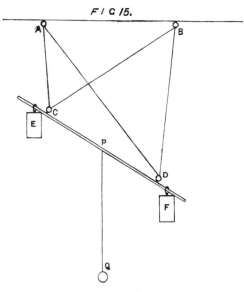

support; C, D similar rings attached to a stout bar, which carries equal heavy weights E, F, attached near its ends, and is supported in a horizontal position at right angles to the beam by a wire passing through the four rings in the manner shewn. When the pendulum is made to vibrate, a point in the rod midway between C and D executes a harmonic motion in a direction parallel to CD, and may be made the point of attachment of another pendulum PQ. If the weights E and F be very great in relation to Q, the upper pendulum swings very nearly in its own proper period, and induces in Q a forced vibration of the same period. When the length PQ is so adjusted that the natural periods of the two pendulums are nearly the same, Q will be thrown into violent motion, even though the vibration of P be of but inconsiderable amplitude. In this case the difference of phase is about a quarter of a period, by which amount the upper pendulum is in advance. If the two periods be very different, the vibrations either agree or are completely opposed in phase, according to equations (4) and (5) of § 46.

63. A very good example of a forced vibration is afforded by a fork under the influence of an intermittent electric current,

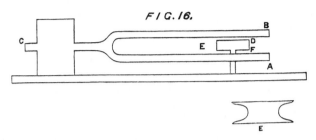

F I G. 16.

whose period is nearly equal to its own. ACB is the fork; E a small electro-magnet, formed by winding insulated wire on an iron core of the shape shewn in E (similar to that known as 'Siemens' armature'), and supported between the prongs of the fork. When an intermittent current is sent through the wire, a periodic force acts upon the fork. This force is not expressible by a simple circular function; but may be expanded by Fourier's theorem in a series of such functions, having periods τ, $\frac{1}{2}\tau$, $\frac{1}{3}\tau$, &c. If any of these, of not too small amplitude, be nearly isochronous with the fork, the latter will be caused to vibrate; otherwise the effect is insignificant. In what follows we will suppose that it is the complete period τ which nearly agrees with that of the fork, and consequently regard the series expressing the periodic force as reduced to its first term.

In order to obtain the maximum vibration, the fork must be carefully tuned by a small sliding piece or by wax, until its natural period (without friction) is equal to that of the force. This is best done by actual trial. When the desired equality is approached, and the fork is allowed to start from rest, the force and complementary free vibration are of nearly equal amplitudes and frequencies, and therefore (§ 48) in the beginning of the motion produce *beats*, whose slowness is a measure of the accuracy of the adjustment. It is not until after the free vibration has had time to subside, that the motion assumes its permanent character. The vibrations of a tuning-fork properly constructed and mounted are subject to very little damping; consequently a very slight deviation from perfect isochronism occasions a marked falling off in the intensity of the resonance.

The amplitude of the forced vibration can be observed with sufficient accuracy by the ear or eye; but the experimental verification of the relations pointed out by theory between its phase and that of the force which causes it, requires a modified arrangement.

Two similar electro-magnets acting on similar forks, and included in the same circuit are excited by the same intermittent current. Under these circumstances it is clear that the systems will be thrown into similar vibrations, because they are acted on by equal forces. This similarity of vibrations refers both to phase and amplitude. Let us suppose now that the vibrations are effected in perpendicular directions, and by means of one of Lissajous' methods are optically compounded. The resulting figure is necessarily a straight line. Starting from the case in which the amplitudes are a maximum, viz. when the natural periods of both forks are the same as that of the force, let one of them be put a little out of tune. It must be remembered that whatever their natural periods may be, the two forks vibrate in perfect unison with the force, and therefore with one another. The principal effect of the difference of the natural periods is to destroy the synchronism of phase. The straight line, which previously represented the compound vibration, becomes an ellipse, and this remains perfectly steady, so long as the forks are not touched. Originally the forks are both a quarter period behind the force. When the pitch of one is slightly lowered, it falls still more behind the force, and at the same time its amplitude diminishes. Let the difference of phase between the two forks be ϵ', and the ratio of amplitudes of vibration $a : a_0$. Then by (6) of § 46

$$a = a_0 \cos \epsilon'.$$

The following table shews the simultaneous values of $a : a_0$ and ϵ'.

$a : a_0$	ϵ'
1·0	0
·9	$25^0 \ 50'$
·8	$36^0 \ 52'$
·7	$45^0 \ 34'$
·6	$53^0 \ 7'$
·5	60^0
·4	$66^0 \ 25'$
·3	$72^0 \ 32'$
·2	$78^0 \ 27'$
·1	$84^0 \ 15'$[1]

[1] *Tonempfindungen*, 3rd edition, p. 190

It appears that a considerable alteration of phase in either direction may be obtained without very materially reducing the amplitude. When one fork is vibrating at its maximum, the other may be made to differ from it on either side by as much as 60^0 in phase, without losing more than half its amplitude, or by as much as 45^0, without losing more than half its *energy*. By allowing one fork to vibrate 45^0 in advance, and the other 45^0 in arrear of the phase corresponding to the case of maximum resonance, we obtain a phase difference of 90^0 in conjunction with an equality of amplitudes. Lissajous' figure then becomes a circle.

[An intermittent electric current may also be applied to regulate the speed of a revolving body. The *phonic wheel,* invented independently by M. La Cour and by the author of this work[1], is of great service in acoustical investigations. It may take various forms; but the essential feature is the approximate closing of the magnetic circuit of an electro-magnet, fed with an intermittent current, by one or more soft iron armatures carried by the wheel and disposed symmetrically round the circumference. If in the revolution of the wheel the closest passage of the armature synchronises with the middle of the time of excitation, the electro-magnetic forces operating upon the armature during its advance and its retreat balance one another. If however the wheel be a little in arrear, the forces promoting advance gain an advantage over those hindering the retreat of the armature, and thus upon the whole the magnetic forces encourage the rotation. In like manner if the phase of the wheel be in advance of that first specified, forces are called into play which retard the motion. By a self-acting adjustment the rotation settles down into such a phase that the driving forces exactly balance the resistances. When the wheel runs lightly, and the electric appliances are moderately powerful, independent driving may not be needed. In this case of course the phase of closest passage must *follow* that which marks the middle of the time of magnetisation. If, as is sometimes advisable, there be an independent driving power, the phase of closest passage may either precede or follow that of magnetisation.

In some cases the oscillations of the motion about the phase into which it should settle down are very persistent and interfere with the applications of the instrument. A remedy may be found in a ring containing water or mercury, revolving concen-

1 *Nature*, May 23, 1878.

trically. When the rotation is uniform, the fluid revolves like a solid body and then exercises no influence. But when from any cause the speed changes, the fluid persists for a time in the former motion, and thus brings into play forces tending to damp out oscillations.]

64. The intermittent current is best obtained by a fork-interrupter invented by Helmholtz. This may consist of a fork and electro-magnet mounted as before. The wires of the magnet are connected, one with one pole of the battery, and the other with a mercury cup. The other pole of the battery is connected with a second mercury cup. A U-shaped rider of insulated wire is carried by the lower prong just over the cups, at such a height that during the vibration the circuit is alternately made and broken by the passage of one end into and out of the mercury. The other end may be kept permanently immersed. By means of the periodic force thus obtained, the effect of friction is compensated, and the vibrations of the fork permanently maintained. In order to set another fork into forced vibration, its associated electro-magnet may be included, either in the same driving-circuit, or in a second, whose periodic interruption is effected by another rider dipping into mercury cups[1].

The *modus operandi* of this kind of self-acting instrument is often imperfectly apprehended. If the force acting on the fork depended only on its position—on whether the circuit were open or closed—the work done in passing through any position would be undone on the return, so that after a complete period there would be nothing outstanding by which the effect of the frictional forces could be compensated. Any explanation which does not take account of the retardation of the current is wholly beside the mark. The causes of retardation are two: irregular contact, and self-induction. When the point of the rider first touches the mercury, the electric contact is imperfect, probably on account of

[1] I have arranged several interrupters on the above plan, all the component parts being of home manufacture. The forks were made by the village blacksmith. The cups consisted of iron thimbles, soldered on one end of copper slips, the further end being screwed down on the base board of the instrument. Some means of adjusting the level of the mercury surface is necessary. In Helmholtz' interrupter a horse-shoe electro-magnet embracing the fork is adopted, but I am inclined to prefer the present arrangement, at any rate if the pitch be low. In some cases a greater motive power is obtained by a horse-shoe magnet acting on a soft iron armature carried horizontally by the upper prong and perpendicular to it. I have usually found a single Smee cell sufficient battery power.

adhering air. On the other hand, in leaving the mercury the contact is prolonged by the adhesion of the liquid in the cup to the amalgamated wire. On both accounts the current is retarded behind what would correspond to the mere position of the fork. But, even if the resistance of the circuit depended only on the position of the fork, the current would still be retarded by its self-induction. However perfect the contact may be, a finite current cannot be generated until after the lapse of a finite time, any more than in ordinary mechanics a finite velocity can be suddenly impressed on an inert body. From whatever causes arising[1], the effect of the retardation is that more work is gained by the fork during the retreat of the rider from the mercury, than is lost during its entrance, and thus a balance remains to be set off against friction.

If the magnetic force depended only on the position of the fork, the phase of its first harmonic component might be considered to be $180°$ in advance of that of the fork's own vibration. The retardation spoken of reduces this advance. If the phase-difference be reduced to $90°$, the force acts in the most favourable manner, and the greatest possible vibration is produced.

It is important to notice that (except in the case just referred to) the actual pitch of the interrupter differs to some extent from that natural to the fork according to the law expressed in (5) of § 46, ϵ being in the present case a prescribed phase-difference depending on the nature of the contacts and the magnitude of the self-induction. If the intermittent current be employed to drive a second fork, the maximum vibration is obtained, when the frequency of the fork coincides, not with the natural, but with the modified frequency of the interrupter.

The deviation of a tuning-fork interrupter from its natural pitch is practically very small; but the fact that such a deviation is possible, is at first sight rather surprising. The explanation (in the case of a small retardation of current) is, that during that half of the motion in which the prongs are the most separated, the electro-magnet acts in aid of the proper recovering power due to rigidity, and so naturally raises the pitch. Whatever the relation of phases may be, the force of the magnet may be divided into

[1] Any desired retardation might be obtained, in default of other means, by attaching the rider, not to the prong itself, but to the further end of a light straight spring carried by the prong and set into forced vibration by the motion of its point of attachment.

two parts respectively proportional to the velocity and displacement (or acceleration). To the first exclusively is due the sustaining power of the force, and to the second the alteration of pitch.

65. The general phenomenon of resonance, though it cannot be exhaustively considered under the head of one degree of freedom, is in the main referable to the same general principles. When a forced vibration is excited in one part of a system, all the other parts are also influenced, a vibration of the same period being excited, whose amplitude depends on the constitution of the system considered as a whole. But it not unfrequently happens that interest centres on the vibration of an outlying part whose connection with the rest of the system is but loose. In such a case the part in question, provided a certain limit of amplitude be not exceeded, is very much in the position of a system possessing one degree of freedom and acted on by a force, which may be regarded as *given*, independently of the natural period. The vibration is accordingly governed by the laws we have already investigated. In the case of approximate equality of periods to which the name of resonance is generally restricted, the amplitude may be very considerable, even though in other cases it might be so small as to be of little account; and the precision required in the adjustment of the periods in order to bring out the effect, depends on the degree of damping to which the system is subjected.

Among bodies which resound without an extreme precision of tuning, may be mentioned stretched membranes, and strings associated with sounding-boards, as in the pianoforte and the violin. When the proper note is sounded in their neighbourhood, these bodies are caused to vibrate in a very perceptible manner. The experiment may be made by singing into a pianoforte the note given by any of its strings, having first raised the corresponding damper. Or if one of the strings belonging to any note be plucked (like a harp string) with the finger, its fellows will be set into vibration, as may immediately be proved by stopping the first.

The phenomenon of resonance is, however, most striking in cases where a very accurate equality of periods is necessary in order to elicit the full effect. Of this class tuning-forks, mounted on resonance boxes, are a conspicuous example. When the unison is perfect the vibration of one fork will be taken up by another across the width of a room, but the slightest deviation of pitch is sufficient to render the phenomenon almost insensible. Forks of 256 vibrations per second are commonly used for the purpose, and it is found that a deviation from unison giving only one beat in a second makes all the difference. When the forks are well tuned and close together, the vibration may be transferred backwards and forwards between them several times, by damping them alternately, with a touch of the finger.

Illustrations of the powerful effects of isochronism must be within the experience of every one. They are often of importance in very different fields from any with which acoustics is concerned. For example, few things are more dangerous to a ship than to lie in the trough of the sea under the influence of waves whose period is nearly that of its own natural rolling.

65 a. It has already (§ 30) been explained how the super-position of two vibrations of equal amplitude and of nearly equal frequency gives rise to a resultant in which the sound rises and falls in beats. If we represent the two components by $\cos 2\pi n_1 t$, $\cos 2\pi n_2 t$, the resultant is

$$2 \cos \pi (n_1 - n_2) t \cdot \cos \pi (n_1 + n_2) t \dots\dots\dots\dots(1);$$

and it may be regarded as a vibration of frequency $\frac{1}{2}(n_1 + n_2)$, and of amplitude $2 \cos \pi (n_1 - n_2) t$. In passing through zero the amplitude changes sign, which is equivalent to a change of phase of 180°, if the amplitude be regarded as always positive. This change of phase is readily detected by measurement in drawings traced by machines for compounding vibrations, and it is a feature of great importance. If a force of this character act upon a system whose natural frequency is $\frac{1}{2}(n_1 + n_2)$, the effect produced is comparatively small. If the system start from rest, the successive impulses cooperate at first, but after a time the later impulses begin to destroy the effect of former ones. The greatest response would be given to forces of frequency n_1 and n_2, and not to a force of frequency $\frac{1}{2}(n_1 + n_2)$.

If, as in some experiments of Prof. A. M. Mayer [1], an otherwise steady sound is rendered intermittent by the periodic interposition of an obstacle, a very different result is arrived at. In this case the phase is resumed after each silence without reversal. If a force of this character act upon an isochronous system, the effect is indeed less than if there were no intermittence; but as all the impulses operate in the same sense without any antagonism, the response is powerful. One kind of intermittent vibration or force is represented by

$$2\,(1 + \cos 2\pi mt)\cos 2\pi nt \ldots\ldots\ldots\ldots(2),$$

in which n is the frequency of the vibration, and m the frequency of intermittence [1]. The amplitude is here always positive, and varies between the values 0 and 4. By ordinary trigonometrical transformation (2) may be put in the form

$$2 \cos 2\pi nt + \cos 2\pi\,(n + m)\,t + \cos 2\pi\,(n - m)\,t\ldots\ldots(3);$$

which shews that the intermittent vibration in question is equivalent to three simple vibrations of frequencies n, $n+m$, $n-m$. This is the explanation of the secondary sounds observed by Mayer.

The form (2) is of course only a particular case. Another in which the intensity of the intermittent sound rises more suddenly to its maximum is given by

$$4 \cos^4 \pi mt \cos 2\pi nt \ldots\ldots\ldots\ldots \ldots(4),$$

which may be transformed into

$$\tfrac{3}{2} \cos 2\pi nt + \cos 2\pi\,(n + m)\,t + \cos 2\pi\,(n - m)\,t$$
$$+ \tfrac{1}{4} \cos 2\pi\,(n + 2m)\,t + \tfrac{1}{4} \cos 2\pi\,(n - 2m)\,t\ldots\ldots(5).$$

There are here *four* secondary sounds, the frequencies of the two new ones differing twice as much as before from that of the primary sound.

The theory of intermittent vibrations is well illustrated by electrically driven forks. A fork interrupter of frequency 128 gave a periodic current, by the passage of which through an electro-magnet a second fork of like pitch could be excited. The action of this current on the second fork could be rendered intermittent by short-circuiting the electro-magnet. This was effected by another interrupter of frequency 4, worked by an *independent* current from a Smee cell. To excite the main current a Grove cell was employed. When the contact of the second interrupter was permanently broken, so that the main current passed continuously through the electro-magnet, the fork was, of course, most powerfully affected when tuned to 128. Scarcely any response was observable when the pitch was changed to 124 or 132. But if the second interrupter were allowed to operate, so as

---

[1] Crum Brown and Tait. *Edin. Proc.* June, 1878. Acoustical Observations II. *Phil. Mag.* April, 1880.

to render the periodic current through the electro-magnet intermittent, then the fork would respond powerfully when tuned to
124 or 132 as well as when tuned to 128, but not when tuned to
intermediate pitches, such as 126 or 130.

The operation of the intermittence in producing a sensitiveness which would not otherwise exist, is easily understood. When
a fork of frequency 124 starts from rest under the influence of a
force of frequency 128, the impulses cooperate at first, but after $\frac{1}{8}$
of a second the new impulses begin to oppose the earlier ones.
After $\frac{1}{4}$ of a second, another series of impulses begins whose effect
agrees with that of the first, and so on. Thus if all these impulses
are allowed to act, the resultant effect is trifling; but if every
alternate series is stopped off, a large vibration accumulates.

Fig. 16 a.

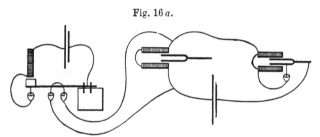

The most general expression for a vibration of frequency n,
whose amplitude and phase are slowly variable with a frequency
m, is

$$\left\{ \begin{aligned} A_0 + A_1 \cos 2\pi mt + A_2 \cos 4\pi mt + A_3 \cos 6\pi mt + \ldots \\ + B_1 \sin 2\pi mt + B_2 \sin 4\pi mt + B_3 \sin 6\pi mt + \ldots \end{aligned} \right\} \cos 2\pi nt$$

$$+ \left\{ \begin{aligned} C_0 + C_1 \cos 2\pi mt + C_2 \cos 4\pi mt + C_3 \cos 6\pi mt + \ldots \\ + D_1 \sin 2\pi mt + D_2 \sin 4\pi mt + D_3 \sin 6\pi mt + \ldots \end{aligned} \right\} \sin 2\pi nt$$

$$\ldots\ldots(6);$$

and this applies both to the case of beats (e.g. if A_1 only be finite)
and to such intermittence as is produced by the interposition of
an obstacle. The vibration in question is accordingly in all cases
equivalent to a combination of simple vibrations of frequencies

$$n, \ n+m, \ n-m, \ n+2m, \ n-2m, \ \&c.$$

It may be well here to emphasise that a simple vibration
implies *infinite* continuance, and does not admit of variations of
phase or amplitude. To suppose, as is sometimes done in optical
speculations, that a train of simple waves may begin at a given
epoch, continue for a certain time involving it may be a large
number of periods, and ultimately cease, is a contradiction in terms.

66. The solution of the equation for free vibration, viz.

$$\ddot{u} + \kappa\dot{u} + n^2 u = 0 \dots\dots\dots\dots\dots (1).$$

may be put into another form by expressing the arbitrary constants of integration A and α in terms of the initial values of u and $\dot{u}$, which we may denote by u_0 and $\dot{u}_0$. We obtain at once

$$u = e^{-\frac{1}{2}\kappa t}\left\{\dot{u}_0\frac{\sin n't}{n'} + u_0\left(\cos n't + \frac{\kappa}{2n'}\sin n't\right)\right\} \dots\dots (2),$$

where $n' = \sqrt{n^2 - \frac{1}{4}\kappa^2}.$

If there be no friction, $\kappa = 0$, and then

$$u = \dot{u}_0\frac{\sin nt}{n} + u_0\cos nt \dots\dots\dots\dots\dots(3).$$

These results may be employed to obtain the solution of the complete equation

$$\ddot{u} + \kappa\dot{u} + n^2 u = U \dots\dots\dots\dots\dots\dots (4),$$

where U is an explicit function of the time; for from (2) we see that the effect at time t of a velocity $\delta\dot{u}$ communicated at time t' is

$$u = \delta\dot{u}\, e^{-\frac{1}{2}\kappa(t-t')}\frac{\sin n'(t-t')}{n'}$$

The effect of U is to generate in time dt' a velocity $U dt'$, whose result at time t will therefore be

$$u = \frac{1}{n'}\, U dt'\, e^{-\frac{1}{2}\kappa(t-t')}\sin n'(t-t'),$$

and thus the solution of (4) will be

$$u = \frac{1}{n'}\int^t e^{-\frac{1}{2}\kappa(t-t')}\sin n'(t-t')\, U\, dt' \dots\dots\dots (5).$$

If there be no friction, we have simply

$$u = \frac{1}{n}\int^t \sin n(t-t')\, U\, dt' \dots\dots\dots\dots (6),$$

U being the force at time t'.

The lower limit of the integrals is so far arbitrary, but it will generally be convenient to make it zero.

On this supposition u and $\dot{u}$ as given by (6) vanish, when $t = 0$, and the complete solution is

$$u = e^{-\frac{1}{2}\kappa t}\left\{\dot{u}_0\frac{\sin n't}{n'} + u_0\left(\cos n't + \frac{\kappa}{2n'}\sin n't\right)\right\}$$
$$+ \frac{1}{n'}\int_0^t e^{-\frac{1}{2}\kappa(t-t')}\sin n'(t-t')\, U\, dt' \dots\dots (7),$$

or if there be no friction

$$u = \dot{u}_0 \frac{\sin nt}{n} + u_0 \cos nt + \frac{1}{n} \int_0^t \sin n (t - t') \, U \, dt' \, \ldots\ldots\ldots (8).$$

When t is sufficiently great, the complementary terms tend to vanish on account of the factor $e^{-\frac{1}{2}\kappa t}$, and may then be omitted.

66 a. In § 66 we have limited the discussion to the case of greatest acoustical importance, that is, we have supposed that n' is real, as happens when n^2 is positive, and κ not too great. But a more general treatment of the problem of free vibrations is not without interest. Whatever may be the values of n^2 and κ, the solution of (1) § 66 may be expressed

$$u = A e^{\mu_1 t} + B e^{\mu_2 t} \ldots\ldots\ldots\ldots\ldots\ldots (1),$$

where μ_1, μ_2 are the roots of

$$\mu^2 + \kappa\mu + n^2 = 0 \ldots\ldots\ldots\ldots\ldots\ldots (2).$$

The case already discussed is that in which the values of μ are imaginary. The motion is then oscillatory, with amplitude which decreases if κ be positive, but increases if κ be negative.

But if n^2, though positive, be less than $\frac{1}{4}\kappa^2$, or if n^2 be negative, n' becomes imaginary, that is μ becomes real. The motion expressed by (1) is then non-oscillatory, and it depends upon the sign of μ whether it increases or diminishes with the time. From the solution of (2), viz.

$$\mu = -\tfrac{1}{2}\kappa \pm \tfrac{1}{2} \sqrt{(\kappa^2 - 4n^2)} \ldots\ldots\ldots\ldots\ldots (3),$$

it is evident that if n^2 be positive (and less than $\frac{1}{4}\kappa^2$) the two values of μ are of the same sign, and that the sign is the opposite of that of κ. Hence if κ be positive, both terms in (1) diminish with the time, so that the system, however disturbed, subsides again into a state of rest. If, on the contrary, κ be negative, the motion increases without limit.

We have still to consider the case of n^2 negative. The real values of μ are then of *opposite* signs. It is possible so to start the system from a displaced position that it shall approach asymptotically the condition of rest in the configuration of equilibrium; but unless a special relation between displacement and velocity is satisfied, the motion tends to increase without limit. Under these circumstances the equilibrium must be regarded as *unstable*. In this sense stability requires that n^2 and κ be both positive.

A word may not be out of place as to the effect of an impressed force upon a statically unstable system. If in § 46 we suppose $\kappa = 0$, the solution (7) does not change its form merely because n^2 becomes negative. The fact that a system is susceptible of purely periodic motion under the operation of an external periodic force is therefore no evidence of stability.

67. For most acoustical purposes it is sufficient to consider the vibrations of the systems, with which we may have to deal, as infinitely small, or rather as similar to infinitely small vibrations. This restriction is the foundation of the important laws of isochronism for free vibrations, and of persistence of period for forced vibrations. There are, however, phenomena of a subordinate but not insignificant character, which depend essentially on the square and higher powers of the motion. We will therefore devote the remainder of this chapter to the discussion of the motion of a system of one degree of freedom, the motion not being so small that the squares and higher powers can be altogether neglected.

The approximate expressions for the kinetic and potential energies will be of the form

$$T = \tfrac{1}{2}\left(m_0 + m_1 u\right)\dot{u}^2, \quad V = \tfrac{1}{2}\left(\mu_0 + \mu_1 u\right)u^2.$$

If the sum of T and V be differentiated with respect to the time, we find as the equation of motion

$$m_0\ddot{u} + \mu_0 u + m_1 u\ddot{u} + \tfrac{1}{2}m_1\dot{u}^2 + \tfrac{3}{2}\mu_1 u^2 = \text{Impressed Force,}$$

which may be treated by the method of successive approximation. For the sake of simplicity we will take the case where $m_1 = 0$, a supposition in no way affecting the essence of the question. The *inertia* of the system is thus constant, while the force of restitution is a composite function of the displacement, partly proportional to the displacement itself and partly proportional to its square—accordingly unsymmetrical with respect to the position of equilibrium. Thus for free vibrations our equation is of the form

$$\ddot{u} + n^2 u + \alpha u^2 = 0 \dots\dots\dots\dots\dots (1),$$

with the approximate solution

$$u = A \cos nt \dots\dots\dots\dots\dots (2),$$

where A—the amplitude—is to be treated as a small quantity.

Substituting the value of u expressed by (2) in the last term, we find

$$\ddot{u} + n^2 u = -\alpha \frac{A^2}{2}\left(1 + \cos 2nt\right),$$

whence for a second approximation to the value of u

$$u = A \cos nt - \frac{\alpha A^2}{2n^2} + \frac{\alpha A^2}{6n^2} \cos 2nt \ldots\ldots\ldots\ldots(3);$$

shewing that the proper tone (n) of the system is accompanied by its *octave* $(2n)$, whose *relative* importance increases with the amplitude of vibration. A trained ear can generally perceive the octave in the sound of a tuning-fork caused to vibrate strongly by means of a bow, and with the aid of appliances, to be explained later, the existence of the octave may be made manifest to any one. By following the same method the approximation can be carried further; but we pass on now to the case of a system in which the recovering power is symmetrical with respect to the position of equilibrium. The equation of motion is then approximately

$$\ddot{u} + n^2 u + \beta u^3 = 0 \ldots\ldots\ldots\ldots\ldots\ldots (4),$$

which may be understood to refer to the vibrations of a heavy pendulum, or of a load carried at the end of a straight spring.

If we take as a first approximation $u = A \cos nt$, corresponding to $\beta = 0$, and substitute in the term multiplied by β, we get

$$\ddot{u} + n^2 u = -\frac{\beta A^3}{4} \cos 3nt - \frac{3\beta A^3}{4} \cos nt.$$

Corresponding to the last term of this equation, we should obtain in the solution a term of the form $t \sin nt$, becoming greater without limit with t. This, as in a parallel case in the Lunar Theory, indicates that our assumed first approximation is not really an approximation at all, or at least does not *continue* to be such. If, however, we take as our starting point $u = A \cos mt$, with a suitable value for m, we shall find that the solution may be completed with the aid of periodic terms only. In fact it is evident beforehand that all we are entitled to assume is that the motion is approximately simple harmonic, with a period *approximately* the same, as if $\beta = 0$. A very slight examination is sufficient to shew that the term varying as u^3, not only may, but *must* affect the period. At the same time it is evident that a solution, in which the period is assumed wrongly, no matter by how little, must at length cease to represent the motion with any approach to accuracy.

We take then for the approximate equation

$$\ddot{u} + n^2 u = -\frac{3\beta A^3}{4} \cos mt - \frac{\beta A^3}{4} \cos 3mt \ldots\ldots\ldots (5),$$

of which the solution will be

$$u = A \cos mt + \frac{\beta A^3}{4} \frac{\cos 3mt}{9m^2 - n^2} \quad \ldots\ldots\ldots\ldots\ldots (6),$$

provided that m be taken so as to satisfy

$$A(-m^2 + n^2) = -\frac{3\beta A^3}{4},$$

or $$m^2 = n^2 + \frac{3\beta A^2}{4} \quad \ldots\ldots\ldots\ldots\ldots\ldots (7).$$

The term in β thus produces two effects. It alters the pitch of the fundamental vibration, and it introduces the *twelfth* as a necessary accompaniment. The alteration of pitch is in most cases exceedingly small—depending on the square of the amplitude, but it is not altogether insensible. Tuning-forks generally rise a little, though very little, in pitch as the vibration dies away. It may be remarked that the same slight dependence of pitch on amplitude occurs when the force of restitution is of the form $n^2 u + \alpha u^2$, as may be seen by continuing the approximation to the solution of (1) one step further than (3). The result in that case is

$$m^2 = n^2 - \frac{5\alpha^2 A^2}{6n^2} \quad \ldots\ldots\ldots\ldots\ldots (8)[1].$$

The difference $m^2 - n^2$ is of the same order in A in both cases; but in one respect there is a distinction worth noting, namely, that in (8) m^2 is always less than n^2, while in (7) it depends on the sign of β whether its effect is to raise or lower the pitch. However, in most cases of the unsymmetrical class the change of pitch would depend partly on a term of the form αu^2 and partly on another of the form βu^3, and then

$$m^2 = n^2 - \frac{5\alpha^2 A^2}{6n^2} + \frac{3\beta A^2}{4} \quad \ldots\ldots\ldots\ldots (9)[1].$$

[In all cases where the period depends upon amplitude, it is necessarily an *even* function thereof, a change of sign in the amplitude being merely equivalent to an alteration in phase of 180°.]

68. We now pass to the consideration of the vibrations forced on an unsymmetrical system by two harmonic forces

$$E \cos pt, \quad F \cos (qt - \epsilon).$$

---

[1] [A correction is here introduced, the necessity for which was pointed out to me by Dr Burton.]

The equation of motion is

$$\ddot{u} + n^2u = -\alpha u^2 + E\cos pt + F\cos(qt-\epsilon) \ \ldots\ldots (1).$$

To find a first approximation we neglect the term containing α. Thus

$$u = e\cos pt + f\cos(qt-\epsilon) \ \ldots\ldots\ldots\ldots (2),$$

where

$$e = \frac{E}{n^2-p^2}, \quad f = \frac{F}{n^2-q^2} \ \ldots\ldots\ldots\ldots (3).$$

Substituting this in the term multiplied by α, we get

$$\ddot{u} + n^2u = E\cos pt + F\cos(qt-\epsilon)$$

$$-\alpha\left[\frac{e^2+f^2}{2} + \frac{e^2}{2}\cos 2pt + \frac{f^2}{2}\cos 2(qt-\epsilon) + ef\cos\{(p-q)t+\epsilon\}\right.$$

$$\left. + ef\cos\{(p+q)t-\epsilon\}\right]$$

whence as a second approximation for u

$$u = e\cos pt + f\cos(qt-\epsilon) - \frac{\alpha(e^2+f^2)}{2n^2} - \frac{\alpha e^2}{2(n^2-4p^2)}\cos 2pt$$

$$- \frac{\alpha f^2}{2(n^2-4q^2)}\cos 2(qt-\epsilon) - \frac{\alpha ef}{n^2-(p-q)^2}\cos\{(p-q)t+\epsilon\}$$

$$- \frac{\alpha ef}{n^2-(p+q)^2}\cos\{(p+q)t-\epsilon\} \ \ldots\ldots\ldots\ldots (4).$$

The additional terms represent vibrations having frequencies which are severally the doubles and the sum and difference of those of the primaries. Of the two latter the amplitudes are proportional to the product of the original amplitudes, shewing that the derived tones increase in relative importance with the intensity of their parent tones.

68a. If an isolated vibrating system be subject to internal dissipative influences, the vibrations cannot be permanent, since they are dependent upon an initial store of energy which suffers gradual exhaustion. In order that the motion may be maintained, the vibrating body must be in connection with a source of energy. We have already considered cases of this kind under the head of forced vibrations, where the system is subject to forces whose amplitude and phase are prescribed, independently of the behaviour of the system. Such forces may have their origin in revolving mechanism (such as electric alternators) governed so as to move at a uniform speed. But more frequently the forces under consideration depend upon the vibrations of other systems, and then the question as to how the vibrations are to be main-- tained represents itself. A good example is afforded by the case already discussed (§§ 63, 65) of a fork maintained in vibration electrically by means of currents governed by a fork interrupter. It has been pointed out that the performance of the latter depends upon the magnetic forces operative upon it differing in phase from the vibrations of the fork itself. With the interrupter may be classed for the present purpose almost all acoustical and musical instruments capable of providing a sustained sound. It may suffice to mention vibrations maintained by wind (organ- pipes, harmonium reeds, æolian harps, &c.), by heat (singing flames, Rijke's tubes, &c.), by friction (violin strings, finger- glasses), and the slower vibrations of clock pendulums and watch balance-wheels.

In considering whether proposed forces are of the right kind for the maintenance or encouragement of a vibration, it is often convenient to regard them as reduced to impulses. Suppose, to take a simple case, that a small horizontal positive impulse acts upon the bob of a vibrating pendulum. The effect depends, of course, upon the phase of the vibration at the instant of the impulse. If the bob be moving positively at the instant in question the vibration is encouraged, and this effect is a maximum when the positive motion is greatest, that is, when the impulse occurs at the moment of positive movement through the position of equilibrium. This is the condition of things aimed at in designing a clock escapement, for the effect of the force is then a maximum in encouraging the vibration, and a minimum (zero to the first order of approximation) in disturbing the period. Of course, if the impulse be half a period earlier or later than is above supposed, the effect is to discourage the vibration, again without altering the period. In like manner we see that if the impulse occur at a moment of maximum elongation the effect is concentrated upon the period, the vibration being neither en- couraged nor discouraged.

In most cases the force acting upon a vibrating system in virtue of its connection with a source of energy may be regarded as harmonic. It may then be divided into two parts, one pro- portional to the displacement u (or to the acceleration $\ddot{u}$), the second proportional to the velocity $\dot{u}$. The inclusion of such forces does not alter the form of the equation of vibration

$$\ddot{u} + \kappa\dot{u} + n^2u = 0 \dots\dots\dots\dots\dots(1).$$

By the first part (proportional to u) the pitch is modified, and by the second the coefficient of decay. If the altered κ be still positive, vibrations gradually die down; but if the effect of the included forces be to render κ negative, vibrations tend on the contrary to increase. The only case in which according to (1) a steady vibration is possible, is when the complete value of κ is zero. If this condition be satisfied, a vibration of any amplitude is permanently maintained.

When κ is negative, so that small vibrations tend to increase, a point is of course soon reached beyond which the approximate equations cease to be applicable. We may form an idea of the state of things which then arises by adding to equation (1) a term proportional to a higher power of the velocity. Let us take

$$\ddot{u} + \kappa\dot{u} + \kappa'\dot{u}^3 + n^2 u = 0 \quad\dots\dots\dots\dots\dots\dots(2),$$

in which κ and κ' are supposed to be small quantities. The approximate solution of (2) is

$$u = A \sin nt + \frac{\kappa' n A^3}{32} \cos 3nt \dots\dots\dots\dots\dots(3),$$

in which A is given by

$$\kappa + \tfrac{3}{4}\kappa' n^2 A^2 = 0 \quad\dots\dots\dots\dots\dots\dots(4).$$

From (4) we see that no steady vibration is possible unless κ and κ' have opposite signs. If κ and κ' be both positive, the vibration in all cases dies down; while if κ and κ' be both negative, the vibration (according to (2)) increases without limit. If κ be negative and κ' positive, the vibration becomes steady and assumes the amplitude determined by (4). A smaller vibration increases up to this point, and a larger vibration falls down to it. If on the other hand κ be positive, while κ' is negative, the steady vibration abstractedly possible is unstable, a departure in either direction from the amplitude given by (4) tending always to increase [1].

68 *b*. We will now consider briefly another and a very curious kind of maintenance, of which the peculiarity is that the maintaining influence operates with a frequency which is the double of that of the vibration maintained. Probably the best known example is that form of Melde's experiment, in which a fine string is maintained in transverse vibration by connecting one of its extremities with the vibrating prong of a massive tuning-fork,

[1] On Maintained Vibrations, *Phil. Mag.*, April, 1883.

the direction of motion of the point of attachment being parallel to the length of the string. The effect of the motion is to render the tension of the string periodically variable ; and at first sight there is nothing to cause the string to depart from its equilibrium condition of straightness. It is known, however, that under these circumstances the equilibrium may become unstable, and that the string may settle down into a state of permanent and vigorous vibration, whose period is the *double* of that of the fork.

As a simpler example, with but one degree of freedom, we may take a pendulum, formed of a bar of soft iron and vibrating upon knife-edges. Underneath is placed symmetrically a vertical bar electro-magnet, through which is caused to pass an electric current rendered intermittent by an interrupter whose frequency is twice that of the pendulum. The magnetic force does not tend to displace the pendulum from its equilibrium position, but produces the same sort of effect as if gravity were subject to a periodic variation of intensity.

A similar result is obtained by causing the point of support of the pendulum to vibrate in a *vertical* path. If we denote this motion by $\eta = \beta \sin 2pt$, the effect is as if gravity were variable by the term $4p^2\beta \sin 2pt$.

Of the same nature are the crispations observed by Faraday[1] and others upon the surface of water which oscillates vertically. Faraday arrived experimentally at the conclusion that there were two complete vibrations of the support for each complete vibration of the liquid.

In the following investigation[2], relative to the case of one degree of freedom, we shall start with the assumption that a steady vibration is in progress, and inquire under what conditions the assumed state of things is possible.

If the force of restitution, or "spring," of a body susceptible of vibration be subject to an imposed periodic variation, the differential equation takes the form

$$\ddot{u} + \kappa\dot{u} + (n^2 - 2\alpha \sin 2pt)\, u = 0 \quad\ldots\ldots\ldots\ldots(1),$$

in which κ and α are supposed to be small. A similar equation would apply approximately to the case of a periodic variation in the effective mass of the body. The motion expressed by the solution of (1) can be regular only when it keeps perfect time

[1] *Phil. Trans.* 1831, p. 299.
[2] *Phil. Mag.,* April, 1883.

with the imposed variations. It will appear that the necessary conditions cannot be satisfied rigorously by any simple harmonic vibration, but we may assume

$$u = A_1 \sin pt + B_1 \cos pt$$
$$+ A_3 \sin 3pt + B_3 \cos 3pt + A_5 \sin 5pt + \ldots\ldots\ldots(2),$$

in which it is not necessary to provide for sines and cosines of even multiples of pt. If the assumption be justifiable, the solution in (2) must be convergent. Substituting in the differential equation, and equating to zero the coefficients of $\sin pt$, $\cos pt$, &c. we find

$$A_1 (n^2 - p^2) - \kappa p B_1 - \alpha B_1 + \alpha B_3 = 0,$$
$$B_1 (n^2 - p^2) + \kappa p A_1 - \alpha A_1 - \alpha A_3 = 0 ;$$
$$A_3 (n^2 - 9p^2) - 3\kappa p B_3 - \alpha B_1 + \alpha B_5 = 0,$$
$$B_3 (n^2 - 9p^2) + 3\kappa p A_3 + \alpha A_1 - \alpha A_5 = 0 ;$$
$$A_5 (n^2 - 25p^2) - 5\kappa p B_5 - \alpha B_3 + \alpha B_7 = 0,$$
$$B_5 (n^2 - 25p^2) + 5\kappa p A_5 + \alpha A_3 - \alpha A_7 = 0 ;$$
$$\ldots\ldots\ldots\ldots\ldots\ldots\ldots\ldots\ldots\ldots\ldots\ldots$$

These equations shew that A_3, B_3 are of the order α relatively to A_1, B_1; that A_5, B_5 are of order α relatively to A_3, B_3, and so on. If we omit A_3, B_3 in the first pair of equations, we find as a first approximation,

$$A_1 (n^2 - p^2) - (\kappa p + \alpha) B_1 = 0,$$
$$A_1 (\kappa p - \alpha) + (n^2 - p^2) B_1 = 0 ;$$

whence
$$\frac{B_1}{A_1} = \frac{n^2 - p^2}{\kappa p + \alpha} = \frac{\alpha - \kappa p}{n^2 - p^2} = \frac{\sqrt{(\alpha - \kappa p)}}{\sqrt{(\alpha + \kappa p)}}\ldots\ldots\ldots\ldots(3),$$

and
$$(n^2 - p^2)^2 = \alpha^2 - \kappa^2 p^2 \ldots\ldots\ldots\ldots\ldots(4).$$

Thus, if α be given, the value of p necessary for a regular motion is definite; and p having this value, the regular motion is

$$u = P \sin (pt + \epsilon),$$

in which ϵ, being equal to $\tan^{-1} (B_1/A_1)$, is also definite. On the other hand, as is evident at once from the linearity of the original equation, there is nothing to limit the amplitude of vibration.

These characteristics are preserved however far it may be necessary to pursue the approximation. If A_{2m+1}, B_{2m+1} may be neglected, the first m pairs of equations determine the *ratios* of all the coefficients, leaving the absolute magnitude open; and they provide further an equation connecting p and α, by which the pitch is determined.

For the second approximation the second pair of equations give

$$A_3 = \frac{\alpha B_1}{n^2 - 9p^2}, \qquad B_3 = -\frac{\alpha A_1}{n^2 - 9p^2},$$

whence

$$u = P \sin (pt + \epsilon) + \frac{\alpha P}{9p^2 - n^2} \cos (3pt + \epsilon) \ldots\ldots\ldots(5),$$

and from the first pair

$$\tan \epsilon = \left\{ n^2 - p^2 - \frac{\alpha^2}{n^2 - 9p^2} \right\} \div (\alpha + \kappa p)\ldots\ldots\ldots\ldots\ldots(6),$$

while p is determined by

$$\left\{ n^2 - p^2 - \frac{\alpha^2}{n^2 - 9p^2} \right\}^2 = \alpha^2 - \kappa^2 p^2\ldots\ldots\ldots\ldots\ldots(7).$$

Returning to the first approximation, we see from (4) that the solution is possible only under the condition that α be not less than κp. If $\alpha = \kappa p$, then $p = n$; that is, the imposed variation in the "spring" must be exactly twice as quick as the natural vibration of the body would be in the absence of friction. From (3) it appears that in this case $\epsilon = 0$, which indicates that the spring is a minimum one-eighth of a period *after* the body has passed its position of equilibrium, and a maximum one-eighth of a period *before* such passage. Under these circumstances the greatest possible amount of energy is communicated to the system; and in the case contemplated it is just sufficient to balance the loss by dissipation, the adjustment being evidently independent of the amplitude.

If $\alpha < \kappa p$ sufficient energy cannot pass to maintain the motion, whatever may be the phase-relation; but if $\alpha > \kappa p$, the balance between energy supplied and energy dissipated may be attained by such an alteration of phase as shall diminish the former quantity to the required amount. The alteration of phase may for this purpose be indifferently in either direction; but if ϵ be positive, we must have

$$p^2 = n^2 - \sqrt{(\alpha^2 - \kappa^2 p^2)} ;$$

while if ϵ be negative

$$p^2 = n^2 + \sqrt{(\alpha^2 - \kappa^2 p^2)}.$$

If α be very much greater than κp, $\epsilon = \pm \frac{1}{4}\pi$. which indicates that when the system passes through its position of equilibrium the spring is at its maximum or at its minimum.

The inference from the equation that the adjustment of pitch must be absolutely rigorous for steady vibration will be subject to some modification in practice; otherwise the experiment could not succeed.   In most cases n^2 is to a certain extent a function of amplitude; so that if n^2 have very nearly the required value, complete coincidence is attainable by the assumption of an amplitude of large and determinate amount without other alterations in the conditions of the system.

The reader who wishes to pursue this subject is referred to a paper by the Author " On the Maintenance of Vibrations by Forces of Double Frequency, and on the Propagation of Waves through a Medium endowed with a Periodic Structure,"[1] in which the analysis of Mr Hill[2] is applied to the present problem.

68 *c*.   The determination of absolute pitch by means of the siren has already been alluded to (§ 17).   In all probability first-rate results might be got by this method if proper provision, with the aid of a phonic wheel for example, were made for uniform speed.   In recent years several experimenters have obtained excellent results by various methods; but a brief notice of these is all that our limits will allow.

One of the most direct determinations is that of Koenig[3], to whom the scientific world has long been indebted for the construction of much excellent apparatus.   This depends upon a special instrument, consisting of a fork of 64 complete vibrations per second, the motion being maintained by a clock movement acting upon an escapement.   A dial is provided marking ordinary time, and serves to record the number of vibrations executed.   The performance of the fork is tested by a comparison between the instrument and any chronometer known to be keeping good time.   The standard fork of 256 complete vibrations was compared with that of the instrument by observing the Lissajous's figure appropriate to the double octave.

M. Koenig has also investigated the influence of resonators upon the pitch of forks.   Thus without a resonator a fork of 256 complete vibrations sounded in a satisfactory manner for about 90 seconds.   A resonator of adjustable pitch was then brought into proximity, and the pitch, originally much graver than that of the

[1] *Phil. Mag.*, August, 1887.

[2] On the Part of the Motion of the Lunar Perigree which is a Function of the Mean Motions of the Sun and Moon, Acta Mathematica 8; 1, 1886.   Mr Hill's work was first published in 1877.

[3] Wied. *Ann.* IX. p. 394, 1880.

fork, was gradually raised. Even when the resonator was still a minor third below the fork, there was observed a slight diminution in the duration of the vibratory movement, and at the same time an augmentation in the frequency of about ·005. As the natural note of the resonator approached nearer to that of the fork, this diminution in the time and this increase in frequency became more pronounced up to the immediate neighbourhood of unison; but at the moment when unison was established, the alteration of pitch suddenly disappeared, and the frequency became exactly the same as in the absence of the resonator. At the same time the sound was powerfully reinforced; but this exaggerated intensity fell off rapidly and the vibration died away after 8 or 10 seconds. The pitch of the resonator being again raised a little, the sound of the fork began to change in the opposite direction, being now as much too grave as before the unison was reached it had been too acute. The displacement then fell away by degrees, as the pitch of the resonator was further raised, and the duration of the vibrations gradually recovered its original value of about 90 seconds. The maximum disturbance in the frequency observed by Koenig was ·035 complete vibrations. For the explanation of these effects see § 117.

The temperature coefficient found by Koenig is ·000112, so that the pitch of a 256 fork falls ·0286 for each degree Cent. by which the temperature rises.

In determinations of absolute pitch[1] by the Author of this work an electrically maintained interrupter fork, whose frequency may for example be 32, was employed to drive a dependent fork of pitch 128. When the apparatus is in good order, there is a fixed relation between the two frequencies, the one being precisely four times the other. The higher is of course readily compared by beats, or by optical methods, with a standard of 128, whose accuracy is to be tested. It remains to determine the frequency of the interrupter fork itself.

For this purpose the interrupter is compared with the pendulum of a standard clock whose rate is known. The comparison may be direct, or the intervention of a phonic wheel (§ 63) may be invoked. In either case the pendulum of the clock is provided with a silvered bead upon which is concentrated the light from a lamp. Immediately in front of the pendulum is placed a screen perforated by a somewhat narrow vertical slit. The bright point of light

[1] *Nature*, xvii. p. 12, 1877 ; *Phil. Trans.* 1883, Part I. p. 316.

reflected by the bead is seen intermittently, either by looking over
the prong of the interrupter or through a hole in the disc of the
phonic wheel. In the first case there are 32 views per second, but
in the latter this number is reduced by the intervention of the
wheel. In the experiments referred to the wheel was so
arranged that one revolution corresponded to four complete vibrations of the interrupter, and there were thus 8 views of the pendulum per second, instead of 32. Any deviation of the period of
the pendulum from a precise multiple of the period of intermittence
shews itself as a cycle of changes in the appearance of the flash
of light, and an observation of the duration of this cycle gives the
data for a precise comparison of frequencies.

The calculation of the results is very simple. Supposing in
the first instance that the clock is correct, let a be the number of
cycles per second (perhaps $\frac{1}{10}$) between the wheel and the clock.
Since the period of a cycle is the time required for the wheel to
gain, or lose, one revolution upon the clock, the frequency of revolution is $8 \pm a$. The frequency of the auxiliary fork is precisely 16
times as great, i.e. $128 \pm 16a$. If b be the number of beats per
second between the auxiliary fork and the standard, the frequency
of the latter is

$$128 \pm 16a \pm b.$$

An error in the mean rate of the clock is readily allowed for;
but care is required to ascertain that the actual rate at the time
of observation does not differ appreciably from the mean rate.
To be quite safe it would be necessary to repeat the determinations at intervals over the whole time required to rate the
clock by observation of the stars. In this case it would probably
be convenient to attach a counting apparatus to the phonic wheel.

In the method of M'Leod and Clarke[1] time, given by a clock,
is recorded automatically upon the revolving drum of a chrono-
graph, which is maintained by a suitable governor in uniform
rotation. The circumference of the drum is marked with a grating
of equidistant lines parallel to the axis, and the comparison between
the drum and the standard fork is effected by observation of the
wavy pattern seen when the revolving grating is looked at past
the edges of the vibrating prongs. These observers made a special
investigation as to the effect of bowing a fork upon previously
existing vibrations. Their conclusion is that in the case of unloaded forks no sensible change of phase occurs.

[1] *Phil. Trans.* 1880, Part I. p. 1.

In the chronographic method of Prof. A. M. Mayer[1] the fork under investigation is armed with a triangular fragment of thin sheet metal, one milligram in weight, and actually traces its vibrations as a curve of sines upon smoked paper. The time is recorded by small electric discharges from an induction apparatus, under the control of a clock, and delivered from the *same tracing.* *point.* Although the disturbance due to the tracing point appears to be very small, it is doubtful whether this method could compete in respect of accuracy with those above described where the comparison with the standard is optical or acoustical. On the other hand, it has the advantage of not requiring a uniform rotation of the drum, and the apparatus lends itself with facility to the determination of small intervals of time after the manner originally proposed by T. Young[2].

68 *d*. The methods hitherto described for the determination of absolute pitch, with the exception of that of Scheibler, may be regarded as rather mechanical in their character, and they depend for the most part upon somewhat special apparatus. It is possible, however, to determine pitch with fair accuracy with no other appliances than a common harmonium and a watch, and as the process is instructive in respect of the theory of overtones, a short account will here be given of it[3].

The fundamental principle is that the absolute frequencies of two musical notes can be deduced from the *interval* between them, i.e. the ratio of their frequencies, and the number of beats which they occasion in a given time when sounded together. For example, if x and y denote the frequencies of two notes whose interval is an equal temperament major third, we know that $y = 1 \cdot 25992\, x$. At the same time the number of beats heard in a second depending upon the deviation of the third from true intonation, is $4y - 5x$. In the case of the notes of a harmonium, which are rich in overtones, these beats are readily counted, and thus two equations are obtained from which the values of x and y are at once found.

Of course in practice the truth of an equal temperament third could not be taken for granted, but the difficulty thence arising would be easily met by including in the counting all the three

[1] National Academy of Sciences, Washington, *Memoirs*, Vol. III. p. 43, 1884.

[2] *Lectures*, Vol. I. p. 191.

[3] *Nature*, Jan. 23, 1879.

major thirds which together make up an octave. Suppose, for example, that the frequencies of c, e, $g\sharp$, c' are respectively x, y, z, $2x$, and that the beats per second between x and y are a, between y and z are b, and between z and $2x$ are c. Then

$$4y - 5x = a, \quad 4z - 5y = b, \quad 8x - 5z = c,$$

from which

$$x = \tfrac{1}{3}(25a + 20b + 16c),$$

$$y = \tfrac{1}{3}(32a + 25b + 20c),$$

$$z = \tfrac{1}{3}(40a + 32b + 25c).$$

In the above statements the octave c—c' is for simplicity supposed to be true. The actual error could readily be allowed for if required; but in practice it is not necessary to use c' at all, inasmuch as the third set of beats can be counted equally well between $g\sharp$ and c.

The principal objection to the method in the above form is that it presupposes the absolute constancy of the notes, for example, that y is the same whether it is being sounded in conjunction with x or in conjunction with z. This condition is very imperfectly satisfied by the notes of a harmonium.

In order to apply the fundamental principle with success, it is necessary to be able to check the accuracy of the interval which is supposed to be known, at the same time that the beats are being counted. If the interval be a major tone (9 : 8), its exactness is proved by the absence of beats between the ninth component of the lower and the eighth of the higher note, and a counting of the beats between the tenth component of the lower and the ninth of the higher note completes the necessary data for determining the absolute pitch.

The equal temperament whole tone (1·12246) is intermediate between the minor tone (1·11111) and the major tone (1·12500), but lies much nearer to the latter. Regarded as a disturbed major tone, it gives slow beats, and regarded as a disturbed minor tone it gives quick ones. Both sets of beats can be heard at the same time, and when counted (by two observers) give the means of calculating the absolute pitch of both notes. If x and y be the frequencies of the two notes, a and b the frequencies of the slow and quick beats respectively,

$$9x - 8y = a, \quad 9y - 10x = b,$$

whence

$$x = 9a + 8b, \quad y = 10a + 9b.$$

The application of this method in no way assumes the truth of the equal temperament whole tone, and in fact it is advantageous to flatten the interval somewhat, so as to make it lie more nearly midway between the major and the minor tone. In this way the rapidity of the quicker beats is diminished, which facilitates the counting.

The course of an experiment is then as follows. The notes C and D are sounded together, and at a given signal the observers begin counting the beats situated at about d'' and e'' on the scale. After the expiration of a measured interval of time a second signal is given, and the number of both sets of beats is recorded.

For further details of the method reference must be made to the original memoir, but one example of the results may be given here. The period being 10 minutes, the number of beats recorded were 2392 and 2341, giving $x = 67 \cdot 09$ as the pitch of C.

CHAPTER IV.

69. WE have now examined in some detail the oscillations of a system possessed of one degree of freedom, and the results, at which we have arrived, have a very wide application. But material systems enjoy in general more than one degree of freedom. In order to define their configuration at any moment several independent variable quantities must be specified, which, by a generalization of language originally employed for a point, are called the *co-ordinates* of the system, the number of independent co-ordinates being the *index of freedom*. Strictly speaking, the displacements possible to a natural system are infinitely various, and cannot be represented as made up of a finite number of displacements of specified type. To the elementary parts of a solid body any arbitrary displacements may be given, subject to conditions of continuity. It is only by a process of abstraction of the kind so constantly practised in Natural Philosophy, that solids are treated as rigid, fluids as incompressible, and other simplifications introduced so that the position of a system comes to depend on a finite number of co-ordinates. It is not, however, our intention to exclude the consideration of systems possessing infinitely various freedom; on the contrary, some of the most interesting applications of the results of this chapter will lie in that direction. But such systems are most conveniently conceived as limits of others, whose freedom is of a more restricted kind. We shall accordingly commence with a system, whose position is specified by a finite number of independent co-ordinates ψ_1, ψ_2, ψ_3, &c.

70. The main problem of Acoustics consists in the investigation of the vibrations of a system about a position of stable equilibrium, but it will be convenient to commence with the statical part of the subject. By the Principle of Virtual Velocities, if we reckon the co-ordinates ψ_1, ψ_2, &c. from the configuration of equilibrium, the potential energy of any other configuration will be a homogeneous quadratic function of the co-ordinates, provided that the displacement be sufficiently small. This quantity is called V, and represents the work that may be gained in passing from the actual to the equilibrium configuration. We may write

$$V = \tfrac{1}{2}c_{11}\psi_1{}^2 + \tfrac{1}{2}c_{22}\psi_2{}^2 + \ldots + c_{12}\psi_1\psi_2 + c_{23}\psi_2\psi_3 + \ldots\ldots(1).$$

Since by supposition the equilibrium is thoroughly stable, the quantities c_{11}, c_{22}, c_{12}, &c. must be such that V is positive for all real values of the co-ordinates.

71. If the system be displaced from the zero configuration by the action of given forces, the new configuration may be found from the Principle of Virtual Velocities. If the work done by the given forces on the hypothetical displacement $\delta\psi_1$, $\delta\psi_2$, &c. be

$$\Psi_1\delta\psi_1 + \Psi_2\delta\psi_2 + \ldots\ldots\ldots\ldots\ldots\ldots\ldots(1),$$

this expression must be equivalent to δV, so that since $\delta\psi_1$, $\delta\psi_2$, &c. are independent, the new position of equilibrium is determined by

$$\frac{dV}{d\psi_1} = \Psi_1, \quad \frac{dV}{d\psi_2} = \Psi_2, \text{ &c.} \ldots\ldots\ldots\ldots\ldots(2),$$

or by (1) of § 70,

$$\left. \begin{aligned} c_{11}\psi_1 + c_{12}\psi_2 + c_{13}\psi_3 + \ldots\ldots = \Psi_1 \\ c_{21}\psi_1 + c_{22}\psi_2 + c_{23}\psi_3 + \ldots\ldots = \Psi_2 \\ \ldots\ldots\ldots\ldots\ldots\ldots\ldots\ldots\ldots\ldots\ldots \end{aligned} \right\} \ldots\ldots\ldots\ldots (3),$$

where there is no distinction in value between c_{rs} and c_{sr}.

From these equations the co-ordinates may be determined in terms of the forces. If ∇ be the determinant

$$\nabla = \begin{vmatrix} c_{11}, & c_{12}, & c_{13}, & \ldots \\ c_{21}, & c_{22}, & c_{23}, & \ldots \\ c_{31}, & c_{32}, & c_{33}, & \ldots \\ \ldots\ldots\ldots\ldots\ldots \end{vmatrix} \ldots\ldots\ldots\ldots\ldots (4),$$

the solution of (3) may be written

$$\left. \begin{aligned} \nabla \cdot \psi_1 = \frac{d\nabla}{dc_{11}}\Psi_1 + \frac{d\nabla}{dc_{12}}\Psi_2 + \ldots.. \\ \nabla \cdot \psi_2 = \frac{d\nabla}{dc_{21}}\Psi_1 + \frac{d\nabla}{dc_{22}}\Psi_2 + \ldots.. \\ \ldots\ldots\ldots\ldots\ldots\ldots\ldots\ldots\ldots \end{aligned} \right\} \ldots\ldots\ldots\ldots(5).$$

These equations determine ψ_1, ψ_2, &c uniquely, since ∇ does not vanish, as appears from the consideration that the equations $dV/d\psi_1 = 0$, &c. could otherwise be satisfied by finite values of the co-ordinates, provided only that the *ratios* were suitable, which is contrary to the hypothesis that the system is thoroughly stable in the zero configuration.

72. If $\psi_1, \ldots \Psi_1, \ldots$ and $\psi_1', \ldots \Psi_1', \ldots$ be two sets of displacements and corresponding forces, we have the following reciprocal relation,

$$\Psi_1\psi_1' + \Psi_2\psi_2' + \ldots = \Psi_1'\psi_1 + \Psi_2'\psi_2 + \ldots\ldots\ldots\ldots(1),$$

as may be seen by substituting the values of the forces, when each side of (1) takes the form,

$$c_{11}\psi_1\psi_1' + c_{22}\psi_2\psi_2' + \ldots\ldots$$
$$\ldots + c_{12}(\psi_2\psi_1' + \psi_2'\psi_1) + c_{23}(\psi_3\psi_2' + \psi_3'\psi_2) + \ldots\ldots$$

Suppose in (1) that all the forces vanish except Ψ_2 and Ψ_1'; then

$$\Psi_2\psi_2' = \Psi_1'\psi_1 \ldots\ldots\ldots\ldots\ldots\ldots\ldots (2).$$

If the forces Ψ_2 and Ψ_1' be of the same kind, we may suppose them equal, and we then recognise that a force of any type acting alone produces a displacement of a second type equal to the displacement of the first type due to the action of an equal force of the second type. For example, if A and B be two points of a rod supported horizontally in any manner, the vertical deflection at A, when a weight W is attached at B, is the same as the deflection at B, when W is applied at A [1].

73. Since V is a homogeneous quadratic function of the co-ordinates,

$$2V = \frac{dV}{d\psi_1}\psi_1 + \frac{dV}{d\psi_2}\psi_2 + \ldots\ldots\ldots\ldots\ldots\ldots(1),$$

or, if Ψ_1, Ψ_2, &c. be the forces necessary to maintain the displacement represented by ψ_1, ψ_2, &c.,

$$2V = \Psi_1\psi_1 + \Psi_2\psi_2 + \ldots\ldots\ldots\ldots\ldots\ldots(2).$$

If $\psi_1 + \Delta\psi_1$, $\psi_2 + \Delta\psi_2$, &c. represent another displacement for which the necessary forces are $\Psi_1 + \Delta\Psi_1$, $\Psi_2 + \Delta\Psi_2$, &c., the

---

[1] On this subject, see *Phil. Mag.*, Dec., 1874, and March, 1875.

corresponding potential energy is given by

$$2 (V + \Delta V) = (\Psi_1 + \Delta \Psi_1)(\psi_1 + \Delta \psi_1) + \ldots$$
$$= 2V + \Psi_1 \Delta \psi_1 + \Psi_2 \Delta \psi_2 + \ldots$$
$$+ \Delta \Psi_1 . \psi_1 + \Delta \Psi_2 . \psi_2 + \ldots$$
$$+ \Delta \Psi_1 . \Delta \psi_1 + \Delta \Psi_2 . \Delta \psi_2 + \ldots,$$

so that we may write

$$2 \Delta V = \Sigma \Psi . \Delta \psi + \Sigma \Delta \Psi . \psi + \Sigma \Delta \Psi . \Delta \psi \ldots\ldots\ldots(3),$$

where ΔV is the difference of the potential energies in the two cases, and we must particularly notice that by the reciprocal relation, § 72 (1),

$$\Sigma \Psi . \Delta \psi = \Sigma \Delta \Psi . \psi \ldots\ldots \ldots\ldots\ldots\ldots(4).$$

From (3) and (4) we may deduce two important theorems, relating to the value of V for a system subjected to given displacements, and to given forces respectively.

74. The first theorem is to the effect that, if given displacements (not sufficient by themselves to determine the configuration) be produced in a system by forces of corresponding types, the resulting value of V for the system so displaced, and in equilibrium, is as small as it can be under the given displacement conditions; and that the value of V for any other configuration exceeds this by the potential energy of the configuration which is the difference of the two. The only difficulty in the above statement consists in understanding what is meant by 'forces of corresponding types.' Suppose, for example, that the system is a stretched string, of which a given point P is to be subject to an obligatory displacement; the force of corresponding type is here a force applied at the point P itself. And generally, the forces, by which the proposed displacement is to be made, must be such as would do no work on the system, provided only that that displacement were *not* made.

By a suitable choice of co-ordinates, the given displacement conditions may be expressed by ascribing given values to the first r co-ordinates $\psi_1, \psi_2, \ldots \psi_r$, and the conditions as to the forces will then be represented by making the forces of the remaining types Ψ_{r+1}, Ψ_{r+2}, &c. vanish. If $\psi + \Delta \psi$ refer to any other configuration of the system, and $\Psi + \Delta \Psi$ be the corresponding forces, we are to suppose that $\Delta \psi_1, \Delta \psi_2$, &c. as far as $\Delta \psi_r$ all vanish. Thus for the first r suffixes $\Delta \psi$ vanishes, and for the remaining suffixes Ψ vanishes.   Accordingly $\Sigma \Psi . \Delta \psi$ is zero, and therefore $\Sigma \Delta \Psi . \psi$ is also zero.   Hence ·

$$2 \Delta V = \Sigma \Delta \Psi . \Delta \psi \dots\dots\dots\dots\dots\dots\dots(I),$$

which proves that if the given displacements be made in any other than the prescribed way, the potential energy is increased by the energy of the difference of the configurations.

By means of this theorem we may trace the effect on V of any relaxation in the stiffness of a system, subject to given displacement conditions.   For, if after the alteration in stiffness the original equilibrium configuration be considered, the value of V corresponding thereto is by supposition less than before; and, as we have just seen, there will be a still further diminution in the value of V when the system passes to equilibrium under the altered conditions.   Hence we conclude that a diminution in V as a function of the co-ordinates entails also a diminution in the actual value of V when a system is subjected to given displacements.   It will be understood that in particular cases the diminution spoken of may vanish[1].

For example, if a point P of a bar clamped at both ends be displaced laterally to a given small amount by a force there applied, the potential energy of the deformation will be diminished by any relaxation (however local) in the stiffness of the bar.

75.   The second theorem relates to a system displaced *by given forces*, and asserts that in this case the value of V in equilibrium is greater than it would be in any other configuration in which the system could be maintained at rest under the given forces, by the operation of mere constraints.   We will shew that the *removal* of constraints increases the value of V.

The co-ordinates may be so chosen that the conditions of constraint are expressed by

$$\psi_1 = 0, \quad \psi_2 = 0, \dots\dots \psi_r = 0 \dots\dots\dots\dots\dots(1).$$

We have then to prove that when Ψ_{r+1}, Ψ_{r+2}, &c. are given, the value of V is least when the conditions (1) hold.   The second configuration being denoted as before by $\psi_1 + \Delta \psi_1$ &c., we see that for suffixes up to r inclusive ψ vanishes, and for higher suffixes $\Delta \Psi$ vanishes.   Hence

$$\Sigma \psi . \Delta \Psi = \Sigma \Delta \psi . \Psi = 0,$$

---

[1] See a paper on General Theorems relating to Equilibrium and Initial and Steady Motions.   *Phil. Mag.*, March. 1875.

and therefore

$$2 \Delta V = \Sigma \Delta \Psi . \Delta \psi \dots\dots\dots\dots\dots(2),$$

shewing that the increase in V due to the removal of the constraints is equal to the potential energy of the difference of the two configurations.

76. We now pass to the investigation of the initial motion of a system which starts from rest under the operation of given impulses. The motion thus acquired is independent of any potential energy which the system may possess when actually displaced, since by the nature of impulses we have to do only with the initial configuration itself. The initial motion is also independent of any forces of a finite kind, whether impressed on the system from without, or of the nature of viscosity.

If P, Q, R be the component impulses, parallel to the axes, on a particle m whose rectangular co-ordinates are x, y, z, we have by D'Alembert's Principle

$$\Sigma m (\dot{x}\delta x + \dot{y}\delta y + \dot{z}\delta z) = \Sigma (P\delta x + Q\delta y + R\delta z)\dots\dots(1),$$

where $\dot{x}$, $\dot{y}$, $\dot{z}$ denote the velocities acquired by the particle in virtue of the impulses, and δx, δy, δz correspond to any arbitrary displacement of the system which does not violate the connection of its parts. It is required to transform (1) into an equation expressed by the independent generalized co-ordinates.

For the first side,

$$\Sigma m (\dot{x}\delta x + \dot{y}\delta y + \dot{z}\delta z) = \delta\psi_1 \Sigma m \left(\dot{x}\frac{dx}{d\psi_1} + \dot{y}\frac{dy}{d\psi_1} + \dot{z}\frac{dz}{d\psi_1} \right)$$

$$+ d\psi_2 \Sigma m \left(\dot{x}\frac{dx}{d\psi_2} + \dot{y}\frac{dy}{d\psi_2} + \dot{z}\frac{dz}{d\psi_2} \right) + \dots\dots$$

$$= \delta\psi_1 \Sigma m \left(\dot{x}\frac{d\dot{x}}{d\dot{\psi}_1} + \dot{y}\frac{d\dot{y}}{d\dot{\psi}_1} + \dot{z}\frac{d\dot{z}}{d\dot{\psi}_1} \right) + \dots\dots$$

$$= \delta\psi_1 . \tfrac{1}{2}\Sigma m \frac{d}{d\dot{\psi}_1} (\dot{x}^2 + \dot{y}^2 + \dot{z}^2) + \dots\dots$$

$$= \delta\psi_1 \frac{dT}{d\dot{\psi}_1} + \delta\psi_2 \frac{dT}{d\dot{\psi}_2} + \dots\dots\dots\dots\dots\dots(2),$$

where T, the kinetic energy of the system is supposed to be expressed as a function of $\dot{\psi}_1$, $\dot{\psi}_2$, &c.

On the second side,

$$\Sigma (P\delta x + Q\delta y + R\delta z) = \delta\psi_1 \Sigma m \left(P\frac{dx}{d\dot\psi_1} + Q\frac{dy}{d\dot\psi_1} + R\frac{dz}{d\dot\psi_1} \right) + \ldots\ldots$$

$$= \xi_1 \delta\psi_1 + \xi_2 \delta\psi_2 + \ldots\ldots\ldots\ldots\ldots\ldots\ldots (3),$$

if
$$\Sigma m \left(P\frac{dx}{d\dot\psi_1} + Q\frac{dy}{d\dot\psi_1} + R\frac{dz}{d\dot\psi_1} \right) = \xi_1, \&c.$$

The transformed equation is therefore

$$\left(\frac{dT}{d\dot\psi_1} - \xi_1 \right) \delta\psi_1 + \left(\frac{dT}{d\dot\psi_2} - \xi_2 \right) \delta\psi_2 + \ldots = 0 \ldots\ldots\ldots\ldots (4),$$

where $\delta\psi_1$, $\delta\psi_2$, &c. are now completely independent. Hence to determine the motion we have

$$\frac{dT}{d\dot\psi_1} = \xi_1, \qquad \frac{dT}{d\dot\psi_2} = \xi_2, \&c. \ldots\ldots\ldots\ldots\ldots\ldots (5),$$

where ξ_1, ξ_2, &c. may be considered as the generalized components of impulse.

77. Since T is a homogeneous quadratic function of the generalized co-ordinates, we may take

$$T = \tfrac{1}{2}a_{11}\dot\psi_1{}^2 + \tfrac{1}{2}a_{22}\dot\psi_2{}^2 + \ldots\ldots + a_{12}\dot\psi_1\dot\psi_2 + a_{23}\dot\psi_2\dot\psi_3 + \ldots\ldots (1),$$

whence

$$\left. \begin{aligned} \xi_1 &= \frac{dT}{d\dot\psi_1} = a_{11}\dot\psi_1 + a_{12}\dot\psi_2 + a_{13}\dot\psi_3 + \ldots\ldots \\ \xi_2 &= \frac{dT}{d\dot\psi_2} = a_{21}\dot\psi_1 + a_{22}\dot\psi_2 + a_{23}\dot\psi_3 + \ldots\ldots \\ &\ldots\ldots\ldots\ldots\ldots\ldots\ldots\ldots\ldots\ldots\ldots\ldots\ldots\ldots\ldots \end{aligned} \right\} \ldots\ldots\ldots\ldots (2),$$

where there is no distinction in value between a_{rs} and a_{sr}.

Again, by the nature of T,

$$2T = \dot\psi_1 \frac{dT}{d\dot\psi_1} + \dot\psi_2 \frac{dT}{d\dot\psi_2} + \ldots\ldots = \xi_1\dot\psi_1 + \xi_2\dot\psi_2 + \ldots\ldots (3).$$

The theory of initial motion is closely analogous to that of the displacement of a system from a configuration of stable equilibrium by steadily applied forces. In the present theory the initial kinetic energy T bears to the velocities and impulses the same relations as in the former V bears to the displacements and forces respectively. In one respect the theory of initial motions is the more complete, inasmuch as T is exactly, while V is in general only approximately, a homogeneous quadratic function of the variables.

If $\dot{\psi}_1, \dot{\psi}_2, ..., \xi_1, \xi_2, ...$ denote one set of velocities and impulses for a system started from rest, and $\dot{\psi}_1', \dot{\psi}_2', ..., \xi_1', \xi_2', ...$ a second set, we may prove, as in § 72, the following reciprocal relation:

$$\xi_1'\dot{\psi}_1 + \xi_2'\dot{\psi}_2 + ... = \xi_1\dot{\psi}_1' + \xi_2\dot{\psi}_2' + ...(4)^1.$$

This theorem admits of interesting application to fluid motion. It is known, and will be proved later in the course of this work, that the motion of a frictionless incompressible liquid, which starts from rest, is of such a kind that its component velocities at any point are the corresponding differential coefficients of a certain function, called the velocity-potential. Let the fluid be set in motion by a prescribed arbitrary deformation of the surface S of a closed space described within it. The resulting motion is deter- mined by the normal velocities of the elements of S, which, being shared by the fluid in contact with them, are denoted by du/dn, if u be the velocity-potential, which interpreted physically denotes the impulsive pressure, if the density be taken as unity. Hence by the theorem, if v be the velocity-potential of a second motion, corres- ponding to another set of arbitrary surface velocities dv/dn,

$$\iint u \frac{dv}{dn} dS = \iint v \frac{du}{dn} dS \quad ....(5),$$

—an equation immediately following from Green's theorem, if besides S there be only fixed solids immersed in the fluid. The present method enables us to attribute to it a much higher gene- rality. For example, the immersed solids, instead of being fixed, may be free, altogether or in part, to take the motion imposed upon them by the fluid pressures.

78. A particular case of the general theorem is worthy of special notice. In the first motion let

$$\dot{\psi}_1 = A, \quad \dot{\psi}_2 = 0, \quad \xi_3 = \xi_4 = \xi_5 \ ...... = 0 ;$$

and in the second,

$$\dot{\psi}_1' = 0, \quad \dot{\psi}_2' = A, \quad \xi_3' = \xi_4' = \xi_5' \ ...... = 0.$$

Then
$$\xi_1 = \xi_2 ....(1)$$

In words, if, by means of a suitable impulse of the correspond- ing type, a given arbitrary velocity of one co-ordinate be impressed on a system, the impulse corresponding to a second co-ordinate necessary in order to prevent it from changing, is the same as would be required for the first co-ordinate, if the given velocity were impressed on the second.

<hr>

[1] Thomson and Tait, § 313 (f).

As a simple example, take the case of two spheres A and B immersed in a liquid, whose centres are free to move along certain lines. If A be set in motion with a given velocity, B will naturally begin to move also. The theorem asserts that the impulse required to prevent the motion of B, is the same as if the functions of A and B were exchanged: and this even though there be other rigid bodies, C, D, &c., in the fluid, either fixed, or free in whole or in part.

The case of electric currents mutually influencing each other by induction is precisely similar. Let there be two circuits A and B, in the neighbourhood of which there may be any number of other wire circuits or solid conductors. If a unit current be suddenly developed in the circuit A, the electromotive impulse induced in B is the same as there would have been in A, had the current been forcibly developed in B.

79. The motion of a system, on which given arbitrary velocities are impressed by means of the necessary impulses of the corresponding types, possesses a remarkable property discovered by Thomson. The conditions are that $\dot{\psi}_1$, $\dot{\psi}_2$, $\dot{\psi}_3$, ... $\dot{\psi}_r$ are given, while ξ_{r+1}, ξ_{r+2}, ... vanish. Let $\dot{\psi}_1$, $\dot{\psi}_2$, ... ξ_1, ξ_2, &c. correspond to the actual motion; and

$$\dot{\psi}_1 + \Delta\dot{\psi}_1, \dot{\psi}_2 + \Delta\dot{\psi}_2, \ldots \xi_1 + \Delta\xi_1, \xi_2 + \Delta\xi_2, \ldots$$

to another motion satisfying the same *velocity* conditions. For each suffix either $\Delta\dot{\psi}$ *or* ξ vanishes. Now for the kinetic energy of the supposed motion,

$$2\,(T + \Delta T) = (\xi_1 + \Delta\xi_1)\,(\dot{\psi}_1 + \Delta\dot{\psi}_1) + \ldots$$
$$= 2T + \xi_1\Delta\dot{\psi}_1 + \xi_2\Delta\dot{\psi}_2 + \ldots$$
$$+ \Delta\xi_1 . \dot{\psi}_1 + \Delta\xi_2 . \dot{\psi}_2 + \ldots + \Delta\xi_1\Delta\dot{\psi}_1 + \Delta\xi_2\Delta\dot{\psi}_2 + \ldots$$

But by the reciprocal relation (4) of § 77

$$\xi_1\Delta\dot{\psi}_1 + \ldots = \Delta\xi_1 . \dot{\psi}_1 + \ldots,$$

of which the former by hypothesis is zero; so that

$$2\Delta T = \Delta\xi_1\Delta\dot{\psi}_1 + \Delta\xi_2\Delta\dot{\psi}_2 + \ldots\ldots\ldots\ldots\ldots(1),$$

shewing that the energy of the supposed motion exceeds that of the actual motion by the energy of that motion which would have to be compounded with the latter to produce the former. The motion actually induced in the system has thus less energy than any other satisfying the same velocity conditions. In a subsequent chapter we shall make use of this property to find a superior limit to the energy of a system set in motion with prescribed velocities.

If any diminution be made in the inertia of any of the parts of a system, the motion corresponding to prescribed velocity conditions will in general undergo a change. The value of T will necessarily be less than before ; for there would be a decrease even if the motion remained unchanged, and therefore *a fortiori* when the motion is such as to make T an absolute minimum. Conversely any increase in the inertia increases the initial value of T.

This theorem is analogous to that of § 74. The analogue for initial motions of the theorem of § 75, relating to the potential energy of a system displaced by given forces, is that of Bertrand, and may be thus stated:—If a system start from rest under the operation of given impulses, the kinetic energy of the actual motion exceeds that of any other motion which the system might have been guided to take with the assistance of mere constraints, by the kinetic energy of the difference of the motions[1].

[The theorems of Kelvin and Bertrand represent different aspects of the same truth. Let us suppose that the prescribed impulse is entirely of the first type ξ_1. Then $T = \frac{1}{2}\xi_1\dot{\psi}_1$, whether the motion be free or be subjected to any constraint. Further, under any given circumstances as to constraint, $\dot{\psi}_1$ is proportional to ξ_1, and the ratio $\xi_1 : \dot{\psi}_1$ may be regarded as the moment of inertia ; so that

$$T = \tfrac{1}{2}\xi_1\dot{\psi}_1 = \tfrac{1}{2}m\dot{\psi}_1{}^2 = \tfrac{1}{2}\xi_1{}^2/m.$$

Kelvin's theorem asserts that the introduction of a constraint can v increase the value of T when $\dot{\psi}_1$ is given. Hence whether $\dot{\psi}_1$ be given or not, the constraint can only increase the ratio of $2T$ to $\dot{\psi}_1{}^2$ or of ξ_1 to $\dot{\psi}_1$. Both theorems are included in the statement that the moment of inertia is increased by the introduction of a constraint.]

80. We will not dwell at any greater length on the mechanics of a system subject to impulses, but pass on to investigate Lagrange's equations for continuous motion. We shall suppose that the connections binding together the parts of the system are not explicit functions of the time ; such cases of forced motion as we shall have to consider will be specially shewn to be within the scope of the investigation.

By D'Alembert's Principle in combination with that of Virtual Velocities,

$$\Sigma m\,(\ddot{x}\delta x + \ddot{y}\delta y + \ddot{z}\delta z) = \Sigma\,(X\delta x + Y\delta y + Z\delta z)\ldots\ldots(1),$$

[1] Thomson and Tait, § 311. *Phil. Mag.* March, 1875.

where δx, δy, δz denote a displacement of the system of the most general kind possible without violating the connections of its parts. Since the displacements of the individual particles of the system are mutually related, $\delta x, \ldots$ are not independent. The object now is to transform to other variables ψ_1, $\psi_2, \ldots$, which shall be independent. We have

$$\ddot{x}\,\delta x = \frac{d}{dt}\,(\dot{x}\,\delta x) - \tfrac{1}{2}\delta \dot{x}^2,$$

so that

$$\Sigma m\,(\ddot{x}\,\delta x + \ddot{y}\,\delta y + \ddot{z}\,\delta z) = \frac{d}{dt}.\,\Sigma m\,(\dot{x}\,\delta x + \dot{y}\,\delta y + \dot{z}\,\delta z) - \delta T.$$

But (§ 76) we have already found that

$$\Sigma m\,(\dot{x}\,\delta x + \dot{y}\,\delta y + \dot{z}\,\delta z) = \frac{dT}{d\dot{\psi}_1}\,\delta\psi_1 + \frac{dT}{d\dot{\psi}_2}\,\delta\psi_2 + \ldots,$$

while $$\delta T = \frac{dT}{d\psi_1}\,\delta\psi_1 + \frac{dT}{d\dot{\psi}_1}\,\delta\dot{\psi}_1 + \ldots,$$

if T be expressed as a quadratic function of $\dot{\psi}_1$, $\dot{\psi}_2, \ldots$, whose coefficients are in general functions of ψ_1, $\psi_2, \ldots$. Also

$$\frac{d}{dt}\left(\frac{dT}{d\dot{\psi}_1}\,\delta\psi_1\right) = \frac{d}{dt}\left(\frac{dT}{d\dot{\psi}_1}\right).\,\delta\psi_1 + \frac{dT}{d\dot{\psi}_1}\,\delta\dot{\psi}_1,$$

inasmuch as $$\frac{d}{dt}\,\delta\psi_1 = \delta\frac{d}{dt}\,\psi_1.$$

Accordingly

$$\Sigma m\,(\ddot{x}\,\delta x + \ddot{y}\,\delta y + \ddot{z}\,\delta z) = \left\{\frac{d}{dt}\left(\frac{dT}{d\dot{\psi}_1}\right) - \frac{dT}{d\psi_1}\right\}\,\delta\psi_1$$

$$+ \left\{\frac{d}{dt}\left(\frac{dT}{d\dot{\psi}_2}\right) - \frac{dT}{d\psi_2}\right\}\,\delta\psi_2 + \ldots\ldots\ldots(2).$$

Thus, if the transformation of the second side of (1) be

$$\Sigma\,(X\,\delta x + Y\,\delta y + Z\,\delta z) = \Psi_1\delta\psi_1 + \Psi_2\delta\psi_2 + \ldots\ldots\ldots(3),$$

we have equations of motion of the form

$$\frac{d}{dt}\left(\frac{dT}{d\dot{\psi}}\right) - \frac{dT}{d\psi} = \Psi\ldots\ldots\ldots\ldots\ldots(4).$$

Since $\Psi\delta\psi$ denotes the work done on the system during a displacement $\delta\psi$, Ψ may be regarded as the generalized component of force.

In the case of a conservative system it is convenient to separate from Ψ those parts which depend only on the configuration of the system. Thus, if V denote the potential energy, we may write

$$\frac{d}{dt}\left(\frac{dT}{d\dot{\psi}}\right) - \frac{dT}{d\psi} + \frac{dV}{d\psi} = \Psi \ldots\ldots\ldots\ldots\ldots (5),$$

where Ψ is now limited to the forces acting on the system which are not already taken account of in the term $dV/d\psi$.

81. There is also another group of forces whose existence it is often advantageous to recognize specially, namely those arising from friction or viscosity. If we suppose that each particle of the system is retarded by forces proportional to its component velocities, the effect will be shewn in the fundamental equation (1) § 80 by the addition to the left-hand member of the terms

$$\Sigma\,(\kappa_x\dot{x}\,\delta x + \kappa_y\dot{y}\,\delta y + \kappa_z\dot{z}\,\delta z),$$

where κ_x, κ_y, κ_z are coefficients independent of the velocities, but possibly dependent on the configuration of the system. The transformation to the independent co-ordinates ψ_1, ψ_2, &c. is effected in a similar manner to that of

$$\Sigma m\,(\dot{x}\,\delta x + \dot{y}\,\delta y + \dot{z}\,\delta z)$$

considered above (§ 80), and gives

$$\frac{dF}{d\dot{\psi}_1}\,\delta\psi_1 + \frac{dF}{d\dot{\psi}_2}\,\delta\psi_2 + \ldots\ldots\ldots\ldots\ldots (1),$$

where
$$F = \tfrac{1}{2}\Sigma\,(\kappa_x\dot{x}^2 + \kappa_y\dot{y}^2 + \kappa_z\dot{z}^2)$$

$$= \tfrac{1}{2}b_{11}\dot{\psi}_1{}^2 + \tfrac{1}{2}b_{22}\dot{\psi}_2{}^2 + \ldots + b_{12}\dot{\psi}_1\dot{\psi}_2 + b_{23}\dot{\psi}_2\dot{\psi}_3 + \ldots\ldots (2).$$

F, it will be observed, is like T a homogeneous quadratic function of the velocities, positive for all real values of the variables. It represents half the rate at which energy is dissipated.

The above investigation refers to retarding forces proportional to the absolute velocities; but it is equally important to consider such as depend on the *relative* velocities of the parts of the system, and fortunately this can be done without any increase of complication. For example, if a force act on the particle x_1 proportional to $(\dot{x}_1 - \dot{x}_2)$, there will be at the same moment an equal and opposite force acting on the particle x_2. The additional terms in the fundamental equation will be of the form

$$\kappa_x\,(\dot{x}_1 - \dot{x}_2)\,\delta\dot{x}_1 + \kappa_x\,(\dot{x}_2 - \dot{x}_1)\,\delta x_2,$$

which may be written

$$\kappa_x\,(\dot{x}_1 - \dot{x}_2)\,\delta(x_1 - x_2) = \delta\psi_1\,\frac{d}{d\dot{\psi}_1}\{\tfrac{1}{2}\kappa_x\,(\dot{x}_1 - \dot{x}_2)^2\} + \ldots,$$

and so on for any number of pairs of mutually influencing particles. The only effect is the addition of new terms to F, which still appears in the form $(2)^1$. We shall see presently that the existence of the function F, which may be called the Dissipation Function, implies certain relations among the coefficients of the generalized equations of vibration, which carry with them important consequences[2].

The equations of motion may now be written in the form

$$\frac{d}{dt}\left(\frac{dT}{d\dot{\psi}}\right) - \frac{dT}{d\psi} + \frac{dF}{d\dot{\psi}} + \frac{dV}{d\psi} = \Psi \ldots\ldots\ldots\ldots (3).$$

82. We may now introduce the condition that the motion takes place in the immediate neighbourhood of a configuration of thoroughly stable equilibrium; T and F are then homogeneous quadratic functions of the velocities with coefficients which are to be treated as constant, and V is a similar function of the co-ordinates themselves, provided that (as we suppose to be the case) the origin of each co-ordinate is taken to correspond with the configuration of equilibrium. Moreover all three functions are essentially positive. Since terms of the form $dT/d\psi$ are of the second order of small quantities, the equations of motion become linear, assuming the form

$$\frac{d}{dt}\left(\frac{dT}{d\dot{\psi}}\right) + \frac{dF}{d\dot{\psi}} + \frac{dV}{d\psi} = \Psi \ldots\ldots\ldots\ldots (1),$$

where under Ψ are to be included all forces acting on the system not already provided for by the differential coefficients of F and V.

The three quadratic functions will be expressed as follows :—

$$\left.\begin{array}{l} T = \tfrac{1}{2}a_{11}\dot{\psi}_1{}^2 + \tfrac{1}{2}a_{22}\dot{\psi}_2{}^2 + \ldots + a_{12}\dot{\psi}_1\dot{\psi}_2 + \ldots \\ F = \tfrac{1}{2}b_{11}\dot{\psi}_1{}^2 + \tfrac{1}{2}b_{22}\dot{\psi}_2{}^2 + \ldots + b_{12}\dot{\psi}_1\dot{\psi}_2 + \ldots \\ V = \tfrac{1}{2}c_{11}\psi_1{}^2 + \tfrac{1}{2}c_{22}\psi_2{}^2 + \ldots + c_{12}\psi_1\psi_2 + \ldots \end{array}\right\} \ldots\ldots\ldots(2),$$

where the coefficients a, b, c are constants.

From equation (1) we may of course fall back on previous results by supposing F and V, or F and T, to vanish.

A third set of theorems of interest in the application to Elec-

---

[1] The differences referred to in the text may of course pass into differential coefficients in the case of a body continuously deformed.

[2] The Dissipation Function appears for the first time, so far as I am aware, in a paper on General Theorems relating to Vibrations, published in the *Proceedings of the Mathematical Society* for June, 1873.

tricity may be obtained by omitting T and V, while F is retained, but it is unnecessary to pursue the subject here.

If we substitute the values of T, F and V, and write D for d/dt, we obtain a system of equations which may be put into the form

$$\left.\begin{aligned}
e_{11}\psi_1 + e_{12}\psi_2 + e_{13}\psi_3 + \ldots &= \Psi_1 \\
e_{21}\psi_1 + e_{22}\psi_2 + e_{23}\psi_3 + \ldots &= \Psi_2 \\
e_{31}\psi_1 + e_{32}\psi_2 + e_{33}\psi_3 + \ldots &= \Psi_3 \\
\ldots\ldots\ldots\ldots\ldots\ldots\ldots\ldots\ldots\ldots\ldots
\end{aligned}\right\}\ldots\ldots\ldots\ldots(3).$$

where e_{rs} denotes the quadratic operator

$$e_{rs} = a_{rs}D^2 + b_{rs}D + c_{rs} \ldots\ldots\ldots\ldots\ldots\ldots (4).$$

It must be particularly remarked that since

$$a_{rs} = a_{sr}, \quad b_{rs} = b_{sr}, \quad c_{rs} = c_{sr},$$

it follows that $\qquad\qquad e_{rs} = e_{sr} \ldots\ldots\ldots\ldots\ldots\ldots\ldots\ldots\ldots(5).$

[The theory of *motional* forces, i.e. forces proportional to the velocities, has been further developed in the second edition of Thomson and Tait's *Natural Philosophy* (1879). In the most general case the equations may be written

$$\left.\begin{aligned}
\frac{d}{dt}\left(\frac{dT}{d\dot{\psi}_1}\right) + \frac{dV}{d\psi_1} + b_{11}\dot{\psi}_1 + (b_{12}+\beta_{12})\,\dot{\psi}_2 + (b_{13}+\beta_{13})\,\dot{\psi}_3 + \ldots &= \Psi_1 \\
\frac{d}{dt}\left(\frac{dT}{d\dot{\psi}_2}\right) + \frac{dV}{d\psi_2} + (b_{21}-\beta_{21})\dot{\psi}_1 + b_{22}\dot{\psi}_2 + (b_{23}-\beta_{23})\,\dot{\psi}_3 + \ldots &= \Psi_2
\end{aligned}\right\}(6),$$

where $\qquad\qquad b_{rs} = b_{sr}, \quad \beta_{rs} = \beta_{sr} \ldots\ldots\ldots\ldots\ldots\ldots(7).$

Of these the terms with the coefficients b can be derived from the dissipation function

$$F = \tfrac{1}{2}b_{11}\dot{\psi}_1{}^2 + \tfrac{1}{2}b_{22}\dot{\psi}_2{}^2 + \ldots + b_{12}\dot{\psi}_1\dot{\psi}_2 + \ldots.$$

The terms in β on the other hand do not represent dissipation, and are called the gyrostatic terms.

If we multiply the first of equations (6) by $\dot{\psi}_1$, the second by $\dot{\psi}_2$, &c., and then add, we obtain

$$\frac{d(T+V)}{dt} + 2F = \Psi_1\dot{\psi}_1 + \Psi_2\dot{\psi}_2 + \ldots\ldots\ldots\ldots(8).$$

In this the first term represents the rate at which energy is being stored in the system; $2F$ is the rate of dissipation; and the two together account for the work done upon the system by the external forces.]

83. Before proceeding further, we may draw an important inference from the *linearity* of our equations. If corresponding respectively to the two sets of forces Ψ_1, Ψ_2,..., Ψ_1', Ψ_2',... two motions denoted by ψ_1, ψ_2, ..., ψ_1', ψ_2', ... be possible, then must also be possible the motion $\psi_1 + \psi_1'$, $\psi_2 + \psi_2'$,... in conjunction with the forces $\Psi_1 + \Psi_1'$, $\Psi_2 + \Psi_2'$, ... Or, as a particular case, when there are no impressed forces, the superposition of any two natural vibrations constitutes also a natural vibration. This is the celebrated principle of the Coexistence of Small Motions, first clearly enunciated by Daniel Bernoulli. It will be understood that its truth depends in general on the justice of the assumption that the motion is so small that its square may be neglected.

[Again, if a system be under the influence of constant forces Ψ_1, &c., which displace it into a new position of equilibrium, the vibrations which may occur about the new position are the same as those which might before have occurred about the old position.]

84. To investigate the free vibrations, we must put Ψ_1, Ψ_2,... equal to zero ; and we will commence with a system on which no frictional forces act, for which therefore the coefficients e_{rs}, &c. are *even* functions of the symbol D. We have

$$\left. \begin{array}{l} e_{11}\psi_1 + e_{12}\psi_2 + \ldots = 0 \\ e_{21}\psi_1 + e_{22}\psi_2 + \ldots = 0 \\ \ldots\ldots\ldots\ldots\ldots\ldots \end{array} \right\} \quad \ldots\ldots\ldots\ldots\ldots\ldots(1).$$

From these equations, of which there are as many (m) as the system possesses degrees of liberty, let all but one of the variables be eliminated. The result, which is of the same form whichever be the co-ordinate retained, may be written

$$\nabla\psi = 0 \quad \ldots\ldots\ldots\ldots\ldots\ldots\ldots\ldots(2),$$

where ∇ denotes the determinant

$$\begin{vmatrix} e_{11}, & e_{12}, & e_{13}, & \cdots \\ e_{21}, & e_{22}, & e_{23}, & \cdots \\ e_{31}, & e_{32}, & e_{33}, & \cdots \\ & \cdots\cdots\cdots \end{vmatrix} \quad \ldots\ldots\ldots\ldots\ldots\ldots(3),$$

and is (if there be no friction) an even function of D of degree $2m$. Let $\pm\lambda_1$, $\pm\lambda_2$, ..., $\pm\lambda_m$ be the roots of $\nabla = 0$ considered as an equation in D. Then by the theory of differential equations the most general value of ψ is

$$\psi = A\epsilon^{\lambda_1 t} + A'\epsilon^{-\lambda_1 t} + B\epsilon^{\lambda_2 t} + B'\epsilon^{-\lambda_2 t} + \ldots\ldots\ldots(4),$$

where the $2m$ quantities A, A', B, B', &c. are arbitrary constants. This form holds good for each of the co-ordinates, but the constants in the different expressions are not independent. In fact if a particular solution be

$$\psi_1 = A_1 \epsilon^{\lambda_1 t}, \ \ \psi_2 = A_2 \epsilon^{\lambda_1 t}, \ \text{&c.},$$

the *ratios* $A_1 : A_2 : A_3 \dots$ are completely determined by the equations

$$\left. \begin{aligned} e_{11}A_1 + e_{12}A_2 + e_{13}A_3 + \dots\dots &= 0 \\ e_{21}A_1 + e_{22}A_2 + e_{23}A_3 + \dots\dots &= 0 \\ \dots\dots\dots\dots\dots\dots\dots\dots\dots\dots\dots\dots\dots\dots \end{aligned} \right\} \dots\dots\dots(5),$$

where in each of the coefficients such as e_{rs}, λ_1 is substituted for D. Equations (5) are necessarily compatible, by the condition that λ_1 is a root of $\nabla = 0$. The ratios $A_1' : A_2' : A_3' \dots$ corresponding to the root $-\lambda_1$ are the same as the ratios $A_1 : A_2 : A_3 \dots$, but for the other pairs of roots λ_2, $-\lambda_2$, &c. there are distinct systems of ratios.

85. The nature of the system with which we are dealing imposes an important restriction on the possible values of λ. If λ_1 were real, either λ_1 or $-\lambda_1$ would be real and positive, and we should obtain a particular solution for which the co-ordinates, and with them the kinetic energy denoted by

$$\lambda_1^2 \left\{ \tfrac{1}{2}a_{11}A_1^2 + \dots a_{12}A_1A_2 + \dots \right\} \epsilon^{\pm 2\lambda_1 t},$$

increase without limit. Such a motion is obviously impossible for a conservative system, whose whole energy can never differ from the sum of the potential and kinetic energies with which it was animated at starting. This conclusion is not evaded by taking λ_1 negative; because we are as much at liberty to trace the motion backwards as forwards. It is as certain that the motion never *was* infinite, as that it never *will be*. The same argument excludes the possibility of a complex value of λ.

We infer that all the values of λ are purely imaginary, corresponding to *real negative* values of λ^2. Analytically, the fact that the roots of $\nabla = 0$, considered as an equation in D^2, are all real and negative, must be a consequence of the relations subsisting between the coefficients $a_{11}, a_{12}, \dots, c_{11}, c_{12}, \dots$ in virtue of the fact that for all real values of the variables T and V are positive. The case of two degrees of liberty will be afterwards worked out in full.

86. The form of the solution may now be advantageously changed by writing in_1 for λ_1, &c. (where $i = \sqrt{-1}$), and taking new arbitrary constants. Thus

$$
\left.
\begin{aligned}
\psi_1 &= A_1 \cos{(n_1 t - \alpha)} + B_1 \cos{(n_2 t - \beta)} + C_1 \cos{(n_3 t - \gamma)} + \dots \\
\psi_2 &= A_2 \cos{(n_1 t - \alpha)} + B_2 \cos{(n_2 t - \beta)} + C_2 \cos{(n_3 t - \gamma)} + \dots \\
\psi_3 &= A_3 \cos{(n_1 t - \alpha)} + B_3 \cos{(n_2 t - \beta)} + C_3 \cos{(n_3 t - \gamma)} + \dots
\end{aligned}
\right\} \dots (1),
$$

where n_1^2, n_2^2, &c. are the m roots of the equation of mth degree in n^2 found by writing $-n^2$ for D^2 in $\nabla = 0$. For each value of n the ratios $A_1 : A_2 : A_3 \dots$ are determinate and real.

This is the complete solution of the problem of the free vibrations of a conservative system. We see that the whole motion may be resolved into m normal harmonic vibrations of (in general) different periods, each of which is entirely independent of the others. If the motion, depending on the original disturbance, be such as to reduce itself to one of these (n_1) we have

$$
\psi_1 = A_1 \cos{(n_1 t - \alpha)}, \quad \psi_2 = A_2 \cos{(n_1 t - \alpha)}, \text{ &c.} \dots \dots (2),
$$

where the ratios $A_1 : A_2 : A_3 \dots$ depend on the constitution of the system, and only the absolute amplitude and phase are arbitrary. The several co-ordinates are always in similar (or opposite) phases of vibration, and the whole system is to be found in the configuration of equilibrium at the same moment.

We perceive here the mechanical foundation of the supremacy of harmonic vibrations. If the motion be sufficiently small, the differential equations become linear with constant coefficients; while circular (and exponential) functions are the only ones which retain their type on differentiation.

87. The m periods of vibration, determined by the equation $\nabla = 0$, are quantities intrinsic to the system, and must come out the same whatever co-ordinates may be chosen to define the configuration. But there is one system of co-ordinates, which is especially suitable, that namely in which the normal types of vibration are defined by the vanishing of all the co-ordinates but one. In the first type the original co-ordinates ψ_1, ψ_2, &c. have given ratios; let the quantity fixing the absolute values be ϕ_1, so that in this type each co-ordinate is a known multiple of ϕ_1. So in the second type each co-ordinate may be regarded as a known multiple of a second quantity ϕ_2, and so on. By a suitable determination of the m quantities ϕ_1, ϕ_2, &c., *any* configuration of the system may be represented as compounded of the m configurations of these types, and thus the quantities ϕ themselves may be looked upon as co-ordinates defining the configuration of the system. They are called the *normal* co-ordinates[1].

When expressed in terms of the normal co-ordinates, T and V are reduced to sums of squares; for it is easily seen that if the products also appeared, the resulting equations of vibration would not be satisfied by putting any $m-1$ of the co-ordinates equal to zero, while the remaining one was finite.

We might have commenced with this transformation, assuming from Algebra that any two homogeneous quadratic functions can be reduced by linear transformations to sums of squares.[2] Thus

$$\left. \begin{aligned} T &= \tfrac{1}{2}a_1\dot{\phi}_1{}^2 + \tfrac{1}{2}a_2\dot{\phi}_2{}^2 + \dots \\ V &= \tfrac{1}{2}c_1\phi_1{}^2 + \tfrac{1}{2}c_2\phi_2{}^2 + \dots \end{aligned} \right\} \dots\dots\dots\dots(1),$$

where the coefficients (in which the double suffixes are no longer required) are necessarily positive if the equilibrium be stable.

Lagrange's equations now become

$$a_1\ddot{\phi}_1 + c_1\phi_1 = 0, \quad a_2\ddot{\phi}_2 + c_2\phi_2 = 0, \text{ &c.} \dots\dots\dots\dots(2),$$

of which the solution is

$$\phi_1 = A\cos(n_1 t - \alpha), \quad \phi_2 = B\cos(n_2 t - \beta), \text{ &c.} \dots\dots(3),$$

where A, $B\dots$, α, β,... are arbitrary constants, and

$$n_1{}^2 = c_1 \div a_1, \quad n_2{}^2 = c_2 \div a_2, \text{ &c.} \dots\dots\dots\dots(4).$$

[The vibrations expressed by the various normal co-ordinates are completely independent of one another, and the energy of the whole motion is the simple sum of the parts corresponding to the several normal vibrations taken separately. In fact by (1)

$$T + V = \tfrac{1}{2}c_1 A_1{}^2 + \tfrac{1}{2}c_2 A_2{}^2 + \dots\dots\dots\dots\dots(5).$$

By the nature of the case the coefficients a are necessarily positive. But if the equilibrium be unstable, some of the coefficients c may be negative. Corresponding to any negative c, n becomes imaginary and the circular functions of the time are replaced by exponentials.

In any motion proportional to $e^{\lambda t}$ the disturbance is equally multiplied in equal times, and the degree of instability may be considered to be measured by λ. If there be more than one

---

[1] Thomson and Tait's *Natural Philosophy*, first edition 1867, § 337.

[2]   See Routh's *Rigid Dynamics*, p. 4○○.

unstable mode, the relative importance is largely determined bv the corresponding values of λ. Thus, if

$$\psi = A e^{\lambda_1 t} + B e^{\lambda_2 t},$$

in which $\lambda_1 > \lambda_2$, then whatever may be the finite ratio of $A : B$, the first term ultimately acquires the preponderance, inasmuch as

$$A e^{\lambda_1 t} : B e^{\lambda_2 t} = (A/B) e^{(\lambda_1 - \lambda_2) t}.$$

In general, unstable equilibrium when disturbed infinitesimally will be departed from according to that mode which is *most unstable*, viz. for which λ is greatest. In a later chapter we shall meet with interesting applications of this principle.

The reduction to·normal co-ordinates allows us readily to trace what occurs when two of the values of n^2 become equal. It is evident that there is no change of form. The spherical pendulum may be referred to as a simple example of equal roots. It is remarkable that both Lagrange and Laplace fell into the error of supposing that equality among roots necessarily implies terms containing t as a factor[1]. The analytical theory of the general case (where the co-ordinates are not normal) has been discussed by Somof[2] and by Routh[3].]

88. The interpretation of the equations of motion leads to a theorem of considerable importance, which may be thus stated[4]. The period of a conservative system vibrating in a constrained type about a position of stable equilibrium is stationary in value when the type is normal. We might prove this from the original equations of vibration, but it will be more convenient to employ the normal co-ordinates. The constraint, which may be supposed to be of such a character as to leave only one degree of freedom, is represented by taking the quantities ϕ in given ratios.

If we put

$$\phi_1 = A_1 \theta, \quad \phi_2 = A_2 \theta, \&\text{c.} \dots\dots\dots\dots(1),$$

θ is a variable quantity, and A_1, A_2, &c. are given for a given constraint.

The expressions for T and V become

$$T = \{\tfrac{1}{2} a_1 A_1{}^2 + \tfrac{1}{2} a_2 A_2{}^2 + \dots\dots\} \dot{\theta}^2,$$
$$V = \{\tfrac{1}{2} c_1 A_1{}^2 + \tfrac{1}{2} c_2 A_2{}^2 + \dots\dots\} \theta^2,$$

---

[1] Thomson and Tait, 2nd edition, § 343 m.

[2] St Petersb. Acad. Sci. Mém. I. 1859.

[3] *Stability of Motion* (Adams Prize Essay for 1877). See also Routh's *Rigid Dynamics*, 5th edition, 1892.

[4] *Proceedings of the Mathematical Society*, June, 1873.

whence, if θ varies as $\cos pt$,

$$p^2 = \frac{c_1 A_1^2 + c_2 A_2^2 + \ldots + c_m A_m^2}{a_1 A_1^2 + a_2 A_2^2 + \ldots + a_m A_m^2} \ldots\ldots\ldots\ldots\ldots(2).$$

This gives the period of the vibration of the constrained type; and it is evident that the period is stationary, when all but one of the coefficients $A_1, A_2, \ldots$ vanish, that is to say, when the type coincides with one of those natural to the system, and no constraint is needed.

[In the foregoing statement the equilibrium is supposed to be thoroughly stable, so that all the quantities c are positive. But the theorem applies equally even though any or all of the c's be negative. Only if p^2 itself be negative, the period becomes imaginary. In this case the stationary character attaches to the coefficients of t in the exponential terms, quantities which measure the *degree* of instability.

Corresponding theorems, of importance in other branches of science, may be stated for systems such that only T and F, or only V and F, are sensible[1].

The stationary property of the roots of Lagrange's determinant (3) § 84, suggests a general method of approximating to their values. Beginning with assumed rough approximations to the ratios $A_1 : A_2 : A_3 \ldots\ldots$ we may calculate a first approximation to p^2 from

$$p^2 = \frac{\frac{1}{2} c_{11} A_1^2 + \frac{1}{2} c_{22} A_2^2 + \ldots + c_{12} A_1 A_2 + \ldots}{\frac{1}{2} a_{11} A_1^2 + \frac{1}{2} a_{22} A_2^2 + \ldots + a_{12} A_1 A_2 + \ldots} \ldots\ldots (3).$$

With this value of p^2 we may recalculate the ratios $A_1 : A_2 \ldots$ from any $(m-1)$ of equations (5) § 84, then again by application of (3) determine an improved value of p^2, and so on.]

By means of the same theorem we may prove that an increase in the mass of any part of a vibrating system is attended by a prolongation of all the natural periods, or at any rate that no period can be diminished. Suppose the increment of mass to be infinitesimal. After the alteration, the types of free vibration will in general be changed; but, by a suitable constraint, the system may be made to retain any one of the former types. If this be done, it is certain that any vibration which involves a motion of the part whose mass has been increased will have its period prolonged. Only as a particular case (as, for example, when a load is placed at the node of a vibrating string) can the period

[1] *Brit. Ass. Rep.* for 1885, p. 911.

remain unchanged. The theorem now allows us to assert that the removal of the constraint, and the consequent change of type, can only affect the period by a quantity of the second order; and that therefore in the limit the free period cannot be less than before the change. By integration we infer that a finite increase of mass must prolong the period of every vibration which involves a motion of the part affected, and that in no case can the period be diminished; but in order to see the correspondence of the two sets of periods, it may be necessary to suppose the alterations made by steps. Conversely, the effect of a removal of part of the mass of a vibrating system must be to shorten the periods of all the free vibrations.

In like manner we may prove that if the system undergo such a change that the potential energy of a given configuration is diminished, while the kinetic energy of a given motion is unaltered, the periods of the free vibrations are all increased, and conversely. This proposition may sometimes be used for tracing the effects of a constraint; for if we suppose that the potential energy of any configuration violating the condition of constraint gradually increases, we shall approach a state of things in which the condition is observed with any desired degree of completeness. During each step of the process every free vibration becomes (in general) more rapid, and a number of the free periods (equal to the degrees of liberty lost) become infinitely small. The same practical result may be reached without altering the potential energy by supposing the *kinetic* energy of any *motion* violating the condition to increase without limit. In this case one or more periods become infinitely large, but the finite periods are ultimately the same as those arrived at when the potential energy is increased, although in one case the periods have been throughout increasing, and in the other diminishing. This example shews the necessity of making the alterations by steps; otherwise we should not understand the correspondence of the two sets of periods. Further illustrations will be given under the head of two degrees of freedom.

By means of the principle that the value of the free periods is stationary, we may easily calculate corrections due to any deviation in the system from theoretical simplicity. If we take as a hypothetical type of vibration that proper to the simple system, the period so found will differ from the truth by quantities depending on the squares of the irregularities. Several examples of such calculations will be given in the course of this work.

89. Another point of importance relating to the period of a system vibrating in an arbitrary type remains to be noticed. It appears from (2) § 88, that the period of the vibration corresponding to any hypothetical type is included between the greatest and least of those natural to the system. In the case of systems like strings and plates which are treated as capable of continuous deformation, there is no least natural period; but we may still assert that the period calculated from any hypothetical type cannot exceed that belonging to the gravest normal type. When therefore the object is to estimate the longest proper period of a system by means of calculations founded on an assumed type, we know *a priori* that the result will come out too small.

In the choice of a hypothetical type judgment must be used, the object being to approach the truth as nearly as can be done without too great a sacrifice of simplicity. Thus the type for a string heavily weighted at one point might suitably be taken from the extreme case of an infinite load, when the two parts of the string would be straight. As an example of a calculation of this kind, of which the result is known, we will take the case of a uniform string of length l, stretched with tension T_1, and inquire what the period would be on certain suppositions as to the type of vibration.

Taking the origin of x at the middle of the string, let the curve of vibration on the positive side be

$$y = \cos pt \left\{ 1 - \left(\frac{2x}{l} \right)^n \right\} \quad \dots\dots\dots\dots(1),$$

and on the negative side the image of this in the axis of y, n being not less than unity. This form satisfies the condition that y vanishes when $x = \pm \frac{1}{2} l$. We have now to form the expressions for T and V, and it will be sufficient to consider the positive half of the string only. Thus, ρ being the longitudinal density,

$$T = \tfrac{1}{2} \int_0^{\frac{1}{2}l} \rho \dot{y}^2 dx = \frac{\rho \, n^2 l \, p^2 \sin^2 pt}{2 \, (n+1) \, (2n+1)},$$

and

$$V = \tfrac{1}{2} T_1 \int_0^{\frac{1}{2}l} \left(\frac{dy}{dx} \right)^2 dx = \frac{n^2 T_1 \cos^2 pt}{(2n-1) \, l}.$$

Hence
$$p^2 = \frac{2(n+1)(2n+1)}{2n-1} \cdot \frac{T_1}{\rho l^2}. \ldots\ldots\ldots\ldots\ldots\ldots(2).$$

If $n = 1$, the string vibrates as if the mass were concentrated in its middle point, and

$$p^2 = \frac{12\,T_1}{\rho l^2}.$$

If $n = 2$, the form is parabolic, and

$$p^2 = \frac{10\,T_1}{\rho l^2}.$$

The true value of p^2 for the gravest type is $\dfrac{\pi^2 T_1}{\rho l^2}$, so that the assumption of a parabolic form gives a period which is too small in the ratio $\pi : \sqrt{10}$ or $\cdot9936 : 1$. The minimum of p^2, as given by (2), occurs when $n = \frac{1}{2}(\sqrt{6}+1) = 1\cdot72474$, and gives

$$p^2 = 9\cdot8990\,\frac{T_1}{\rho l^2}.$$

The period is now too small in the ratio

$$\pi : \sqrt{9\cdot8990} = \cdot99851 : 1.$$

It will be seen that there is considerable latitude in the choice of a type, even the violent supposition that the string vibrates as two straight pieces giving a period less than ten per cent. in error. And whatever type we choose to take, the period calculated from it cannot be greater than the truth.

[In the above applications it is assumed that there are no unstable modes. When unstable modes exist, the statement is that a constrained mode if stable possesses a frequency of vibration less than that of the highest normal mode, and if unstable has a degree of instability less than that of the most unstable normal mode.]

90. The rigorous determination of the periods and types of vibration of a given system is usually a matter of great difficulty, arising from the fact that the functions necessary to express the modes of vibration of most continuous bodies are not as yet recognised in analysis. It is therefore often necessary to fall back on methods of approximation, referring the proposed system to some other of a character more amenable to analysis, and calculating corrections depending on the supposition that the difference between the two systems is small. The problem of approximately simple systems is thus one of great importance, more especially as it is impossible in practice actually to realise the simple forms about which we can most easily reason.

Let us suppose then that the vibrations of a simple system are thoroughly known, and that it is required to investigate those of a system derived from it by introducing small variations in the mechanical functions. If ϕ_1, ϕ_2, &c. be the normal co-ordinates of the original system,

$$T = \tfrac{1}{2} a_1 \dot{\phi}_1^2 + \tfrac{1}{2} a_2 \dot{\phi}_2^2 + \ldots,$$
$$V = \tfrac{1}{2} c_1 \phi_1^2 + \tfrac{1}{2} c_2 \phi_2^2 + \ldots,$$

and for the varied system, referred to the same co-ordinates, which are now only approximately normal,

$$\left. \begin{aligned} T + \delta T &= \tfrac{1}{2} (a_1 + \delta a_{11}) \dot{\phi}_1^2 + \ldots + \delta a_{12} \dot{\phi}_1 \dot{\phi}_2 + \ldots \\ V + \delta V &= \tfrac{1}{2} (c_1 + \delta c_{11}) \phi_1^2 + \ldots + \delta c_{12} \phi_1 \phi_2 + \ldots \end{aligned} \right\} \ldots (1),$$

in which δa_{11}, δa_{12}, δc_{11}, δc_{12}, &c. are to be regarded as small quantities. In certain cases new co-ordinates may appear, but if so their coefficients must be small. From (1) we obtain for the Lagrangian equations of motion,

$$\left. \begin{aligned} (a_1 + \delta a_{11} D^2 + c_1 + \delta c_{11}) \phi_1 + (\delta a_{12} D^2 + \delta c_{12}) \phi_2 \\ + (\delta a_{13} D^2 + \delta c_{13}) \phi_3 + \ldots = 0 \\ (\delta a_{21} D^2 + \delta c_{21}) \phi_1 + (a_2 + \delta a_{22} D^2 + c_2 + \delta c_{22}) \phi_2 \\ + (\delta a_{23} D^2 + \delta c_{23}) \phi_3 + \ldots = 0 \\ \ldots\ldots\ldots\ldots\ldots\ldots\ldots\ldots\ldots\ldots\ldots\ldots\ldots\ldots\ldots\ldots\ldots\ldots\ldots\ldots\ldots\ldots\ldots \end{aligned} \right\} \ldots\ldots(2).$$

In the original system the fundamental types of vibration are those which correspond to the variation of but a single co-ordinate at a time. Let us fix our attention on one of them, involving say a variation of ϕ_r, while all the remaining co-ordinates vanish. The change in the system will in general entail an alteration in the fundamental or normal types; but under the circumstances contemplated the alteration is small. The new normal type is expressed by the synchronous variation of the other co-ordinates in addition to ϕ_r; but the ratio of any other ϕ_s to ϕ_r is small. When these ratios are known, the normal mode of the altered system will be determined.

Since the whole motion is simple harmonic, we may suppose that each co-ordinate varies as $\cos p_r t$, and substitute in the differential equations $-p_r^2$ for D^2. In the sth equation ϕ_s occurs with the finite coefficient

$$- a_s p_r^2 - \delta a_{ss} p_r^2 + c_s + \delta c_{ss}.$$

The coefficient of ϕ_r is

$$- \delta a_{rs} p_r^2 + \delta c_{rs}.$$

The other terms are to be neglected in a first approximation, since both the co-ordinate (relatively to ϕ_r) and its coefficient are small quantities. Hence

$$\phi_s : \phi_r = - \frac{\delta c_{rs} - p_r^2 \delta a_{rs}}{c_s - p_r^2 a_s} \quad \ldots\ldots\ldots\ldots\ldots (3).$$

Now

$$- a_s p_s^2 + c_s = 0,$$

and thus

$$\phi_s : \phi_r = \frac{p_r^2 \delta a_{rs} - \delta c_{rs}}{a_s (p_s^2 - p_r^2)} \ldots\ldots\ldots\ldots\ldots (4),$$

the required result.

If the kinetic energy alone undergo variation,

$$\phi_s : \phi_r = \frac{p_r^2}{p_s^2 - p_r^2} \frac{\delta a_{rs}}{a_s} \ldots\ldots\ldots\ldots\ldots (5).$$

The corrected value of the period is determined by the rth equation of (2), not hitherto used. We may write it,

$$\phi_r \{ - p_r^2 a_r - p_r^2 \delta a_{rr} + c_r + \delta c_{rr} \} + \Sigma \phi_s (- p_r^2 \delta a_{rs} + \delta c_{rs}) = 0.$$

Substituting for $\phi_s : \phi_r$ from (4), we get

$$p_r^2 = \frac{c_r + \delta c_{rr}}{a_r + \delta a_{rr}} - \Sigma \frac{(\delta c_{rs} - p_r^2 \delta a_{rs})^2}{a_s a_r (p_s^2 - p_r^2)} \ldots\ldots\ldots\ldots (6).$$

The first term gives the value of p_r^2 calculated without allowance for the change of type, and is sufficient, as we have already proved, when the square of the alteration in the system may be neglected. The terms included under the symbol Σ, in which the summation extends to all values of s other than r, give the correction due to the change of type and are of the second order. Since a_s and a_r are positive, the sign of any term depends upon that of $p_s^2 - p_r^2$. If $p_s^2 > p_r^2$, that is, if the mode s be more acute than the mode r, the correction is negative, and makes the calculated note graver than before; but if the mode s be the graver, the correction raises the note. If r refer to the gravest mode of the system, the whole correction is negative; and if r refer to the acutest mode, the whole correction is positive, as we have already seen by another method.

91. As an example of the use of these formulæ, we may take the case of a stretched string, whose longitudinal density ρ is not quite constant. If x be measured from one end, and y

be the transverse displacement, the configuration at any time t will be expressed by

$$y = \phi_1 \sin \frac{\pi x}{l} + \phi_2 \sin \frac{2\pi x}{l} + \phi_3 \sin \frac{3\pi x}{l} + \ldots\ldots\ldots(1),$$

l being the length of the string. ϕ_1, ϕ_2, ... are the normal co-ordinates for $\rho = $ constant, and though here ρ is not strictly constant, the configuration of the system may still be expressed by means of the same quantities. Since the potential energy of any configuration is the same as if $\rho = $ constant, $\delta V = 0$. For the kinetic energy we have

$$T + \delta T = \tfrac{1}{2} \int_0^l \rho \left(\dot{\phi}_1 \sin \frac{\pi x}{l} + \dot{\phi}_2 \sin \frac{2\pi x}{l} + \ldots \right)^2 dx$$

$$= \tfrac{1}{2} \dot{\phi}_1{}^2 \int_0^l \rho \sin^2 \frac{\pi x}{l} \, dx + \tfrac{1}{2} \dot{\phi}_2{}^2 \int_0^l \rho \sin^2 \frac{2\pi x}{l} \, dx + \ldots$$

$$+ \dot{\phi}_1 \dot{\phi}_2 \int_0^l \rho \sin \frac{\pi x}{l} \sin \frac{2\pi x}{l} \, dx + \ldots.$$

If ρ were constant, the products of the velocities would disappear, since ϕ_1, ϕ_2, &c. are, on that supposition, the normal co-ordinates. As it is, the integral coefficients, though not actually evanescent, are small quantities. Let $\rho = \rho_0 + \delta\rho$; then in our previous notation

$$a_r = \tfrac{1}{2} l\rho_0, \quad \delta a_{rr} = \int_0^l \delta\rho \sin^2 \frac{r\pi x}{l} \, dx, \quad \delta a_{rs} = \int_0^l \delta\rho \sin \frac{r\pi x}{l} \sin \frac{s\pi x}{l} \, dx.$$

Thus the type of vibration is expressed by

$$\phi_s : \phi_r = \frac{p_r{}^2}{p_s{}^2 - p_r{}^2} \cdot \frac{2}{l\rho_0} \int_0^l \delta\rho \sin \frac{r\pi x}{l} \sin \frac{s\pi x}{l} \, dx;$$

or, since

$$p_r{}^2 : p_s{}^2 = r^2 : s^2,$$

$$\phi_s : \phi_r = \frac{r^2}{s^2 - r^2} \int_0^l \frac{2\delta\rho}{l\rho_0} \sin \frac{r\pi x}{l} \sin \frac{s\pi x}{l} \, dx \ldots\ldots(2).$$

Let us apply this result to calculate the displacement of the nodal point of the second mode ($r = 2$), which would be in the middle, if the string were uniform. In the neighbourhood of this point, if $x = \tfrac{1}{2}l + \delta x$, the approximate value of y is

$$y = \phi_1 \sin \frac{\pi}{2} + \phi_2 \sin \frac{2\pi}{2} + \phi_3 \sin \frac{3\pi}{2} + \ldots$$

$$+ \delta x \left\{ \frac{\pi}{l} \phi_1 \cos \frac{\pi}{2} + \frac{2\pi}{l} \phi_2 \cos \frac{2\pi}{2} + \ldots \right\}$$

$$= \phi_1 - \phi_3 + \phi_5 - \ldots + \frac{\pi}{l} \delta x \left\{ -2\phi_2 + 2\phi_4 + \ldots \right\}.$$

Hence when $y = 0$,

$$\delta x = \frac{l}{2\pi\phi_2} \{\phi_1 - \phi_3 + \phi_5 - \dots \} \dots\dots\dots\dots (3)$$

approximately, where

$$\phi_s : \phi_2 = \frac{4}{s^2 - 4} \int_0^l \frac{2\delta\rho}{l\rho_0} \sin \frac{2\pi x}{l} \sin \frac{s\pi x}{l} \, dx \dots\dots\dots (4).$$

To shew the application of these formulæ, we may suppose the irregularity to consist in a small load of mass $\rho_0\lambda$ situated at $x = \frac{1}{4} l$, though the result might be obtained much more easily directly. We have

$$\delta x = \frac{2\lambda}{\pi\sqrt{2}} \left\{ \frac{2}{1^2 - 4} - \frac{2}{3^2 - 4} - \frac{2}{5^2 - 4} + \frac{2}{7^2 - 4} + \dots\dots \right\},$$

from which the value of δx may be calculated by approximation. The real value of δx is, however, very simple. The series within brackets may be written

$$-\left\{ 1 + \frac{1}{3} - \frac{1}{5} - \frac{1}{7} + \frac{1}{9} + \frac{1}{11} - \&c. \right\},$$

which is equal to

$$-\int_0^1 \frac{1 + x^2}{1 + x^4} \, dx.$$

The value of the definite integral is

$$\pi \div 4 \sin \frac{\pi}{4} \,[1],$$

and thus

$$\delta x = -\frac{2\lambda}{\pi\sqrt{2}} \cdot \frac{\pi\sqrt{2}}{4} = -\frac{\lambda}{2},$$

as may also be readily proved by equating the periods of vibration of the two parts of the string, that of the loaded part being calculated approximately on the assumption of unchanged type.

As an example of the formula (6) § 90 for the period, we may take the case of a string carrying a small load $\rho_0\lambda$ at its middle point. We have

$$a_r = \frac{1}{2} l\rho_0, \quad \delta a_{rr} = \rho_0\lambda \sin^2 \frac{r\pi}{2}, \quad \delta a_{rs} = \rho_0\lambda \sin \frac{r\pi}{2} \sin \frac{s\pi}{2},$$

and thus, if P_r be the value corresponding to $\lambda = 0$, we get when r is even, $p_r = P_r$, and when r is odd,

$$p_r^2 = P_r^2 \left\{ \frac{1}{1 + 2\lambda/l} - \Sigma \frac{4r^2}{s^2 - r^2} \frac{\lambda^2}{l^2} \right\} \dots\dots\dots\dots (5),$$

[1] Todhunter's *Int. Calc.* § 255.

where the summation is to be extended to all the odd values of s other than r. If $r = 1$,

$$p_1{}^2 = P_1{}^2 \left\{ 1 - \frac{2\lambda}{l} + \frac{4\lambda^2}{l^2} - \Sigma \frac{4}{s^2 - 1} \frac{\lambda^2}{l^2} \right\}.$$

Now

$$2\Sigma \frac{1}{s^2 - 1} = \Sigma \frac{1}{s - 1} - \Sigma \frac{1}{s + 1},$$

in which the values of s are 3, 5, 7, 9.... Accordingly

$$\Sigma \frac{1}{s^2 - 1} = \frac{1}{4}$$

and

$$p_1{}^2 = P_1{}^2 \left\{ 1 - \frac{2\lambda}{l} + \frac{3\lambda^2}{l^2} + \ldots \right\} \ldots \ldots \ldots (6),$$

giving the pitch of the gravest tone accurately as far as the square of the ratio $\lambda : l$.

In the general case the value of $p_r{}^2$, correct as far as the first order in $\delta\rho$, will be

$$p_r{}^2 = P_r{}^2 \left\{ 1 - \frac{\delta a_{rr}}{a_r} \right\} = P_r{}^2 \left\{ 1 - \frac{2}{l} \int_0^l \frac{\delta\rho}{\rho_0} \sin^2 \frac{r\pi x}{l} \, dx \right\} \ldots (7).$$

92. The theory of vibrations throws great light on expansions of arbitrary functions in series of other functions of specified types. The best known example of such expansions is that generally called after Fourier, in which an arbitrary periodic function is resolved into a series of harmonics, whose periods are submultiples of that of the given function. It is well known that the difficulty of the question is confined to the proof of the *possibility* of the expansion; if this be assumed, the determination of the coefficients is easy enough. What I wish now to draw attention to is, that in this, and an immense variety of similar cases, the possibility of the expansion may be inferred from physical considerations.

To fix our ideas, let us consider the small vibrations of a uniform string stretched between fixed points. We know from the general theory that the whole motion, whatever it may be, can be analysed into a series of component motions, each represented by a harmonic function of the time, and capable of existing by itself. If we can discover these normal types, we shall be in a position to represent the most general vibration possible by combining them, assigning to each an arbitrary amplitude and phase.

Assuming that a motion is harmonic with respect to time, we get to determine the type an equation of the form

$$\frac{d^2y}{dx^2} + k^2y = 0,$$

whence it appears that the normal functions are

$$y = \sin\frac{\pi x}{l}, \quad y = \sin\frac{2\pi x}{l}, \quad y = \sin\frac{3\pi x}{l}, \text{ \&c.}$$

We infer that the most general position which the string can assume is capable of representation by a series of the form

$$A_1 \sin\frac{\pi x}{l} + A_2 \sin\frac{2\pi x}{l} + A_3 \sin\frac{3\pi x}{l} + \ldots\ldots,$$

which is a particular case of Fourier's theorem. There would be no difficulty in proving the theorem in its most general form.

So far the string has been supposed uniform. But we have only to introduce a variable density, or even a single load at any point of the string, in order to alter completely the expansion whose possibility may be inferred from the dynamical theory. It is unnecessary to dwell here on this subject, as we shall have further examples in the chapters on the vibrations of particular systems, such as bars, membranes, and confined masses of air.

92 a. In § 88 we have a formula for the frequency of vibration applicable when by the imposition of given constraints the original system is left with only one degree of freedom. It is of interest to trace also the effect of less complete constraints, such as may be expressed by linear relations among the normal co-ordinates of number less by at least two than that of the (original) degrees of freedom. Thus we may suppose that

$$\left.\begin{array}{l} f_1\phi_1 + f_2\phi_2 + f_3\phi_3 + \ldots = 0 \\ g_1\phi_1 + g_2\phi_2 + g_3\phi_3 + \ldots = 0 \\ h_1\phi_1 + h_2\phi_2 + h_3\phi_3 + \ldots = 0 \\ \ldots\ldots\ldots\ldots\ldots\ldots\ldots\ldots\ldots\ldots \end{array}\right\} \ldots\ldots\ldots\ldots (1).$$

If the number of equations (r) fall short of the number of the degrees of freedom by unity, the ratios $\phi_1 : \phi_2 : \phi_3 \ldots$ are fully determined, and the case is that of but one outstanding degree of freedom discussed in § 88.

This problem may be treated in more than one way, but the

most instructive procedure is to trace the effect of additions to T and V. We will suppose that equations (1) § 87 are altered to

$$T = \tfrac{1}{2}a_1\dot{\phi}_1^2 + \tfrac{1}{2}a_2\dot{\phi}_2^2 + \dots + \tfrac{1}{2}\alpha\,(f_1\dot{\phi}_1 + f_2\dot{\phi}_2 + \dots\,)^2 \dots\dots\dots(2),$$

$$V = \tfrac{1}{2}c_1\phi_1^2 + \tfrac{1}{2}c_2\phi_2^2 + \dots + \tfrac{1}{2}\gamma\,(f_1\phi_1 + f_2\phi_2 + \dots\,)^2 \dots\dots\dots(3),$$

and that F, not previously existent, is now

$$F = \tfrac{1}{2}\beta\,(f_1\dot{\phi}_1 + f_2\dot{\phi}_2 + \dots\,)^2 \dots\dots\dots\dots\dots(4).$$

The connection with the proposed problem will be understood by supposing for instance that $\alpha = 0$, $\beta = 0$, while $\gamma = \infty$. By (3) the potential energy of any displacement violating the condition

$$f_1\phi_1 + f_2\phi_2 + \dots = 0 \dots\dots\dots\dots\dots(5)$$

is then infinite, and this is tantamount to the imposition of the constraint represented by (5).

Lagrange's equations with λ written for D now become

$$\left.\begin{aligned}(a_1\lambda^2 + c_1)\,\phi_1 + f_1\,(\alpha\lambda^2 + \beta\lambda + \gamma)\,(f_1\phi_1 + f_2\phi_2 + \dots\,) &= 0\\(a_2\lambda^2 + c_2)\,\phi_2 + f_2\,(\alpha\lambda^2 + \beta\lambda + \gamma)\,(f_1\phi_1 + f_2\phi_2 + \dots\,) &= 0\\\dots\dots\dots\dots\dots\dots\dots\dots\dots\dots\dots\dots\dots\dots\dots\dots\dots&\end{aligned}\right\}\dots(6).$$

If we multiply the first of these by $f_1/(a_1\lambda^2 + c_1)$, the second by $f_2/(a_2\lambda^2 + c_2)$, and so on, and add the results together, the factor $(f_1\phi_1 + f_2\phi_2 + \dots\,)$ will divide out, and the determinant takes the form

$$\frac{f_1^2}{a_1\lambda^2 + c_1} + \frac{f_2^2}{a_2\lambda^2 + c_2} + \dots\dots\dots + \frac{1}{\alpha\lambda^2 + \beta\lambda + \gamma} = 0 \dots\dots(7).$$

If any one of the quantities α, β, γ become infinite while the others remain finite, the effect is equivalent to the imposition of the constraint (5), and the result may be written

$$\Sigma f^2/(a\lambda^2 + c) = 0\dots\dots\dots\dots\dots\dots(8)[1].$$

When multiplied out this equation is of degree $(m-1)$ in λ^2, one degree of freedom having been lost.

If we put $\beta = 0$, (7) is an equation of the mth degree in λ^2, and the coefficients α, γ enter in the same way as do a_1, c_1; a_2, c_2; &c.

In order to refer more directly to the case of vibrations about stable equilibrium, we will write p^2 for $-\lambda^2$. The values of p^2 belonging to the unaltered system, viz. n_1^2, n_2^2,..., are given as before by

$$c_1 - a_1 n_1^2 = 0; \qquad c_2 - a_2 n_2^2 = 0, \text{ &c.,} \dots\dots\dots\dots(9);$$

and we will also write

$$\gamma - \alpha\nu^2 = 0 \dots\dots\dots\dots\dots\dots(10),$$

[1] Routh's *Rigid Dynamics*, 5th edition, 1892, § 67.

where ν^2 relates to the supposed additions to T and V considered as belonging to an independent vibrator. Let the order of magnitude of these quantities be

$$n_1{}^2,\ n_2{}^2,\ldots\ldots\ldots n_r{}^2,\ \nu^2,\ n_{r+1}{}^2,\ldots\ldots\ldots n_m{}^2 \ldots\ldots\ldots (11).$$

We shall see that there is a root of (7) between each consecutive pair of the quantities (11).

Our equation may be written

$$f_1{}^2 (\gamma - \alpha p^2)(c_2 - a_2 p^2)(c_3 - a_3 p^2)\ldots\ldots$$
$$+ f_2{}^2 (\gamma - \alpha p^2)(c_1 - a_1 p^2)(c_3 - a_3 p^2)\ldots\ldots$$
$$+ \ldots\ldots\ldots\ldots\ldots\ldots\ldots\ldots\ldots\ldots\ldots\ldots\ldots$$
$$+ (c_1 - a_1 p^2)(c_2 - a_2 p^2)\ldots\ldots\ldots = 0 \ldots\ldots\ldots\ldots (12).$$

When p^2 coincides with any of the quantities (11), all but one of the terms in (12) vanish, and the sign of the expression is the same as that of the term which remains over. When $p^2 < n_1{}^2$, all the terms are positive, so that there is no root less than $n_1{}^2$. When $p^2 = n_1{}^2$, the expression (12) reduces to the positive quantity

$$f_1{}^2 (\gamma - \alpha n_1{}^2)(c_2 - a_2 n_1{}^2)(c_3 - a_3 n_1{}^2)\ldots\ldots$$

When p^2 rises to $n_2{}^2$, (12) becomes

$$f_2{}^2 (\gamma - \alpha n_2{}^2)(c_1 - a_1 n_2{}^2)(c_3 - a_3 n_2{}^2)\ldots\ldots ;$$

and this is *negative*, since the factor $(c_1 - a_1 n_2{}^2)$ is now negative. Hence there is a root of (12) between $n_1{}^2$ and $n_2{}^2$. When $p^2 = n_3{}^2$, the expression is again positive, and thus there is a root between $n_2{}^2$ and $n_3{}^2$. This argument may be continued, and it proves that there is a root of (12) between any consecutive two of the $(m+1)$ quantities (11). The m roots of (12) are now accounted for, and there is none greater than $n_m{}^2$. If we compare the values of the roots before and after the change, we see that the effect is to cause a movement which is in every case *towards* ν^2.[1] Considered absolutely the movement is in one direction for those roots that are greater than ν^2 and in the opposite direction for those that are less than ν^2. Accordingly the interval from $n_r{}^2$ to $n_{r+1}{}^2$, in which ν^2 lies, contains after the change *two* roots, one on either side of ν^2.

If ν^2 be less than any of the quantities n^2, as happens when $\gamma = 0$, one root lies between ν^2 and $n_1{}^2$, one between $n_1{}^2$ and $n_2{}^2$, and so on. Thus every root is depressed. On the other hand if $\nu^2 > n_m{}^2$, every root is increased. This happens if $\alpha = 0$. (§ 88.)

---

[1] It will be understood that in particular cases the movement may vanish.

The results now arrived at are of course independent of the special machinery of normal co-ordinates used in the investigation. If to any part of a system $(n_1{}^2, n_2{}^2 \ldots\ldots)$ be attached a vibrator (ν^2) having a single degree of freedom, the effect is to displace all the quantities $n_1{}^2, \ldots$ in the direction of ν^2. Let us now suppose that a second change is made in the vibrator whereby α becomes $\alpha + \alpha'$, and γ becomes $\gamma + \gamma'$. Every root of the determinantal equation moves towards ν'^2, where $\gamma' - \alpha'\nu'^2 = 0$. If we suppose that $\nu'^2 = \nu^2$, the movements are in all cases in the same directions as before. Going back now to the original system, and supposing that α, γ grow from zero to their actual values in such a manner that ν^2 remains constant, we see that during this process the roots move without regression in the direction of closer agreement with ν^2.

As α and γ become infinite, one root of (12) moves to coincidence with ν^2, while the remaining $(m-1)$ roots, corresponding to the constrained system, are given by

$$\Sigma f^2/(c - ap^2) = 0 \ \ldots\ldots\ldots\ldots\ldots\ldots\ldots\ldots(13),$$

and are independent of the value of ν^2.

Particular cases are obtained by supposing either $\nu^2 = 0$, or $\nu^2 = \infty$. Whether the constraint is effected by making infinite the kinetic energy of any motion, or the potential energy of any displacement, which violates it, makes no difference to the vibrations which remain. In the first case one vibration becomes infinitely slow, and in the second case one becomes infinitely quick. However the constraint be arrived at, the $(m-1)$ frequencies of vibration of the constrained system *separate*[1] the m frequencies of the original system.

Any number of examples of this theorem may be invented without difficulty. Consider the case of a uniform stretched string, held at both ends and vibrating transversely. This is the original system. Now introduce a constraint by holding at rest a point which divides the length in the proportion (say) of 3 : 2. The two parts vibrate independently, and the frequencies for each part form an arithmetical progression. If the frequencies proper to the undivided string be 1, 2, 3, 4 ......; those for the parts are

---

[1] But in particular cases the "separation" may vanish. The theorem in the text was proved for two degrees of freedom in the first edition of this work. In its generality it appears to be due to Routh.

$\frac{5}{2}(1, 2, 3, \ldots)$ and $\frac{5}{3}(1, 2, 3, \ldots)$. The beginning of each series is shewn in the accompanying scheme;

and it will be seen that between any consecutive numbers in the first row there is a number to be found *either* in the second *or* in the third row. In the case of 5 and 10 we have an extreme condition of things; but the slightest displacement of the point at which the constraint is applied will displace one of the fives, tens &c. to the left and the other to the right.

The coincidences may be avoided by dividing the string incommensurably. Thus, if x be an incommensurable number less than unity, one of the series of quantities m/x, $m/(1-x)$, where m is a whole number, can be found which shall lie between any given consecutive integers, and but one such quantity can be found.

Again, let us suppose that a system is referred to co-ordinates which are not normal (§ 84), and let the constraint represented by $\psi_1 = 0$ be imposed. The determinant of the altered system is formed from that of the original system by erasing the first row and the first column. It may be called ∇_1, and from this again may be formed in like manner a new determinant ∇_2, and so on. These determinants form a series of functions of p^2, regularly decreasing in degree; and we conclude that the roots of each separate the roots of that immediately preceding[1].

It may be remarked that while for the sake of simplicity of statement we have supposed that the equilibrium of the original system was thoroughly stable, as also that of the vibration brought into connection therewith, these restrictions may easily be dispensed with. In any case the series of positive and negative quantities, n_1^2, n_2^2, ... and ν^2, may be arranged in algebraic order, and the effect of the vibrator is to cause a movement of every value of p^2 in the direction of ν^2.

In order to extend the above theory we will now suppose that the addition to T is

$$\tfrac{1}{2}\alpha_f (f_1\dot{\phi}_1 + f_2\dot{\phi}_2 + \ldots)^2 + \tfrac{1}{2}\alpha_g (g_1\dot{\phi}_1 + g_2\dot{\phi}_2 + \ldots)^2$$
$$+ \tfrac{1}{2}\alpha_h (h_1\dot{\phi}_1 + h_2\dot{\phi}_2 + \ldots)^2 + \ldots\ldots\ldots(14)$$

[1] Routh's *Rigid Dynamics*, 5th edition, Part II. § 58.

and the addition to V

$$\tfrac{1}{2}\gamma_f\,(f_1\phi_1+f_2\phi_2+\ldots)^2 + \tfrac{1}{2}\gamma_g\,(g_1\phi_1+g_2\phi_2+\ldots)^2 + \ldots\ldots(15).$$

If we set

$$\alpha_f\lambda^2 + \gamma_f = F',\;\; \alpha_g\lambda^2 + \gamma_g = G',\;\ldots\ldots\ldots\ldots(16),$$

and so on, Lagrange's equations become

$$(a_1\lambda^2 + c_1)\,\phi_1 + F'f_1\,(f_1\phi_1+f_2\phi_2+\ldots)$$
$$+\,G'g_1\,(g_1\phi_1+g_2\phi_2+\ldots) + H'h_1\,(h_1\phi_1+h_2\phi_2+\ldots) + \ldots = 0\ldots(17)$$

$$(a_2\lambda^2 + c_2)\,\phi_2 + F'f_2\,(f_1\phi_1+f_2\phi_2+\ldots)$$
$$+\,G'g_2\,(g_1\phi_1+g_2\phi_2+\ldots) + H'h_2\,(h_1\phi_1+h_2\phi_2+\ldots) + \ldots = 0\ldots(18),$$

and so on, the number of equations being equal to the number (m) of co-ordinates $\phi_1,\ \phi_2\ \ldots.$ The number of additions (r), corresponding to the letters $f,\ g,\ h,\ \ldots$, is supposed to be less than m.

From the above m equations let r new ones be formed, as follows. For the first multiply (17) by $f_1/(a_1\lambda^2+c_1)$, (18) by $f_2/(a_2\lambda^2+c_2)$, and so on, and add the results together. For the second proceed in the same manner, using the multipliers $g_1/(a_1\lambda^2+c_1)$, $g_2/(a_2\lambda^2+c_2)$, &c. In like manner for the third equation use h instead of g, and so on. In this way we obtain r equations which may be written

$$F'\,(f_1\phi_1+f_2\phi_2+\ldots)\,\{1/F' + F_1^2 + F_2^2 + F_3^2+\ldots\}$$
$$+\,G'\,(g_1\phi_1+g_2\phi_2+\ldots)\,\{F_1G_1 + F_2G_2 + \ldots\}$$
$$+\,H'\,(h_1\phi_1+h_2\phi_2+\ldots)\,\{F_1H_1 + F_2H_2 + \ldots\} + \ldots\ldots = 0\ \ldots(19),$$

$$F'\,(f_1\phi_1+f_2\phi_2+\ldots)\,\{G_1F_1 + G_2F_2 + \ldots\}$$
$$+\,G'\,(g_1\phi_1+g_2\phi_2+\ldots)\,\{1/G' + G_1^2 + G_2^2 + \ldots\}$$
$$+\,H'\,(h_1\phi_1+h_2\phi_2+\ldots)\,\{G_1H_1 + G_2H_2 + \ldots\} + \ldots\ldots = 0\ldots(20),$$

and so on, where for brevity

$$\left.\begin{array}{l} F_1^2 = f_1^2/(a_1\lambda^2 + c_1),\;\; F_2^2 = f_2^2/(a_2\lambda^2 + c_2),\ \&c.,\\[4pt] G_1^2 = g_1^2/(a_1\lambda^2 + c_1),\;\; G_2^2 = g_2^2/(a_2\lambda^2 + c_2),\ \&c.\\[4pt] F_1G_1 = f_1g_1/(a_1\lambda^2 + c_1),\ \&c. \end{array}\right\}\ \ldots.(21).$$

The determinantal equation, of the rth order, is thus

$$\begin{vmatrix} 1/F' + \Sigma F^2, & \Sigma FG, & \Sigma FH,\ldots \\[4pt] \Sigma FG,\ 1/G' + \Sigma G^2, & \Sigma GH, \\[4pt] \Sigma FH, & \Sigma GH,\ 1/H' + \Sigma H^2,\ldots \\[4pt] \multicolumn{3}{c}{\ldots\ldots\ldots\ldots\ldots\ldots\ldots} \end{vmatrix} = 0\ldots\ldots(22).$$

If, for example, there be two additions to T and V of the kind prescribed, the equation is

$$\frac{1}{F'G'} + \frac{\Sigma F^2}{G'} + \frac{\Sigma G^2}{F'} + \Sigma F^2 . \Sigma G^2 - \{\Sigma FG\}^2 = 0 \quad \ldots\ldots\ldots(23),$$

and herein

$$(F_1{}^2 + F_2{}^2 + \ldots)(G_1{}^2 + G_2{}^2 + \ldots) - (F_1 G_1 + F_2 G_2 + \ldots)^2$$
$$= \Sigma\Sigma (F_1 G_2 - F_2 G_1)^2 \ldots\ldots\ldots\ldots(24).$$

Equation (23) is in general of the mth degree in λ^2, and determines the frequencies of vibration. In the extreme case where F' and G' are made infinite, the system is subject to the two constraints

$$\left. \begin{array}{l} f_1\phi_1 + f_2\phi_2 + \ldots = 0 \\ g_1\phi_1 + g_2\phi_2 + \ldots = 0 \end{array} \right\} \quad \ldots\ldots\ldots\ldots\ldots(25),$$

and the equation [1] giving the $(m-2)$ outstanding roots is

$$\frac{(f_1 g_2 - f_2 g_1)^2}{(a_1\lambda^2 + c_1)(a_2\lambda^2 + c_2)} + \frac{(f_1 g_3 - f_3 g_1)^2}{(a_1\lambda^2 + c_1)(a_3\lambda^2 + c_3)} + \ldots\ldots = 0 \ldots\ldots(26).$$

In general if the system be subject to the r constraints (1), the determinantal equation is

$$\begin{vmatrix} \Sigma FF, & \Sigma FG, & \Sigma FH, \ldots \\ \Sigma FG, & \Sigma GG, & \Sigma GH, \ldots \\ \Sigma FH, & \Sigma GH, & \Sigma HH, \ldots \\ \ldots\ldots & \ldots\ldots\ldots & \end{vmatrix} = 0 \quad \ldots\ldots\ldots (27).$$

If r be less than m, this determinant can be resolved [2] into a sum of squares of determinants of the same order (r). Thus if there be three constraints, the first of these squares is

$$\begin{vmatrix} F_1 & F_2 & F_3 \\ G_1 & G_2 & G_3 \\ H_1 & H_2 & H_3 \end{vmatrix}^2 \ldots\ldots\ldots\ldots\ldots(28),$$

and the others are to be found by including every combination of the m suffixes taken three together. To fall back upon the original notation we have merely in (28) to replace the capital letters $F, G, \ldots$ by $f, g, \ldots$, and to introduce the denominator

$$(a_1\lambda^2 + c_1)(a_2\lambda^2 + c_2)(a_3\lambda^2 + c_3).$$

The determinantal equation for a system originally of m degrees of freedom and subjected to r constraints is thus found. Its form

---

[1] This result is due to Routh, *loc. cit.* § 67.

[2] Salmon, *Lessons on Higher Algebra*, § 24.

is largely determined by the consideration that it must remain unaffected by interchanges either of the letters or of the suffixes.
That it would become nugatory if two of the conditions of constraint coincided, could also have been foreseen. If $r = m - 1$,
the system is reduced to one degree of freedom, and the equation
is

$$\begin{vmatrix} f_2 & f_3 & f_4 \cdots \\ g_2 & g_3 & g_4 \cdots \\ h_2 & h_3 & h_4 \cdots \\ \cdots\cdots\cdots \end{vmatrix}^2 (a_1\lambda^2 + c_1) + \begin{vmatrix} f_1 & f_3 & f_4 \cdots \\ g_1 & g_3 & g_4 \cdots \\ h_1 & h_3 & h_4 \cdots \\ \cdots\cdots\cdots \end{vmatrix}^2 (a_2\lambda^2 + c_2) + \ldots = 0 \ldots\ldots(29),$$

in agreement with § (88).

There are theories, parallel to the foregoing, for systems in
which T and F, or V and F, are alone sensible. In these cases, if
the functions be intrinsically positive, the normal motions are
proportional to exponential functions of the time such as $e^{-t/\tau}$.
The quantities τ_1, τ_2,... are called the time-constants, or persistences, of the motions, being the times occupied by the motions in
subsiding in the ratio of $e : 1$. The new persistences, after the
introduction of a constraint, will separate the original values.

The best illustrations of this theory are electrical, where the
motions are not restricted to be small. Suppose (to take an
electro-magnetic example) that in one branch of a net-work of
conductors there is introduced a coil of persistence (when closed
upon itself) equal to τ', the original persistences being τ_1, τ_2,....
Then the new persistences lie in all cases nearer to τ', and they
separate the quantities τ', τ_1, τ_2.... If τ' be made infinite as by
increasing the self-induction of the additional coil without limit,
or be made to vanish as by breaking the contact in the branch,
the result is a constraint, and the new values of the persistences
separate the former ones.

93. The determination of the coefficients to suit arbitrary
initial conditions may always be readily effected by the fundamental property of the normal functions, and it may be convenient
to sketch the process here for systems like strings, bars, membranes, plates, &c. in which there is only one dependent variable
ζ to be considered. If $u_1, u_2 \ldots$ be the normal functions, and
$\phi_1, \phi_2 \ldots$ the corresponding co-ordinates,

$$\zeta = \phi_1 u_1 + \phi_2 u_2 + \phi_3 u_3 + \ldots\ldots\ldots\ldots(1).$$

The equations of free motion are

$$\ddot{\phi}_1 + n_1^2\phi_1 = 0, \quad \ddot{\phi}_2 + n_2^2\phi_2 = 0, \&c. \ \dots\dots\dots\dots(2),$$

of which the solutions are

$$\left. \begin{aligned} \phi_1 &= A_1 \sin n_1 t + B_1 \cos n_1 t \\ \phi_2 &= A_2 \sin n_2 t + B_2 \cos n_2 t \\ &\dots\dots\dots\dots\dots\dots\dots \end{aligned} \right\} \ \dots\dots\dots\dots(3).$$

The initial values of ζ and $\dot{\zeta}$ are therefore

$$\left. \begin{aligned} \zeta_0 &= B_1 u_1 + B_2 u_2 + B_3 u_3 + \dots \\ \dot{\zeta}_0 &= n_1 A_1 u_1 + n_2 A_2 u_2 + n_3 A_3 u_3 + \dots \end{aligned} \right\} \ \dots\dots\dots (4).$$

and the problem is to determine $A_1, A_2, \dots B_1, B_2 \dots$ so as to correspond with arbitrary values of ζ_0 and $\dot{\zeta}_0$.

If $\rho\, dx$ be the mass of the element dx, we have from (1)

$$T = \tfrac{1}{2} \int \rho \dot{\zeta}^2 dx$$

$$= \tfrac{1}{2} \dot{\phi}_1^2 \int \rho\, u_1^2 dx + \tfrac{1}{2} \dot{\phi}_2^2 \int \rho\, u_2^2 dx + \dots + \dot{\phi}_1 \dot{\phi}_2 \int \rho u_1 u_2 dx + \dots.$$

But the expression for T in terms of $\dot{\phi}_1$, $\dot{\phi}_2$, &c. cannot contain the products of the normal generalized velocities, and therefore every integral of the form

$$\int \rho\, u_r u_s dx = 0 \ \dots\dots\dots\dots\dots\dots(5).$$

Hence to determine B_r we have only to multiple the first of equations (4) by ρu_r and integrate over the system. We thus obtain

$$B_r \int \rho\, u_r^2 dx = \int \rho\, u_r \zeta_0 dx \ \dots\dots\dots\dots\dots(6).$$

Similarly, $\quad n_r A_r \int \rho\, u_r^2 dx = \int \rho\, u_r \dot{\zeta}_0 dx \ \dots\dots\dots \ \dots\dots(7).$

The process is just the same whether the element dx be a line area, or volume.

The conjugate property, expressed by (5), depends upon the fact that the functions u are normal. As soon as this is known by the solution of a differential equation or otherwise, we may infer the conjugate property without further proof, but the property itself is most intimately connected with the fundamental variational equation of motion § 94.

94. If V be the potential energy of deformation, ζ the displacement, and ρ the density of the (line, area, or volume) element dx, the equation of virtual velocities gives immediately

$$\delta V + \int \rho \ddot{\zeta}\,\delta\zeta\,dx = 0 \ldots\ldots\ldots\ldots\ldots\ldots(1).$$

In this equation δV is a symmetrical function of ζ and $\delta\zeta$, as may be readily proved from the expression for V in terms of generalized co-ordinates. In fact if

$$V = \tfrac{1}{2}c_{11}\psi_1^2 + \ldots + c_{12}\psi_1\psi_2 + \ldots,$$
$$\delta V = c_{11}\psi_1\delta\psi_1 + c_{22}\psi_2\delta\psi_2 + \ldots$$
$$+ c_{12}(\psi_1\delta\psi_2 + \psi_2\delta\psi_1) + \ldots.$$

Suppose now that ζ refers to the motion corresponding to a normal function u_r, so that $\ddot{\zeta} + n_r^2\zeta = 0$, while $\delta\zeta$ is identified with another normal function u_s; then

$$\delta V = n_r^2 \int \rho\, u_r u_s\, dx.$$

Again, if we suppose, as we are equally entitled to do, that ζ varies as u_s and $\delta\zeta$ as u_r, we get for the same quantity δV,

$$\delta V = n_s^2 \int \rho\, u_r u_s\, dx;$$

and therefore $\qquad (n_r^2 - n_s^2)\int \rho\, u_r u_s\, dx = 0 \ldots\ldots\ldots\ldots\ldots(2),$

from which the conjugate property follows, if the motions represented respectively by u_r and u_s have different periods.

A good example of the connection of the two methods of treatment will be found in the chapter on the transverse vibrations of bars.

95. Professor Stokes[1] has drawn attention to a very general law connecting those parts of the free motion which depend on the initial *displacements* of a system not subject to frictional forces, with those which depend on the initial *velocities*. If a velocity of any type be communicated to a system at rest, and then after a small interval of time the opposite velocity be communicated, the effect in the limit will be to start the system without velocity, but with a displacement of the corresponding type. We may readily prove from this that in order

[1] *Dynamical Theory of Diffraction, Cambridge Trans.* Vol. IX. p. 1, 1856.

to deduce the motion depending on initial displacements from that depending on the initial velocities, it is only necessary to differentiate with respect to the time, and to replace the arbitrary constants (or functions) which express the initial velocities by those which express the corresponding initial displacements.

Thus, if ϕ be any normal co-ordinate satisfying the equation

$$\ddot{\phi} + n^2\phi = 0,$$

the solution in terms of the initial values of ϕ and $\dot{\phi}$ is

$$\phi = \phi_0 \cos nt + \frac{1}{n} \dot{\phi}_0 \sin nt \dots\dots\dots\dots\dots(1),$$

of which the first term may be obtained from the second by Stokes' rule.

CHAPTER V.

96. WHEN dissipative forces act upon a system, the character of the motion is in general more complicated. If two only of the functions T, F, and V be finite, we may by a suitable linear transformation rid ourselves of the products of the co-ordinates, and obtain the normal types of motion. In the preceding chapter we have considered the case of $F = 0$. The same theory with obvious modifications will apply when $T = 0$, or $V = 0$, but these cases though of importance in other parts of Physics, such as Heat and Electricity, scarcely belong to our present subject.

The presence of friction will not interfere with the reduction of T and V to sums of squares; but the transformation proper for them will not in general suit also the requirements of F. The general equation can then only be reduced to the form

$$a_1\ddot{\phi}_1 + b_{11}\,\dot{\phi}_1 + b_{12}\,\dot{\phi}_2 + \ldots + c_1\phi_1 = \Phi_1, \quad \&\text{c.} \ldots\ldots\ldots(1),$$

and not to the simpler form applicable to a system of one degree of freedom, viz.

$$a_1\phi_1 + b_1\dot{\phi}_1 + c_1\phi_1 = \Phi_1, \quad \&\text{c.} \ldots\ldots\ldots\ldots(2).$$

We may, however, choose which pair of functions we shall reduce, though in Acoustics the choice would almost always fall on T and V.

97. There is, however, a not unimportant class of cases in which the reduction of all three functions may be effected; and the theory then assumes an exceptional simplicity. Under this head the most important are probably those when F is of the same form as T or V. The first case occurs frequently, in books at any rate, when the motion of each part of the system is resisted by a retarding force, proportional both to the mass and velocity of the part. The same exceptional reduction is possible when F is a linear function of T and V, or when T is itself of the same form as V. In any of these cases, the equations of motion are of the same form as for a system of one degree of freedom, and the theory possesses certain peculiarities which make it worthy of separate consideration.

The equations of motion are obtained at once from T, F and V:—

$$\left. \begin{array}{l} a_1\ddot{\phi}_1 + b_1\dot{\phi}_1 + c_1\phi_1 = \Phi_1, \\ a_2\ddot{\phi}_2 + b_2\dot{\phi}_2 + c_2\phi_2 = \Phi_2, \text{ \&c.} \end{array} \right\} \dots\dots\dots\dots(1),$$

in which the co-ordinates are separated.

For the free vibrations we have only to put $\Phi_1 = 0$, &c., and the solution is of the form

$$\phi = e^{-\frac{1}{2}\kappa t}\left\{\dot{\phi}_0\,\frac{\sin n't}{n'} + \phi_0\left(\cos n't + \frac{\kappa}{2n'}\sin n't\right)\right\}\dots\dots(2),$$

where $\qquad \kappa = b/a, \quad n^2 = c/a, \quad n' = \surd(n^2 - \tfrac{1}{4}\kappa^2),$

and ϕ_0 and $\dot{\phi}_0$ are the initial values of ϕ and $\dot{\phi}$.

The whole motion may therefore be analysed into component motions, each of which corresponds to the variation of but one normal co-ordinate at a time. And the vibration in each of these modes is altogether similar to that of a system with only one degree of liberty. After a certain time, greater or less according to the amount of dissipation, the free vibrations become insignificant, and the system returns sensibly to rest.

[If F be of the same form as T, all the values of κ are equal, viz. all vibrations die out at the same rate.]

Simultaneously with the free vibrations, but in perfect independence of them, there may exist forced vibrations depending on the quantities Φ. Precisely as in the case of one degree of freedom, the solution of

$$a\ddot{\phi} + b\dot{\phi} + c\phi = \Phi \dots\dots\dots\dots\dots\dots(3)$$

may be written

$$\phi = \frac{1}{n'}\int_0^t e^{-\frac{1}{2}\kappa(t-t')}\sin n'(t-t')\,\Phi\,dt'\dots\dots\dots(4),$$

where as above

$$\kappa = b/a, \quad n^2 = c/a, \quad n' = \surd(n^2 - \tfrac{1}{4}\kappa^2).$$

To obtain the complete expression for ϕ we must add to the right-hand member of (4), which makes the initial values of ϕ and $\dot{\phi}$ vanish, the terms given in (2) which represent the residue at time t of the initial values ϕ_0 and $\dot{\phi}_0$. If there be no friction, the value of ϕ in (4) reduces to

$$\phi = \frac{1}{n}\int_0^t \sin n\,(t-t')\,\Phi\,dt' \dots\dots\dots\dots (5).$$

98. The complete independence of the normal co-ordinates leads to an interesting theorem concerning the relation of the subsequent motion to the initial disturbance. For if the forces which act upon the system be of such a character that they do no work on the displacement indicated by $\delta\phi_1$, then $\Phi_1 = 0$. No such forces, however long continued, can produce any effect on the motion ϕ_1. If it exist, they cannot destroy it; if it do not exist, they cannot generate it. The most important application of the theorem is when the forces applied to the system act at a node of the normal component ϕ_1, that is, at a point which the component vibration in question does not tend to set in motion. Two extreme cases of such forces may be specially noted, (1) when the force is an impulse, starting the system from rest, (2) when it has acted so long that the system is again at rest under its influence in a disturbed position. So soon as the force ceases, natural vibrations set in, and in the absence of friction would continue for an indefinite time. We infer that whatever in other respects their character may be, they contain no component of the type ϕ_1. This conclusion is limited to cases where T, F, V admit of simultaneous reduction, including of course the case of no friction.

99. The formulæ quoted in § 97 are applicable to any kind of force, but it will often happen that we have to deal only with the effects of impressed forces of the harmonic type, and we may then advantageously employ the more special formulæ applicable to such forces. In using normal co-ordinates, we have first to calculate the forces Φ_1, Φ_2, &c. corresponding to each period, and thence deduce the values of the co-ordinates themselves. If among the natural periods (calculated without allowance for friction) there be any nearly agreeing in magnitude with the period of an impressed force, the corresponding component vibrations will be abnormally large, unless indeed the force itself be greatly attenuated in the preliminary resolution. Suppose, for example, that a transverse force of harmonic type and given period acts at a single point of a stretched string. All the normal modes of vibration will, in general, be excited, not however in their own proper periods, but in the period of the impressed force; but any normal component, which has a node at the point of application, will not be excited. The magnitude of each component thus depends on two things: (1) on the situation of its nodes with respect to the point at which the force is applied, and (2) on the degree of agreement between its own proper period and that of the force. It is important to remember that in response to a simple harmonic force, the system will vibrate in general in *all* its modes, although in particular cases it may sometimes be sufficient to attend to only one of them as being of paramount importance.

100. When the periods of the forces operating are very long relatively to the free periods of the system, an equilibrium theory is sometimes adequate, but in such a case the solution could generally be found more easily without the use of the normal co-ordinates. Bernoulli's theory of the Tides is of this class, and proceeds on the assumption that the free periods of the masses of water found on the globe are small relatively to the periods of the operative forces, in which case the inertia of the water might be left out of account. As a matter of fact this supposition is only very roughly and partially applicable, and we are consequently still in the dark on many important points relating to the tides. The principal forces have a semi-diurnal period, which is not sufficiently long in relation to the natural periods concerned, to allow of the inertia of the water being neglected. But if the rotation of the earth had been much slower, the equilibrium theory of the tides might have been adequate.

A corrected equilibrium theory is sometimes useful, when the period of the impressed force is sufficiently long in comparison with most of the natural periods of a system, but not so in the case of one or two of them. It will be sufficient to take the case where there is no friction. In the equation

$$a\ddot{\phi} + c\phi = \Phi, \quad \text{or} \quad \ddot{\phi} + n^2\phi = \Phi/a,$$

suppose that the impressed force varies as $\cos pt$. Then

$$\phi = \Phi \div a\,(n^2 - p^2) \dots\dots\dots\dots\dots (1).$$

The equilibrium theory neglects p^2 in comparison with n^2, and takes

$$\phi = \Phi \div an^2 \dots\dots\dots\dots\dots\dots (2).$$

Suppose now that this course is justifiable, except in respect of the single normal co-ordinate ϕ_1. We have then only to add to the result of the equilibrium theory, the difference between the true and the there assumed value of ϕ_1, viz.

$$\phi_1 = \frac{\Phi_1}{a_1(n_1^2 - p^2)} - \frac{\Phi_1}{a_1 n_1^2} = \frac{p^2}{n_1^2 - p^2} \cdot \frac{\Phi_1}{a} \dots\dots(3).$$

The other extreme case ought also to be noticed. If the forced vibrations be extremely rapid, they may become nearly independent of the potential energy of the system. Instead of neglecting p^2 in comparison with n^2, we have then to neglect n^2 in comparison with p^2, which gives

$$\phi = - \Phi \div ap^2 \dots\dots\dots(4).$$

If there be one or two co-ordinates to which this treatment is not applicable, we may supplement the result, calculated on the hypothesis that V is altogether negligible, with corrections for these particular co-ordinates.

101. Before passing on to the general theory of the vibrations of systems subject to dissipation, it may be well to point out some peculiarities of the free vibrations of continuous systems, started by a force applied at a single point. On the suppositions and notations of § 93, the configuration at any time is determined by

$$\zeta = \phi_1 u_1 + \phi_2 u_2 + \phi_3 u_3 + \dots\dots\dots(1),$$

where the normal co-ordinates satisfy equations of the form

$$a_r \ddot{\phi}_r + c_r \phi_r = \Phi_r \dots\dots\dots(2).$$

Suppose now that the system is held at rest by a force applied at the point Q. The value of Φ_r is determined by the consideration that $\Phi_r \delta\phi_r$ represents the work done upon the system by the impressed forces during a hypothetical displacement $\delta\zeta = \delta\phi_r u_r$, that is

$$\delta\phi_r \int Z u_r \, dx \, ;$$

thus $$\Phi_r = \int Z u_r \, dx = u_r(Q) \int Z \, dx \, ;$$

so that initially by (2)

$$c_r \phi_r = u_r(Q) \int Z \, dx \dots\dots\dots(3).$$

If the system be let go from this configuration at $t = 0$, we have at any subsequent time t,

$$\phi_r = \cos n_r t \frac{u_r(Q)\int Z dx}{c_r} = \cos n_r t \frac{u_r(Q)\int Z dx}{n_r^2 \int \rho \, u_r^2 dx} \dots\dots(4),$$

and at the point P

$$\zeta = \Sigma \cos n_r t \frac{u_r(P) \, u_r(Q)\int Z dx}{n_r^2 \int \rho \, u_r^2 dx} \dots\dots\dots(5).$$

At particular points $u_r(P)$ and $u_r(Q)$ vanish, but on the whole

$$u_r(P) \, u_r(Q) \div \int \rho \, u_r^2 dx$$

neither converges, nor diverges, with r. The series for ζ therefore converges with n_r^{-2}.

Again, suppose that the system is started by an impulse from the configuration of equilibrium. In this case initially

$$a_r \dot{\phi}_r = \int \Phi_r dt = u_r(Q) \int Z_1 dx,$$

whence at time t

$$\phi_r = \frac{\sin n_r t}{a_r n_r} . u_r(Q) . \int Z_1 dx = \frac{\sin n_r t . u_r(Q)}{n_r \int \rho u_r^2 dx} \int Z_1 dx \dots\dots(6).$$

This gives

$$\zeta = \Sigma \sin n_r t \frac{u_r(P) \, u_r(Q)\int Z_1 dx}{n_r \int \rho u_r^2 dx} \dots\dots\dots\dots(7),$$

shewing that in this case the series converges with n_r^{-1}, that is more slowly than in the previous case.

In both cases it may be observed that the value of ζ is symmetrical with respect to P and Q, proving that the displacement at time t for the point P when the force or impulse is applied at Q, is the same as it would be at Q if the force or impulse had been applied at P. This is an example of a very general reciprocal theorem, which we shall consider at length presently.

As a third case we may suppose the body to start from rest as deformed by a force *uniformly distributed*, over its length, area, or volume. We readily find

$$\zeta = \Sigma \cos n_r t \; \frac{u_r(P) . Z . \int u_r dx}{n_r{}^2 \int \rho u_r{}^2 dx} \dots\dots\dots\dots(8).$$

The series for ζ will be more convergent than when the force is concentrated in a single point.

In exactly the same way we may treat the case of a continuous body whose motion is subject to dissipation, provided that the three functions T, F, V be simultaneously reducible, but it is not necessary to write down the formulæ.

102. If the three mechanical functions T, F and V of any system be not simultaneously reducible, the natural vibrations (as has already been observed) are more complicated in their character. When, however, the dissipation is small, the method of reduction is still useful; and this class of cases besides being of some importance in itself will form a good introduction to the more general theory. We suppose then that T and V are expressed as sums of squares

$$\left. \begin{array}{l} T = \tfrac{1}{2} a_1 \dot{\phi}_1{}^2 + \tfrac{1}{2} a_2 \dot{\phi}_2{}^2 + \dots \\ V = \tfrac{1}{2} c_1 \phi_1{}^2 + \tfrac{1}{2} c_2 \phi_2{}^2 + \dots \end{array} \right\} \dots\dots\dots (1),$$

while F still appears in the more general form

$$F = \tfrac{1}{2} b_{11} \dot{\phi}_1{}^2 + \tfrac{1}{2} b_{22} \dot{\phi}_2{}^2 + \dots + b_{12} \dot{\phi}_1 \dot{\phi}_2 + \dots\dots\dots\dots(2).$$

The equations of motion are accordingly

$$\left. \begin{array}{l} a_1 \ddot{\phi}_1 + b_{11} \dot{\phi}_1 + b_{12} \dot{\phi}_2 + b_{13} \dot{\phi}_3 + \dots + c_1 \phi_1 = 0 \\ a_2 \ddot{\phi}_2 + b_{21} \dot{\phi}_1 + b_{22} \dot{\phi}_2 + b_{23} \dot{\phi}_3 + \dots + c_2 \phi_2 = 0 \\ \dots\dots\dots\dots\dots\dots\dots\dots\dots\dots\dots\dots\dots \end{array} \right\} \dots\dots\dots(3),$$

in which the coefficients b_{11}, b_{12}, &c. are to be treated as small. If there were no friction, the above system of equations would be satisfied by supposing one co-ordinate ϕ_r to vary suitably, while the other co-ordinates vanish. In the actual case there will be a corresponding solution in which the value of any other co-ordinate ϕ_s will be small relatively to ϕ_r.

Hence, if we omit terms of the second order, the rth equation becomes,

$$a_r \ddot{\phi}_r + b_{rr} \dot{\phi}_r + c_r \phi_r = 0 \dots\dots\dots\dots(4),$$

from which we infer that ϕ_r varies approximately as if there were no change due to friction in the type of vibration. If ϕ_r vary as e^{prt}, we obtain to determine p_r

$$a_r p_r^2 + b_{rr} p_r + c_r = 0 \quad \dots\dots\dots\dots\dots\dots (5).$$

The roots of this equation are complex, but the real part is small in comparison with the imaginary part. [The character of the motion represented by (5) has already been discussed (§ 45). The rate at which the vibrations die down is proportional to b_{rr}, and the period, if the term be still admitted, is approximately the same as if there were no dissipation.]

From the sth equation, if we introduce the supposition that all the co-ordinates vary as e^{prt}, we get

$$(p_s^2 a_s + c_s)\, \phi_s + b_{rs} p_r \phi_r = 0,$$

terms of the second order being omitted; whence

$$\phi_s : \phi_r = -\frac{b_{rs} p_r}{p_r^2 a_s + c_s} = \frac{b_{rs} p_r}{a_s (p_s^2 - p_r^2)} \quad \dots\dots\dots (6).$$

This equation determines approximately the altered type of vibration. Since the chief part of p_r is imaginary, we see that the co-ordinates ϕ_s are approximately in the same phase, *but that that phase differs by a quarter period from the phase of ϕ_r.* Hence when the function F does not reduce to a sum of squares, the character of the elementary modes of vibration is less simple than otherwise, and the various parts of the system are no longer simultaneously in the same phase.

We proved above that, when the friction is small, the value of p_r may be calculated approximately without allowance for the change of type; but by means of (6) we may obtain a still closer approximation, in which the squares of the small quantities are retained. The rth equation (3) gives

$$a_r p_r^2 + c_r + b_{rr} p_r + \Sigma \frac{p_r^2 b_{rs}^2}{a_s (p_s^2 - p_r^2)} = 0 \quad \dots\dots\dots\dots (7).$$

The leading part of the terms included under Σ being real, the correction has no effect on the real part of p_r on which the rate of decay depends.

102 a. Following the electrical analogy we may conveniently describe the forces expressed by F as forces of *resistance*. In § 102 we have seen that if the resistances be small, the periods are independent of them. We may therefore extend to this case the application of the theorems with regard to the effect upon the periods of additions to T and V, which have been already proved when there are no resistances.

By (5) § 102, if the forces of resistance be increased, the rates of subsidence of all the normal motions are in general increased with them; but in particular cases it may happen that there is no change in a rate of subsidence.

It is natural to inquire whether this conclusion is limited to *small* resistances, for at first sight it would appear likely to hold good generally. An argument sufficient to decide this question may be founded upon a particular case. Consider a system formed by attaching two loads at any points of a stretched string vibrating transversely. If the mass of the string itself be neglected, there are two degrees of freedom and two periods of vibration corresponding to two normal modes. In each of these modes both loads in general vibrate. Now suppose that a force of resistance is introduced retarding the motion of *one* of the loads, and that this force gradually increases. At first the effect is to cause both kinds of vibration to die out and that at an increasing rate, but afterwards the law changes. For when the resistance becomes infinite, it is equivalent to a *constraint,* holding at rest the load upon which it acts. The remaining vibration is then unaffected by resistance, and maintains itself indefinitely. Thus the rate of subsidence of one of the normal modes has decreased to evanescence in spite of a continual increase in the forces of resistance F. This case is of course sufficient to disprove the suggested general theorem.

103. We now return to the consideration of the general equations of § 84.

If ψ_1, ψ_2, &c. be the co-ordinates and Ψ_1, Ψ_2, &c. the forces, we have

$$\left. \begin{array}{l} e_{11}\psi_1 + e_{12}\psi_2 + \ldots = \Psi_1 \\ e_{21}\psi_1 + e_{22}\psi_2 + \ldots = \Psi_2, \&c. \end{array} \right\} \quad \ldots\ldots\ldots\ldots\ldots(1),$$

where
$$e_{rs} = a_{rs}D^2 + b_{rs}D + c_{rs} \ldots\ldots\ldots\ldots\ldots\ldots(2).$$

For the free vibrations Ψ_1, &c. vanish. If ∇ be the determinant

$$\nabla = \left| \begin{array}{l} e_{11}, \ e_{12}, \ldots \\ e_{21}, \ e_{22}, \ldots \\ \ldots\ldots\ldots\ldots \end{array} \right| \quad \ldots\ldots\ldots\ldots\ldots\ldots\ldots(3)$$

the result of eliminating from (1) all the co-ordinates but one, is

$$\nabla \psi = 0 \dots\dots\dots\dots\dots\dots\dots(4).$$

Since ∇ now contains odd powers of D, the $2m$ roots of the equation $\nabla = 0$ no longer occur in equal positive and negative pairs, but contain a real as well as an imaginary part. The complete integral may however still be written

$$\psi = A e^{\mu_1 t} + A' e^{\mu_1' t} + B e^{\mu_2 t} + B' e^{\mu_2' t} + \dots\dots\dots(5),$$

where the pairs of conjugate roots are μ_1, μ_1'; μ_2, μ_2'; &c. Corresponding to each root, there is a particular solution such as

$$\psi_1 = A_1 e^{\mu_1 t}, \quad \psi_2 = A_2 e^{\mu_1 t}, \quad \psi_3 = A_3 e^{\mu_1 t}, \text{ &c.,}$$

in which the *ratios* $A_1 : A_2 : A_3 \dots$ are determined by the equations of motion, and only the absolute value remains arbitrary. In the present case however (where ∇ contains odd powers of D) these ratios are not in general real, and therefore the variations of the co-ordinates ψ_1, ψ_2, &c. are not synchronous in phase. If we put $\mu_1 = \alpha_1 + i\beta_1$, $\mu_1' = \alpha_1 - i\beta_1$, &c., we see that none of the quantities α can be positive, since in that case the energy of the motion would increase with the time, as we know it cannot do.

103 a. The general argument (§§ 85, 103) from considerations of energy as to the nature of the roots of the determinantal equation (Thomson and Tait's *Natural Philosophy*, 1st edition 1867) has been put into a more mathematical form by Routh[1]. His investigation relates to the most general form of the equation in which the relations § 82

$$a_{rs} = a_{sr}, \qquad b_{rs} = b_{sr}, \qquad c_{rs} = c_{sr} \dots\dots\dots\dots (1),$$

are not assumed. But for the sake of brevity and as sufficient for almost all acoustical problems, these relations will here be supposed to hold.

We shall have occasion to consider two solutions corresponding to two roots μ, ν of the equation. For the first we have

$$\psi_1 = M_1 e^{\mu t}, \ \psi_2 = M_2 e^{\mu t}, \ \psi_3 = M_3 e^{\mu t}, \text{ &c.} \dots\dots\dots(2),$$

and for the second

$$\psi_1 = N_1 e^{\nu t}, \ \psi_2 = N_2 e^{\nu t}, \ \psi_3 = N_3 e^{\nu t}, \text{ &c.} \dots\dots\dots (3).$$

In either of these solutions, for example (2), the ratios

$$M_1 : M_2 : M_3 : \dots\dots$$

[1] *Rigid Dynamics*, 5th edition, Ch. VII.

are determinate when μ has been chosen. They are real when μ is real; and when μ is complex $(\alpha \pm i\beta)$, they take the form $P \pm iQ$.

If now we substitute the values of ψ from (2) in the equations of motion, we get

$$\left. \begin{aligned} (a_{11}\mu^2 + b_{11}\mu + c_{11}) M_1 + (a_{12}\mu^2 + b_{12}\mu + c_{12}) M_2 + \ldots\ldots = 0 \\ (a_{12}\mu^2 + b_{12}\mu + c_{12}) M_1 + (a_{22}\mu^2 + b_{22}\mu + c_{22}) M_2 + \ldots\ldots = 0 \\ \ldots\ldots\ldots\ldots\ldots\ldots\ldots\ldots\ldots\ldots\ldots\ldots\ldots\ldots\ldots\ldots\ldots\ldots\ldots \end{aligned} \right\} \ldots(4).$$

The first result is obtained by multiplying these equations in order by M_1, M_2, &c. and adding. It may be written

$$A\mu^2 + B\mu + C = 0, \ldots\ldots\ldots\ldots\ldots (5),$$

where

$$A = \tfrac{1}{2}a_{11}M_1^2 + \tfrac{1}{2}a_{22}M_2^2 + a_{12}M_1M_2 + \ldots\ldots\ldots (6),$$
$$B = \tfrac{1}{2}b_{11}M_1^2 + \tfrac{1}{2}b_{22}M_2^2 + b_{12}M_1M_2 + \ldots\ldots\ldots (7),$$
$$C = \tfrac{1}{2}c_{11}M_1^2 + \tfrac{1}{2}c_{22}M_2^2 + c_{12}M_1M_2 + \ldots\ldots\ldots (8).$$

The functions A, B, C, are, it will be seen, the same as we have already denoted by T, F, and V respectively; but the varied notation may be useful as reminding us that there is as yet no limitation upon the nature of these quadratic functions.

The following inferences from (5) are drawn by Routh:—

(α) If A, B, C either be zero, or be one-signed functions of the same sign, the fundamental determinant cannot have a real positive root. For if μ were real, the coefficients M_1, M_2,...... would be real. We should thus have the sum of three positive quantities equal to zero.

(β) If there be no forces of resistance, i.e. if the term B be absent, and if A and C be one-signed and have the same sign, the fundamental determinant cannot have a real root, positive or negative.

(γ) If A, B, C be one-signed functions, but if the sign of B be opposite to that of A and C, the fundamental determinant cannot have a real negative root.

The second equation is obtained as before from (4), except that now the multipliers are N_1, N_2,... appropriate to the root ν. The result may be written

$$A(\mu, \nu)\mu^2 + B(\mu, \nu)\mu + C(\mu, \nu) = 0 \ldots\ldots\ldots\ldots(9),$$

where

$$2A(\mu, \nu) = a_{11}M_1N_1 + a_{22}M_2N_2 + \ldots\ldots$$
$$+ a_{12}(M_1N_2 + M_2N_1) + \ldots\ldots\ldots\ldots(10),$$

with similar suppositions for $B(\mu, \nu)$ and $C(\mu, \nu)$. $A(\mu, \nu)$ is thus a symmetrical function of the M's and N's, so that

$$A(\mu, \nu) = A(\nu, \mu) \quad\dots\dots\dots\dots\dots\dots(11).$$

It will be observed that according to this notation $A(\mu, \mu)$ is the same as A in (6).

In like manner

$$A(\mu, \nu)\,\nu^2 + B(\mu, \nu)\,\nu + C(\mu, \nu) = 0 \quad\dots\dots\dots\dots(12),$$

shewing that μ, ν are both roots of the quadratic, whose co-efficients are $A(\mu, \nu)$, $B(\mu, \nu)$, $C(\mu, \nu)$. Accordingly

$$u + \nu = -\frac{B(\mu, \nu)}{A(\mu, \nu)}, \qquad \mu\nu = \frac{C(\mu, \nu)}{A(\mu, \nu)} \quad\dots\dots\dots\dots(13).$$

We will now suppose that μ, ν are two conjugate complex roots, so that

$$\mu = \alpha + i\beta, \qquad \nu = \alpha - i\beta,$$

where α, β are real. Under these circumstances if M_1, M_2, ... be $P_1 + iQ_1$, $P_2 + iQ_2$, ..., then $N_1, N_2, ...$ will be $P_1 - iQ_1$, $P_2 - iQ_2$, ..., the P's and Q's being real. Thus by (10)

$$2A(\mu, \nu) = a_{11}(P_1^2 + Q_1^2) + a_{22}(P_2^2 + Q_2^2) + \dots\dots$$
$$+ 2a_{12}(P_1P_2 + Q_1Q_2) + \dots\dots$$
$$= 2A(P) + 2A(Q) \quad\dots\dots\dots\dots\dots\dots(14).$$

In (14) $A(P)$, $A(Q)$ are functions, such as (6), of real variables. From (13) we now find

$$2\alpha = -\frac{B(P) + B(Q)}{A(P) + A(Q)} \quad\dots\dots\dots\dots(15),$$

$$\alpha^2 + \beta^2 = \frac{C(P) + C(Q)}{A(P) + A(Q)} \quad\dots\dots\dots\dots(16).$$

From these Routh deduces the following conclusions:—

(δ) If A and B be one-signed and have the same sign (whether C be a one-signed function or not), then the real part α of every imaginary root must be negative and not zero. But if B be absent, then the real part of every imaginary root is zero.

(ϵ) If A and C be one-signed and have opposite signs, then whatever may be the character of B, there can be no imaginary roots.

It may be remarked that if B do not occur, and if μ^2 and ν^2 be different roots of the determinant, it follows from (9), (12) that

$$A(\mu, \nu) = C(\mu, \nu) = 0\dots\dots\dots\dots\dots\dots(17).$$

When the number of degrees of freedom is finite, the fundamental determinant may be expanded in powers of μ, giving an equation $f(\mu) = 0$ of degree $2m$. The condition of stability is that all the real roots and the real parts of all the complex roots should be negative. If, as usual, complex quantities $x + iy$ be represented by points whose co-ordinates are x, y, the condition is that all points representing roots should lie to the left of the axis of y. The application of Cauchy's rule relative to the number of roots within any contour, by taking as the contour the infinite semi-circle on the positive side of the axis of y, is very fully discussed by Routh[1], who has thrown the results into forms convenient for practical application to particular cases.

103 b. The theorems of § 103 a do not exhaust all that general mechanical principles would lead us to expect as to the character of the roots of the fundamental determinant, and it may be well to pursue the question a little further. We will suppose throughout that A is one-signed and *positive*.

If B and C be both one-signed and positive, we see that the equilibrium is thoroughly stable; for from (α) it follows that there can be no positive root, and from (δ) that no complex root can have its real part positive.

In like manner the equations of § 103 a suffice for the case where C is one-signed and positive, B one-signed and negative. By (5) every real root is positive, and by (15) the real part of every complex root. Hence the equilibrium is unstable in *every* mode.

When C is one-signed and negative, all the roots are real (ϵ); but (5) does not tell us whether they are positive or negative. When $B = 0$, we know (§ 87) that the roots occur in pairs of equal numerical value and of opposite sign. In this case therefore there are m positive and m negative roots. We will prove that this state of things cannot be disturbed by B. For if the determinant be expanded, the coefficient of μ^{2m} is the discriminant of A, and the coefficient of μ^0 is the discriminant of C. By supposition neither of these quantities is zero, and thus no root of the equation can be other than finite. Hence as B increases from zero to its actual magnitude as a function of the variables, no root of the equation can change sign, and accordingly there remain m

---

[1] Adams Prize Essay 1877; *Rigid Dynamics* § 290.

positive and m negative roots. It should be noticed that in this argument there is no restriction upon the character of B.

In the case of a real root the values of M_1, M_2, ... are real, and thus the motion is such as might take place under a constraint reducing the system to one degree of freedom. But if this constraint were actually imposed, there would be *two* corresponding values of μ, being the values given by (5). In general only one of these is applicable to the question in hand. Otherwise it would be possible to define m kinds of constraint, one or other of which would be consistent with any of the $2m$ roots. But this could only happen when the *three* functions A, B, C are simultaneously reducible to sums of squares (§ 97).

When $B = 0$, there are m modes of motion, and two roots for each mode. In the present application to the case where C is one-signed and negative, each of the m modes for $B = 0$ gives one positive and one negative root. The positive root denotes instability, and although the negative root gives a motion which diminishes without limit, the character of instability is considered to attach to the mode as a whole, and all the m modes are said to be unstable. But when B is finite, there are in general $2m$ distinct modes with one root corresponding to each. Of the $2m$ modes m are unstable, but the remaining m modes must be reckoned as stable. On the whole, however, the equilibrium is unstable, so that the influence of B, even when positive, is insufficient to obviate the instability due to the character of C.

We must not prolong much further our discussion of unstable systems, but there is one theorem respecting real roots too fundamental to be passed over. It may be regarded as an extension of that of § 88.

The value of μ corresponding to a given constraint $M_1 : M_2 : ...$ is one of the roots of (5): and it follows from (4) that the value of μ is stationary when the imposed constraint coincides with one of the modes of free motion. The effect of small changes in A, B, C may thus be calculated from (5) without allowance for the accompanying change of type.

Let C, being negative for the mode under consideration, undergo numerical increase, while A and B remain unchanged as functions of the co-ordinates. The latter condition requires that the roots of (5), one of which is positive and one negative, should move either both towards zero or both away from zero; and the first condition excludes the former alternative. Whether it be the positive or the negative root of (5) which is the root of the determinant, we infer that the change in question causes the latter to move away from zero.

In like manner if A increase, while B and C remain unchanged, the movement of the root, whether positive or negative, is necessarily towards zero.

Again, if A and C be given, while B increases algebraically as a function of the variables, the movement of the root of the determinant must be in the negative direction.

An algebraic increase in B thus increases the stability, or decreases the instability, in every mode. A numerical increase in C or decrease in A on the other hand promotes the stability of the stable modes and the instability of the unstable modes.

We can do little more than allude to the theorem relating to the effect of a single constraint upon a system for which C is one-signed and negative. Whatever be the nature of B, the $(m-1)$ positive roots of the determinant, appropriate to the system after the constraint has been applied, will separate the m positive roots of the original determinant, and a like proposition will hold for the negative roots. Upon this we may found a generalization of the foregoing conclusions analogous to that of § 92 *a*. Consider an independent vibrator of one degree of freedom for which C is positive, and let the roots of the frequency equation be ν_1, ν_2, one negative and one positive. If we regard this as forming part of the system, we have in all $(2m + 2)$ roots. The effect of a constraint by which the two parts of the system are connected will be to reduce the $(2m + 2)$ back to $2m$. Of these the m positive will separate the $(m + 1)$ quantities formed of the m positive roots of the original equation together with (the positive) ν_2, and a similar proposition will hold for the negative roots. The effect of the vibrator upon the original system is thus to cause a movement of the positive roots towards ν_2, and a movement of the negative roots towards ν_1. This conclusion covers all the previous statements as to the effect of changes in A, B, C upon the values of the roots.

Enough has now been said on the subject of the free vibrations of a system in general. Any further illustration that it may require will be afforded by the discussion of the case of two degrees of freedom, § 112, and by the vibrations of strings and other special bodies with which we shall soon be occupied. We resume the equations (1) § 103, with the view of investigating further the nature of *forced vibrations*.

104. In order to eliminate from the equations all the co-ordinates but one (ψ_1), operate on them in succession with the minor determinants

$$\frac{d\nabla}{de_{11}}, \quad \frac{d\nabla}{de_{21}}, \quad \frac{d\nabla}{de_{31}}, \quad \&c.,$$

and add the results together; and in like manner for the other co-ordinates. We thus obtain as the equivalent of the original system of equations

$$\left. \begin{aligned}
\nabla\psi_1 &= \frac{d\nabla}{de_{11}}\Psi_1 + \frac{d\nabla}{de_{21}}\Psi_2 + \frac{d\nabla}{de_{31}}\Psi_3 + \dots \\
\nabla\psi_2 &= \frac{d\nabla}{de_{12}}\Psi_1 + \frac{d\nabla}{de_{22}}\Psi_2 + \frac{d\nabla}{de_{32}}\Psi_3 + \dots \\
\nabla\psi_3 &= \frac{d\nabla}{de_{13}}\Psi_1 + \frac{d\nabla}{de_{23}}\Psi_2 + \frac{d\nabla}{de_{33}}\Psi_3 + \dots \\
&\dots\dots\dots\dots\dots\dots\dots\dots\dots\dots\dots
\end{aligned} \right\} \quad \dots\dots\dots (1),$$

in which the differentiations of ∇ are to be made without recognition of the equality subsisting between e_{rs} and e_{sr}.

The forces Ψ_1, Ψ_2, &c. are any whatever, subject, of course, to the condition of not producing so great a displacement or motion that the squares of the small quantities become sensible. If, as is often the case, the forces operating be made up of two parts, one constant with respect to time, and the other periodic, it is convenient to separate in imagination the two classes of effects produced. The effect due to the constant forces is exactly the same as if they acted alone, and is found by the solution of a statical problem. It will therefore generally be sufficient to suppose the forces periodic, the effects of any constant forces, such as gravity, being merely to alter the configuration about which the vibrations proper are executed. We may thus without any real loss of generality confine ourselves to periodic, and therefore by Fourier's theorem to harmonic forces.

We might therefore assume as expressions for Ψ_1, &c. circular functions of the time; but, as we shall have frequent occasion to recognise in the course of this work, it is usually more convenient to employ an imaginary exponential function, such as $E\,e^{ipt}$, where E is a constant which may be complex. When the corresponding symbolical solution is obtained, its real and imaginary parts may be separated, and belong respectively to the real and imaginary parts of the data. In this way the analysis gains considerably in brevity, inasmuch as differentiations and alterations of phase are expressed by merely modifying the complex coefficient without changing the form of the function. We therefore write

$$\Psi_1 = E_1 e^{ipt}, \quad \Psi_2 = E_2 e^{ipt}, \; \&c.$$

The minor determinants of the type $\dfrac{d\nabla}{de_{rs}}$ are rational integral functions of the symbol D, and operate on Ψ_1, &c. according to the law

$$f(D)\, e^{ipt} = f(ip)\, e^{ipt} \; \dotfill (2).$$

Our equations therefore assume the form

$$\nabla\psi_1 = A_1 e^{ipt}, \quad \nabla\psi_2 = A_2 e^{ipt}, \; \&c. \dotfill (3),$$

where A_1, A_2, &c. are certain complex constants. And the symbolical solutions are

$$\psi_1 = A_1 \nabla^{-1} e^{ipt}, \; \&c.,$$

or by (2),

$$\psi_1 = A_1 \frac{e^{ipt}}{\nabla(ip)}, \; \&c. \dotfill (4),$$

where $\nabla(ip)$ denotes the result of substituting ip for D in ∇.

Consider first the case of a system exempt from friction.

∇ and its differential coefficients are then *even* functions of D, so that $\nabla(ip)$ is real. Throwing away the imaginary part of the solution, writing $R_1 e^{i\theta_1}$ for A_1, &c., we have

$$\psi_1 = \frac{R_1}{\nabla(ip)} \cos(pt + \theta_1), \; \&c. \dotfill (5).$$

If we suppose that the forces Ψ_1, &c. (in the case of more than one generalized component) have all the same phase, they may be expressed by

$$E_1 \cos(pt + \alpha), \quad E_2 \cos(pt + \alpha), \; \&c.;$$

and then, as is easily seen, the co-ordinates themselves agree in phase with the forces:

$$\psi_1 = \frac{R_1}{\nabla(ip)} \cos(pt + \alpha) \dotfill (6).$$

The amplitudes of the vibrations depend among other things on the magnitude of $\nabla(ip)$. Now, if the period of the forces be the same as one of those belonging to the free vibrations, $\nabla (ip) = 0$, and the amplitude becomes infinite. This is, of course, just the case in which it is essential to introduce the consideration of friction, from which no natural system is really exempt.

If there be friction, $\nabla (ip)$ is complex; but it may be divided into two parts—one real and the other purely imaginary, of which the latter depends entirely on the friction. Thus, if we put

$$\nabla (ip) = \nabla_1 (ip) + ip \, \nabla_2 (ip) \dots\dots\dots\dots\dots (7),$$

∇_1, ∇_2 are even functions of ip, and therefore real. If as before $A_1 = R_1 e^{i\theta_1}$, our solution takes the form

$$\psi_1 = \frac{R_1 e^{i\theta_1} e^{i\gamma} e^{ipt}}{\{|\nabla_1 (ip)|^2 + p^2 |\nabla_2 (ip)|^2\}^{\frac{1}{2}}},$$

or, on throwing away the imaginary part,

$$\psi_1 = \frac{R_1 \cos (pt + \theta_1 + \gamma)}{\{|\nabla_1 (ip)|^2 + p^2 |\nabla_2 (ip)|^2\}^{\frac{1}{2}}} \dots\dots\dots\dots\dots (8),$$

where

$$\tan \gamma = -\frac{p \, \nabla_2 (ip)}{\nabla_1 (ip)} \dots\dots\dots\dots\dots (9).$$

We have said that $\nabla_2 (ip)$ depends entirely on the friction; but it is not true, on the other hand, that $\nabla_1 (ip)$ is exactly the same, as if there had been no friction. However, this is approximately the case, if the friction be small; because any part of $\nabla (ip)$, which depends on the first power of the coefficients of friction, is necessarily imaginary. Whenever there is a coincidence between the period of the force and that of one of the free vibrations, $\nabla_1 (ip)$ vanishes, and we have $\tan \gamma = -\infty$, and therefore

$$\psi_1 = \frac{R_1 \sin (pt + \theta_1)}{p \, \nabla_2 (ip)} \dots\dots\dots\dots\dots (10),$$

indicating a vibration of large amplitude, only limited by the friction.

On the hypothesis of small friction, θ is in general small, and so also is γ, except in case of approximate equality of periods. With certain exceptions, therefore, the motion has nearly the same (or opposite) phase with the force that excites it.

When a force expressed by a harmonic term acts on a system, the resulting motion is everywhere harmonic, and retains the original period, provided always that the squares of the displacements and velocities may be neglected. This important principle was enunciated by Laplace and applied by him to the theory of the tides. Its great generality was also recognised by Sir John Herschel, to whom we owe a formal demonstration of its truth[1].

If the force be not a harmonic function of the time, the types of vibration in different parts of the system are in general different from each other and from that of the force. The harmonic functions are thus the only ones which preserve their type unchanged, which, as was remarked in the Introduction, is a strong reason for anticipating that they correspond to simple tones.

105. We now turn to a somewhat different kind of forced vibration, where, instead of given *forces* as hitherto, given inexorable *motions* are prescribed.

If we suppose that the co-ordinates ψ_1, ψ_2, ... ψ_r are given functions of the time, while the forces of the remaining types Ψ_{r+1}, Ψ_{r+2}, ... Ψ_m vanish, the equations of motion divide themselves into two groups, viz.

$$\left. \begin{aligned} e_{11}\psi_1 + e_{12}\psi_2 + \ldots + e_{1m}\psi_m &= \Psi_1 \\ e_{21}\psi_1 + e_{22}\psi_2 + \ldots + e_{2m}\psi_m &= \Psi_2 \\ \cdots\cdots\cdots\cdots\cdots\cdots\cdots\cdots\cdots\cdots\cdots \\ e_{r1}\psi_1 + e_{r2}\psi_2 + \ldots + e_{rm}\psi_m &= \Psi_r \end{aligned} \right\} \quad \ldots\ldots\ldots\ldots (1);$$

and

$$\left. \begin{aligned} e_{r+1,1}\,\psi_1 + e_{r+1,2}\,\psi_2 + \ldots + e_{r+1,m}\,\psi_m &= 0 \\ \cdots\cdots\cdots\cdots\cdots\cdots\cdots\cdots\cdots\cdots\cdots\cdots\cdots \\ e_{m1}\quad\psi_1 + e_{m2}\quad\psi_2 + \ldots + e_{mm}\quad\psi_m &= 0 \end{aligned} \right\} \ldots\ldots\ldots(2).$$

In each of the $m - r$ equations of the latter group, the first r terms are known explicit functions of the time, and have the same effect as known forces acting on the system. The equations of this group are therefore sufficient to determine the unknown quantities; after which, if required, the forces necessary to maintain the prescribed motion may be determined from the first group. It is obvious that there is no essential difference between the two classes of problems of forced vibrations.

106. The motion of a system devoid of friction and executing simple harmonic vibrations in consequence of prescribed variations of some of the co-ordinates, possesses a peculiarity parallel to those considered in §§ 74, 79. Let

$$\psi_1 = A_1 \cos pt, \quad \psi_2 = A_2 \cos pt, \quad \&c.,$$

---

[1] *Encyc. Metrop.* art. 323. Also *Outlines of Astronomy*, § 650.

in which the quantities $A_1 \ldots A_r$ are regarded as given, while the remaining ones are arbitrary. We have from the expressions for T and V, § 82,

$$2(T + V) = \tfrac{1}{2}(c_{11} + p^2 a_{11}) A_1{}^2 + \ldots + (c_{12} + p^2 a_{12}) A_1 A_2 + \ldots$$
$$+ \{\tfrac{1}{2}(c_{11} - p^2 a_{11}) A_1{}^2 + \ldots + (c_{12} - p^2 a_{12}) A_1 A_2 + \ldots\} \cos 2pt,$$

from which we see that the equations of motion express the condition that E, the variable part of $T + V$, which is proportional to

$$\tfrac{1}{2}(c_{11} - p^2 a_{11}) A_1{}^2 + \ldots + (c_{12} - p^2 a_{12}) A_1 A_2 + \ldots \quad \ldots \ldots (1),$$

shall be stationary in value, for all variations of the quantities $A_{r+1} \ldots A_m$. Let p'^2 be the value of p^2 natural to the system when vibrating under the restraint defined by the ratios

$$A_1 : A_2 \ldots A_r : A_{r+1} : \ldots A_m ;$$

then

$$p'^2 = \{\tfrac{1}{2}c_{11}A_1{}^2 + \ldots + c_{12}A_1 A_2 + \ldots\} \div \{\tfrac{1}{2}a_{11}A_1{}^2 + \ldots + a_{12}A_1 A_2 + \ldots\},$$

so that

$$E = (p'^2 - p^2) \{\tfrac{1}{2}a_{11}A_1{}^2 + \ldots + a_{12}A_1 A_2 + \ldots\} \ldots \ldots \ldots (2).$$

From this we see that if p^2 be certainly less than p'^2; that is, if the prescribed period be greater than any of those natural to the system under the partial constraint represented by

$$A_1 : A_2 \ldots A_r,$$

then E is necessarily positive, and the stationary value—there can be but one—is an absolute minimum. For a similar reason, if the prescribed period be *less* than any of those natural to the partially constrained system, E is an absolute maximum algebraically, but arithmetically an absolute minimum. But when p^2 lies within the range of possible values of p'^2, E may be positive or negative, and the actual value is not the greatest or least possible. Whenever a natural vibration is consistent with the imposed conditions, that will be the vibration assumed. The variable part of $T + V$ is then zero.

For convenience of treatment we have considered apart the two great classes of forced vibrations and free vibrations; but there is, of course, nothing to prevent their coexistence. After the lapse of a sufficient interval of time, the free vibrations always disappear, however small the friction may be. The case of absolutely no friction is purely ideal.

There is one caution, however, which may not be superfluous in respect to the case where given *motions* are forced on the system. Suppose, as before, that the co-ordinates $\psi_1, \psi_2, \ldots \psi_r$ are given. Then the free vibrations, whose existence or non-existence is a matter of indifference so far as the forced motion is concerned, must be understood to be such as the system is capable of, when the co-ordinates $\psi_1 \ldots \psi_r$ *are not allowed to vary from zero.* In order to prevent their varying, forces of the corresponding types must be introduced; so that from one point of view the motion in question may be regarded as forced. But the applied forces are merely of the nature of a constraint; and their effect is the same as a limitation on the freedom of the motion.

106 a. The principles of the last sections shew that if ψ_1, $\psi_2 \ldots \psi_r$ be given harmonic functions of the time $A_1 \cos pt$, $A_2 \cos pt, \ldots$, the forces of the other types vanishing, then the motion is determinate, unless p is so chosen as to coincide with one of the values proper to the system when ψ_1, $\psi_2 \ldots \psi_r$ are maintained at zero. As an example, consider the case of a membrane capable of vibrating transversely. If the displacement ψ at every point of the contour be given (proportional to $\cos pt$), then in general the value in the interior is determinate; but an exception occurs if p have one of the values proper to the membrane when vibrating with the contour held at rest. This problem is considered by M. Duhem[1] on the basis of a special analytical investigation by Schwartz. It will be seen that it may be regarded as a particular case of a vastly more general theorem.

A like result may be stated for an elastic solid of which the surface motion (proportional to $\cos pt$) is given at every point. Of course, the motion at the boundary need not be more than partially given. Thus for a mass of air we may suppose given the motion *normal* to a closed surface. The internal motion is then determinate, unless the frequency chosen is one of those proper to the mass, when the surface is made unyielding.

107. Very remarkable reciprocal relations exist between the forces and motions of different types, which may be regarded as extensions of the corresponding theorems for systems in which only V or T has to be considered (§ 72 and §§ 77, 78). If we suppose that all the component forces, except two—Ψ_1 and Ψ_2—are zero, we obtain from § 104,

$$\left. \begin{aligned} \nabla \psi_1 &= \frac{d\nabla}{de_{11}} \Psi_1 + \frac{d\nabla}{de_{21}} \Psi_2 \\ \nabla \psi_2 &= \frac{d\nabla}{de_{12}} \Psi_1 + \frac{d\nabla}{de_{22}} \Psi_2 \end{aligned} \right\} \quad \ldots\ldots\ldots\ldots\ldots (1).$$

[1] *Cours de Physique Mathématique*, Tome Second, p. 190, Paris, 1891.

We now consider two cases of motion for the same system; first when Ψ_2 vanishes, and secondly (with dashed letters) when Ψ_1' vanishes. If $\Psi_2 = 0$,

$$\psi_2 = \nabla^{-1} \frac{d\nabla}{de_{12}} \Psi_1 \quad \dots \dots \dots \dots \dots \dots (2).$$

Similarly, if $\Psi_1' = 0$,

$$\psi_1' = \nabla^{-1} \frac{d\nabla}{de_{21}} \Psi_2' \quad \dots \dots \dots \dots \dots \dots (3).$$

In these equations ∇ and its differential coefficients are rational integral functions of the symbol D; and since in every case $e_{rs} = e_{sr}$, ∇ is a symmetrical determinant, and therefore

$$\frac{d\nabla}{de_{rs}} = \frac{d\nabla}{de_{sr}} \quad \dots \dots \dots \dots \dots \dots (4).$$

Hence we see that if a force Ψ_1 act on the system, the co-ordinate ψ_2 is related to it in the same way as the co-ordinate ψ_1' is related to the force Ψ_2', when this latter force is supposed to act alone.

In addition to the motion here contemplated, there may be free vibrations dependent on a disturbance already existing at the moment subsequent to which all new sources of disturbance are included in Ψ; but these vibrations are themselves the effect of forces which acted previously. However small the dissipation may be, there must be an interval of time after which free vibrations die out, and beyond which it is unnecessary to go in taking account of the forces which have acted on a system. If therefore we include under Ψ forces of sufficient remoteness, there are no independent vibrations to be considered, and in this way the theorem may be extended to cases which would not at first sight appear to come within its scope. Suppose, for example, that the system is at rest in its position of equilibrium, and then begins to be acted on by a force of the first type, gradually increasing in magnitude from zero to a finite value Ψ_1, at which point it ceases to increase. If now at a given epoch of time the force be suddenly destroyed and remain zero ever afterwards, free vibrations of the system will set in, and continue until destroyed by friction. At any time t subsequent to the given epoch, the co-ordinate ψ_2 has a value dependent upon t proportional to Ψ_1. The theorem allows us to assert that this value ψ_2 bears the same relation to Ψ_1 as ψ_1' would at the same moment have borne to Ψ_2', if the original cause of the vibrations had been a force of the second type increasing gradually from zero to Ψ_2', and then suddenly vanishing at the given epoch of time. We have already had an example of this in § 101, and a like result obtains when the cause of the original disturbance is an impulse, or, as in the problem of the pianoforte-string, a variable force of finite though short duration. In these applications of our theorem we obtain results relating to free vibrations, considered as the residual effect of forces whose actual operation may have been long before.

108. In an important class of cases the forces Ψ_1 and Ψ_2' are harmonic, and of the same period. We may represent them by $A_1 e^{ipt}$, $A_2' e^{ipt}$, where A_1 and A_2' may be assumed to be *real*, if the forces be in the same phase at the moments compared. The results may then be written

$$\left. \begin{aligned} \psi_2 &= A_1 \frac{d \log \nabla (ip)}{de_{12}} e^{ipt} \\ \psi_1' &= A_2' \frac{d \log \nabla (ip)}{de_{21}} e^{ipt} \end{aligned} \right\} \quad \dots\dots\dots\dots\dots(1),$$

where ip is written for D. Thus,

$$A_2' \psi_2 = A_1 \psi_1' \dots\dots\dots\dots\dots\dots (2).$$

Since the ratio $A_1 : A_2'$ is by hypothesis real, the same is true of the ratio $\psi_1' : \psi_2$; which signifies that the motions represented by those symbols are in the same phase. Passing to real quantities we may state the theorem thus:—

If a force $\Psi_1 = A_1 \cos pt$, *acting on the system give rise to the motion* $\psi_2 = \theta A_1 \cos (pt - \epsilon)$; *then will a force* $\Psi_2' = A_2' \cos pt$ *produce the motion* $\psi_1' = \theta A_2' \cos (pt - \epsilon)$.

If there be no friction, ϵ will be zero.

If $A_1 = A_2'$, then $\psi_1' = \psi_2$. But it must be remembered that the forces Ψ_1 and Ψ_2' are not necessarily comparable, any more than the co-ordinates of corresponding types, one of which for example may represent a linear and another an angular displacement.

The reciprocal theorem may be stated in several ways, but before proceeding to these we will give another investigation, not requiring a knowledge of determinants.

If $\Psi_1, \Psi_2, \dots \psi_1, \psi_2, \dots$ and $\Psi_1', \Psi_2', \dots \psi_1', \psi_2', \dots$ be two sets

of forces and corresponding displacements, the equations of
motion, § 103, give

$$\Psi_1\psi_1' + \Psi_2\psi_2' + \ldots = \psi_1'(e_{11}\psi_1 + e_{12}\psi_2 + e_{13}\psi_3 + \ldots)$$
$$+ \psi_2'(e_{21}\psi_1 + e_{22}\psi_2 + e_{23}\psi_3 + \ldots) + \ldots.$$

Now, if all the forces vary as e^{ipt}, the effect of a symbolic
operator such as e_{rs} on any of the quantities ψ is merely to
multiply that quantity by the constant found by substituting
ip for D in e_{rs}. Supposing this substitution made, and having
regard to the relations $e_{rs} = e_{sr}$, we may write

$$\Psi_1\psi_1' + \Psi_2\psi_2' + \ldots = e_{11}\psi_1\psi_1' + e_{22}\psi_2\psi_2' + \ldots$$
$$+ e_{12}(\psi_1'\psi_2 + \psi_2'\psi_1) + \ldots \ldots\ldots\ldots\ldots (3).$$

Hence by the symmetry

$$\Psi_1\psi_1' + \Psi_2\psi_2' + \ldots = \Psi_1'\psi_1 + \Psi_2'\psi_2 + \ldots \ldots\ldots (4),$$

which is the expression of the reciprocal relation.

109. In the applications that we are about to make it
will be supposed throughout that the forces of all types but
two (which we may as well take as the first and second) are
zero. Thus

$$\Psi_1\psi_1' + \Psi_2\psi_2' = \Psi_1'\psi_1 + \Psi_2'\psi_2 \ldots\ldots\ldots\ldots (1).$$

The consequences of this equation may be exhibited in three
different ways. In the first we suppose that

$$\Psi_2 = 0, \quad \Psi_1' = 0,$$

whence
$$\psi_2 : \Psi_1 = \psi_1' : \Psi_2' \ldots\ldots\ldots\ldots\ldots (2),$$

shewing, as before, that the relation of ψ_2 to Ψ_1 in the first
case when $\Psi_2 = 0$ is the same as the relation of ψ_1' to Ψ_2' in
the second case, when $\Psi_1 = 0$, the identity of relationship extending to phase as well as amplitude.

A few examples may promote the comprehension of a law,
whose extreme generality is not unlikely to convey an impression
of vagueness.

If P and Q be two points of a horizontal bar supported in
any manner (e.g. with one end clamped and the other free), a
given harmonic transverse force applied at P will give at any
moment the same vertical deflection at Q as would have been
found at P, had the force acted at Q.

If we take angular instead of linear displacements, the

theorem will run:—A given harmonic *couple* at P will give the same *rotation* at Q as the couple at Q would give at P.

Or if one displacement be linear and the other angular, the result may be stated thus: Suppose for the first case that a harmonic couple acts at P, and for the second that a vertical force of the same period and phase acts at Q, then the linear displacement at Q in the first case has at every moment the same phase as the rotatory displacement at P in the second, and the amplitudes of the two displacements are so related that the maximum couple at P would do the same work in acting over the maximum rotation at P due to the force at Q, as the maximum force at Q would do in acting through the maximum displacement at Q due to the couple at P. In this case the statement is more complicated, as the forces, being of different kinds, cannot be taken equal.

If we suppose the period of the forces to be excessively long, the momentary position of the system tends to coincide with that in which it would be maintained at rest by the then acting forces, and the equilibrium theory becomes applicable. Our theorem then reduces to the statical one proved in § 72.

As a second example, suppose that in a space occupied by air, and either wholly, or partly, confined by solid boundaries, there are two spheres A and B, whose centres have one degree of freedom. Then a periodic force acting on A will produce the same motion in B, as if the parts were interchanged; and this, whatever membranes, strings, forks on resonance cases, or other bodies capable of being set into vibration, may be present in their neighbourhood.

Or, if A and B denote two points of a solid elastic body of any shape, a force parallel to OX, acting at A, will produce the same motion of the point B parallel to OY as an equal force parallel to OY acting at B would produce in the point A, parallel to OX.

Or again, let A and B be two points of a space occupied by air, between which are situated obstacles of any kind. Then a sound originating at A is perceived at B with the same intensity as that with which an equal sound originating at B would be perceived at A.[1] The obstacle, for instance, might consist of a rigid

---

[1] Helmholtz, *Crelle*, Bd. LVII., 1859. The sounds must be such as in the absence of obstacles would diffuse themselves equally in all directions.

wall pierced with one or more holes. This example corresponds to the optical law that if by any combination of reflecting or refracting surfaces one point can be seen from a second, the second can also be seen from the first. In Acoustics the sound shadows are usually only partial in consequence of the not insignificant value of the wave-length in comparison with the dimensions of ordinary obstacles: and the reciprocal relation is of considerable interest.

A further example may be taken from electricity. Let there be two circuits of insulated wire A and B, and in their neighbourhood any combination of wire-circuits or solid conductors in communication with condensers. A periodic electro-motive force in the circuit A will give rise to the same current in B as would be excited in A if the electro-motive force operated in B.

Our last example will be taken from the theory of conduction and radiation of heat, Newton's law of cooling being assumed as a basis. The temperature at any point A of a conducting and radiating system due to a steady (or harmonic) source of heat at B is the same as the temperature at B due to an equal source at A. Moreover, if at any time the source at B be removed, the whole subsequent course of temperature at A will be the same as it would be at B if the parts of B and A were interchanged.

110. The second way of stating the reciprocal theorem is arrived at by taking in (1) of § 109,

$$\psi_1 = 0, \quad \psi_2' = 0 \, ;$$

whence

$$\Psi_1 \psi_1' = \Psi_2' \psi_2 \dots\dots\dots\dots\dots\dots (1),$$

or

$$\Psi_1 : \psi_2 = \Psi_2' : \psi_1' \dots\dots\dots\dots\dots (2),$$

shewing that the relation of Ψ_1 to ψ_2 in the first case, when $\psi_1 = 0$, is the same as the relation of Ψ_2' to ψ_1' in the second case, when $\psi_2' = 0$.

Thus in the example of the rod, if the point P be held at rest while a given vibration is imposed upon Q (by a force there applied), the reaction at P is the same both in amplitude and phase as it would be at Q if that point were held at rest and the given vibration were imposed upon P.

So if A and B be two electric circuits in the neighbourhood of any number of others, $C, D, \dots$ whether closed or terminating in condensers, and a given periodic current be excited in A by the necessary electro-motive force, the induced electro-motive force in B is the same as it would be in A, if the parts of A and B were interchanged.

The third form of statement is obtained by putting in (1) of § 109,

$$\Psi_1 = 0, \quad \psi_2' = 0 ;$$

whence

$$\Psi_1' \psi_1 + \Psi_2' \psi_2 = 0 \dots\dots\dots\dots\dots (3),$$

or

$$\psi_1 : \psi_2 = - \Psi_2' : \Psi_1' \dots\dots\dots\dots\dots (4),$$

proving that the ratio of ψ_1 to ψ_2 in the first case, when Ψ_2 acts alone, is the negative of the ratio of Ψ_2' to Ψ_1' in the second case, when the forces are so related as to keep ψ_2' equal to zero.

Thus if the point P of the rod be held at rest while a periodic force acts at Q, the reaction at P bears the same numerical ratio to the force at Q as the displacement at Q would bear to the displacement at P, if the rod were caused to vibrate by a force applied at P.

111. The reciprocal theorem has been proved for all systems in which the frictional forces can be represented by the function F, but it is susceptible of a further and an important generalization. We have indeed proved the existence of the function F for a large class of cases where the motion is resisted by forces proportional to the absolute or relative velocities, but there are other sources of dissipation not to be brought under this head, whose effects it is equally important to include; for example, the dissipation due to the conduction or radiation of heat. Now although it be true that the forces in these cases are not *for all possible motions* in a constant ratio to the velocities or displacements, yet in any actual case of periodic motion (τ) they are necessarily periodic, and therefore, whatever their phase, expressible by a sum of two terms, one proportional to the displacement (absolute or relative) and the other proportional to the velocity of the part of the system affected. If the coefficients be the same, not necessarily for all motions whatever, *but for all motions of the period* τ, the function F exists in the only sense required for our present purpose. In fact since it is exclusively with motions of period τ that the theorem is concerned, it is plainly a matter of indifference whether the functions T, F, V are dependent upon τ or not. Thus extended, the theorem is perhaps sufficiently general to cover the whole field of dissipative forces.

It is important to remember that the Principle of Reciprocity is limited to systems which vibrate about a configuration of *equilibrium*, and is therefore not to be applied without reservation to such a problem as that presented by the transmission of sonorous waves through the atmosphere when disturbed by wind. The vibrations must also be of such a character that the square of the motion can be neglected throughout; otherwise our demonstration would not hold good. Other apparent exceptions depend on a misunderstanding of the principle itself. Care must be taken to observe a proper correspondence between the forces and displacements, the rule being that the action of the force over the displacement is to represent *work done*. Thus *couples* correspond to *rotations*, *pressures* to increments of *volume*, and so on.

111 *a*. The substance of the preceding sections is taken from a paper by the Author [1], in which the action of dissipative forces appears first to have been included. Reciprocal theorems of a special character, and with exclusion of dissipation, had been previously given by other writers. One, due to von Helmholtz, has already been quoted. Reference may also be made to the reciprocal theorem of Betti [2], relating to a uniform isotropic elastic solid, upon which bodily and surface forces act. Lamb [3] has shewn that these results and more recent ones of von Helmholtz [4] may be deduced from a very general equation established by Lagrange in the *Mécanique Analytique*.

111 *b*. In many cases of practical interest the external force, in response to which a system vibrates harmonically, is applied at a single point. This may be called the driving-point, and it becomes important to estimate the reaction of the system upon it. When T and F only are sensible, or F and V only, certain general conclusions may be stated, of which a specimen will here be given. For further details reference must be made to a paper by the Author [5].

[1] " Some General Theorems relating to Vibrations," *Proc. Math. Soc.*, 1873.

[2] *Il Nuovo Cimento*, 1872.

[3] *Proc. Math. Soc.*, Vol. xix., p. 144, Jan. 1888.

[4] *Crelle*, t. 100, pp. 137, 213.   1886.

[5] "The Reaction upon the Driving-point of a System executing Forced Harmonic Oscillations of various Periods." *Phil. Mag.*, May, 1886.

Consider a system, devoid of potential energy, in which the co-ordinate ψ_1 is made to vary by the operation of the harmonic force Ψ_1, proportional to e^{ipt}. The other co-ordinates may be chosen arbitrarily, and it will be very convenient to choose them so that no product of them enters into the expressions for T and F. They would be in fact the normal co-ordinates of the system on the supposition that ψ_1 is constrained (by a suitable force of its own type) to remain zero. The expressions for T and F thus take the following forms :—

$$T = \tfrac{1}{2} a_{11} \dot{\psi}_1{}^2 + \tfrac{1}{2} a_{22} \dot{\psi}_2{}^2 + \tfrac{1}{2} a_{33} \dot{\psi}_3{}^2 + \ldots$$
$$+ a_{12} \dot{\psi}_1 \dot{\psi}_2 + a_{13} \dot{\psi}_1 \dot{\psi}_3 + a_{14} \dot{\psi}_1 \dot{\psi}_4 + \ldots \ldots \ldots \ldots (1).$$

$$F = \tfrac{1}{2} b_{11} \dot{\psi}_1{}^2 + \tfrac{1}{2} b_{22} \dot{\psi}_2{}^2 + \tfrac{1}{2} b_{33} \dot{\psi}_3{}^2 + \ldots$$
$$+ b_{12} \dot{\psi}_1 \dot{\psi}_2 + b_{13} \dot{\psi}_1 \dot{\psi}_3 + b_{14} \dot{\psi}_1 \dot{\psi}_4 + \ldots, \ldots \ldots \ldots (2).$$

The equations for a force Ψ_1, proportional to e^{ipt}, are accordingly

$$(ipa_{11} + b_{11})\, \psi_1 + (ipa_{12} + b_{12})\, \psi_2 + (ipa_{13} + b_{13})\, \dot{\psi}_3 + \ldots = \Psi_1,$$
$$(ipa_{12} + b_{12})\, \dot{\psi}_1 + (ipa_{22} + b_{22})\, \dot{\psi}_2 = 0,$$
$$(ipa_{13} + b_{13})\, \dot{\psi}_1 + (ipa_{33} + b_{33})\, \dot{\psi}_3 = 0,$$
$$\ldots\ldots\ldots\ldots\ldots\ldots\ldots\ldots\ldots\ldots\ldots\ldots\ldots\ldots$$

By means of the second and following equations $\dot{\psi}_2$, $\dot{\psi}_3 \ldots$ are expressed in terms of $\dot{\psi}_1$. Introducing these values into the first equation, we get

$$\Psi_1 / \dot{\psi}_1 = ipa_{11} + b_{11} - \frac{(ipa_{12} + b_{12})^2}{ipa_{22} + b_{22}} - \frac{(ipa_{13} + b_{13})^2}{ipa_{33} + b_{33}} - \ldots \quad \ldots\ldots\ldots (3).$$

The ratio $\Psi_1 / \dot{\psi}_1$ is a complex quantity, of which the real part corresponds to the work done by the force in a complete period and dissipated in the system. By an extension of electrical language we may call it the *resistance* of the system and denote it by the letter R'. The other part of the ratio is imaginary. If we denote it by $ipL'\dot{\psi}_1$, or $L'\dot{\psi}_1$, L' will be the moment of inertia, or self-induction of electrical theory. We write therefore

$$\Psi_1 = (R' + ipL')\, \dot{\psi}_1 \ldots\ldots\ldots\ldots\ldots\ldots (4);$$

and the values of R' and L' are to be deduced by separation of the real and the imaginary parts of the right-hand member of (3). In this way we get

$$R' = b_{11} - \Sigma\, \frac{b_{12}{}^2}{b_{22}} + p^2 \Sigma\, \frac{(a_{12}b_{22} - a_{22}b_{12})^2}{b_{22}\,(b_{22}{}^2 + p^2 a_{22}{}^2)} \ldots\ldots\ldots\ldots (5).$$

This is the value of the resistance as determined by the constitution of the system, and by the frequency of the imposed vibration. Each component of the latter series (which alone involves p) is of the form $\alpha p^2/(\beta + \gamma p^2)$, where α, β, γ are all positive, and (as may be seen most easily by considering its reciprocal) increases continually as p^2 increases from zero to infinity. We conclude that as the frequency of vibration increases, the value of R' increases continuously with it. At the lower limit the motion is determined sensibly by the quantities b (the resistances) only, and the corresponding resultant resistance R' is an absolute minimum, whose value is

$$b_{11} - \Sigma \, (b_{12}{}^2/b_{22}) \quad\dots\dots\dots\dots\dots\dots (6).$$

At the upper limit the motion is determined by the inertia of the component parts without regard to resistances, and the value of R' is

$$b_{11} - \Sigma \frac{b_{12}{}^2}{b_{22}} + \Sigma \frac{(a_{12}b_{22} - a_{22}b_{12})^2}{b_{22}a_{22}{}^2},$$

or

$$b_{11} + \Sigma \left(b_{22} \frac{a_{12}{}^2}{a_{22}{}^2} - 2b_{12} \frac{a_{12}}{a_{22}} \right) \dots\dots\dots\dots\dots (7).$$

When p is either very large or very small, all the co-ordinates are in the same phase, and (6), (7) may be identified with $2F/\dot{\psi}_1{}^2$.

Also

$$L' = a_{11} - \Sigma \frac{a_{12}{}^2}{a_{22}} + \Sigma \frac{(a_{12}b_{22} - a_{22}b_{12})^2}{a_{22}(b_{22}{}^2 + p^2 a_{22}{}^2)} \quad \dots\dots\dots (8).$$

In the latter series every term is positive, and continually diminishes as p^2 increases. Hence every increase of frequency is attended by a diminution of the moment of inertia, which tends ultimately to the minimum corresponding to the disappearance of the dissipative terms.

If p be either very large or very small, (8) identifies itself with $2T/\dot{\psi}_1{}^2$.

As a simple example take the problem of the reaction upon the primary circuit of the electric currents generated in a neighbouring secondary circuit. In this case the co-ordinates (or rather their rates of increase) are naturally taken to be the currents themselves, so that $\dot{\psi}_1$ is the primary, and $\dot{\psi}_2$ the secondary current. In usual electrical notation we represent the coefficients of self-induction by L, N, and of mutual induction by M, so that

$$T = \tfrac{1}{2}L\dot{\psi}_1{}^2 + M\dot{\psi}_1\dot{\psi}_2 + \tfrac{1}{2}N\dot{\psi}_2{}^2,$$

and the resistances by R and S. Thus

$$a_{11} = L, \quad a_{12} = M, \quad a_{22} = N \, ;$$
$$b_{11} = R, \quad b_{12} = 0, \quad b_{22} = S \, ;$$

and (5) and (8) become at once

$$R' = R + \frac{p^2 M^2 S}{S^2 + p^2 N^2} \dots\dots\dots\dots\dots (9),$$

$$L' = L - \frac{p^2 M^2 N}{S^2 + p^2 N^2} \dots\dots\dots\dots (10).$$

These formulæ were given originally by Maxwell, who remarked that the reaction of the currents in the secondary has the effect of increasing the effective resistance and diminishing the effective self-induction of the primary circuit.

If the rate of alternation be very slow, the secondary circuit is without influence. If, on the other hand, the rate be very rapid,

$$R' = R + M^2 S / N^2, \qquad L' = L - M^2 / N.$$

112. In Chapter III. we considered the vibrations of a system with one degree of freedom. The remainder of the present Chapter will be devoted to some details of the case where the degrees of freedom are two.

If x and y denote the two co-ordinates, the expressions for T and V are of the form

$$\left. \begin{aligned} 2T &= L\dot{x}^2 + 2M\dot{x}\dot{y} + N\dot{y}^2 \\ 2V &= Ax^2 + 2Bxy + Cy^2 \end{aligned} \right\} \dots\dots\dots\dots (1);$$

so that, in the absence of friction, the equations of motion are

$$\left. \begin{aligned} L\ddot{x} + M\ddot{y} + Ax + By &= X \\ M\ddot{x} + N\ddot{y} + Bx + Cy &= Y \end{aligned} \right\} \dots\dots\dots\dots (2).$$

When there are no impressed forces, we have for the natural vibrations

$$\left. \begin{aligned} (LD^2 + A)x + (MD^2 + B)y &= 0 \\ (MD^2 + B)x + (ND^2 + C)y &= 0 \end{aligned} \right\} \dots\dots\dots\dots (3),$$

D being the symbol of differentiation with respect to time.

If a solution of (3) be $x = l\, e^{\lambda t}$, $y = m\, e^{\lambda t}$, λ^2 is one of the roots of

$$(L\lambda^2 + A)(N\lambda^2 + C) - (M\lambda^2 + B)^2 = 0 \dots\dots\dots (4),$$

or

$$\lambda^4 (LN - M^2) + \lambda^2 (LC + NA - 2MB) + AC - B^2 = 0 \dots\dots (5).$$

The constants L, M, N; A, B, C, are not entirely arbitrary. Since T and V are essentially positive, the following inequalities must be satisfied:—

$$LN > M^2, \quad AC > B^2 \dots\dots\dots\dots\dots (6).$$

Moreover, L, N, A, C must themselves be positive.

We proceed to examine the effect of these restrictions on the roots of (5).

In the first place the three coefficients in the equation are positive. For the first and third, this is obvious from (6). The coefficient of λ^2

$$= (\sqrt{LC} - \sqrt{NA})^2 + 2\sqrt{LNAC} - 2MB,$$

in which, as is seen from (6), $\sqrt{LNAC}$ is necessarily greater than MB. We conclude that the values of λ^2, if real, are both negative.

It remains to prove that the roots are in fact real. The condition to be satisfied is that the following quantity be not negative :—

$$(LC + NA - 2MB)^2 - 4(LN - M^2)(AC - B^2).$$

After reduction this may be brought into the form

$$4(\sqrt{LN}.B - \sqrt{AC}.M)^2$$
$$+ (\sqrt{LC} - \sqrt{NA})^2\{(\sqrt{LC} - \sqrt{NA})^2 + 4(\sqrt{LNAC} - MB)\},$$

which shews that the condition is satisfied, since $\sqrt{LNAC} - MB$ is positive. This is the analytical proof that the values of λ^2 are both real and negative; a fact that might have been anticipated without any analysis from the physical constitution of the system whose vibrations they serve to express.

The two values of λ^2 are different, unless *both*

$$\left. \begin{array}{c} \sqrt{LN}.B - \sqrt{AC}.M = 0 \\ \sqrt{LC} - \sqrt{NA} = 0 \end{array} \right\},$$

which require that

$$L : M : N = A : B : C \ \dots\dots\dots\dots(7).$$

The common spherical pendulum is an example of this case.

By means of a suitable force Y the co-ordinate y may be prevented from varying. The system then loses one degree of freedom, and the period corresponding to the remaining one is in general different from either of those possible before the introduction of Y. Suppose that the types of the motions obtained by thus preventing in turn the variation of y and x are respectively $e^{\mu_1 t}$, $e^{\mu_2 t}$. Then μ_1^2, μ_2^2 are the roots of the equation

$$(L\lambda^2 + A)(N\lambda^2 + C) = 0,$$

being that obtained from (4) by suppressing M and B. Hence (4) may itself be put into the form

$$LN\,(\lambda^2 - \mu_1^{\,2})\,(\lambda^2 - \mu_2^{\,2}) = (M\lambda^2 + B)^2 \ldots\ldots\ldots\ldots(8),$$

which shews at once that neither of the roots of λ^2 can be intermediate in value between $\mu_1^{\,2}$ and $\mu_2^{\,2}$. A little further examination will prove that one of the roots is greater than both the quantities $\mu_1^{\,2}$, $\mu_2^{\,2}$, and the other less than both. For if we put

$$f(\lambda^2) = LN\,(\lambda^2 - \mu_1^{\,2})\,(\lambda^2 - \mu_2^{\,2}) - (M\lambda^2 + B)^2,$$

we see that when λ^2 is very small, f is positive $(AC - B^2)$; when λ^2 decreases (algebraically) to $\mu_1^{\,2}$, f changes sign and becomes negative. Between 0 and $\mu_1^{\,2}$ there is therefore a root; and also by similar reasoning between $\mu_2^{\,2}$ and $-\infty$. We conclude that the tones obtained by subjecting the system to the two kinds of constraint in question are both intermediate in pitch between the tones given by the natural vibrations of the system. In particular cases $\mu_1^{\,2}$, $\mu_2^{\,2}$ may be equal, and then

$$\lambda^2 = \frac{\sqrt{LN}\mu^2 \pm B}{\sqrt{LN} \mp M} = \frac{-\sqrt{AC} \pm B}{\sqrt{LN} \mp M} \ldots\ldots\ldots\ldots(9).$$

This proposition may be generalized. *Any* kind of constraint which leaves the system still in possession of one degree of freedom may be regarded as the imposition of a forced relation between the co-ordinates, such as

$$\alpha x + \beta y = 0 \ldots\ldots\ldots\ldots\ldots\ldots\ldots(10).$$

Now if $\alpha x + \beta y$, and any other homogeneous linear function of x and y, be taken as new variables, the same argument proves that the single period possible to the system after the introduction of the constraint, is intermediate in value between those two in which the natural vibrations were previously performed. Conversely, the two periods which become possible when a constraint is removed, lie one on each side of the original period.

If the values of λ^2 be equal, which can only happen when

$$L : M : N = A : B : C,$$

the introduction of a constraint has no effect on the period; for instance, the limitation of a spherical pendulum to one vertical plane.

113. As a simple example of a system with two degrees of freedom, we may take a stretched string of length l, itself without inertia, but carrying two equal masses m at distances a and b from one end (Fig. 17). Tension $= T_1$.

Fig. 17.

If x and y denote the displacements,

$$2T = m\,(\dot{x}^2 + \dot{y}^2),$$

$$2V = T_1 \left\{ \frac{x^2}{a} + \frac{(x-y)^2}{b-a} + \frac{y^2}{l-b} \right\}$$

Since T and V are not of the same form, it follows that the two periods of vibration are in every case unequal.

If the loads be symmetrically attached, the character of the two component vibrations is evident. In the first, which will have the longer period, the two weights move together, so that x and y remain equal throughout the vibration. In the second x and y are numerically equal, but opposed in sign. The middle point of the string then remains at rest, and the two masses are always to be found on a straight line passing through it. In the first case $x - y = 0$, and in the second $x + y = 0$; so that $x - y$ and $x + y$ are the new variables which must be assumed in order to reduce the functions T and V simultaneously to a sum of squares.

For example, if the masses be so attached as to divide the string into three equal parts,

$$\left. \begin{aligned} 2T &= \frac{m}{2}\, \{(\dot{x}+\dot{y})^2 + (\dot{x}-\dot{y})^2\} \\ 2V &= \frac{3T_1}{2l}\, \{(x+y)^2 + 3\,(x-y)^2\} \end{aligned} \right\} \dots\dots\dots\dots(1),$$

from which we obtain as the complete solution,

$$\left. \begin{aligned} x + y &= A \cos\left(\sqrt{\frac{3T_1}{lm}}\,.\,t + \alpha\right) \\ x - y &= B \cos\left(\sqrt{\frac{9T_1}{lm}}\,.\,t + \beta\right) \end{aligned} \right\} \dots\dots\dots\dots(2),$$

where, as usual, the constants A, α, B, β are to be determined by the initial circumstances.

114. When the two natural periods of a system are nearly equal, the phenomenon of intermittent vibration sometimes presents itself in a very curious manner. In order to illustrate this, we may recur to the string loaded, we will now suppose, with two equal masses at distances from its ends equal to one-fourth of the length. If the middle point of the string were absolutely fixed, the two similar systems on either side of it would be completely independent, or, if the whole be considered as one system, the two periods of vibration would be equal. We now suppose that instead of being absolutely fixed, the middle point is attached to springs, or other machinery, destitute of inertia, so that it is capable of yielding *slightly*. The reservation as to inertia is to avoid the introduction of a third degree of freedom.

From the symmetry it is evident that the fundamental vibrations of the system are those represented by $x+y$ and $x-y$. Their periods are slightly different, because, on account of the yielding of the centre, the potential energy of a displacement when x and y are equal, is less than that of a displacement when x and y are opposite; whereas the kinetic energies are the same for the two kinds of vibration. In the solution

$$\left. \begin{aligned} x+y &= A \cos (n_1 t + \alpha) \\ x-y &= B \cos (n_2 t + \beta) \end{aligned} \right\} \dots\dots\dots\dots\dots(1),$$

we are therefore to regard n_1 and n_2 as nearly, but not quite, equal. Now let us suppose that initially x and $\dot{x}$ vanish. The conditions are

$$\left. \begin{aligned} A \cos \alpha + B \cos \beta &= 0 \\ n_1 A \sin \alpha + n_2 B \sin \beta &= 0 \end{aligned} \right\},$$

which give approximately

$$A + B = 0, \quad \alpha = \beta.$$

Thus

$$\left. \begin{aligned} x &= A \sin \frac{n_2 - n_1}{2} t \; \sin \left(\frac{n_1 + n_2}{2} t + \alpha \right) \\ y &= A \cos \frac{n_2 - n_1}{2} t \; \cos \left(\frac{n_1 + n_2}{2} t + \alpha \right) \end{aligned} \right\} \dots\dots\dots(2).$$

The value of the co-ordinate x is here approximately expressed by a harmonic term, whose amplitude, being proportional to $\sin \frac{1}{2} (n_2 - n_1) t$, is a slowly varying harmonic function of the time. The vibrations of the co-ordinates are therefore intermittent, and so adjusted that each amplitude vanishes at the moment that the other is at its maximum.

This phenomenon may be prettily shewn by a tuning fork of very low pitch, heavily weighted at the ends, and firmly held by

screwing the stalk into a massive support. When the fork vibrates
in the normal manner, the rigidity, or want of rigidity, of the
stalk does not come into play; but if the displacements of the two
prongs be in the same direction, the slight yielding of the stalk
entails a small change of period. If the fork be excited by striking
one prong, the vibrations are intermittent, and appear to transfer
themselves backwards and forwards between the prongs. Unless,
however, the support be very firm, the abnormal vibration, which
involves a motion of the centre of inertia, is soon dissipated; and
then, of course, the vibration appears to become steady. If the
fork be merely held in the hand, the phenomenon of intermittence
cannot be obtained at all.

115. The stretched string with two attached masses may be
used to illustrate some general principles. For example, the period
of the vibration which remains possible when one mass is held
at rest, is intermediate between the two free periods. Any increase in either load depresses the pitch of both the natural
vibrations, and conversely. If the new load be situated at a point
of the string not coinciding with the places where the other loads
are attached, nor with the node of one of the two previously
possible free vibrations (the other has no node), the effect is still
to prolong both the periods already present. With regard to the
third finite period, which becomes possible for the first time after
the addition of the new load, it must be regarded as derived from
one of infinitely small magnitude, of which an indefinite number
may be supposed to form part of the system. It is instructive
to trace the effect of the introduction of a new load and its gradual
increase from zero to infinity, but for this purpose it will be
simpler to take the case where there is but one other. At the
commencement there is one finite period τ_1, and another of infinitesimal magnitude τ_2. As the load increases τ_2 becomes finite,
and both τ_1 and τ_2 continually increase. Let us now consider
what happens when the load becomes very great. One of the
periods is necessarily large and capable of growing beyond all
limit. The other must approach a fixed finite limit. The first
belongs to a motion in which the larger mass vibrates nearly as
if the other were absent; the second is the period of the vibration
of the smaller mass, taking place much as if the larger were fixed.
Now since τ_1 and τ_2 can never be equal, τ_1 must be always the
greater; and we infer, that as the load becomes continually larger,

it is τ_1 that increases indefinitely, and τ_2 that approaches a finite limit.

We now pass to the consideration of forced vibrations.

116. The general equations for a system of two degrees of freedom including friction are

$$\left. \begin{array}{l} (LD^2 + \alpha D + A)\, x + (MD^2 + \beta D + B)\, y = X \\ (MD^2 + \beta D + B)\, x + (ND^2 + \gamma D + C)\, y = Y \end{array} \right\} \dots\dots\dots(1).$$

In what follows we shall suppose that $Y = 0$, and that $X = e^{ipt}$. The solution for y is

$$y = -\frac{(B - p^2 M + i\beta p)\, e^{ipt}}{(A - p^2 L + i\alpha p)(C - p^2 N + i\gamma p) - (B - p^2 M + i\beta p)^2} \dots(2).$$

If the connection between x and y be of a loose character, the constants M, β, B are small, so that the term $(B - p^2 M + i\beta p)^2$ in the denominator may in general be neglected. When this is permissible, the co-ordinate y is the same as if x had been prevented from varying, and a force Y had been introduced whose magnitude is independent of N, γ, and C. But if, in consequence of an approximate isochronism between the force and one of the motions which become possible when x or y is constrained to be zero, either $A - p^2 L + i\alpha p$ or $C - p^2 N + i\gamma p$ be small, then the term in the denominator containing the coefficients of mutual influence must be retained, being no longer *relatively* unimportant; and the solution is accordingly of a more complicated character.

Symmetry shews that if we had assumed $X = 0$, $Y = e^{ipt}$, we should have found the same value for x as now obtains for y. This is the Reciprocal Theorem of § 108 applied to a system capable of two independent motions. The string and two loads may again be referred to as an example.

117. So far for an imposed force. We shall next suppose that it is a *motion* of one co-ordinate ($x = e^{ipt}$) that is prescribed, while $Y = 0$; and for greater simplicity we shall confine ourselves to the case where $\beta = 0$. The value of y is

$$y = -\frac{(B - Mp^2)\, e^{ipt}}{C - Np^2 + i\gamma p} \dots\dots\dots\dots\dots\dots (1).$$

Let us now inquire into the reaction of this motion on x. We have

$$(MD^2 + B)\, y = -\frac{(B - Mp^2)^2\, e^{ipt}}{C - Np^2 + i\gamma p} \dots\dots\dots\dots(2).$$

If the real and imaginary parts of the coefficient of e^{ipt} be respectively A' and $i\alpha'p$, we may put

$$(MD^2 + B) y = A'x + \alpha'\dot{x} \quad \dots\dots\dots\dots(3),$$

and
$$A' = -\frac{(B - Mp^2)^2 (C - Np^2)}{(C - Np^2)^2 + \gamma^2 p^2} \quad \dots\dots\dots\dots(4),$$

$$\alpha' = \frac{(B - Mp^2)^2 \gamma}{(C - Np^2)^2 + \gamma^2 p^2} \quad \dots\dots\dots\dots\dots\dots(5).$$

It appears that the effect of the reaction of y (over and above what would be caused by holding $y = 0$) is represented by changing A into $A + A'$, and α into $\alpha + \alpha'$, where A' and α' have the above values, and is therefore equivalent to the effect of an alteration in the coefficients of spring and friction. These alterations, however, are not constants, *but functions of the period of the motion contemplated*, whose character we now proceed to consider.

Let n be the value of p corresponding to the natural frictionless period of y (x being maintained at zero); so that $C - n^2 N = 0$. Then

$$\left. \begin{aligned} A' &= (B - Mp^2)^2 \, \frac{N (p^2 - n^2)}{N^2 (p^2 - n^2)^2 + \gamma^2 p^2} \\ \alpha' &= (B - Mp^2)^2 \, \frac{\gamma}{N^2 (p^2 - n^2)^2 + \gamma^2 p^2} \end{aligned} \right\} \dots\dots\dots(6).$$

In most cases with which we are practically concerned γ is small, and interest centres mainly on values of p not much differing from n. We shall accordingly leave out of account the variations of the positive factor $(B - Mp^2)^2$, and in the small term $\gamma^2 p^2$, substitute for p its approximate value n. When p is not nearly equal to n, the term in question is of no importance.

As might be anticipated from the general principle of work, α' is always positive. Its maximum value occurs when $p = n$ nearly, and is then proportional to $1/\gamma n^2$, which varies *inversely* with γ. This might not have been expected on a superficial view of the matter, for it seems rather a paradox that, the greater the friction, the less should be its result. But it must be remembered that γ is only the *coefficient* of friction, and that when γ is small the maximum motion is so much increased that the whole work spent against friction is greater than if γ were more considerable.

But the point of most interest is the dependence of A' on p. If p be less than n, A' is negative. As p passes through the value

n, A' vanishes, and changes sign. When A' is negative, the influence of y is to diminish the recovering power of the vibration x, and we see that this happens when the forced vibration is slower than that natural to y. The tendency of the vibration y is thus to retard the vibration x, if the latter be already the slower, but to accelerate it, if it be already the more rapid, only vanishing in the critical case of perfect isochronism. The attempt to make x vibrate at the rate determined by n is beset with a peculiar difficulty, analogous to that met with in balancing a heavy body with the centre of gravity above the support. On whichever side a slight departure from precision of adjustment may occur the influence of the dependent vibration is always to increase the error. Examples of the instability of pitch accompanying a strong resonance will come across us hereafter; but undoubtedly the most interesting application of the results of this section is to the explanation of the anomalous refraction, by substances possessing a very marked selective absorption, of the two kinds of light situated (in a normal spectrum) immediately on either side of the absorption band[1]. It was observed by Christiansen and Kundt, the discoverers of this remarkable phenomenon, that media of the kind in question (for example, *fuchsine* in alcoholic solution) refract the ray immediately *below* the absorption-band abnormally *in excess*, and that *above* it *in defect*. If we suppose, as on other grounds it would be natural to do, that the intense absorption is the result of an agreement between the vibrations of the kind of light affected, and some vibration proper to the molecules of the absorbing agent, our theory would indicate that for light of somewhat greater period the effect must be the same as a relaxation of the natural elasticity of the ether, manifesting itself by a slower propagation and increased refraction. On the other side of the absorption-band its influence must be in the opposite direction.

In order to trace the law of connection between A' and p, take for brevity, $\gamma n = a$, $N(p^2 - n^2) = x$, so that

$$A' \propto \frac{x}{x^2 + a^2}.$$

When the sign of x is changed, A' is reversed with it, but preserves its numerical value. When $x = 0$, or $\pm \infty$, A' vanishes.

[1] *Phil. Mag.*, May, 1872. Also Sellmeier, *Pogg. Ann.* t. cxliii. p. 272, 1871.

Fig. 18.

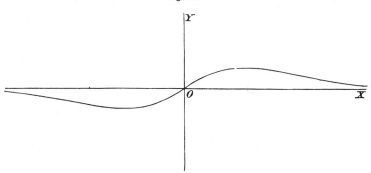

Hence the origin is on the representative curve (Fig. 18), and the axis of x is an asymptote. The maximum and minimum values of A' occur when x is respectively equal to $+a$, or $-a$; and then

$$\frac{x}{x^2 + a^2} = \pm \frac{1}{2a}.$$

The corresponding values of p are given by

$$p^2 = n^2 \pm \frac{\gamma n}{N} \quad\dots\dots\dots\dots\dots\dots\dots (7).$$

Hence, the smaller the value of a or γ, the greater will be the maximum alteration of A, and the corresponding value of p will approach nearer and nearer to n. It may be well to repeat, that in the optical application a diminished γ is attended by an *increased* maximum absorption. When the adjustment of periods is such as to favour A' as much as possible, the corresponding value of α' is one half of *its* maximum.

CHAPTER VI.

118. AMONG vibrating bodies there are none that occupy a more prominent position than Stretched Strings. From the earliest times they have been employed for musical purposes, and in the present day they still form the essential parts of such important instruments as the pianoforte and the violin. To the mathematician they must always possess a peculiar interest as the battle-field on which were fought out the controversies of D'Alembert, Euler, Bernoulli and Lagrange, relating to the nature of the solutions of partial differential equations. To the student of Acoustics they are doubly important. In consequence of the comparative simplicity of their theory, they are the ground on which difficult or doubtful questions, such as those relating to the nature of simple tones, can be most advantageously faced; while in the form of a Monochord or Sonometer, they afford the most generally available means for the comparison of pitch.

The 'string' of Acoustics is a perfectly uniform and flexible filament of solid matter stretched between two fixed points—in fact an ideal body, never actually realized in practice, though closely approximated to by most of the strings employed in music. We shall afterwards see how to take account of any small deviations from complete flexibility and uniformity.

The vibrations of a string may be divided into two distinct classes, which are practically independent of one another, if the amplitudes do not exceed certain limits. In the first class the displacements and motions of the particles are *longitudinal*, so that the string always retains its straightness. The potential energy of a displacement depends, not on the whole tension, but on the *changes* of tension which occur in the various parts of the string, due to the increased or diminished extension. In order to calculate it we must know the relation between the extension of a string and the stretching force. The approximate law (given by Hooke) may be expressed by saying that the extension varies as the tension, so that if l and l' denote the natural and the stretched lengths of a string, and T the tension,

$$\frac{l'-l}{l} = \frac{T}{E} \quad \dots\dots\dots\dots\dots\dots\dots\dots (1),$$

where E is a constant, depending on the material and the section, which may be interpreted to mean the tension that would be necessary to stretch the string to twice its natural length, if the law applied to so great extensions, which, in general, it is far from doing.

119. The vibrations of the second kind are *transverse;* that is to say, the particles of the string move sensibly in planes perpendicular to the line of the string. In this case the potential energy of a displacement depends upon the general tension, and the small variations of tension accompanying the additional stretching due to the displacement may be left out of account. It is here assumed that the stretching due to the motion may be neglected in comparison with that to which the string is already subject in its position of equilibrium. Once assured of the fulfilment of this condition, we do not, in the investigation of transverse vibrations, require to know anything further of the law of extension.

The most general vibration of the transverse, or lateral, kind may be resolved, as we shall presently prove, into two sets of component normal vibrations, executed in perpendicular planes. Since it is only in the initial circumstances that there can be any distinction, pertinent to the question, between one plane and another, it is sufficient for most purposes to regard the motion as entirely confined to a single plane passing through the line of the string.

In treating of the theory of strings it is usual to commence with two particular solutions of the partial differential equation, representing the transmission of waves in the positive and negative directions, and to combine these in such a manner as to suit the case of a finite string, whose ends are maintained at rest; neither of the solutions taken by itself being consistent with the existence of *nodes*, or places of permanent rest. This aspect of the question is very important, and we shall fully consider it; but it seems scarcely desirable to found the solution in the first instance on a property so peculiar to a *uniform* string as the undisturbed transmission of waves. We will proceed by the more general method of assuming (in conformity with what was proved in the last chapter) that the motion may be resolved into normal components of the harmonic type, and determining their periods and character by the special conditions of the system.

Towards carrying out this design the first step would naturally be the investigation of the partial differential equation, to which the motion of a continuous string is subject. But in order to throw light on a point, which it is most important to understand clearly,—the connection between finite and infinite freedom, and the passage corresponding thereto between arbitrary constants and arbitrary functions, we will commence by following a somewhat different course.

120. In Chapter III. it was pointed out that the fundamental vibration of a string would not be entirely altered in character, if the mass were concentrated at the middle point. Following out this idea, we see that if the whole string were divided into a number of small parts and the mass of each concentrated at its centre, we might by sufficiently multiplying the number of parts arrive at a system, still of finite freedom, but capable of representing the continuous string with any desired accuracy, so far at least as the lower component vibrations are concerned. If the analytical solution for any number of divisions can be obtained, its limit will give the result corresponding to a uniform string. This is the method followed by Lagrange.

Let l be the length, ρl the whole mass of the string, so that ρ denotes the mass per unit length, T_1 the tension.

Fig. 19.

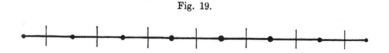

The length of the string is divided into $m + 1$ equal parts (a), so that

$$(m + 1)\, a = l \dots\dots\dots\dots\dots(1).$$

At the m points of division equal masses (μ) are supposed concentrated, which are the representatives of the mass of the portions (a) of the string, which they severally bisect. The mass of each terminal portion of length $\frac{1}{2}a$ is supposed to be concentrated at the final points. On this understanding, we have

$$(m+1)\,\mu = \rho l \quad\text{......................} (2).$$

We proceed to investigate the vibrations of a string, itself devoid of inertia, but loaded at each of m points equidistant (a) from themselves and from the ends, with a mass μ.

If ψ_1, ψ_2 ...... ψ_{m+2} denote the lateral displacements of the loaded points, including the initial and final points, we have the following expressions for T and V,

$$T = \tfrac{1}{2}\,\mu\,\{\dot{\psi}_1{}^2 + \dot{\psi}_2{}^2 + \ldots + \dot{\psi}^2{}_{m+1} + \dot{\psi}^2{}_{m+2}\} \quad\text{...................} (3)$$

$$V = \frac{T_1}{2a}\,\{(\psi_2 - \psi_1)^2 + (\psi_3 - \psi_2)^2 + \ldots + (\psi_{m+2} - \psi_{m+1})^2\}\ldots(4),$$

with the conditions that ψ_1 and ψ_{m+2} vanish. These give by Lagrange's Method the m equations of motion,

$$\left.\begin{array}{l}
B\psi_1 + A\psi_2 + B\psi_3 = 0 \\
B\psi_2 + A\psi_3 + B\psi_4 = 0 \\
B\psi_3 + A\psi_4 + B\psi_5 = 0 \\
\text{.................................} \\
B\psi_m + A\psi_{m+1} + B\psi_{m+2} = 0
\end{array}\right\} \quad\text{...............} (5),$$

where $\qquad A = \mu D^2 + \dfrac{2T_1}{a}, \qquad B = -\dfrac{T_1}{a} \text{....................} (6).$

Supposing now that the vibration under consideration is one of normal type, we assume that ψ_1, ψ_2, &c. are all proportional to $\cos(nt - \epsilon)$, where n remains to be determined. A and B may then be regarded as constants, with a substitution of $-n^2$ for D^2.

If for the sake of brevity we put

$$C = A \div B = -2 + \frac{\mu a n^2}{T_1} \quad\text{...................} (7),$$

the determinantal equation, which gives the values of n^2, assumes the form

$$\begin{vmatrix} C, & 1, & 0, & 0, & 0...... \\ 1, & C, & 1, & 0, & 0...... \\ 0, & 1, & C, & 1, & 0...... \\ 0, & 0, & 1, & C, & 1...... \\ 0, & 0, & 0, & 1, & C...... \\ .... \\ .... \end{vmatrix} \begin{array}{l} m \text{ rows} \end{array} = 0............(8).$$

From this equation the values of the roots might be found. It may be proved that, if $C = 2 \cos \theta$, the determinant is equivalent to $\sin(m+1)\theta \div \sin \theta$; but we shall attain our object with greater ease directly from (5) by acting on a hint derived from the known results relating to a continuous string, and assuming for trial a particular type of vibration. Thus let a solution be

$$\psi_r = P \sin(r-1)\beta \, \cos(nt - \epsilon) \, ... (9),$$

a form which secures that $\psi_1 = 0$. In order that ψ_{m+2} may vanish,

$$(m+1)\beta = s\pi ....(10),$$

where s is an integer. Substituting the assumed values of ψ in the equations (5), we find that they are satisfied, provided that

$$2B \cos \beta + A = 0 \, ...(11);$$

so that the value of n in terms of β is

$$n = 2 \sin \frac{\beta}{2} \sqrt{\frac{T_1}{\mu a}} \, ...(12).$$

A normal vibration is thus represented by

$$\psi_r = P_s \sin \frac{(r-1)s\pi}{m+1} \cos(n_s t - \epsilon_s)...........(13),$$

where

$$n_s = 2 \sqrt{\frac{T_1}{\mu a}} \, \sin \frac{s\pi}{2(m+1)} \, ...(14),$$

and P_s, ϵ_s denote arbitrary constants independent of the general constitution of the system. The m admissible values of n are found from (14) by ascribing to s in succession the values 1, 2, 3...m, and are all different. If we take $s = m+1$, ψ_r vanishes, so that this does not correspond to a possible vibration. Greater values of s give only the same periods over again. If $m+1$ be even, one of the values of n—that, namely, corresponding to

$s = \frac{1}{2}(m+1)$,—is the same as would be found in the case of only a single load $(m = 1)$. The interpretation is obvious. In the kind of vibration considered every alternate particle remains at rest, so that the intermediate ones really move as though they were attached to the centres of strings of length $2a$, fastened at the ends.

The most general solution is found by putting together all the possible particular solutions of normal type

$$\psi_r = \sum_{s=1}^{s=m} P_s \sin \frac{(r-1)s\pi}{m+1} \cos(n_s t - \epsilon_s)\ldots\ldots(15),$$

and, by ascribing suitable values to the arbitrary constants, can be identified with the vibration resulting from arbitrary initial circumstances.

Let x denote the distance of the particle r from the end of the string, so that $(r-1)a = x$; then by substituting for μ and a from (1) and (2), our solution may be written,

$$\psi(x) = P_s \sin s \frac{\pi x}{l} \cos(n_s t - \epsilon_s) \ldots\ldots\ldots (16),$$

$$n_s = \frac{2(m+1)}{l} \sqrt{\frac{T_1}{\rho}} \sin \frac{s\pi}{2(m+1)} \ldots\ldots\ldots (17).$$

In order to pass to the case of a continuous string, we have only to put m infinite. The first equation retains its form, and specifies the displacement at any point x. The limiting form of the second is simply

$$n = \frac{s\pi}{l} \sqrt{\frac{T_1}{\rho}} \ldots\ldots\ldots\ldots\ldots (18),$$

whence for the periodic time,

$$\tau = \frac{2\pi}{n} = \frac{2l}{s} \sqrt{\frac{\rho}{T_1}} \ldots\ldots\ldots\ldots\ldots (19).$$

The periods of the component tones are thus aliquot parts of that of the gravest of the series, found by putting $s = 1$. The whole motion is in all cases periodic; and the period is $2l\sqrt{(\rho/T_1)}$. This statement, however, must not be understood as excluding a shorter period; for in particular cases any number of the lower components may be absent. All that is asserted is that the above-mentioned interval of time is *sufficient* to bring about a complete recurrence. We defer for the present any further discussion of the important formula (19), but it is interesting to observe the approach to a limit in (17), as m is made successively greater and greater. For this purpose it will be sufficient to take the gravest tone for which $s = 1$, and accordingly to trace the variation of

$$\frac{2\,(m+1)}{\pi}\,\sin\frac{\pi}{2\,(m+1)}.$$

The following are a series of simultaneous values of the function and variable :—

m	1	2	3	4	9	19	39
$\dfrac{2\,(m+1)}{\pi}\,\sin\dfrac{\pi}{2\,(m+1)}$	·9003	·9549	·9745	·9836	·9959	·9990	·9997

It will be seen that for very moderate values of m the limit is closely approached. Since m is the number of (moveable) loads, the case $m = 1$ corresponds to the problem investigated in Chapter III., but in comparing the results we must remember that we there supposed the *whole* mass of the string to be concentrated at the centre. In the present case the load at the centre is only half as great; the remainder being supposed concentrated at the ends, where it is without effect.

From the fact that our solution is general, it follows that any initial form of the string can be represented by

$$\psi\,(x) = \sum_{s=1}^{s=\infty} (P\cos\epsilon)_s\,\sin s\frac{\pi x}{l} \quad \dots\dots\dots\dots (20).$$

And, since any form possible for the string at all may be regarded as initial, we infer that any finite single valued function of x, which vanishes at $x = 0$ and $x = l$, can be expanded within those limits in a series of sines of $\pi x/l$ and its multiples,—which is a case of Fourier's theorem. We shall presently shew how the more general form can be deduced.

121. We might now determine the constants for a continuous string by integration as in § 93, but it is instructive to solve the problem first in the general case (m finite), and afterwards to proceed to the limit. The initial conditions are

$$\psi(a) = A_1 \sin \frac{\pi a}{l} \quad + A_2 \sin 2\frac{\pi a}{l} \quad + \ldots + A_m \sin m\frac{\pi a}{l},$$

$$\psi(2a) = A_1 \sin 2\frac{\pi a}{l} + A_2 \sin 4\frac{\pi a}{l} + \ldots + A_m \sin 2m\frac{\pi a}{l},$$

....

$$\psi(ma) = A_1 \sin m\frac{\pi a}{l} + A_2 \sin 2m\frac{\pi a}{l} + \ldots + A_m \sin mm\frac{\pi a}{l};$$

where, for brevity, $A_s = P_s \cos \epsilon_s$, and $\psi(a)$, $\psi(2a) \ldots \psi(ma)$ are the initial displacements of the m particles.

To determine any constant A_s, multiply the first equation by $\sin(s\pi a/l)$, the second by $\sin(2s\pi a/l)$, &c., and add the results. Then, by Trigonometry, the coefficients of all the constants, except A_s, vanish, while that of $A_s = \frac{1}{2}(m+1)^1$. Hence

$$A_s = \frac{2}{m+1} \Sigma_{r=1}^{r=m} \psi(ra) \sin rs \frac{\pi a}{l} \ldots\ldots\ldots\ldots(1).$$

We need not stay here to write down the values of B_s (equal to $P_s \sin \epsilon_s$) as depending on the initial velocities. When a becomes infinitely small, ra under the sign of summation ranges by infinitesimal steps from zero to l. At the same time $\frac{1}{m+1} = \frac{a}{l}$, so that writing $ra = x$, $a = dx$, we have ultimately

$$A_s = \frac{2}{l} \int_0^l \psi(x) \sin\left(\frac{s\pi x}{l}\right) dx \ldots\ldots\ldots\ldots (2),$$

expressing A_s in terms of the initial displacements.

122. We will now investigate independently the partial differential equation governing the transverse motion of a perfectly flexible string, on the suppositions (1) that the magnitude of the tension may be considered constant, (2) that the square of the inclination of any part of the string to its initial direction may be neglected. As before, ρ denotes the linear density at any point, and T_1 is the constant tension. Let rectangular co-ordinates be taken parallel, and perpendicular to the string, so that x gives the equilibrium and x, y, z the displaced position of any particle at time t. The forces acting on the element dx are the tensions at

[1] Todhunter's *Int. Calc.*, p. 267.

its two ends, and any impressed forces $Y\rho\,dx$, $Z\rho\,dx$. By D'Alembert's Principle these form an equilibrating system with the reactions against acceleration, $-\rho\,d^2y/dt^2$, $-\rho\,d^2z/dt^2$. At the point x the components of tension are

$$T_1\frac{dy}{dx}, \quad T_1\frac{dz}{dx},$$

if the squares of dy/dx, dz/dx be neglected; so that the forces acting on the element dx arising out of the tension are

$$T_1\frac{d}{dx}\left(\frac{dy}{dx}\right)dx, \quad T_1\frac{d}{dx}\left(\frac{dz}{dx}\right)dx.$$

Hence for the equations of motion,

$$\left.\begin{aligned}\frac{d^2y}{dt^2} &= \frac{T_1}{\rho}\frac{d^2y}{dx^2} + Y \\ \frac{d^2z}{dt^2} &= \frac{T_1}{\rho}\frac{d^2z}{dx^2} + Z\end{aligned}\right\} \quad \ldots\ldots\ldots\ldots\ldots\ldots (1),$$

from which it appears that the dependent variables y and z are altogether independent of one another.

The student should compare these equations with the corresponding equations of finite differences in § 120. The latter may be written

$$\mu\frac{d^2}{dt^2}\psi(x) = \frac{T_1}{a}\{\psi(x-a) + \psi(x+a) - 2\psi(x)\}.$$

Now in the limit, when a becomes infinitely small,

$$\psi(x-a) + \psi(x+a) - 2\psi(x) = \psi''(x)\,a^2,$$

while $\mu = \rho a$; and the equation assumes ultimately the form

$$\frac{d^2}{dt^2}\psi(x) = \frac{T_1}{\rho}\frac{d^2}{dx^2}\psi(x),$$

agreeing with (1).

In like manner the limiting forms of (3) and (4) of § 120 are

$$T' = \tfrac{1}{2}\int\rho\left(\frac{dy}{dt}\right)^2 dx \ldots\ldots\ldots\ldots\ldots\ldots (2),$$

$$V = \tfrac{1}{2}T_1\int\left(\frac{dy}{dx}\right)^2 dx \ldots\ldots\ldots\ldots\ldots\ldots (3),$$

which may also be proved directly.

The first is obvious from the definition of T. To prove the second, it is sufficient to notice that the potential energy in any configuration is the work required to produce the necessary stretching against the tension T_1. Reckoning from the configuration of equilibrium, we have

$$V = T_1 \int \left(\frac{ds}{dx} - 1 \right) dx ;$$

and, so far as the third power of $\frac{dy}{dx}$,

$$\frac{ds}{dx} - 1 = \tfrac{1}{2} \left(\frac{dy}{dx} \right)^2 .$$

123. In most of the applications that we shall have to make, the density ρ is constant, there are no impressed forces, and the motion may be supposed to take place in one plane. We may then conveniently write

$$\frac{T_1}{\rho} = a^2 \dotfill (1),$$

and the differential equation is expressed by

$$\frac{d^2 y}{d\,(at)^2} = \frac{d^2 y}{dx^2} \dotfill (2).$$

If we now assume that y varies as $\cos mat$, our equation becomes

$$\frac{d^2 y}{dx^2} + m^2 y = 0 \dotfill (3),$$

of which the most general solution is

$$y = (A \sin mx + C \cos mx) \cos mat \dotfill (4),$$

This, however, is not the most general harmonic motion of the period in question. In order to obtain the latter, we must assume

$$y = y_1 \cos mat + y_2 \sin mat \dotfill (5),$$

where y_1, y_2 are functions of x, not necessarily the same. On substitution in (2) it appears that y_1 and y_2 are subject to equations of the form (3), so that finally

$$y = \begin{aligned} &(A \sin mx + C \cos mx) \cos mat \\ &+ (B \sin mx + D \cos mx) \sin mat \end{aligned} \Bigg\} \dotfill (6),$$

an expression containing four arbitrary constants. For any continuous length of string satisfying without interruption the differential equation, this is the most general solution possible, under the condition that the motion at every point shall be simple harmonic. But whenever the string forms part of a system vibrating freely and without dissipation, we know from former chapters that all parts are simultaneously in the same phase, which requires that

$$A : B = C : D \dots\dots\dots\dots\dots (7);$$

and then the most general vibration of simple harmonic type is

$$y = \{\alpha \sin mx + \beta \cos mx\} \cos (mat - \epsilon) \dots\dots\dots (8).$$

124. The most simple as well as the most important problem connected with our present subject is the investigation of the free vibrations of a finite string of length l held fast at both its ends. If we take the origin of x at one end, the terminal conditions are that when $x = 0$, and when $x = l$, y vanishes for all values of t. The first requires that in (6) of § **123**

$$C = 0, \quad D = 0 \dots\dots\dots\dots\dots (1);$$

and the second that

$$\sin ml = 0 \dots\dots\dots\dots\dots (2),$$

or that $ml = s\pi$, where s is an integer. We learn that the only harmonic vibrations possible are such as make

$$m = \frac{s\pi}{l} \dots\dots\dots\dots\dots (3),$$

and then

$$y = \sin \frac{s\pi x}{l} \left(A \cos \frac{s\pi at}{l} + B \sin \frac{s\pi at}{l} \right) \dots\dots\dots (4).$$

Now we know *a priori* that whatever the motion may be, it can be represented as a sum of simple harmonic vibrations, and we therefore conclude that the most general solution for a string, fixed at 0 and l, is

$$y = \sum_{s=1}^{s=\infty} \sin \frac{s\pi x}{l} \left(A_s \cos \frac{s\pi at}{l} + B_s \sin \frac{s\pi at}{l} \right) \dots\dots (5).$$

The slowest vibration is that corresponding to $s = 1$. Its period (τ_1) is given by

$$\tau_1 = \frac{2l}{a} = 2l \sqrt{\frac{\rho}{T_1}} \dots\dots\dots\dots\dots (6).$$

The other components have periods which are aliquot parts of τ_1 :—

$$\tau_s = \tau_1 \div s \dots\dots\dots\dots\dots (7);$$

so that, as has been already stated, the whole motion is under all circumstances periodic in the time τ_1. The sound emitted constitutes in general a musical *note*, according to our definition of that term, whose pitch is fixed by τ_1, the period of its gravest component. It may happen, however, in special cases that the gravest vibration is absent, and yet that the whole motion is not periodic in any shorter time. This condition of things occurs, if $A_1^2 + B_1^2$ vanish, while, for example. $A_2^2 + B_2^2$ and $A_3^2 + B_3^2$ are finite. In such cases the sound could hardly be called a note; but it usually happens in practice that, when the gravest tone is absent, some other takes its place in the character of fundamental, and the sound still constitutes a note in the ordinary sense, though, of course, of elevated pitch. A simple case is when all the odd components beginning with the first are missing. The whole motion is then periodic in the time $\frac{1}{2}\tau_1$, and if the second component be present, the sound presents nothing unusual.

The pitch of the note yielded by a string (6), and the character of the fundamental vibration, were first investigated on mechanical principles by Brook Taylor in 1715; but it is to Daniel Bernoulli (1755) that we owe the general solution contained in (5). He obtained it, as we have done, by the synthesis of particular solutions, permissible in accordance with his Principle of the Coexistence of Small Motions. In his time the generality of the result so arrived at was open to question; in fact, it was the opinion of Euler, and also, strangely enough, of Lagrange[1], that the series of sines in (5) was not capable of representing an arbitrary function; and Bernoulli's argument on the other side, drawn from the infinite number of the disposable constants, was certainly inadequate[2].

Most of the laws embodied in Taylor's formula (6) had been discovered experimentally long before (1636) by Mersenne. They may be stated thus:—

[1] See Riemann's *Partielle Differential Gleichungen*, § 78.

[2] Dr Young, in his memoir of 1800, seems to have understood this matter quite correctly. He says, "At the same time, as M. Bernoulli has justly observed, since every figure may be infinitely approximated, by considering its ordinates as composed of the ordinates of an infinite number of trochoids of different magnitudes, it may be demonstrated that all these constituent curves would revert to their initial state, in the same time that a similar chord bent into a trochoidal curve would perform a single vibration; and this is in some respects a convenient and compendious method of considering the problem."

(1) For a given string and a given tension, the time varies as the length.

This is the fundamental principle of the monochord, and appears to have been understood by the ancients[1].

(2) When the length of the string is given, the time varies inversely as the square root of the tension.

(3) Strings of the same length and tension vibrate in times, which are proportional to the square roots of the linear density.

These important results may all be obtained by the method of dimensions, if it be assumed that τ depends only on l, ρ, and T_1.

For, if the units of length, time and mass be denoted respectively by $[L]$, $[T]$, $[M]$, the dimensions of these symbols are given by

$$l = [L], \quad \rho = [ML^{-1}], \quad T_1 = [MLT^{-2}],$$

and thus (see § 52) the only combination of them capable of representing a time is $T_1^{-\frac{1}{2}} . \rho^{\frac{1}{2}} . l$. The only thing left undetermined is the numerical factor.

125. Mersenne's laws are exemplified in all stringed instruments. In playing the violin different notes are obtained from the same string by shortening its efficient length. In tuning the violin or the pianoforte, an adjustment of pitch is effected with a constant length by varying the tension; but it must be remembered that ρ is not quite invariable.

To secure a prescribed pitch with a string of given material, it is requisite that one relation only be satisfied between the length, the thickness, and the tension; but in practice there is usually no great latitude. The length is often limited by considerations of convenience, and its curtailment cannot always be compensated by an increase of thickness, because, if the tension be not increased proportionally to the section, there is a loss of flexibility, while if the tension be so increased, nothing is effected towards lowering the pitch. The difficulty is avoided in the lower strings of the pianoforte and violin by the addition of a coil of fine wire, whose effect is to impart inertia without too much impairing flexibility.

---

[1] Aristotle "knew that a pipe or a chord of double length produced a sound of which the vibrations occupied a double time; and that the properties of concords depended on the proportions of the times occupied by the vibrations of the separate sounds."—Young's *Lectures on Natural Philosophy*, Vol. 1. p. 404.

For quantitative investigations into the laws of strings, the sonometer is employed. By means of a weight hanging over a pulley, a catgut, or a metallic wire, is stretched across two bridges mounted on a resonance case. A moveable bridge, whose position is estimated by a scale running parallel to the wire, gives the means of shortening the efficient portion of the wire to any desired extent. The vibrations may be excited by plucking, as in the harp, or with a bow (well supplied with rosin), as in the violin.

If the moveable bridge be placed half-way between the fixed ones, the note is raised an octave; when the string is reduced to one-third the note obtained is the twelfth.

By means of the law of lengths, Mersenne determined for the first time the frequencies of known musical notes. He adjusted the length of a string until its note was one of assured position in the musical scale, and then prolonged it under the same tension until the vibrations were slow enough to be counted.

For experimental purposes it is convenient to have two, or more, strings mounted side by side, and to vary in turn their lengths, their masses, and the tensions to which they are subjected. Thus in order that two strings of equal length may yield the interval of the octave, their tensions must be in the ratio of 1 : 4, if the masses be the same; or, if the tensions be the same the masses must be in the reciprocal ratio.

The sonometer is very useful for the numerical determination of pitch. By varying the tension, the string is tuned to unison with a fork, or other standard of known frequency, and then by adjustment of the moveable bridge, the length of the string is determined, which vibrates in unison with any note proposed for measurement. The law of lengths then gives the means of effecting the desired comparison of frequencies.

Another application by Scheibler to the determination of absolute pitch is important. The principle is the same as that explained in Chapter III., and the method depends on deducing the absolute pitch of two notes from a knowledge of both the *ratio* and the *difference* of their frequencies. The lengths of the sonometer string when in unison with a fork, and when giving with it four beats per second, are carefully measured. The ratio of the lengths is the inverse ratio of the frequencies, and the difference of the frequencies is four. From these data the absolute pitch of the fork can be calculated.

The pitch of a string may be calculated also by Taylor's formula from the mechanical elements of the system, but great precautions are necessary to secure accuracy. The tension is produced by a weight, whose mass (expressed with the same unit as ρ) may be called P; so that $T_1 = gP$, where $g = 32\cdot2$, if the units of length and time be the foot and the second. In order to secure that the whole tension acts on the vibrating segment, no bridge must be interposed, a condition only to be satisfied by suspending the string vertically. After the weight is attached, a portion of the string is isolated by clamping it firmly at two points, and the length is measured. The mass of the unit of length ρ refers to the stretched state of the string, and may be found indirectly by observing the elongation due to a tension of the same order of magnitude as T_1, and calculating what would be produced by T_1 according to Hooke's law, and by weighing a known length of the string in its normal state. After the clamps have been secured great care is required to avoid fluctuations of temperature, which would seriously influence the tension. In this way Seebeck obtained very accurate results.

126. When a string vibrates in its gravest normal mode, the excursion is at any moment proportional to $\sin(\pi x/l)$, increasing numerically from either end towards the centre; no intermediate point of the string remains permanently at rest. But it is otherwise in the case of the higher normal components. Thus, if the vibration be of the mode expressed by

$$y = \sin\frac{s\pi x}{l}\left(A_s \cos\frac{s\pi at}{l} + B_s \sin\frac{s\pi at}{l}\right),$$

the excursion is proportional to $\sin(s\pi x/l)$, which vanishes at $s-1$ points, dividing the string into s equal parts. These points of no motion are called nodes, and may evidently be touched or held fast without in any way disturbing the vibration. The production of 'harmonics' by lightly touching the string at the points of aliquot division is a well-known resource of the violinist. All component modes are excluded which have not a node at the point touched; so that, as regards pitch, the effect is the same as if the string were securely fastened there.

127. The constants, which occur in the general value of y, § 124, depend on the special circumstances of the vibration, and may be expressed in terms of the initial values of y and $\dot{y}$.

Putting $t = 0$, we find

$$y_0 = \Sigma_{s=1}^{s=\infty} A_s \sin \frac{s\pi x}{l}; \quad \dot{y}_0 = \frac{\pi a}{l} \Sigma_{s=1}^{s=\infty} s B_s \sin \frac{s\pi x}{l} \ \ldots\ldots (1).$$

Multiplying by $\sin \frac{s\pi x}{l}$, and integrating from 0 to l, we obtain

$$A_s = \frac{2}{l} \int_0^l y_0 \sin \frac{s\pi x}{l}\, dx; \quad B_s = \frac{2}{\pi a s} \int_0^l \dot{y}_0 \sin \frac{s\pi x}{l}\, dx \ \ldots\ldots (2).$$

These results exemplify Stokes' law, § 95; for that part of y, which depends on the initial velocities, is

$$y = \Sigma_{s=1}^{s=\infty} \frac{2}{\pi a s} \sin \frac{s\pi x}{l} \ \sin \frac{s\pi a t}{l} \int_0^l \dot{y}_0 \sin \frac{s\pi x}{l}\, dx,$$

and from this the part depending on initial displacements may be inferred, by differentiating with respect to the time, and substituting y_0 for $\dot{y}_0$.

When the condition of the string at some one moment is thoroughly known, these formulæ allow us to calculate the motion for all subsequent time. For example, let the string be initially at rest, and so displaced that it forms two sides of a triangle. Then $B_s = 0$; and

Fig. 20.

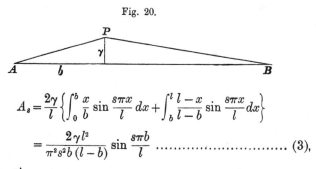

$$A_s = \frac{2\gamma}{l} \left\{ \int_0^b \frac{x}{b} \sin \frac{s\pi x}{l}\, dx + \int_b^l \frac{l-x}{l-b} \sin \frac{s\pi x}{l}\, dx \right\}$$

$$= \frac{2\gamma l^2}{\pi^2 s^2 b\, (l-b)} \sin \frac{s\pi b}{l} \ \ldots\ldots\ldots\ldots\ldots\ldots\ldots\ldots (3),$$

on integration.

We see that A_s vanishes, if $\sin (s\pi b/l) = 0$, that is, if there be a node of the component in question situated at P. A more comprehensive view of the subject will be afforded by another mode of solution to be given presently.

128.   In the expression for y the coefficients of $\sin(s\pi x/l)$ are the normal co-ordinates of Chapters IV. and V. We will denote them therefore by ϕ_s, so that the configuration and motion of the system at any instant are defined by the values of ϕ_s and $\dot{\phi}_s$ according to the equations

$$\left. \begin{aligned} y &= \phi_1 \sin\frac{\pi x}{l} + \phi_2 \sin\frac{2\pi x}{l} + \ldots + \phi_s \sin\frac{s\pi x}{l} + \ldots \\ \dot{y} &= \dot{\phi}_1 \sin\frac{\pi x}{l} + \dot{\phi}_2 \sin\frac{2\pi x}{l} + \ldots + \dot{\phi}_s \sin\frac{s\pi x}{l} + \ldots \end{aligned} \right\} \ \ldots\ldots (1).$$

We proceed to form the expressions for T and V, and thence to deduce the normal equations of vibration.

For the kinetic energy,

$$T = \tfrac{1}{2}\rho \int_0^l \dot{y}^2 dx = \tfrac{1}{2}\rho \int_0^l \left\{ \Sigma_{s=1}^{s=\infty} \dot{\phi}_s \sin\frac{s\pi x}{l} \right\}^2 dx$$

$$= \tfrac{1}{2}\rho \int_0^l \Sigma_{s=1}^{s=\infty} \dot{\phi}_s^2 \sin^2\frac{s\pi x}{l}\, dx,$$

the product of every pair of terms vanishing by the general property of normal co-ordinates.   Hence

$$T = \tfrac{1}{4}\rho l\, \Sigma_{s=1}^{s=\infty} \dot{\phi}_s^2 \ \ldots\ldots\ldots\ldots\ldots\ldots (2).$$

In like manner,

$$V = \tfrac{1}{2}T_1 \int_0^l \left(\frac{dy}{dx}\right)^2 dx = \tfrac{1}{2}T_1 \int_0^l \left\{ \Sigma_{s=1}^{s=\infty} \phi_s \frac{s\pi}{l}\cos\frac{s\pi x}{l} \right\}^2 dx$$

$$= \tfrac{1}{4}T_1 l . \Sigma_{s=1}^{s=\infty} \frac{s^2\pi^2}{l^2}\, \phi_s^2 \ \ldots\ldots\ldots\ldots\ldots\ldots\ldots\ldots\ldots\ldots\ldots (3).$$

These expressions do not presuppose any particular motion, either natural, or otherwise; but we may apply them to calculate the whole energy of a string vibrating naturally, as follows:—If M be the whole mass of the string (ρl), and its equivalent $(a^2\rho)$ be substituted for T_1, we find for the sum of the energies,

$$T + V = \tfrac{1}{4}M . \Sigma_{s=1}^{s=\infty} \left\{ \dot{\phi}_s^2 + \frac{s^2\pi^2 a^2}{l^2}\phi_s^2 \right\} \ \ldots\ldots\ldots\ldots (4),$$

or, in terms of A_s and B_s of § 126,

$$T + V = \pi^2 M . \Sigma_{s=1}^{s=\infty} \frac{A_s^2 + B_s^2}{\tau_s^2} \ \ldots\ldots\ldots\ldots (5).$$

If the motion be not confined to the plane of xy, we have merely to add the energy of the vibrations in the perpendicular plane.

Lagrange's method gives immediately the equation of motion

$$\ddot{\phi}_s + \left(\frac{s\pi a}{l}\right)^2 \phi_s = \frac{2}{l\rho}\Phi_s \quad\dots\dots\dots\dots (6),$$

which has been already considered in § 66. If ϕ_0 and $\dot{\phi}_0$ be the initial values of ϕ and $\dot{\phi}$, the general solution is

$$\phi = \dot{\phi}_0 \frac{\sin nt}{n} + \phi_0 \cos nt$$

$$+ \frac{2}{l\rho n}\int_0^t \sin n\,(t-t')\,\Phi\,dt' \dots\dots\dots\dots\dots(7),$$

where n is written for $s\pi a/l$.

By definition Φ_s is such that $\Phi_s\,\delta\phi_s$ represents the work done by the impressed forces on the displacement $\delta\phi_s$. Hence, if the force acting at time t on an element of the string $\rho\,dx$ be $\rho\,Ydx$,

$$\Phi_s = \int_0^l \rho\,Y \sin\frac{s\pi x}{l}\,dx \dots\dots\dots\dots\dots(8).$$

In these equations ϕ_s is a linear quantity, as we see from (1); and Φ_s is therefore a force of the ordinary kind.

129. In the applications that we have to make, the only impressed force will be supposed to act in the immediate neighbourhood of one point $x = b$, and may usually be reckoned as a whole, so that

$$\Phi_s = \sin\frac{s\pi b}{l}\int \rho\,Ydx \dots\dots\dots\dots\dots (1).$$

If the point of application of the force coincide with a node of the mode (s), $\Phi_s = 0$, and we learn that the force is altogether without influence on the component in question. This principle is of great importance; it shews, for example, that if a string be at rest in its position of equilibrium, no force applied at its centre, whether in the form of plucking, striking, or bowing, can generate any of the even normal components[1]. If after the operation of the force, its point of application be damped, as by touching it

---

[1] The observation that a harmonic is not generated, when one of its nodal points is plucked, is due to Young.

with the finger, all motion must forthwith cease; for those components which have not a node at the point in question are stopped by the damping, and those which have, are absent from the beginning[1]. More generally, by damping any point of a sounding string, we stop all the component vibrations which have not, and leave entirely unaffected those which have a node at the point touched.

The case of a string pulled aside at one point and afterwards let go from rest may be regarded as included in the preceding statements. The complete solution may be obtained thus. Let the motion commence at the time $t = 0$; from which moment $\Phi_s = 0$. The value of ϕ_s at time t is

$$\phi_s = (\phi_s)_0 \cos nt + \frac{1}{n} (\dot{\phi}_s)_0 \sin nt \dots\dots\dots\dots (2),$$

where $(\phi_s)_0$, $(\dot{\phi}_s)_0$ denote the initial values of the quantities affected with the suffix s. Now in the problem in hand $(\dot{\phi}_s)_0 = 0$, and $(\phi_s)_0$ is determined by

$$n^2 (\phi_s)_0 = \frac{2}{l\rho} \Phi_s = \frac{2}{l\rho} Y' \sin \frac{s\pi b}{l} \dots\dots\dots\dots (3),$$

if Y' denote the force with which the string is held aside at the point b. Hence at time t

$$\phi_s = \frac{2}{l\rho n^2} Y' \sin \frac{s\pi b}{l} \cos nt \dots\dots\dots\dots (4),$$

and by (1) of § 128

$$y = \frac{2}{l\rho} Y' . \sum_{s=1}^{s=\infty} \sin \frac{s\pi b}{l} \sin \frac{s\pi x}{l} \frac{\cos nt}{n^2} \dots\dots\dots (5),$$

where $n = s\pi a/l$.

The symmetry of the expression (5) in x and b is an example of the principle of § 107.

The problem of determining the subsequent motion of a string set into vibration by an impulse acting at the point b, may be treated in a similar manner. Integrating (6) of § 128 over the duration of the impulse, we find ultimately, with the same notation as before,

$$(\dot{\phi}_s)_0 = \frac{2}{l\rho} \sin \frac{s\pi b}{l} Y_1,$$

---

[1] A like result ensues when the point which is damped is at the same distance from one end of the string as the point of excitation is from the other end.

if $\int Y' dt$ be denoted by Y_1. At the same time $(\phi_s)_0 = 0$, so that by (2) at time t

$$y = \frac{2Y_1}{l\rho} \sum_{s=1}^{s=\infty} \sin \frac{s\pi b}{l} \sin \frac{s\pi x}{l} \frac{\sin nt}{n} \dots\dots\dots (6).$$

The series of component vibrations is less convergent for a struck than for a plucked string, as the preceding expressions shew. The reason is that in the latter case the initial value of y is continuous, and only dy/dx discontinuous, while in the former it is $\dot{y}$ itself that makes a sudden spring. See §§ 32, 101.

The problem of a string set in motion by an impulse may also be solved by the general formulæ (7) and (8) of § 128. The force finds the string at rest at $t = 0$, and acts for an infinitely short time from $t = 0$ to $t = \tau'$. Thus $(\phi_s)_0$ and $(\dot{\phi}_s)_0$ vanish, and (7) of § 128 reduces to

$$\phi_s = \frac{2}{l\rho n} \sin nt \int_0^{\tau'} \Phi_s dt',$$

while by (8) of § 128

$$\int_0^{\tau'} \Phi_s dt' = \sin \frac{s\pi b}{l} \int_0^{\tau'} Y' dt' = \sin \frac{s\pi b}{l} Y_1.$$

Hence, as before,

$$\phi_s = \frac{2}{l\rho n} Y_1 \sin \frac{s\pi b}{l} \sin nt \dots\dots\dots\dots (7).$$

Hitherto we have supposed the disturbing force to be concentrated at a single point. If it be distributed over a distance β on either side of b, we have only to integrate the expressions (6) and (7) with respect to b, substituting, for example, in (7) in place of $Y_1 \sin (s\pi b/l)$,

$$\int_{b-\beta}^{b+\beta} Y_1' \sin \frac{s\pi b}{l} db.$$

If Y_1' be constant between the limits, this reduces to

$$Y_1' \frac{2l}{s\pi} \sin \frac{s\pi \beta}{l} \sin \frac{s\pi b}{l} \dots\dots\dots\dots (8).$$

The principal effect of the distribution of the force is to render the series for y more convergent.

130. The problem which will next engage our attention is that of the pianoforte wire. The cause of the vibration is here the blow of a hammer, which is projected against the string, and after the impact rebounds. But we should not be justified in assuming, as in the last section, that the mutual action occupies so short a time that its duration may be neglected. Measured by the standards of ordinary life the duration of the contact is indeed very small, but here the proper comparison is with the natural periods of the string. Now the hammers used to strike the wires of a pianoforte are covered with several layers of cloth for the express purpose of making them more yielding, with the effect of prolonging the contact. The rigorous treatment of the problem would be difficult, and the solution, when obtained, probably too complicated to be of use; but by introducing a certain simplification Helmholtz has obtained a solution representing all the essential features of the case. He remarks that since the actual yielding of the string must be slight in comparison with that of the covering of the hammer, the law of the force called into play during the contact must be nearly the same as if the string were absolutely fixed, in which case the force would vary very nearly as a circular function. We shall therefore suppose that at the time $t = 0$, when there are neither velocities nor displacements, a force $F \sin pt$ begins to act on the string at $x = b$, and continues through half a period of the circular function, that is, until $t = \pi/p$, after which the string is once more free. The magnitude of p will depend on the mass and elasticity of the hammer, but not to any great extent on the velocity with which it strikes the string.

The required solution is at once obtained by substituting for Φ_s in the general formula (7) of § 128 its value given by

$$\Phi_s = F \sin \frac{s\pi b}{l} \sin pt' \quad \dots\dots\dots\dots\dots(1),$$

the range of the integration being from 0 to π/p. We find $(t > \pi/p)$

$$\phi_s = \frac{2F}{ln\rho} \sin \frac{s\pi b}{l} \int_0^{\frac{\pi}{p}} \sin n(t - t') \sin pt' \, dt'$$

$$= \frac{4p \cos \frac{n\pi}{2p}}{l\rho n(p^2 - n^2)} . F \sin \frac{s\pi b}{l} . \sin n\left(t - \frac{\pi}{2p}\right) \quad \dots\dots\dots(2),$$

and the final solution for y becomes, if we substitute for n and ρ their values,

$$y = \frac{4apl^2 F}{\pi T_1} \sum_{s=1}^{s=\infty} \frac{\cos \frac{s\pi^2 a}{2pl} . \sin \frac{s\pi b}{l}}{s\left(l^2 p^2 - s^2 a^2 \pi^2\right)} \sin \frac{s\pi x}{l} \sin \frac{s\pi a}{l}\left(t - \frac{\pi}{2p}\right) \dots(3).$$

We see that all components vanish which have a node at the point of excitement, but this conclusion does not depend on any particular law of force. The interest of the present solution lies in the information that may be elicited from it as to the dependence of the resulting vibrations on the duration of contact. If we denote the ratio of this quantity to the fundamental period of the string by ν, so that $\nu = \pi a : 2pl$, the expression for the amplitude of the component s is

$$\frac{8Fl}{\pi^2 T_1} \cdot \frac{\nu \cos (s\pi\nu)}{s (1 - 4s^2\nu^2)} \sin \frac{s\pi b}{l} \dots\dots\dots\dots\dots (4).$$

We fall back on the case of an impulse by putting $\nu = 0$, and

$$Y_1 = \int_{0}^{\pi/p} F \sin pt\, dt = \frac{2F}{p}.$$

When ν is finite, those components disappear, whose periods are $\frac{2}{3}, \frac{2}{5}, \frac{2}{7}, \dots$ of the duration of contact; and when s is very great, the series converges with s^{-3}. Some allowance must also be made for the finite breadth of the hammer, the effect of which will also be to favour the convergence of the series.

The laws of the vibration of strings may be verified, at least in their main features, by optical methods of observation—either with the vibration-microscope, or by a tracing point recording the character of the vibration on a revolving drum. This character depends on two things,—the mode of excitement, and the point whose motion is selected for observation. Those components do not appear which have nodes either at the point of excitement, or at the point of observation. The former are not generated, and the latter do not manifest themselves. Thus the simplest motion is obtained by plucking the string at the centre, and observing one of the points of trisection, or *vice versa*. In this case the first harmonic which contaminates the purity of the principal vibration is the fifth component, whose intensity is usually insufficient to produce much disturbance.

[The dynamical theory of the vibration of strings may be employed to test the laws of hearing, and the necessary experiments are easily carried out upon a grand pianoforte. Having freed a string, say c, from its damper by pressing the digital, pluck it at one-third of its length. According to Young's theorem the third component vibration is not excited then, and in correspondence with that fact the ear fails to detect the component g'. A slight displacement of the point plucked brings g' in again; and if a resonator (g') be used to assist the ear, it is only with difficulty that the point can be hit with such precision as entirely to extinguish the tone. Experiments of this kind shew that the ear analyses the sound of a string into precisely the same constituents as are found by sympathetic resonance, that is, into simple tones, according to Ohm's definition of this conception. Such experiments are also well adapted to shew that it is not a mere play of imagination when we hear overtones, as some people believe it is on hearing them for the first time[1].

If, after the string has been sounded loudly by striking the digital, it be touched with the finger at one of the points of trisection, all components are stopped except the 3rd, 6th, &c., so that these are left isolated. The inexperienced observer is usually surprised by the loudness of the residual sound, and begins to appreciate the large part played by overtones.]

131.   The case of a periodic force is included in the general solution of § 128, but we prefer to follow a somewhat different method, in order to make an extension in another direction. We have hitherto taken no account of dissipative forces, but we will now suppose that the motion of each element of the string is resisted by a force proportional to its velocity. The partial differential equation becomes

$$\frac{d^2y}{dt^2} + \kappa \frac{dy}{dt} = a^2 \frac{d^2y}{dx^2} + Y \dots\dots\dots\dots(1),$$

by means of which the subject may be treated. But it is still simpler to avail ourselves of the results of the last chapter, remarking that in the present case the dissipation-function F is of the same form as T. In fact

$$F = \tfrac{1}{4} \rho \kappa l \, . \, \Sigma_{s=1}^{s=\infty} \dot{\phi}_s{}^2 \dots\dots\dots\dots\dots(2),$$

where ϕ_1, ϕ_2,... are the normal co-ordinates, by means of which T and V are reduced to sums of squares. The equations of motion are therefore simply

$$\ddot{\phi}_s + \kappa \dot{\phi}_s + n^2 \phi_s = \frac{2}{l\rho} \Phi_s \quad \dots\dots\dots\dots(3),$$

[1] Helmholtz, Ch. iv.; Brandt, *Pogg. Ann.*, Vol. cxii. p. 324, 1861.

of the same form as obtains for systems with but one degree of freedom. It is only necessary to add to what was said in Chapter III., that since κ is independent of s, the natural vibrations subside in such a manner that the amplitudes maintain their relative values.

If a periodic force $F \cos pt$ act at a single point, we have

$$\Phi_s = F \sin \frac{s\pi b}{l} \cos pt \dots\dots\dots\dots\dots\dots (4),$$

and § 46

$$\phi_s = \frac{2F \sin \epsilon}{l\rho\, p\kappa} \sin \frac{s\pi b}{l} \cos (pt - \epsilon) \dots\dots\dots\dots (5),$$

where

$$\tan \epsilon = \frac{p\kappa}{n^2 - p^2} \dots\dots\dots\dots\dots\dots\dots (6).$$

If among the natural vibrations there be any one nearly isochronous with $\cos pt$, then a large vibration of that type will be forced, unless indeed the point of excitement should happen to fall near a node. In the case of exact coincidence, the component vibration in question vanishes; for no force applied at a node can generate it, under the present law of friction, which however, it may be remarked, is very special in character. If there be no friction, $\kappa = 0$, and

$$l\rho\,\phi_s = \frac{2F}{n^2 - p^2} \sin \frac{s\pi b}{l} \cos pt \dots\dots\dots\dots (7),$$

which would make the vibration infinite, in the case of perfect isochronism, unless $\sin (s\pi b/l) = 0$.

The value of y is here, as usual.

$$y = \phi_1 \sin \frac{\pi x}{l} + \phi_2 \sin \frac{2\pi x}{l} + \phi_3 \sin \frac{3\pi x}{l} + \dots\dots\dots\dots (8).$$

132. The preceding solution is an example of the use of normal co-ordinates in a problem of forced vibrations. It is of course to free vibrations that they are more especially applicable, and they may generally be used with advantage throughout, whenever the system after the operation of various forces is ultimately left to itself. Of this application we have already had examples.

In the case of vibrations due to periodic forces, one advantage of the use of normal co-ordinates is the facility of comparison with the *equilibrium theory*, which it will be remembered is the theory of the motion on the supposition that the inertia of the system may be left out of account. If the value of the normal co-ordinate ϕ_s on the equilibrium theory be $A_s \cos pt$, then the actual value will be given by the equation

$$\phi_s = \frac{n^2 A_s}{n^2 - p^2} \cos pt \dots\dots\dots\dots\dots(1),$$

so that, when the result of the equilibrium theory is known and can readily be expressed in terms of the normal co-ordinates, the true solution with the effects of inertia included can at once be written down.

In the present instance, if a force $F \cos pt$ of very long period act at the point b of the string, the result of the equilibrium theory, in accordance with which the string would at any moment consist of two straight portions, will be

$$l\rho \phi_s = \frac{2F}{n^2} \sin \frac{s\pi b}{l} \cos pt \dots\dots\dots\dots\dots(2).$$

from which the actual result for all values of p is derived by simply writing $(n^2 - p^2)$ in place of n^2.

The value of y in this and similar cases may however be expressed in finite terms, and the difficulty of obtaining the finite expression is usually no greater than that of finding the form of the normal functions when the system is free. Thus in the equation of motion

$$\frac{d^2y}{dt^2} = a^2 \frac{d^2y}{dx^2} + Y,$$

suppose that Y varies as $\cos mat$. The forced vibration will then satisfy

$$\frac{d^2y}{dx^2} + m^2 y = - \frac{1}{a^2} Y \dots\dots\dots\dots\dots(3).$$

If $Y = 0$, the investigation of the normal functions requires the solution of

$$\frac{d^2y}{dx^2} + m^2 y = 0,$$

and a subsequent determination of m to suit the boundary conditions. In the problem of forced vibrations m is given, and we have only to supplement any particular solution of (3) with the complementary function containing two arbitrary constants. This function, apart from the value of m and the ratio of the constants, is of the same form as the normal functions; and all that remains to be effected is the determination of the two constants in accordance with the prescribed boundary conditions which the complete solution must satisfy. Similar considerations apply in the case of any continuous system.

133. If a periodic force be applied at a single point, there are two distinct problems to be considered; the first, when at the point $x = b$, a given periodic force acts; the second, when it is the actual motion of the point b that is obligatory. But it will be convenient to treat them together.

The usual differential equation

$$\frac{d^2y}{dt^2} + \kappa \frac{dy}{dt} = a^2 \frac{d^2y}{dx^2} \dots\dots\dots\dots(1),$$

is satisfied over both the parts into which the string is divided at b, but is violated in crossing from one to the other.

In order to allow for a change in the arbitrary constants, we must therefore assume distinct expressions for y, and afterwards introduce the two conditions which must be satisfied at the point of junction. These are

(1) That there is no discontinuous change in the value of y;

(2) That the resultant of the tensions acting at b balances the impressed force.

Thus, if $F \cos pt$ be the force, the second condition gives

$$T_1 \Delta \left(\frac{dy}{dx}\right) + F \cos pt = 0 \dots\dots\dots\dots(2),$$

where $\Delta(dy/dx)$ denotes the alteration in the value of dy/dx incurred in crossing the point $x = b$ in the positive direction.

We shall, however, find it advantageous to replace $\cos pt$ by the complex exponential e^{ipt}, and finally discard the imaginary part, when the symbolical solution is completed. On the assumption that y varies as e^{ipt}, the differential equation becomes

$$\frac{d^2y}{dx^2} + \lambda^2 y = 0 \dots\dots\dots\dots(3);$$

where λ^2 is the complex constant,

$$\lambda^2 = \frac{1}{a^2}(p^2 - ip\kappa) \dots\dots\dots\dots(4).$$

The most general solution of (3) consists of two terms, proportional respectively to $\sin \lambda x$, and $\cos \lambda x$; but the condition to be satisfied at $x = 0$ shews that the second does not occur here. Hence if γe^{ipt} be the value of y at $x = b$,

$$y = \gamma \frac{\sin \lambda x}{\sin \lambda b} \cdot e^{ipt} \dotfill (5),$$

is the solution applying to the first part of the string from $x = 0$ to $x = b$. In like manner it is evident that for the second part we shall have

$$y = \gamma \frac{\sin \lambda (l - x)}{\sin \lambda (l - b)} e^{ipt} \dotfill (6).$$

If γ be given, these equations constitute the symbolical solution of the problem; but if it be the force that is given, we require further to know the relation between it and γ.

Differentiation of (5) and (6) and substitution in the equation analogous to (2) gives

$$\gamma = \frac{F}{T_1} \frac{\sin \lambda b \; \sin \lambda (l - b)}{\lambda \sin \lambda l} \dotfill (7).$$

Thus

$$y = \frac{F}{T_1} \frac{\sin \lambda x \; \sin \lambda (l - b)}{\lambda \sin \lambda l} e^{ipt}$$
$$\text{from } x = 0 \text{ to } x = b$$
$$y = \frac{F}{T_1} \frac{\sin \lambda (l - x) \; \sin \lambda b}{\lambda \sin \lambda l} e^{ipt}$$
$$\text{from } x = b \text{ to } x = l$$
$$\left. \right\} \dotsc (8)^1.$$

These equations exemplify the general law of reciprocity proved in the last chapter; for it appears that the motion at x due to the force at b is the same as would have been found at b, had the force acted at x.

In discussing the solution we will take first the case in which there is no friction. The coefficient κ is then zero; while λ is real, and equal to p/a. The real part of the solution, corresponding to the force $F \cos pt$, is found by simply putting $\cos pt$ for e^{ipt} in (8), but it seems scarcely necessary to write the equations again for the sake of so small a change. The same remark applies to the forced motion given in terms of γ.

It appears that the motion becomes infinite in case the force

[1] Donkin's *Acoustics*, p. 121.

is isochronous with one of the natural vibrations of the entire string, unless the point of application be a node; but in practice it is not easy to arrange that a string shall be subject to a force of given magnitude. Perhaps the best method would be to attach a small mass of iron, attracted periodically by an electro-magnet, whose coils are traversed by an intermittent current. But unless some means of compensation were devised, the mass would have to be very small in order to avoid its inertia introducing a new complication.

A better approximation may be obtained to the imposition of an obligatory motion. A massive fork of low pitch, excited by a bow or sustained in permanent operation by electro-magnetism, executes its vibrations in approximate independence of the re-actions of any light bodies which may be connected with it. In order therefore to subject any point of a string to an obligatory transverse motion, it is only necessary to attach it to the extremity of one prong of such a fork, whose plane of vibration is perpendicular to the length of the string. This method of exhibiting the forced vibrations of a string appears to have been first used by Melde[1].

Another arrangement, better adapted for aural observation, has been employed by Helmholtz. The end of the stalk of a powerful tuning-fork, set into vibration with a bow, or otherwise, is pressed against the string. It is advisable to file the surface, which comes into contact with the string, into a suitable (saddle-shaped) form, the better to prevent slipping and jarring.

Referring to (5) we see that, if sin λb vanished, the motion (according to this equation) would become infinite, which may be taken to prove that in the case contemplated, the motion would really become great,—so great that corrections, previously insignificant, rise into importance. Now sin λb vanishes, when the force is isochronous with one of the natural vibrations of the first part of the string, supposed to be held fixed at 0 and b.

When a fork is placed on the string of a monochord, or other instrument properly provided with a sound-board, it is easy to find by trial the places of maximum resonance. A very slight displacement on either side entails a considerable falling off in the volume of the sound. The points thus determined divide the string into a number of equal parts, of such length that the natural note of any one of them (when fixed at both ends) is

[1] Pogg. *Ann.* cix. p. 193, 1859.

the same as the note of the fork, as may readily be verified. The important applications of resonance which Helmholtz has made to purify a simple tone from extraneous accompaniment will occupy our attention later.

134. Returning now to the general case where λ is complex, we have to extract the real parts from (5), (6), (8) of § 133. For this purpose the sines which occur as factors, must be reduced to the form $Re^{i\epsilon}$. Thus let

$$\sin \lambda x = R_x e^{i\epsilon_x} \dots\dots\dots\dots\dots\dots\dots(1),$$

with a like notation for the others. From (5) § 133 we shall thus obtain

$$y = \gamma \frac{R_x}{R_b} \cos (pt + \epsilon_x - \epsilon_b)\dots\dots\dots\dots\dots\dots(2),$$

$$\text{from } x = 0 \text{ to } x = b,$$

and from (6) § 133

$$y = \gamma \frac{R_{l-x}}{R_{l-b}} \cos (pt + \epsilon_{l-x} - \epsilon_{l-b}),$$

$$\text{from } x = b \text{ to } x = l,$$

corresponding to the obligatory motion $y = \gamma \cos pt$ at b.

By a similar process from (8) § 133, if

$$\lambda = \alpha + i\beta\dots\dots\dots\dots\dots\dots\dots (3),$$

we should obtain

$$\left. \begin{aligned} y &= \frac{F}{T_1} \frac{R_x . R_{l-b}}{\sqrt{(\alpha^2 + \beta^2)} . R_l} \cos \left(pt + \epsilon_x + \epsilon_{l-b} - \epsilon_l - \tan^{-1}(\beta/\alpha) \right) \\ &\qquad\qquad\qquad \text{from } x = 0 \text{ to } x = b \\ y &= \frac{F}{T_1} \frac{R_{l-x} . R_b}{\sqrt{(\alpha^2 + \beta^2)} . R_l} \cos \left(pt + \epsilon_{l-x} + \epsilon_b - \epsilon_l - \tan^{-1}(\beta/\alpha) \right) \\ &\qquad\qquad\qquad \text{from } x = b \text{ to } x = \iota \end{aligned} \right\} \dots (4),$$

corresponding to the impressed force $F \cos pt$ at b. It remains to obtain the forms of R_x, ϵ_x, &c.

The values of α and β are determined by

$$\alpha^2 - \beta^2 = \frac{p^2}{a^2}, \quad 2\alpha\beta = -\frac{p\kappa}{a^2}\dots\dots\dots\dots (5),$$

and $\sin \lambda x = \sin \alpha x \cos i\beta x + \cos \alpha x \sin i\beta x$

$$= \sin \alpha x \frac{e^{\beta x} + e^{-\beta x}}{2} + i \cos \alpha x \frac{e^{\beta x} - e^{-\beta x}}{2},$$

so that

$$R_x{}^2 = \sin^2 \alpha x \left(\frac{e^{\beta x} + e^{-\beta x}}{2}\right)^2 + \cos^2 \alpha x \left(\frac{e^{\beta x} - e^{-\beta x}}{2}\right)^2 \dots (6),$$

$$\tan \epsilon_x = \frac{e^{\beta x} - e^{-\beta x}}{e^{\beta x} + e^{-\beta x}} \cot \alpha x \dots\dots\dots\dots\dots (7),$$

while

$$\sqrt{(\alpha^2 + \beta^2)} = \frac{1}{a} \sqrt[4]{(p^4 + p^2 \kappa^2)} \dots\dots\dots\dots (8).$$

This completes the solution.

If the friction be very small, the expressions may be simplified. For instance in this case, to a sufficient approximation.

$$\alpha = p/a, \quad \beta = -\kappa/2a, \quad \sqrt{(\alpha^2 + \beta^2)} = p/a,$$

$$\tfrac{1}{2}(e^{\beta x} + e^{-\beta x}) = 1, \quad \tfrac{1}{2}(e^{\beta x} - e^{-\beta x}) = -\kappa x/2a ;$$

so that, corresponding to the obligatory motion at b $y = \gamma \cos pt$, the amplitude of the motion between $x = 0$ and $x = b$ is, approximately

$$\gamma \left\{\frac{\sin^2 \dfrac{px}{a} + \dfrac{\kappa^2 x^2}{4a^2}\cos^2 \dfrac{px}{a}}{\sin^2 \dfrac{pb}{a} + \dfrac{\kappa^2 b^2}{4a^2}\cos^2 \dfrac{pb}{a}}\right\}^{\frac{1}{2}} \dots\dots\dots\dots (9),[1]$$

which becomes great, but not infinite, when $\sin (pb/a) = 0$, or the point of application is a node.

If the imposed force, or motion, be not expressed by a single harmonic term, it must first be resolved into such. The preceding solution may then be applied to each component separately, and the results added together. The extension to the case of more than one point of application of the impressed forces is also obvious. To obtain the most general solution satisfying the conditions, the expression for the natural vibrations must also be added; but these become reduced to insignificance after the motion has been in progress for a sufficient time.

The law of friction assumed in the preceding investigation is the only one whose results can be easily followed deductively, and it is sufficient to give a general idea of the effects of dissipative forces on the motion of a string. But in other respects the conclusions drawn from it possess a fictitious simplicity, depending on the fact that F—the dissipation-function—is similar in form to T, which makes the normal co-ordinates independent of each other.

[1] Reference may be made to a paper by Morton & Vinycomb, *Phil. Mag.*

In almost any other case (for example, when but a single point of the string is retarded by friction) there are no normal co-ordinates properly so called.   There exist indeed elementary types of vibration into which the motion may be resolved, and which are perfectly independent, but these are essentially different in character from those with which we have been concerned hitherto, for the various parts of the system (as affected by one elementary vibration) are not simultaneously in the same phase.   Special cases excepted, no linear transformation of the co-ordinates (with real coefficients) can reduce T, F, and V together to a sum of squares.

If we suppose that the string has no inertia, so that $T = 0$, F and V may then be reduced to sums of squares.   This problem is of no acoustical importance, but it is interesting as being mathematically analogous to that of the conduction and radiation of heat in a bar whose ends are maintained at a constant temperature.

135.   Thus far we have supposed that at two fixed points, $x = 0$ and $x = l$, the string is held at rest.   Since absolute fixity cannot be attained in practice, it is not without interest to inquire in what manner the vibrations of a string are liable to be modified by a yielding of the points of attachment; and the problem will furnish occasion for one or two remarks of importance.   For the sake of simplicity we shall suppose that the system is symmetrical with reference to the centre of the string, and that each extremity is attached to a mass M (treated as unextended in space), and is urged by a spring (μ) towards the position of equilibrium.   If no frictional forces act, the motion is necessarily resolvable into normal vibrations.   Assume

$$y = \{\alpha \sin mx + \beta \cos mx\} \cos (mat - \epsilon)\ldots\ldots\ldots\ldots(1).$$

The conditions at the ends are that

$$\text{when} \quad x = 0, \quad M\ddot{y} + \mu y = T_1\frac{dy}{dx} \left.\right\}$$

$$\text{when} \quad x = l, \quad M\ddot{y} + \mu y = -T_1\frac{dy}{dx} \left.\right\} \quad \ldots\ldots\ldots\ldots(2),$$

which give

$$\frac{\alpha}{\beta} = \frac{\beta \tan ml - \alpha}{\alpha \tan ml + \beta} = \frac{\mu - Ma^2m^2}{mT_1} \quad \ldots\ldots\ldots\ldots(3),$$

two equations, sufficient to determine m, and the ratio of β to α. Eliminating the latter ratio, we find

$$\tan ml = \frac{2\nu}{1-\nu^2} \dots\dots\dots(4),$$

if for brevity we write ν for $\dfrac{\mu - Ma^2m^2}{mT_1}$.

Equation (3) has an infinite number of roots, which may be found by writing $\tan\theta$ for ν, so that $\tan ml = \tan 2\theta$, and the result of adding together *all* the corresponding particular solutions, each with its two arbitrary constants α and ϵ, is necessarily the most general solution of which the problem is capable, and is therefore adequate to represent the motion due to an arbitrary initial distribution of displacement and velocity. We infer that any function of x may be expanded between $x=0$ and $x=l$ in a series of terms

$$\phi_1(\nu_1\sin m_1 x + \cos m_1 x) + \phi_2(\nu_2\sin m_2 x + \cos m_2 x) + \dots\dots(5),$$

m_1, m_2, &c. being the roots of (3) and ν_1, ν_2, &c. the corresponding values of ν. The quantities ϕ_1, ϕ_2, &c. are the *normal* co-ordinates of the system.

From the symmetry of the system it follows that in each normal vibration the value of y is numerically the same at points equally distant from the middle of the string, for example, at the two ends, where $x=0$ and $x=l$. Hence $\nu_s\sin m_s l + \cos m_s l = \pm 1$, as may be proved also from (4).

The kinetic energy T of the whole motion is made up of the energy of the string, and that of the masses M. Thus

$$T = \tfrac{1}{2}\rho\int_0^l \{\Sigma\phi(\nu\sin mx + \cos mx)\}^2 dx$$

$$+ \tfrac{1}{2}M\{\dot\phi_1 + \dot\phi_2 + \dots\}^2 + \tfrac{1}{2}M\{\dot\phi_1(\nu_1\sin m_1 l + \cos m_1 l) + \dots\}^2.$$

But by the characteristic property of normal co-ordinates, terms containing their products cannot be really present in the expression for T, so that

$$\rho\int_0^l (\nu_r\sin m_r x + \cos m_r x)(\nu_s\sin m_s x + \cos m_s x)\,dx$$

$$+ M + M(\nu_r\sin m_r l + \cos m_r l)(\nu_s\sin m_s l + \cos m_s l) = 0\dots\dots(6),$$

if r and s be different.

This theorem suggests how to determine the arbitrary constants, so that the series (5) may represent an arbitrary function y. Take the expression

$$\rho \int_0^l y (\nu_s \sin m_s x + \cos m_s x) dx + M y_0 + M y_l (\nu_s \sin m_s l + \cos m_s l) \ldots (7),$$

and substitute in it the series (5) expressing y. The result is a series of terms of the type

$$\rho \int_0^l \phi_r (\nu_r \sin m_r x + \cos m_r x)(\nu_s \sin m_s x + \cos m_s x)\, dx$$

$$+ M \phi_r + M \phi_r (\nu_r \sin m_r l + \cos m_r l)(\nu_s \sin m_s l + \cos m_s l),$$

all of which vanish by (6), except the one for which $r = s$. Hence ϕ_s is equal to the expression (7) divided by

$$\rho \int_0^l (\nu_s \sin m_s x + \cos m_s x)^2\, dx + M + M (\nu_s \sin m_s l + \cos m_s l)^2 \ldots (8),$$

and thus the coefficients of the series are determined. If $M = 0$, even although μ be finite, the process is of course much simpler, but the unrestricted problem is instructive. So much stress is often laid on special proofs of Fourier's and Laplace's series, that the student is apt to acquire too contracted a view of the nature of those important results of analysis.

We shall now shew how Fourier's theorem in its general form can be deduced from our present investigation. Let $M = 0$; then if $\mu = \infty$, the ends of the string are fast, and the equation determining m becomes $\tan ml = 0$, or $ml = s\pi$, as we know it must be. In this case the series for y becomes

$$y = A_1 \sin \frac{\pi x}{l} + A_2 \sin \frac{2\pi x}{l} + A_3 \sin \frac{3\pi x}{l} + \ldots \ldots (9),$$

which must be general enough to represent any arbitrary functions of x, vanishing at 0 and l, between those limits. But now suppose that μ is zero, M still vanishing. The ends of the string may be supposed capable of sliding on two smooth rails perpendicular to its length, and the terminal condition is the vanishing of dy/dx. The equation in m is the *same as before;* and we learn that any function y' whose rates of variation vanish at $x = 0$ and $x = l$, can be expanded in a series

$$y' = B_1 \cos \frac{\pi x}{l} + B_2 \cos \frac{2\pi x}{l} + B_3 \cos \frac{3\pi x}{l} + \ldots \ldots (10).$$

This series remains unaffected when the sign of x is changed, and the first series merely changes sign without altering its numerical magnitude. If therefore y' be an even function of x, (10) represents it from $-l$ to $+l$. And in the same way, if y be an odd function of x, (9) represents it between the same limits.

Now, whatever function of x $\phi(x)$ may be, it can be divided into two parts, one of which is even, and the other odd, thus:

$$\phi(x) = \frac{\phi(x) + \phi(-x)}{2} + \frac{\phi(x) - \phi(-x)}{2};$$

so that, if $\phi(x)$ be such that $\phi(-l) = \phi(+l)$ and $\phi'(-l) = \phi'(+l)$, it can be represented between the limits $\pm l$ by the mixed series

$$A_1 \sin\frac{\pi x}{l} + B_1 \cos\frac{\pi x}{l} + A_2 \sin\frac{2\pi x}{l} + B_2 \cos\frac{2\pi x}{l} + \ldots \ldots (11).$$

This series is periodic, with the period $2l$. If therefore $\phi(x)$ possess the same property, no matter what in other respects its character may be, the series is its complete equivalent. This is Fourier's theorem[1].

We now proceed to examine the effects of a slight yielding of the supports, in the case of a string whose ends are approximately fixed. The quantity ν may be great, either through μ or through M. We shall confine ourselves to the two principal cases, (1) when μ is great and M vanishes, (2) when μ vanishes and M is great.

In the first case $\nu = \dfrac{\mu}{T_1 m}$,

and the equation in m is approximately

$$\tan ml = -\frac{2}{\nu} = -\frac{2T_1 m}{\mu}.$$

Assume $ml = s\pi + x$, where x is small; then

$$x = \tan x = -\frac{2T_1 \cdot s\pi}{\mu l} \quad \text{approximately,}$$

and $ml = s\pi\left(1 - \dfrac{2T_1}{\mu l}\right)$ ....(12).

---

[1] The best 'system' for proving Fourier's theorem from dynamical considerations is an endless chain stretched round a smooth cylinder (§ 139), or a thin re-entrant column of air enclosed in a ring-shaped tube.

To this order of approximation the tones do not cease to form a harmonic scale, but the pitch of the whole is slightly lowered. The effect of the yielding is in fact the same as that of an increase in the length of the string in the ratio $1 : 1 + \dfrac{2T_1}{\mu l}$, as might have been anticipated.

The result is otherwise if μ vanish, while M is great. Here

$$\nu = -\frac{Ma^2m}{T_1},$$

and $$\tan ml = \frac{2T_1}{Ma^2m} \quad \text{approximately.}$$

Hence $$ml = s\pi + \frac{2T_1l}{Ma^2 . s\pi} \quad\ldots\ldots\ldots\ldots\ldots\ldots(13).$$

The effect is thus equivalent to a decrease in l in the ratio

$$1 : 1 - \frac{2T_1l}{Ma^2 . s^2\pi^2},$$

and consequently there is a rise in pitch, the rise being the greater the lower the component tone. It might be thought that any kind of yielding would depress the pitch of the string, but the preceding investigation shews that this is not the case. Whether the pitch will be raised or lowered, depends on the sign of ν, and this again depends on whether the natural note of the mass M urged by the spring μ is lower or higher than that of the component vibration in question.

136. The problem of an otherwise uniform string carrying a finite load M at $x = b$ can be solved by the formulæ investigated in § 133. For, if the force $F \cos pt$ be due to the reaction against acceleration of the mass M,

$$F = \gamma\, p^2 M \ldots\ldots\ldots\ldots\ldots\ldots\ldots\ldots(1),$$

which combined with equation (7) of § 133 gives, to determine the possible values of λ (or $p : a$),

$$a^2 M \lambda \sin \lambda b \sin \lambda (l - b) = T_1 \sin \lambda l \ldots\ldots\ldots\ldots(2).$$

The value of y for any normal vibration corresponding to λ is

$$\left. \begin{array}{c} y = P \sin \lambda x \sin \lambda (l - b) \cos (a\lambda t - \epsilon) \\ \text{from } x = 0 \text{ to } x = b \\ y = P \sin \lambda (l - x) \sin \lambda b \cos (a\lambda t - \epsilon) \\ \text{from } x = b \text{ to } x = l \end{array} \right\} \quad\ldots\ldots\ldots (3),$$

where P and ϵ are arbitrary constants.

It does not require analysis to prove that any normal components which have a node at the point of attachment are unaffected by the presence of the load. For instance, if a string be weighted at the centre, its component vibrations of even orders remain unchanged, while all the odd components are depressed in pitch. Advantage may sometimes be taken of this effect of a load, when it is desired for any purpose to disturb the harmonic relation of the component tones.

If M be very great, the gravest component is widely separated in pitch from all the others. We will take the case when the load is at the centre, so that $b = l - b = \frac{1}{2} l$. The equation in λ then becomes

$$\sin \frac{\lambda l}{2} \cdot \left\{ \frac{\lambda l}{2} \tan \frac{\lambda l}{2} - \frac{\rho l}{M} \right\} = 0 \dots\dots\dots\dots(4),$$

where $\rho l : M$, denoting the ratio of the masses of the string and the load, is a small quantity which may be called α^2. The first root corresponding to the tone of lowest pitch occurs when $\frac{1}{2}\lambda l$ is small, and such that

$$(\tfrac{1}{2}\lambda l)^2 \left\{ 1 + \tfrac{1}{3} (\tfrac{1}{2} \lambda l)^2 \right\} = \alpha^2 \text{ nearly,}$$

whence $\qquad\qquad \tfrac{1}{2}\lambda l = \alpha (1 - \tfrac{1}{6} \alpha^2),$

and the periodic time is given by

$$\tau = \pi \sqrt{\frac{Ml}{T_1}} \left(1 + \frac{\rho l}{6M} \right) \dots\dots\dots\dots\dots(5).$$

The second term constitutes a correction to the rough value obtained in a previous chapter (§ 52), by neglecting the inertia of the string altogether. That it would be additive might have been expected, and indeed the formula as it stands may be obtained from the consideration that in the actual vibration the two parts of the string are nearly straight, and may be assumed to be exactly so in computing the kinetic and potential energies, without entailing any appreciable error in the calculated period. On this supposition the retention of the inertia of the string increases the kinetic energy corresponding to a given velocity of the load in the ratio of $M : M + \tfrac{1}{3} \rho l$, which leads to the above result. This method has indeed the advantage in one respect, as it might be applied when ρ is not uniform, or nearly uniform. All that is necessary is that the load M should be sufficiently predominant.

There is no other root of (4), until $\sin \tfrac{1}{2}\lambda l = 0$, which gives the second component of the string,—a vibration independent of the load. The roots after the first occur in closely contiguous pairs: for one set is given by $\frac{1}{2}\lambda l = s\pi$, and the other approximately by $\frac{1}{2}\lambda l = s\pi + \dfrac{\rho l}{s\pi M}$, in which the second term is small. The two types of vibration for $s = 1$ are shewn in the figure.

Fig 21.

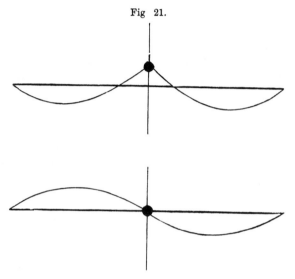

The general formula (2) may also be applied to find the effect of a small load on the pitch of the various components.

137. Actual strings and wires are not perfectly flexible. They oppose a certain resistance to bending, which may be divided into two parts, producing two distinct effects. The first is called viscosity, and shews itself by damping the vibrations. This part produces no sensible effect on the periods. The second is conservative in its character, and contributes to the potential energy of the system, with the effect of shortening the periods. A complete investigation cannot conveniently be given here, but the case which is most interesting in its application to musical instruments, admits of a sufficiently simple treatment.

When rigidity is taken into account, something more must be specified with respect to the terminal conditions than that y vanishes. Two cases may be particularly noted:—

(i)   When the ends are clamped, so that $dy/dx = 0$ at the ends.

(ii) When the terminal directions are perfectly free, in which case $d^2y/dx^2 = 0$.

It is the latter which we propose now to consider.

If there were no rigidity, the type of vibration would be

$$y \propto \sin\frac{s\pi x}{l}, \text{ satisfying the second condition.}$$

The effect of the rigidity might be slightly to disturb the type; but whether such a result occur or not, the period calculated from the potential and kinetic energies on the supposition that the type remains unaltered is necessarily correct as far as the first order of small quantities (§ 88).

Now the potential energy due to the stiffness is expressed by

$$\delta V = \tfrac{1}{2} B \int_0^l \left(\frac{d^2y}{dx^2}\right)^2 dx \dots\dots\dots\dots\dots (1),$$

where B is a quantity depending on the nature of the material and on the form of the section in a manner that we are not now prepared to examine. The *form* of δV is evident, because the force required to bend any element ds is proportional to ds, and to the amount of bending already effected, that is to ds/ρ. The whole work which must be done to produce a curvature $1/\rho$ in ds is therefore proportional to ds/ρ^2; while to the approximation to which we work $ds = dx$, and $1/\rho = d^2y/dx^2$.

Thus, if $y = \phi \sin\dfrac{s\pi x}{l}$,

$$T = \tfrac{1}{4}\rho l \, \dot\phi^2; \quad V = \tfrac{1}{4} T_1 l \cdot \frac{s^2\pi^2}{l^2} \, \phi^2 \left(1 + \frac{B}{T_1}\frac{s^2\pi^2}{l^2}\right),$$

and the period of ϕ is given by

$$\tau = \tau_0 \left(1 - \frac{B}{2T_1}\frac{s^2\pi^2}{l^2}\right) \dots\dots\dots\dots\dots (2),$$

if τ_0 denote what the period would become if the string were endowed with perfect flexibility. It appears that the effect of the stiffness increases rapidly with the order of the component vibrations, which cease to belong to a harmonic scale. However, in the strings employed in music, the tension is usually sufficient to reduce the influence of rigidity to insignificance.

The method of this section cannot be applied without modification to the other case of terminal condition, namely, when the ends are clamped. In their immediate neighbourhood the type of vibration must differ from that assumed by a perfectly flexible string by a quantity, which is no longer small, and whose square therefore cannot be neglected. We shall return to this subject, when treating of the transverse vibrations of rods.

138. There is one problem relating to the vibrations of strings which we have not yet considered, but which is of some practical interest, namely, the character of the motion of a violin (or cello) string under the action of the bow. In this problem the *modus operandi* of the bow is not sufficiently understood to allow us to follow exclusively the *a priori* method: the indications of theory must be supplemented by special observation. By a dexterous combination of evidence drawn from both sources Helmholtz has succeeded in determining the principal features of the case, but some of the details are still obscure.

Since the note of a good instrument, well handled, is musical, we infer that the vibrations are strictly periodic, or at least that strict periodicity is the ideal. Moreover—and this is very important—the note elicited by the bow has nearly, or quite, the same pitch as the natural note of the string. The vibrations, although forcèd, are thus in some sense free. They are wholly dependent for their maintenance on the energy drawn from the bow, and yet the bow does not determine, or even sensibly modify, their periods. We are reminded of the self-acting electrical interrupter, whose motion is indeed forced in the technical sense, but has that kind of freedom which consists in determining (wholly, or in part) under what influences it shall come.

But it does not at once follow from the fact that the string vibrates with its natural periods, that it conforms to its natural types. If the coefficients of the Fourier expansion.

$$y = \phi_1 \sin \frac{\pi x}{l} + \phi_2 \sin \frac{2\pi x}{l} + \ldots\ldots$$

be taken as the independent co-ordinates by which the configuration of the system is at any moment defined, we know that when there is no friction, or friction such that $F \propto T$, the natural vibrations are expressed by making each co-ordinate a *simple* harmonic (or quasi-harmonic) function of the time; while, for all that has hitherto appeared to the contrary, each co-ordinate in the present case might be *any* function of the time periodic in time τ. But a little examination will shew that the vibrations must be sensibly natural in their types as well as in their periods.

The force exercised by the bow at its point of application may be expressed by

$$Y = \Sigma A_r \cos \left(\frac{2r\pi t}{\tau} - \epsilon_r \right);$$

so that the equation of motion for the co-ordinate ϕ_s is

$$\ddot{\phi}_s + \kappa \dot{\phi}_s + \frac{s^2 \pi^2 a^2}{l^2} \phi_s = \frac{2}{l\rho} \sin \frac{s\pi b}{l} . \Sigma A_r \cos \left(\frac{2r\pi t}{\tau} - \epsilon_r \right),$$

b being the point of application. Each of the component parts of Φ_s will give a corresponding term of its own period in the solution, but the one whose period is the same as the natural period of ϕ_s will rise enormously in relative importance. Practically then, if the damping be small, we need only retain that part of ϕ_s which depends on $A_s \cos \left(\frac{2s\pi t}{\tau} - \epsilon_s \right)$, that is to say, we may regard the vibrations as natural in their types.

Another material fact, supported by evidence drawn both from theory and aural observation, is this. All component vibrations are absent which have a node at the point of excitation. "In order, however, to extinguish these tones, it is necessary that the coincidence of the point of application of the bow with the node should be very *exact*. A very small deviation reproduces the missing tones with considerable strength[1]."

The remainder of the evidence on which Helmholtz' theory rests, was derived from direct observation with the vibration-microscope. As explained in Chapter II., this instrument affords a view of the curve representing the motion of the point under observation, as it would be seen traced on the surface of a transparent cylinder. In order to deduce the representative curve in its ordinary form, the imaginary cylinder must be conceived to be unrolled, or developed, into a plane.

The simplest results are obtained when the bow is applied at a node of one of the higher components, and the point observed is one of the other nodes of the same system. If the bow work fairly so as to draw out the fundamental tone clearly and strongly, the representative curve is that shewn in figure 22; where the

[1] Donkin's *Acoustics*, p. 131.

abscissæ correspond to the time (AB being a complete period), and the ordinates represent the displacement. The remarkable

Fig. 22.

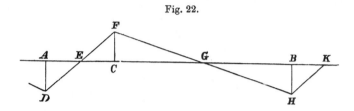

fact is disclosed that the whole period τ may be divided into two parts τ_0 and $\tau - \tau_0$, during each of which the velocity of the observed point is constant; but the velocities to and fro are in general unequal.

We have now to represent this curve by a series of harmonic terms. If the origin of time correspond to the point A, and $AD = FC = \gamma$, Fourier's theorem gives

$$y = \frac{2\gamma\tau^2}{\pi^2\tau_0(\tau - \tau_0)} \; \Sigma_{s=1}^{s=\infty} \frac{1}{s^2} \sin\frac{s\pi\tau_0}{\tau} \sin\frac{2s\pi}{\tau}\left(t - \frac{\tau_0}{2}\right) \dots\dots(1).$$

With respect to the value of τ_0, we know that all those components of y must vanish for which $\sin(s\pi x_0/l) = 0$ (x_0 being the point of observation), because under the circumstances of the case the bow cannot generate them. There is therefore reason to suppose that $\tau_0 : \tau = x_0 : l$; and in fact observation proves that $AC : CB$ (in the figure) is equal to the ratio of the two parts into which the string is divided by the point of observation.

Now the free vibrations of the string are represented in general by

$$y = \Sigma_{s=1}^{s=\infty} \sin\frac{s\pi x}{l}\left\{A_s \cos\frac{2s\pi t}{\tau} + B_s \sin\frac{2s\pi t}{\tau}\right\};$$

and this at the point $x = x_0$ must agree with (1). For convenience of comparison, we may write

$$A_s \cos\frac{2s\pi t}{\tau} + B_s \sin\frac{2s\pi t}{\tau} = C_s \cos\frac{2s\pi}{\tau}\left(t - \frac{\tau_0}{2}\right)$$

$$+ D_s \sin\frac{2s\pi}{\tau}\left(t - \frac{\tau_0}{2}\right),$$

and it then appears that $C_s = 0$.

We find also to determine D_s

$$\sin \frac{s\pi x_0}{l} \cdot D_s = \frac{2\gamma\tau^2}{\pi^2\tau_0(\tau-\tau_0)} \frac{1}{s^2} \sin \frac{s\pi x_0}{l},$$

whence

$$D_s = \frac{2\gamma\tau^2}{\pi^2\tau_0(\tau-\tau_0)} \frac{1}{s^2} \dots\dots\dots\dots (2),$$

unless $\sin(s\pi x_0/l) = 0$.

In the case reserved, the comparison leaves D_s undetermined, but we know on other grounds that D_s then vanishes. However, for the sake of simplicity, we shall suppose for the present that D_s is always given by (2). If the point of application of the bow do not coincide with a node of any of the lower components, the error committed will be of no great consequence.

On this understanding the complete solution of the problem is

$$y = \frac{2\gamma\tau^2}{\pi^2\tau_0(\tau-\tau_0)} \sum_{s=1}^{s=\infty} \frac{1}{s^2} \sin \frac{s\pi x}{l} \sin \frac{2s\pi}{\tau}\left(t - \frac{\tau_0}{2}\right) \dots\dots (3).$$

The amplitudes of the components are therefore proportional to s^{-2}. In the case of a plucked string we found for the corresponding function $s^{-2}\sin(s\pi b/l)$, which is somewhat similar. If the string be plucked at the middle, the even components vanish, but the odd ones follow the same law as obtains for a violin string. The equation (3) indicates that the string is always in the form of two straight lines meeting at an angle. In order more conveniently to shew this, let us change the origin of the time, and the constant multiplier so that

$$y = \frac{8P}{\pi^2} \sum \frac{1}{s^2} \sin \frac{s\pi x}{l} \sin \frac{2s\pi t}{\tau} \dots\dots\dots\dots (4)$$

will be the equation expressing the form of the string at any time

Now we know (§ 127) that the equation of the pair of lines proceeding from the fixed ends of the string, and meeting at a point whose co-ordinates are α, β, is

$$y = \frac{2\beta l^2}{\pi^2\alpha(l-\alpha)} \sum \frac{1}{s^2} \sin \frac{s\pi\alpha}{l} \sin \frac{s\pi x}{l}$$

Thus at the time t, (4) represents such a pair of lines, meeting at the point whose co-ordinates are given by

$$\frac{\beta l^2}{\alpha(l-\alpha)} = \pm 4P,$$

$$\sin \frac{s\pi\alpha}{l} = \pm \sin \frac{2s\pi t}{\tau}.$$

These equations indicate that the projection on the axis of x of the point of intersection moves uniformly backwards and forwards between $x = 0$ and $x = l$, and that the point of intersection itself is situated on one or other of two parabolic arcs, of which the equilibrium position of the string is a common chord.

Since the motion of the string as thus defined by that of the point of intersection of its two straight parts, has no especial relation to x_0 (the point of observation), it follows that, according to these equations, the same kind of motion might be observed at any other point. And this is approximately true. But the theoretical result, it will be remembered, was only obtained by assuming the presence in certain proportions of component vibrations having nodes at x_0, though in fact their absence is required by mechanical laws. The presence or absence of these components is a matter of indifference when a node is the point of observation, but not in any other case. When the node is departed from, the vibration curve shews a series of ripples, due to the absence of the components in question. Some further details will be found in Helmholtz and Donkin.

The sustaining power of the bow depends upon the fact that solid friction is less at moderate than at small velocities, so that when the part of the string acted upon is moving with the bow (not improbably at the same velocity), the mutual action is greater than when the string is moving in the opposite direction with a greater relative velocity. The accelerating effect in the first part of the motion is thus not entirely neutralised by the subsequent retardation, and an outstanding acceleration remains capable of maintaining the vibration in spite of other losses of energy. A curious effect of the same peculiarity of solid friction has been observed by W. Froude, who found that the vibrations of a pendulum swinging from a shaft might be maintained or even increased by causing the shaft to rotate.

[Another case in which the vibrations of a string are maintained is that of the Aeolian Harp. It has often been suggested that the action of the wind is analogous to that of a bow; but the analogy is disproved by the observation[1] that the vibrations are executed in a plane *transverse* to the direction of the wind. The true explanation involves hydrodynamical theory not yet developed.]

[1] *Phil. Mag.*, March, 1879, p. 161.

139. A string stretched on a smooth curved surface will in equilibrium lie along a geodesic line, and, subject to certain conditions of stability, will vibrate about this configuration, if displaced. The simplest case that can be proposed is when the surface is a cylinder of any form, and the equilibrium position of the string is perpendicular to the generating lines. The student will easily prove that the motion is independent of the curvature of the cylinder, and that the vibrations are in all essential respects the same as if the surface were developed into a plane. The case of an endless string, forming a necklace round the cylinder, is worthy of notice.

In order to illustrate the characteristic features of this class of problems, we will take the comparatively simple example of a string stretched on the surface of a smooth sphere, and lying, when in equilibrium, along a great circle. The co-ordinates to which it will be most convenient to refer the system are the latitude θ measured from the great circle as equator, and the longitude ϕ measured along it. If the radius of the sphere be a, we have

$$T = \frac{1}{2} \int \rho \, (a\dot{\theta})^2 a \, d\phi = \frac{a^3 \rho}{2} \int \dot{\theta}^2 d\phi \ldots\ldots\ldots\ldots (1).$$

The extension of the string is denoted by

$$\int (ds - a\, d\phi) = a \int \left(\frac{ds}{a\, d\phi} - 1 \right) d\phi.$$

Now

$$ds^2 = (a\, d\theta)^2 + (a \cos \theta \, d\phi)^2 \, ;$$

so that

$$\frac{ds}{a\, d\phi} - 1 = \left\{ \left(\frac{d\theta}{d\phi} \right)^2 + \cos^2 \theta \right\}^{\frac{1}{2}} - 1 = \frac{1}{2} \left(\frac{d\theta}{d\phi} \right)^2 - \frac{\theta^2}{2}, \text{ approximately.}$$

Thus

$$V = \tfrac{1}{2} a T_1 \int \left\{ \left(\frac{d\theta}{d\phi} \right)^2 - \theta^2 \right\} d\phi \ldots\ldots\ldots\ldots (2);\, [1]$$

and

$$\delta V = a T_1 \cdot \delta\theta \left[\frac{d\theta}{d\phi} \right]_0^l - a T_1 \int_0^l \delta\theta \left(\frac{d^2\theta}{d\phi^2} + \theta \right) d\phi.$$

If the ends be fixed,

$$\delta\theta \left[\frac{d\theta}{d\phi} \right]_0^l = 0,$$

[1] Cambridge Mathematical Tripos Examination, 1876.

and the equation of virtual velocities is

$$a^3 \rho \int_0^l \ddot{\theta}\, \delta\theta\, d\phi - aT_1 \int_0^l \delta\theta \left(\frac{d^2\theta}{d\phi^2} + \theta\right) d\phi = 0,$$

whence, since $\delta\theta$ is arbitrary,

$$a^2 \rho\, \ddot{\theta} = T_1 \left(\frac{d^2\theta}{d\phi^2} + \theta\right) \dots\dots\dots\dots\dots (3).$$

This is the equation of motion.

If we assume $\theta \propto \cos pt$, we get

$$\frac{d^2\theta}{d\phi^2} + \theta + \frac{a^2\rho}{T_1} p^2\theta = 0 \dots\dots\dots\dots\dots (4),$$

of which the solution, subject to the condition that θ vanishes with ϕ, is

$$\theta = A \sin\left\{\frac{a^2\rho}{T_1} p^2 + 1\right\}^{\frac{1}{2}}\phi \cdot \cos pt \dots\dots\dots\dots (5).$$

The remaining condition to be satisfied is that θ vanishes when $a\phi = l$, or $\phi = \alpha$, if $\alpha = l/a$.

This gives

$$p^2 = \frac{T_1}{a^2\rho}\left(\frac{m^2\pi^2}{\alpha^2} - 1\right) = \frac{T_1}{\rho}\left(\frac{m^2\pi^2}{l^2} - \frac{1}{a^2}\right) \dots\dots\dots (6),$$

where m is an integer.

The normal functions are thus of the same form as for a straight string, viz.

$$\theta = A \sin\frac{m\pi\phi}{\alpha} \cos pt. \dots\dots\dots\dots (7),$$

but the series of periods is different. The effect of the curvature is to make each tone graver than the corresponding tone of a straight string. If $\alpha > \pi$, one at least of the values of p^2 is negative, indicating that the corresponding modes are unstable. If $\alpha = \pi$, p_1 is zero, the string being of the same length in the displaced position, as when $\theta = 0$.

A similar method might be applied to calculate the motion of a string stretched round the equator of any surface of revolution[1].

140. The approximate solution of the problem for a vibrating string of nearly but not quite uniform longitudinal density has been fully considered in Chapter IV. § 91, as a convenient example of

---

[1] [For a more general treatment of this question see Michell, *Messenger of Mathematics*, vol. XIX. p. 87, 1890.]

the general theory of approximately simple systems. It will be sufficient here to repeat the result. If the density be $\rho_0 + \delta\rho$, the period τ_r of the rth component vibration is given by

$$\tau_r^2 = \frac{4l^2\rho_0}{T_1}\left\{1 + \frac{2}{l}\int_0^l \frac{\delta\rho}{\rho_0}\sin^2\frac{r\pi x}{l}dx\right\} \dots\dots\dots\dots (1).$$

If the irregularity take the form of a small load of mass m at the point $x = b$, the formula may be written

$$\tau_r^2 = \frac{4l^2\rho_0}{T_1}\left\{1 + \frac{2m}{l\rho_0}\sin^2\frac{r\pi b}{l}\right\} \dots\dots\dots\dots (2).$$

These values of τ^2 are correct as far as the first power of the small quantities $\delta\rho$ and m, and give the means of calculating a correction for such slight departures from uniformity as must always occur in practice.

As might be expected, the effect of a small load vanishes at nodes, and rises to a maximum at the points midway between consecutive nodes. When it is desired merely to make a rough estimate of the effective density of a nearly uniform string, the formula indicates that attention is to be given to the neighbourhood of loops rather than to that of nodes.

[The effect of a small variation of density upon the period is the same whether it occur at a distance x from one end of the string, or at an equal distance from the other end. The *mean* variation at points equidistant from the centre is all that we need regard, and thus no generality will be lost if we suppose that the density remains symmetrically distributed with respect to the centre. Thus we may write

$$\tau_r^2 = \frac{4l^2\rho_0}{T_1}(1 + \alpha_r) \dots\dots\dots\dots\dots(3)$$

where
$$\alpha_r = \frac{2}{l}\int_0^{\frac{1}{2}l} \frac{\delta\rho}{\rho_0}\left(1 - \cos\frac{2\pi r x}{l}\right)dx \dots\dots\dots\dots (4).$$

In this equation $\delta\rho$ may be expanded from 0 to $\frac{1}{2}l$ in the series

$$\frac{\delta\rho}{\rho_0} = A_0 + A_1\cos\frac{2\pi x}{l} + \dots + A_r\cos\frac{2\pi r x}{l} + \dots\dots\dots\dots(5),$$

where
$$A_0 = \frac{2}{l}\int_0^{\frac{1}{2}l} \frac{\delta\rho}{\rho_0}dx \dots\dots\dots\dots\dots\dots(6),$$

$$A_r = \frac{4}{l}\int_0^{\frac{1}{2}l} \frac{\delta\rho}{\rho_0}\cos\frac{2\pi r x}{l}dx \dots\dots\dots\dots(7).$$

Accordingly,

$$\alpha_r = A_0 - \tfrac{1}{2}A_r \quad \dots\dots\dots\dots\dots\dots\dots\dots(8).$$

This equation, as it stands, gives the changes in period in terms of the changes of density supposed to be known. And it shews conversely that a variation of density may always be found which will give prescribed arbitrary displacements to all the periods. This is a point of some interest.

In order to secure a reasonable continuity in the density, it is necessary to suppose that $\alpha_1, \alpha_2 \dots$ are so prescribed that α_r assumes ultimately a constant value when r is increased indefinitely. If this condition be satisfied, we may take $A_0 = \alpha_\infty$, and then A_r tends to zero as r increases.

As a simple example, suppose that it be required so to vary the density of a string that, while the pitch of the fundamental tone is displaced, all other tones shall remain unaltered. The conditions give

$$\alpha_2 = \alpha_3 = \alpha_4 \dots\dots = \alpha_\infty = 0.$$

Accordingly

$$A_0 = A_2 = A_3 = \dots\dots = 0,$$

and

$$A_1 = -2\alpha_1.$$

Thus by (5)

$$\delta\rho/\rho_0 = -2\alpha_1 \cos(2\pi x/l).]$$

141. The differential equation determining the motion of a string, whose longitudinal density ρ is variable, is

$$\rho \frac{d^2y}{dt^2} = T_1 \frac{d^2y}{dx^2} \quad \dots\dots\dots\dots\dots\dots (1),$$

from which, if we assume $y \propto \cos pt$, we obtain to determine the normal functions

$$\frac{d^2y}{dx^2} + \nu^2\rho y = 0 \quad \dots\dots\dots\dots\dots\dots (2),$$

where ν^2 is written for p^2/T_1. This equation is of the second order and linear, but has not hitherto been solved in finite terms. Considered as defining the curve assumed by the string in the normal mode under consideration, it determines the *curvature* at any point, and accordingly embodies a rule by which the curve can be constructed graphically. Thus in the application to a string fixed at both ends, if we start from either end at an arbitrary inclination, and with zero curvature, we are always directed by the equation with what curvature to proceed, and in this way we may trace out the entire curve.

If the assumed value of ν^2 be right, the curve will cross the axis of x at the required distance, and the law of vibration will be completely determined. If ν^2 be not known, different values may be tried until the curve ends rightly; a sufficient approximation to the value of ν^2 may usually be arrived at by a calculation founded on an assumed type (§§ 88, 90).

Whether the longitudinal density be uniform or not, the periodic time of any simple vibration varies *cœteris paribus* as the square root of the density and inversely as the square root of the tension under which the motion takes place.

The converse problem of determining the density, when the period and the type of vibration are given, is always soluble. For this purpose it is only necessary to substitute the given value of y, and of its second differential coefficient in equation (2). Unless the density be infinite, the extremities of a string are points of zero curvature.

When a given string is shortened, every component tone is raised in pitch. For the new state of things may be regarded as derived from the old by introduction, at the proposed point of fixture, of a spring (without inertia), whose stiffness is gradually increased without limit. At each step of the process the potential energy of a given deformation is augmented, and therefore (§ 88) the pitch of every tone is raised. In like manner an addition to the length of a string depresses the pitch, even though the added part be destitute of inertia.

142. Although a general integration of equation (2) of § 141 is beyond our powers, we may apply to the problem some of the many interesting properties of the solution of the linear equation of the second order, which have been demonstrated by MM. Sturm and Liouville[1]. It is impossible in this work to give anything like a complete account of their investigations; but a sketch, in which the leading features are included, may be found interesting, and will throw light on some points connected with the general

---

[1] The memoirs referred to in the text are contained in the first volume of Liouville's *Journal* (1836).

theory of the vibrations of continuous bodies. I have not thought it necessary to adhere very closely to the methods adopted in the original memoirs.

At no point of the curve satisfying the equation

$$\frac{d^2y}{dx^2} + \nu^2\rho\, y = 0 \dots\dots\dots\dots\dots\dots (1),$$

can both y and dy/dx vanish together. By successive differentiations of (1) it is easy to prove that, if y and dy/dx vanish simultaneously, all the higher differential coefficients d^2y/dx^2, d^3y/dx^3, &c. must also vanish at the same point, and therefore by Taylor's theorem the curve must coincide with the axis of x.

Whatever value be ascribed to ν^2, the curve satisfying (1) is *sinuous*, being concave throughout towards the axis of x, since ρ is everywhere positive. If at the origin y vanish, and dy/dx be positive, the ordinate will remain positive for all values of x below a certain limit dependent on the value ascribed to ν^2. If ν^2 be very small, the curvature is slight, and the curve will remain on the positive side of the axis for a great distance. We have now to prove that as ν^2 increases, all the values of x which satisfy the equation $y = 0$ gradually diminish in magnitude.

Let y' be the ordinate of a second curve satisfying the equation

$$\frac{d^2y'}{dx^2} + \nu'^2\rho\, y' = 0 \dots\dots\dots\dots\dots\dots (2),$$

as well as the condition that y' vanishes at the origin, and let us suppose that ν'^2 is somewhat greater than ν^2. Multiplying (2) by y, and (1) by y', subtracting, and integrating with respect to x between the limits 0 and x, we obtain, since y and y' both vanish with x,

$$y'\frac{dy}{dx} - y\frac{dy'}{dx} = (\nu'^2 - \nu^2)\int_0^x \rho\, y y'\, dx \dots\dots\dots\dots (3).$$

If we further suppose that x corresponds to a point at which y vanishes, and that the difference between ν'^2 and ν^2 is very small, we get ultimately

$$y'\frac{dy}{dx} = \delta\nu^2 \int_0^x \rho\, y^2\, dx \dots\dots\dots\dots\dots (4).$$

The right-hand member of (4) being essentially positive, we earn that y' and dy/dx are of the same sign, and therefore that, whether dy/dx be positive or negative, y' is already of the same sign as that to which y is changing, or in other words, the value of x for which y' vanishes is less than that for which y vanishes.

If we fix our attention on the portion of the curve lying between $x = 0$ and $x = l$, the ordinate continues positive throughout as the value of ν^2 increases, until a certain value is attained, which we will call $\nu_1{}^2$. The function y is now identical in form with the first normal function u_1 of a string of density ρ fixed at 0 and l, and has no root except at those points. As ν^2 again increases, the first root moves inwards from $x = l$ until, when a second special value $\nu_2{}^2$ is attained, the curve again crosses the axis at the point $x = l$, and then represents the second normal function u_2. This function has thus one internal root, and one only. In like manner corresponding to a higher value $\nu_3{}^2$ we obtain the third normal function u_3 with two internal roots, and so on. The nth function u_n has thus exactly $n-1$ internal roots, and since its first differential coefficient never vanishes simultaneously with the function, it changes sign each time a root is passed.

From equation (3) it appears that if u_r and u_s be two different normal functions,

$$\int_0^l \rho\, u_r u_s\, dx = 0 \quad\dots\dots\dots\dots\dots\dots (5).$$

A beautiful theorem has been discovered by Sturm relating to the number of the roots of a function derived by addition from a finite number of normal functions. If u_m be the component of lowest order, and u_n the component of highest order, the function

$$f(x) = \phi_m u_m + \phi_{m+1} u_{m+1} + \dots\dots + \phi_n u_n \dots\dots\dots (6),$$

where ϕ_m, ϕ_{m+1}, &c. are arbitrary coefficients, has *at least* $m-1$ internal roots, and *at most* $n-1$ internal roots. The extremities at $x = 0$ and at $x = l$ correspond of course to roots in all cases. The following demonstration bears some resemblance to that given by Liouville, but is considerably simpler, and, I believe, not less rigorous.

If we suppose that $f(x)$ has exactly μ internal roots (any number of which may be equal), the derived function $f'(x)$ cannot have less than $\mu + 1$ internal roots, since there must be at least one root of $f'(x)$ between each pair of consecutive roots of $f(x)$, and the whole number of roots of $f(x)$ concerned is $\mu + 2$. In like manner, we see that there must be at least μ roots of $f''(x)$,

besides the extremities, which themselves necessarily correspond
to roots; so that in passing from $f(x)$ to $f''(x)$ it is impossible
that any roots can be lost.   Now

$$f''(x) = \phi_m u_m'' + \phi_{m+1} u''_{m+1} + \ldots\ldots + \phi_n u_n''$$
$$= -\rho\,(\nu_m^2\,\phi_m u_m + \nu^2_{m+1}\,\phi_{m+1} u_{m+1} + \ldots\ldots + \nu_n^2\,\phi_n u_n)\ldots(7),$$

as we see by (1); and therefore, since ρ is always positive, we
infer that

$$\nu_m^2\,\phi_m u_m + \nu^2_{m+1}\,\phi_{m+1} u_{m+1} + \ldots\ldots + \nu_n^2\,\phi_n u_n \ldots\ldots(8),$$

has at least μ roots.

. Again, since (8) is an expression of the same form as $f(x)$,
similar reasoning proves that

$$\nu_m^4\,\phi_m u_m + \nu^4_{m+1}\,\phi_{m+1} u_{m+1} + \ldots\ldots + \nu_n^4\,\phi_n u_n$$

has at least μ internal roots; and the process may be continued
to any extent.   In this way we obtain a series of functions, all
with μ internal roots at least, which differ from the original
function $f(x)$ by the continually increasing relative importance of
the components of the higher orders.   When the process has been
carried sufficiently far, we shall arrive at a function, whose form
differs as little as we please from that of the normal function of
highest order, viz. u_n, and which has therefore $n-1$ internal roots.
It follows that, since no roots can be lost in passing down the
series of functions, the number of internal roots of $f(x)$ cannot
exceed $n-1$.

The other half of the theorem is proved in a similar manner
by continuing the series of functions backwards from $f(x)$.   In
this way we obtain

$$\phi_m u_m + \qquad \phi_{m+1} u_{m+1} + \ldots\ldots + \qquad \phi_n u_n$$
$$\nu_m^{-2}\,\phi_m u_m + \nu^{-2}_{m+1}\,\phi_{m+1} u_{m+1} + \ldots\ldots + \nu_n^{-2}\,\phi_n u_n$$
$$\nu_m^{-4}\,\phi_m u_m + \nu^{-4}_{m+1}\,\phi_{m+1} u_{m+1} + \ldots\ldots + \nu_n^{-4}\,\phi_n u_n$$
$$\ldots\ldots\ldots\ldots\ldots\ldots\ldots\ldots\ldots\ldots\ldots\ldots\ldots\ldots\ldots,$$

arriving at last at a function sensibly coincident in form with the
normal function of *lowest* order, viz. u_m, and having therefore
$m-1$ internal roots.   Since no roots can be lost in passing up the
series from this function to $f(x)$, it follows that $f(x)$ cannot have
fewer internal roots than $m-1$; but it must be understood that
any number of the $m-1$ roots may be equal.

We will now prove that $f(x)$ cannot be identically zero, unless all the coefficients ϕ vanish. Suppose that ϕ_r is not zero. Multiply (6) by ρu_r, and integrate with respect to x between the limits 0 and l. Then by (5)

$$\int_0^l \rho\, u_r f(x)\, dx = \phi_r \int_0^l \rho\, u_r{}^2 dx \quad \ldots\ldots\ldots\ldots (9);$$

from which, since the integral on the right-hand side is finite, we see that $f(x)$ cannot vanish for all values of x included within the range of integration.

Liouville has made use of Sturm's theorem to shew how a series of normal functions may be compounded so as to have an arbitrary sign at all points lying between $x = 0$ and $x = l$. His method is somewhat as follows.

The values of x for which the function is to change sign being a, b, c, ..., quantities which without loss of generality we may suppose to be all different, let us consider the series of determinants,

$$\begin{vmatrix} u_1(a), & u_1(x) \\ u_2(a), & u_2(x) \end{vmatrix}, \qquad \begin{vmatrix} u_1(a), & u_1(b), & u_1(x) \\ u_2(a), & u_2(b), & u_2(x) \\ u_3(a), & u_3(b), & u_3(x) \end{vmatrix}, \&c.$$

The first is a linear function of $u_1(x)$ and $u_2(x)$, and by Sturm's theorem has therefore one internal root at most, which root is evidently a. Moreover the determinant is not identically zero, since the coefficient of $u_2(x)$, viz. $u_1(a)$, does not vanish, whatever be the value of a. We have thus obtained a function, which changes sign at an arbitrary point a, and there only internally.

The second determinant vanishes when $x = a$, and when $x = b$, and, since it cannot have more than two internal roots, it changes sign, when x passes through these values, and there only. The coefficient of $u_3(x)$ is the value assumed by the first determinant when $x = b$, and is therefore finite. Hence the second determinant is not identically zero.

Similarly the third determinant in the series vanishes and changes sign when $x = a$, when $x = b$, and when $x = c$, and at these internal points only. The coefficient of $u_4(x)$ is finite, being the value of the second determinant when $x = c$.

It is evident that by continuing this process we can form functions compounded of the normal functions, which shall vanish and change sign for any arbitrary values of x, and not elsewhere internally; or, in other words, we can form a function whose sign is arbitrary over the whole range from $x = 0$ to $x = l$.

On this theorem Liouville founds his demonstration of the possibility of representing an arbitrary function between $x = 0$ and $x = l$ by a series of normal functions. If we assume the possibility of the expansion and take

$$f(x) = \phi_1 u_1(x) + \phi_2 u_2(x) + \phi_3 u_3(x) + \ldots\ldots\ldots (10),$$

the necessary values of ϕ_1, ϕ_2, &c. are determined by (9), and we find

$$f(x) = \Sigma \left\{ u_r(x) \int_0^l \rho\, u_r(x) f(x)\, dx \div \int_0^l \rho\, u_r^2(x)\, dx \right\} \ldots\ldots (11).$$

If the series on the right be denoted by $F(x)$, it remains to establish the identity of $f(x)$ and $F(x)$.

If the right-hand member of (11) be multiplied by $\rho u_r(x)$ and integrated with respect to x from $x = 0$ to $x = l$, we see that

$$\int_0^l \rho\, u_r(x)\, F(x)\, dx = \int_0^l \rho\, u_r(x) f(x)\, dx,$$

or, as we may also write it,

$$\int_0^l \{F(x) - f(x)\}\, \rho u_r(x)\, dx = 0 \ldots\ldots\ldots\ldots\ldots (12),$$

where $u_r(x)$ is *any* normal function. From (12) it follows that

$$\int_0^l \{F(x) - f(x)\} \{A_1 u_1(x) + A_2 u_2(x) + A_3 u_3(x) + \ldots\} \rho\, dx = 0 \ldots (13),$$

where the coefficients A_1, A_2, &c. are arbitrary.

Now if $F(x) - f(x)$ be not identically zero, it will be possible so to choose the constants A_1, A_2, &c. that $A_1 u_1(x) + A_2 u_2(x) + \ldots$ has throughout the same sign as $F(x) - f(x)$, in which case every element of the integral would be positive, and equation (13) could not be true. It follows that $F(x) - f(x)$ cannot differ from zero, or that the series of normal functions forming the right-hand member of (11) is identical with $f(x)$ for all values of x from $x = 0$ to $x = l$.

The arguments and results of this section are of course applicable to the particular case of a uniform string for which the normal functions are circular.

[As a particular case of variable density the supposition that

$\rho = \sigma x^{-2}$ is worthy of notice, § 148 b. In the notation there adopted

$$m^2 + \tfrac{1}{4} = n^2 = p^2 \sigma / T_1 \dots\dots\dots\dots\dots(14),$$

and the general solution is

$$y = A x^{\frac{1}{2}+in} + B x^{\frac{1}{2}-in} \dots\dots\dots\dots\dots(15).$$

If the string be fixed at two points, whose abscissæ x_1, x_2 are as r to 1, the frequency equation is $r^{2im} = 1$, or

$$n^2 = \tfrac{1}{4} + \frac{s^2 \pi^2}{(\log r)^2} \dots\dots\dots\dots\dots(16),$$

where s denotes an integer. The proper frequencies thus depend only upon the *ratio* of the terminal abscissæ. By supposing r nearly equal to unity we may fall back upon the usual formula (§ 124) applicable to a uniform string.

The general form of the normal function is

$$y = x^{\frac{1}{2}} \sin \frac{s\pi \log (x/x_1)}{\log (x_2/x_1)} \dots\dots\dots\dots(17).]$$

142 *a.* The points where the string remains at rest, or nodes, are of course determined by the roots of the normal functions, when the vibrations are free. In this case the frequency is limited to certain definite values; but when the vibrations are forced, they may be of any frequency, and it becomes possible to trace the motion of the nodal points as the frequency increases continuously.

For example, suppose that the imposed force acts at a single point P of a string AB, whose density may be variable. So long as the frequency is less than that of either of the two parts AP, PB (supposed to be held at rest at both extremities) into which the string is divided, there can be no (interior) node (Q). Otherwise, that part of the string AQ between the node Q and one extremity (A), which does not include P, would be vibrating freely, and more slowly than is possible for the longer length AP, included between the point P and the same extremity. When the frequency is raised, so as to coincide with the smaller of those proper to AP, PB, say AP, a node enters at P and then advances towards A. At each coincidence of the frequency with one of those proper to the whole string AB, the vibration identifies itself with the corresponding free vibration, and at each coincidence with a frequency proper to AP, or BP, a new node appears at P, and advances in the first case towards A and in the second towards B. And throughout the whole sequence of events all the nodes move *outwards* from P towards A or B.

Thus, if the string be uniform and be bisected at P, there is no node until the pitch rises to the octave (c') of the note (c) of the string. At this stage two nodes enter at P, and move outwards symmetrically. When g' is reached, the mode of vibration is that of the free vibration of the same pitch, and the nodes are at the two points of trisection. At c'' these nodes have moved outwards so far as to bisect AP, BP, and two new nodes enter at P.

143. When the vibrations of a string are not confined to one plane, it is usually most convenient to resolve them into two sets executed in perpendicular planes, which may be treated independently. There is, however, one case of this description worth a passing notice, in which the motion is most easily conceived and treated without resolution.

Suppose that

$$y = \sin\frac{s\pi x}{l}\cos\frac{2s\pi t}{\tau} \left.\begin{array}{c}\\\\\end{array}\right\}\quad\quad(1).$$
$$z = \sin\frac{s\pi x}{l}\sin\frac{2s\pi t}{\tau}$$

Then

$$r = \sqrt{(y^2 + z^2)} = \sin\frac{s\pi x}{l}\quad\quad(2),$$

and

$$z : y = \tan(2s\pi t/\tau)\quad\quad(3),$$

shewing that the whole string is at any moment in one plane, which revolves uniformly, and that each particle describes a circle with radius $\sin(s\pi x/l)$. In fact, the whole system turns without relative displacement about its position of equilibrium, completing each revolution in the time τ/s. The mechanics of this case is quite as simple as when the motion is confined to one plane, the resultant of the tensions acting at the extremities of any small portion of the string's length being balanced by the centrifugal force.

144. The general differential equation for a uniform string, viz.

$$\frac{d^2y}{dt^2} = a^2\frac{d^2y}{dx^2}\quad\quad(1),$$

may be transformed by a change of variables into

$$\frac{d^2 y}{du\,dv} = 0 \dots\dots\dots\dots\dots\dots\dots(2),$$

where $u = x - at$, $v = x + at$. The general solution of (2) is

$$y = f(u) + F(v) = f(x - at) + F(x + at)\dots\dots\dots(3)^1,$$

f, F being two arbitrary functions.

Let us consider first the case in which F vanishes. When t has any particular value, the equation

$$y = f(x - at)\dots\dots\dots\dots\dots\dots\dots\dots(4),$$

expressing the relation between x and y, represents the form of the string. A change in the value of t is merely equivalent to an alteration in the origin of x, so that (4) indicates that a certain *form* is propagated along the string with uniform velocity a in the positive direction. Whatever the value of y may be at the point x and at the time t, the same value of y will obtain at the point $x + a\,\Delta t$ at the time $t + \Delta t$.

The form thus perpetuated may be any whatever, so long as it does not violate the restrictions on which (1) depends.

When the motion consists of the propagation of a wave in the positive direction, a certain relation subsists between the inclination and the velocity at any point. Differentiating (4) we find

$$\frac{dy}{dt} = -a\,\frac{dy}{dx} \dots\dots\dots\dots\dots\dots\dots(5).$$

Initially, dy/dt and dy/dx may both be given arbitrarily, but if the above relation be not satisfied, the motion cannot be represented by (4).

In a similar manner the equation

$$y = F(x + at) \dots\dots\dots\dots\dots\dots(6)$$

denotes the propagation of a wave in the *negative* direction, and the relation between dy/dt and dy/dx corresponding to (5) is

$$\frac{dy}{dt} = a\,\frac{dy}{dx} \dots\dots\dots\dots\dots\dots(7).$$

In the general case the motion consists of the simultaneous propagation of two waves with velocity a, the one in the positive,

---

[1] [Equations (1) and (3) are due to D'Alembert (1750).]

and the other in the negative direction; and these waves are entirely independent of one another. In the first $dy/dt = -a\,dy/dx$, and in the second $dy/dt = a\,dy/dx$. The initial values of dy/dt and dy/dx must be conceived to be divided into two parts, which satisfy respectively the relations (5) and (7). The first constitutes the wave which will advance in the positive direction without change of form; the second, the negative wave. Thus, initially,

$$f'(x) + F'(x) = \frac{dy}{dx} \\ f'(x) - F'(x) = -\frac{1}{a}\frac{dy}{dt} \Bigg\},$$

whence

$$f'(x) = \tfrac{1}{2}\left(\frac{dy}{dx} - \frac{1}{a}\frac{dy}{dt}\right) \\ F'(x) = \tfrac{1}{2}\left(\frac{dy}{dx} + \frac{1}{a}\frac{dy}{dt}\right) \Bigg\} \dots\dots\dots\dots(8),$$

equations which determine the functions f' and F' for all values of the argument from $x = -\infty$ to $x = \infty$, if the initial values of dy/dx and dy/dt be known.

If the disturbance be originally confined to a finite portion of the string, the positive and negative waves separate after the interval of time required for each to traverse half the disturbed portion.

Fig. 23.

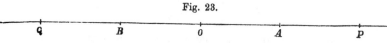

Suppose, for example, that AB is the part initially disturbed. A point P on the positive side remains at rest until the positive wave has travelled from A to P, is disturbed during the passage of the wave, and ever after remains at rest. The negative wave never affects P at all. Similar statements apply, *mutatis mutandis*, to a point Q on the negative side of AB. If the character of the original disturbance be such that $a\,dy/dx - dy/dt$ vanishes initially, there is no positive wave, and the point P is never disturbed at all; and if $a\,dy/dx + dy/dt$ vanish initially, there is no negative wave. If dy/dt vanish initially, the positive and the negative waves are similar and equal, and then neither can vanish. In cases where either wave vanishes, its evanescence may be considered to be due to the mutual destruction of two component waves, one depending on the initial displacements, and the other on the initial velocities. On the one side these two waves conspire, and on the other they destroy one another. This explains the apparent paradox, that P can fail to be affected sooner or later after AB has been disturbed.

The subsequent motion of a string that is initially displaced without velocity, may be readily traced by graphical methods. Since the positive and the negative waves are equal, it is only necessary to divide the original disturbance into two equal parts, to displace these, one to the right, and the other to the left, through a space equal to at, and then to recompound them. We shall presently apply this method to the case of a plucked string of finite length.

145. Vibrations are called *stationary*, when the motion of each particle of the system is proportional to some function of the time, the same for all the particles. If we endeavour to satisfy

$$\frac{d^2y}{dt^2} = a^2 \frac{d^2y}{dx^2} \dots\dots\dots\dots\dots\dots\dots(1),$$

by assuming $y = XT$, where X denotes a function of x only, and T a function of t only, we find

$$\frac{1}{T}\frac{d^2T}{d(at)^2} = \frac{1}{X}\frac{d^2X}{dx^2} = m^2 \quad \text{(a constant)},$$

so that

$$\left. \begin{array}{l} T = A\, \cos mat + B \sin mat \\ X = C\, \cos mx\ + D \sin mx \end{array} \right\} \quad \dots\dots\dots\dots(2),$$

proving that the vibrations must be simple harmonic, though of arbitrary period. The value of y may be written

$$y = P \cos (mat - \epsilon)\ \cos (mx - \alpha)$$
$$= \tfrac{1}{2} P \cos (mat + mx - \epsilon - \alpha) + \tfrac{1}{2} P \cos (mat - mx - \epsilon + \alpha)\dots(3),$$

shewing that the most general kind of stationary vibration may be regarded as due to the superposition of equal progressive vibrations, whose directions of propagation are opposed. Conversely, two stationary vibrations may combine into a progressive one.

The solution $y = f(x - at) + F(x + at)$ applies in the first instance to an infinite string, but may be interpreted so as to give the solution of the problem for a finite string in certain cases. Let us suppose, for example, that the string terminates at $x = 0$, and is held fast there, while it extends to infinity in the positive direction only. Now so long as the point $x = 0$ actually remains at rest, it is a matter of indifference whether the string be prolonged on the negative side or not. We are thus led to regard the given string as forming part of one doubly infinite, and to seek whether and how the initial displacements and velocities on the negative side can be taken, so that on the whole there shall be no displacement at $x = 0$ throughout the subsequent motion. The initial values of y and $\dot{y}$ on the positive side determine the corresponding parts of the positive and negative waves, into which we know that the whole motion can be resolved. The former has no influence at the point $x = 0$. On the negative side the positive and the negative waves are initially at our disposal, but with the latter we are not concerned. The problem is to determine the positive wave on the negative side, so that in conjunction with the given negative wave on the positive side of the origin, it shall leave that point undisturbed.

Let $OPQRS...$ be the line (of any form) representing the wave in OX, which advances in the negative direction. It is

Fig. 24.

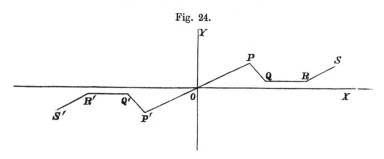

evident that the requirements of the case are met by taking on the other side of O what may be called the *contrary* wave, so that O is the geometrical centre, bisecting every chord (such as PP') which passes through it. Analytically, if $y = f(x)$ is the equation of $OPQRS......$, $-y = f(-x)$ is the equation of $OP'Q'R'S'......$ When after a time t the curves are shifted to the left and to the right respectively through a distance at, the co-ordinates corresponding to $x = 0$ are necessarily equal and opposite, and therefore when compounded give zero resultant displacement.

The effect of the constraint at O may therefore be represented by supposing that the negative wave moves through undisturbed, but that a positive wave at the same time emerges from O. This reflected wave may at any time be found from its parent by the following rule:

Let $APQRS...$ be the position of the parent wave. Then the reflected wave is the position which this would assume, if it were

Fig. 25.

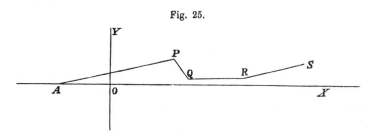

turned through two right angles, first about OX as an axis of rotation, and then through the same angle about OY. In other words, the return wave is the image of $APQRS$ formed by successive optical reflection in OX and OY, regarded as plane mirrors.

The same result may also be obtained by a more analytical process. In the general solution

$$y = f(x - at) + F(x + at),$$

the functions $f(z)$, $F(z)$ are determined by the initial circumstances for all positive values of z. The condition at $x = 0$ requires that

$$f(-at) + F(at) = 0$$

for all positive values of t, or

$$f(-z) = -F(z)$$

for positive values of z. The functions f and F are thus determined for all positive values of x and t.

There is now no difficulty in tracing the course of events when *two* points of the string A and B are held fast. The initial disturbance in AB divides itself into positive and negative waves, which are reflected backwards and forwards between the fixed points, changing their character from positive to negative, and *vice versâ*, at each reflection. After an even number of reflections in each case the original form and motion is completely recovered. The process is most easily followed in imagination when the initial disturbance is confined to a small part of the string, more particularly when its character is such as to give rise to a wave propagated in one direction only. The *pulse* travels with uniform velocity (*a*) to and fro along the length of the string, and after it has returned *a second time* to its starting point the original condition of things is exactly restored. The period of the motion is thus the time required for the pulse to traverse the length of the string twice, or

$$\tau = 2l/a \quad \dots\dots\dots\dots\dots\dots\dots\dots(1).$$

The same law evidently holds good whatever may be the character of the original disturbance, only in the general case it may happen that the *shortest* period of recurrence is some aliquot part of τ.

146. The method of the last few sections may be advantageously applied to the case of a plucked string. Since the initial velocity vanishes, half of the displacement belongs to the positive and half to the negative wave. The manner in which the wave must be completed so as to produce the same effect as the constraint, is shewn in the figure, where the upper curve represents

Fig. 26.

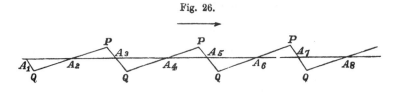

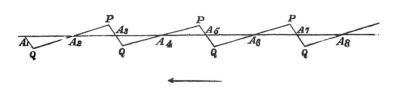

the positive, and the lower the negative wave in their initial positions. In order to find the configuration of the string at any future time, the two curves must be superposed, after the upper has been shifted to the right and the lower to the left through a space equal to *at*.

The resultant curve, like its components, is made up of straight pieces. A succession of six at intervals of a twelfth of the period,

Fig. 27.

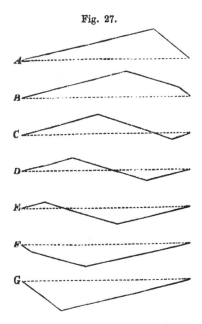

shewing the course of the vibration, is given in the figure (Fig. 27), taken from Helmholtz. From G the string goes back again to A through the same stages[1].

It will be observed that the inclination of the string at the points of support alternates between two constant values.

147. If a small disturbance be made at the time t at the point x of an infinite stretched string, the effect will not be felt at O until after the lapse of the time x/a, and will be in all respects the same as if a like disturbance had been made at the point $x + \Delta x$ at time $t - \Delta x/a$. Suppose that similar disturbances are communicated to the string at intervals of time τ at points whose distances from O increase each time by $a\,\delta\tau$, then it is evident that the result at O will be the same as if the disturbances were all made at the same point, provided that the time-intervals be increased from τ to $\tau + \delta\tau$. This remark con-

---

[1] This method of treating the vibration of a plucked string is due to Young. *Phil. Trans.*, 1800. The student is recommended to make himself familiar with it by actually constructing the forms of Fig. 27.

tains the theory of the alteration of pitch due to motion of the source of disturbance; a subject which will come under our notice again in connection with aerial vibrations.

148. When one point of an infinite string is subject to a forced vibration, trains of waves proceed from it in both directions according to laws, which are readily investigated. We shall suppose that the origin is the point of excitation, the string being there subject to the forced motion $y = A e^{ipt}$; and it will be sufficient to consider the positive side. If the motion of each element ds be resisted by the frictional force $\kappa \rho \dot{y} ds$, the differential equation is

$$\frac{d^2 y}{dt^2} + \kappa \frac{dy}{dt} = a^2 \frac{d^2 y}{dx^2} \dots\dots\dots\dots\dots(1);$$

or since $y \propto e^{ipt}$,

$$\frac{d^2 y}{dx^2} = \left(\frac{i \kappa p}{a^2} - \frac{p^2}{a^2} \right) y = \lambda^2 y \dots\dots\dots\dots(2),$$

if for brevity we write λ^2 for the coefficient of y.

The general solution is

$$y = \{ C e^{-\lambda x} + D e^{+\lambda x} \} e^{ipt} \dots\dots\dots\dots(3).$$

Now since y is supposed to vanish at an infinite distance, D must vanish, if the real part of λ be taken positive. Let

$$\lambda = \alpha + i\beta,$$

where α is positive.

Then the solution is

$$y = A e^{-(\alpha + i\beta) x + ipt} \dots\dots\dots\dots(4),$$

or, on throwing away the imaginary part,

$$y = A e^{-\alpha x} \cos (pt - \beta x) \dots\dots\dots\dots(5),$$

corresponding to the forced motion at the origin

$$y = A \cos pt \dots\dots\dots\dots(6).$$

An arbitrary constant may, of course, be added to t.

To determine α and β, we have

$$\alpha^2 - \beta^2 = -\frac{p^2}{a^2}; \qquad 2\alpha\beta = \frac{\kappa p}{a^2} \dots\dots\dots\dots(7).$$

If we suppose that κ is small,

$$\beta = p/a, \qquad \alpha = \kappa/2a \quad \text{nearly},$$

and

$$y = A e^{-\kappa x/2a} \cos \left(pt - \frac{p}{a} x \right) \dots\dots\dots\dots(8).$$

This solution shews that there is propagated along the string a wave, whose amplitude slowly diminishes on account of the exponential factor. If $\kappa = 0$, this factor disappears, and we have simply

$$y = A \cos \left(pt - \frac{px}{a}\right) \dots\dots\dots\dots\dots(9).$$

This result stands in contradiction to the general law that, when there is no friction, the forced vibrations of a system (due to a single simple harmonic force) must be synchronous in phase throughout. According to (9), on the contrary, the phase varies continuously in passing from one point to another along the string. The fact is, that we are not at liberty to suppose $\kappa = 0$ in (8), inasmuch as that equation was obtained on the assumption that the real part of λ in (3) is positive, and not zero. However long a finite string may be, the coefficient of friction may be taken so small that the vibrations are not damped before reaching the further end. After this point of smallness, reflected waves begin to complicate the result, and when the friction is diminished indefinitely, an infinite series of such must be taken into account, and would give a resultant motion of the same phase throughout.

This problem may be solved for a string whose mass is supposed to be concentrated at equidistant points, by the method of § 120. The co-ordinate ψ_1 may be supposed to be given ($= He^{ipt}$), and it will be found that the system of equations (5) of § 120 may all be satisfied by taking

$$\psi_r = \theta^{r-1}\psi_1 \dots\dots\dots\dots\dots(10),$$

where θ is a complex constant determined by a quadratic equation. The result for a continuous string may be afterwards deduced.

[In the notation of § 120 the quadratic equation is

$$B\theta^2 + A\theta + B = 0 \dots\dots\dots\dots(11),$$

where $\quad A = -\mu p^2 + \frac{2T_1}{a}, \quad B = -\frac{T_1}{a} \dots\dots\dots(12).$

The roots of (11) are

$$\theta = \frac{-A \pm \sqrt{(A^2 - 4B^2)}}{2B} \dots\dots\dots(13),$$

and are imaginary if $4B^2 > A^2$, that is, if

$$p^2 < \frac{4T_1}{\mu a} \dots\dots\dots\dots(14),$$

a condition always satisfied in passing to the limit where α and μ are infinitely small. In any case when (14) is satisfied the modulus of θ is unity, so that (10) represents wave propagation.

If, however, (14) be not satisfied, the values of θ are real. In this case all the motions are in the same phase, and no wave is propagated. The vibration impressed upon ψ_1 is imitated upon a reduced scale by ψ_2, ψ_3......, with amplitudes which form a geometrical progression. In the first case the motion is propagated to an infinite distance, but in the second it is practically confined to a limited region round the source.]

148 a. So long as the conditions of § 144 are satisfied, a positive, or a negative, wave is propagated undisturbed. If however there be any want of uniformity, such (for example) as that caused by a load attached at a particular point, reflection will ensue when that point is reached. The most interesting problem under this head is that of two strings of different longitudinal densities, attached to one another, and vibrating transversely under the common tension T_1. Or, if we regard the string as single, the density may be supposed to vary discontinuously from one uniform value (ρ_1) to another (ρ_2). If a_1, a_2 denote the corresponding velocities of propagation,

$$a_1{}^2 = T_1/\rho_1, \qquad a_2{}^2 = T_1/\rho_2 \dots\dots\dots\dots\dots(1),$$

and

$$\mu = a_1/a_2 = \sqrt{(\rho_2/\rho_1)} \dots\dots\dots\dots\dots\dots(2).$$

The conditions to be satisfied at the junction of the two parts are (i) the continuity of the displacement y, and (ii) the continuity of dy/dx. If the two parts met at a finite angle, an infinitely small element at the junction would be subject to a finite force.

Let us suppose that a positive wave of harmonic type, travelling in the first part (ρ_1), impinges upon the second (ρ_2). In the latter the motion will be adequately represented by a positive wave, but in the former we must provide for a negative reflected wave. Thus we may take for the two parts respectively

$$y = H e^{ik_1(a_1t-x)} + K e^{ik_1(a_1t+x)} \dots\dots\dots\dots\dots(3),$$

$$y = L e^{ik_2(a_2t-x)} \dots\dots\dots\dots\dots\dots\dots(4),$$

where

$$k_1 = 2\pi/\lambda_1, \qquad k_2 = 2\pi/\lambda_2,$$

so that

$$k_1 a_1 = k_2 a_2 \dots\dots\dots\dots\dots\dots\dots(5).$$

The conditions at the junction ($x = 0$) give

$$H + K = L \quad \dots\dots\dots\dots\dots\dots(6),$$

$$k_1 H - k_1 K = k_2 L \quad \dots\dots\dots\dots\dots(7)$$

whence
$$\frac{K}{H} = \frac{k_1 - k_2}{k_1 + k_2} = -\frac{\mu - 1}{\mu + 1} \quad \dots\dots\dots\dots(8).$$

Since the ratio K/H is real, we may suppose that both quantities are real; and if we throw away the imaginary parts from (3) and (4) we get as the solution in terms of real quantities

$$y = H \cos k_1 (a_1 t - x) + K \cos k_1 (a_1 t + x) \dots\dots\dots\dots(9);$$

$$y = (H + K) \cos k_2 (a_2 t - x) \quad \dots\dots\dots\dots(10).$$

The ratio of amplitudes of the reflected and the incident waves expressed by (8) is that first obtained by T. Young for the corresponding problem in Optics.

148 b. The expression for the intensity of reflection established in § 148 a depends upon the assumption that the transition from the one density to the other is sudden, that is occupies a distance which is small in comparison with a wave length. If the transition be gradual, the reflection may be expected to fall off, and in the limit to disappear altogether.

The problem of gradual transition includes, of course, that of a variable medium, and would in general be encumbered with great difficulties. There is, however, one case for which the solution may be readily expressed, and this it is proposed to consider in the present section. The longitudinal density is supposed to vary as the inverse square of the abscissa. If y, denoting the transverse displacement be proportional to e^{ipt}, the equation which it must satisfy as a function of x, is (§ 141),

$$\frac{d^2 y}{dx^2} + n^2 x^{-2} y = 0 \quad \dots\dots\dots\dots\dots(1),$$

where n^2 is some positive constant, of the nature of an abstract number.

The solution of (1) is $y = A x^{\frac{1}{2} + im} + B x^{\frac{1}{2} - im} \dots\dots\dots\dots(2),$

where
$$m^2 = n^2 - \tfrac{1}{4} \quad \dots\dots\dots\dots\dots(3).$$

If m be real, that is, if $n > \tfrac{1}{2}$, we may obtain, by supposing $A = 0$, as a final solution in real quantities,

$$y = C x^{\frac{1}{2}} \cos (pt - m \log x + \epsilon) \quad \dots\dots\dots\dots(4),$$

which represents a positive progressive wave, in many respects similar to those propagated in uniform media.

Let us now suppose that, to the left of the point $x = x_1$, the variable medium is replaced by one of uniform constitution, such that there is no discontinuity of density at the point of transition; and let us inquire what reflection a positive progressive wave in the uniform medium will undergo on arrival at the variable medium. It will be sufficient to consider the case where m is real, that is, where the change of density is but moderately rapid.

By supposition, there is no negative wave in the variable medium, so that $A = 0$ in (2). Thus

$$y = Bx^{\frac{1}{2}-im}, \qquad \frac{dy}{dx} = (\tfrac{1}{2} - im)Bx^{-\frac{1}{2}-im};$$

and, when $x = x_1$,

$$\frac{dy}{y\,dx} = \frac{\tfrac{1}{2} - im}{x_1} \qquad \dotfill (5).$$

The general solution for the uniform medium, satisfying the equation $d^2y/dx^2 + n^2x_1^{-2}y = 0$, may be written

$$y = He^{-in\frac{x-x_1}{x_1}} + Ke^{+in\frac{x-x_1}{x_1}} \qquad \dotfill (6),$$

from which, when $x = x_1$,

$$\frac{dy}{y\,dx} = -\frac{in}{x_1}\frac{H-K}{H+K} \qquad \dotfill (7).$$

In equation (6), H represents the amplitude of the incident positive wave, and K the amplitude of the reflected negative wave. The condition to be satisfied at $x = x_1$ is expressed by equating the values of $\dfrac{dy}{y\,dx}$ given by (5) and (7). Thus

$$\frac{K}{H} = \frac{i(n-m)+\tfrac{1}{2}}{i(n+m)-\tfrac{1}{2}} \qquad \dotfill (8),$$

which gives, in symbolical form, the ratio of the reflected to the incident vibration.

Having regard to (3), we may write (8) in the form

$$\frac{K}{H} = \frac{-i}{2(n+m)} \qquad \dotfill (9);$$

so that the amplitude of the reflected wave is $\tfrac{1}{2}(n+m)^{-1}$ of that of the incident. Thus, as was to be expected, when n and m are great, i.e., when the density changes slowly in the variable medium, there is but little reflection. As regards phase, the result embodied in (9) may be represented by supposing that the reflection occurs at $x = x_1$, and involves a change of phase amounting to a quarter period.

Passing on now to the more important problem, we will suppose that the variable medium extends only so far as the point $x = x_2$, beyond which the density retains uniformly its value at that point. A positive wave travelling at first in a uniform medium of density proportional to x_1^{-2}, passes at the point $x = x_1$ into a variable medium of density proportional to x^{-2}, and again, at the point $x = x_2$, into a uniform medium of density proportional to x_2^{-2}. The velocities of propagation are inversely proportional to the square roots of the densities, so that, if μ be the refractive index between the extreme media,

$$\mu = \frac{x_1}{x_2} \quad \dots\dots\dots\dots\dots\dots\dots(10).$$

The thickness (d) of the layer of transition is

$$d = x_2 - x_1 \quad \dots\dots\dots\dots\dots\dots(11).$$

The wave-lengths in the two media are given by

$$\lambda_1 = \frac{2\pi x_1}{n}, \qquad \lambda_2 = \frac{2\pi x_2}{n};$$

so that

$$n = \frac{2\pi d}{\lambda_2 - \lambda_1} = \frac{2\pi d}{(\mu^{-1} - 1)\,\lambda_1} \quad \dots\dots\dots\dots(12).$$

For the first medium we take, as before,

$$y = He^{-in\frac{x - x_1}{x_1}} + Ke^{+in\frac{x - x_1}{x_1}} \quad \dots\dots\dots\dots(6).$$

giving, when $x = x_1$,

$$\frac{dy}{y\,dx} = -\frac{in}{x_1}\frac{H - K}{H + K} = -\frac{in\theta}{x_1} \quad \dots\dots\ \dots\dots\dots(7).$$

if, for brevity, we write θ for $\dfrac{H - K}{H + K}$.

For the variable medium,

$$y = Ax^{\frac{1}{2}+im} + Bx^{\frac{1}{2}-im} \dots\dots\dots\dots\dots\dots(2),$$

giving, when $x = x_1$,

$$\frac{dy}{y\,dx} = x_1^{-1}\frac{(\frac{1}{2} + im)\,Ax_1^{im} + (\frac{1}{2} - im)\,Bx_1^{-im}}{Ax_1^{im} + Bx_1^{-im}} \quad \dots\dots(13).$$

Hence the condition to be satisfied at $x = x_1$ gives

$$\tfrac{1}{2} + im \frac{Ax_1{}^{im} - Bx_1{}^{-im}}{Ax_1{}^{im} + Bx_1{}^{-im}} = -in\theta;$$

whence

$$\frac{A}{B} = x_1{}^{-2im} \frac{im - in\theta - \tfrac{1}{2}}{im + in\theta + \tfrac{1}{2}} \dots \dots \dots \dots (14).$$

The condition to be satisfied at $x = x_2$ may be deduced from (14), by substituting x_2 for x_1, putting at the same time $\theta = 1$ in virtue of the supposition that in the second medium there is no negative wave. Hence, equating the two values of $A : B$, we get

$$x_1{}^{-2im} \frac{im - in\theta - \tfrac{1}{2}}{im + in\theta + \tfrac{1}{2}} = x_2{}^{-2im} \frac{im - in - \tfrac{1}{2}}{im + in + \tfrac{1}{2}} \dots \dots (15),$$

as the equation from which the reflected wave in the first medium is to be found. Having regard to (3), we get

$$\theta = \frac{H - K}{H + K} = \frac{m + n + \tfrac{1}{2}i + \mu^{2im}(m - n - \tfrac{1}{2}i)}{m + n - \tfrac{1}{2}i + \mu^{2im}(m - n + \tfrac{1}{2}i)},$$

so that

$$\frac{K}{H} = \frac{-i + \mu^{2im}i}{2(m + n) + 2\mu^{2im}(m - n)} \dots \dots \dots (16).$$

This is the symbolical solution. To interpret it in real quantities, we must distinguish the cases of m real and m imaginary. If the transition be not too sudden, m is real, and (16) may be written

$$\frac{K}{H} = \frac{i}{2} \frac{-1 + \cos(2m \log \mu) + i \sin(2m \log \mu)}{m + n + (m - n)\cos(2m \log \mu) + i(m - n)\sin(2m \log \mu)}$$

Thus the expression for the ratio of the *intensities* of the reflected and the incident waves is, after reduction,

$$\frac{\sin^2(m \log \mu)}{4m^2 + \sin^2(m \log \mu)} \dots \dots \dots (17).$$

If m be imaginary, we may write $im = m'$; (16) then gives for the ratio of intensities,

$$\frac{(\mu^{m'} - \mu^{-m'})^2}{(\mu^{m'} - \mu^{-m'})^2 + 16m'^2} \dots \dots \dots (18);$$

or, if we introduce the notation of hyperbolic trigonometry § 170,

$$\frac{\sinh^2(m' \log \mu)}{\sinh^2(m' \log \mu) + 4m'^2} \dots \dots \dots (19).$$

For the critical value $m = 0$, we get, from (17) or (19),

$$\frac{(\log \mu)^2}{4 + (\log \mu)^2} \dots \dots \dots (20).$$

These expressions allow us to trace the effect of a more or less gradual transition between media of given indices. If the transition be absolutely abrupt, $n = 0$, by (12); so that $m' = \frac{1}{2}$. In this case, (18) gives us (§ 148 a) Young's well-known formula

$$\left(\frac{\mu - 1}{\mu + 1}\right)^2 \quad \dots\dots\dots\dots\dots\dots\dots(21).$$

Since $\dfrac{\sinh x}{x}$ increases continually from $x = 0$, the ratio (19) increases continually from $m' = 0$ to $m' = \frac{1}{2}$, i.e., diminishes continually from the case of sudden transition $m' = \frac{1}{2}$, when its value is (21), to the critical case $m' = 0$, when its value is (20), after which this form no longer holds good. When $m' = 0$, $n = \frac{1}{2}$, and, by (12), $d = (\lambda_2 - \lambda_1)/4\pi$.

When $n > \frac{1}{2}$, (17) is the appropriate form. We see from it that with increasing n the reflection diminishes, until it vanishes, when $m \log \mu = \pi$, i.e. when

$$n^2 = \frac{1}{4} + \frac{\pi^2}{(\log \mu)^2} \dots\dots\dots\dots\dots\dots\dots(22).$$

With a still more gradual transition the reflection revives, reaches a maximum, again vanishes when $m \log \mu = 2\pi$, and so on[1].

148 c. In the problem of connected strings, vibrating under the influence of tension alone, the velocity in each uniform part is independent of wave length, and there is nothing corresponding to optical *dispersion*. This state of things will be departed from if we introduce the consideration of stiffness, and it may be of interest to examine in a simple case how far the problem of reflection is thereby modified. As in § 148 a, we will suppose that at $x = 0$ the density changes discontinuously from ρ_1 to ρ_2, but that now the vibrations of the second part occur under the influence of sensible stiffness. The differential equation applicable in this case is, § 188,

$$\beta^2 \frac{d^4y}{dx^4} - a_2^2 \frac{d^2y}{dx^2} + \frac{d^2y}{dt^2} = 0,$$

or, if y vary as e^{int},

$$-\beta^2 \frac{d^4y}{dx^4} + a_2^2 \frac{d^2y}{dx^2} + n^2y = 0 \dots\dots\dots\dots (1),$$

so that, if y vary as e^{ikx},

$$\beta^2 k^4 + a_2^2 k^2 - n^2 = 0 \dots\dots\dots\dots\dots (2).$$

---

[1] *Proc. Math. Soc.*, vol. xi. February, 1880 ; where will also be found a numerical example illustrative of optical conditions.

In consequence of the stiffness represented by β^2 the velocity of propagation deviates from a_2, and must be found from (2). The two values of k^2 given by this equation are real, one being positive and the other negative. The four admissible values of k may thus be written $\pm k_2$, $\pm ih_2$, so that the complete solution of (1) will be

$$y = Ae^{ik_2x} + Be^{-ik_2x} + Ce^{-h_2x} + De^{h_2x} \dots\dots (3),$$

h_2, k_2 being real and positive. The velocity of propagation is n/k_2

In the application which we have to make the disturbance of the imperfectly flexible second part is due to a positive wave entering it from the first part. When x is great and positive, (3) must reduce to its second term. Thus

$$A = 0, \quad D = 0 ;$$

and we are left with

$$y = Be^{-ik_2x} + Ce^{-h_2x} \dots\dots\dots\dots (4).$$

This holds when x is positive. When x is negative, corresponding to the perfectly flexible first part, we have

$$y = He^{-ik_1x} + Ke^{ik_1x} \dots\dots\dots\dots (5),$$

in which $\quad k_1 = n/a_1 \dots\dots\dots\dots\dots\dots(6).$

The "refractive index" is given by

$$\mu = k_2/k_1 \dots\dots\dots\dots\dots\dots(7).$$

The conditions at the junction are first the continuity of y and dy/dx. Further, d^2y/dx^2 in (4) must vanish at this place, inasmuch as curvature implies a couple (§ 162), and this could not be transmitted by the first part. Hence

$$H + K = B + C \dots\dots\dots\dots\dots(8),$$

$$k_1 (H - K) = k_2 B - ih_2 C \dots\dots\dots (9),$$

$$- k_2^2 B + h_2^2 C = 0 \dots\dots\dots\dots (10).$$

From these we deduce

$$\frac{H + K}{H - K} = \frac{k_1 (h_2 + ik_2)}{k_2 h_2} \dots\dots\dots (11),$$

$$\frac{K}{H} = \frac{h_2 (k_1 - k_2) + ik_1 k_2}{h_2 (k_1 + k_2) + ik_1 k_2} \dots\dots\dots (12);$$

and thence for the *intensity* of reflection, equal to Mod². (K/H),

$$\frac{(k_1 - k_2)^2 + k_1{}^2k_2{}^2/h_2{}^2}{(k_1 + k_2)^2 + k_1{}^2k_2{}^2/h_2{}^2} \quad\dots\dots\dots\dots\dots\dots (13).$$

If the second part, as well as the first, be perfectly flexible, $\beta = 0$, $h_2 = \infty$, and we fall back on Young's formula. In general, the intensity of reflection is not accurately given by this formula, even though we employ therein the value of the refractive index appropriate to the waves actually under propagation.

CHAPTER VII.

149. THE next system to the string in order of simplicity is the bar, by which term is usually understood in Acoustics a mass of matter of uniform substance and elongated cylindrical form. At the ends the cylinder is cut off by planes perpendicular to the generating lines. The centres of inertia of the transverse sections lie on a straight line which is called the *axis*.

The vibrations of a bar are of three kinds—longitudinal, torsional, and lateral. Of these the last are the most important, but at the same time the most difficult in theory. They are considered by themselves in the next chapter, and will only be referred to here so far as is necessary for comparison and contrast with the other two kinds of vibrations.

Longitudinal vibrations are those in which the axis remains unmoved, while the transverse sections vibrate to and fro in the direction perpendicular to their planes. The moving power is the resistance offered by the rod to extension or compression.

One peculiarity of this class of vibrations is at once evident. Since the force necessary to produce a given extension in a bar is proportional to the area of the section, while the mass to be moved is also in the same proportion, it follows that for a bar of given length and material the periodic times and the modes of vibration are independent of the area and of the form of the transverse section. A similar law obtains, as we shall presently see, in the case of torsional vibrations.

It is otherwise when the vibrations are lateral. The periodic times are indeed independent of the thickness of the bar in the direction perpendicular to the plane of flexure, but the motive power in this case, viz. the resistance to bending, increases more rapidly than the thickness in that plane, and therefore an increase in thickness is accompanied by a rise of pitch.

In the case of longitudinal and lateral vibrations, the mechanical constants concerned are the density of the material and the value of Young's modulus. For small extensions (or compressions) Hooke's law, according to which the tension varies as the extension, holds good. If the extension, viz. $\dfrac{\text{actual length} - \text{natural length}}{\text{natural length}}$, be called ϵ, we have $T = q\epsilon$, where q is Young's modulus, and T is the tension per unit area necessary to produce the extension ϵ. Young's modulus may therefore be defined as the force which would have to be applied to a bar of unit section, in order to double its length, if Hooke's law continued to hold good for so great extensions; its dimensions are accordingly those of a force divided by an area.

The torsional vibrations depend also on a second elastic constant μ, whose interpretation will be considered in the proper place.

Although in theory the three classes of vibrations, depending respectively on resistance to extension, to torsion, and to flexure are quite distinct, and independent of one another so long as the squares of the strains may be neglected, yet in actual experiments with bars which are neither uniform in material nor accurately cylindrical in figure it is often found impossible to excite longitudinal or torsional vibrations without the accompaniment of some measure of lateral motion. In bars of ordinary dimensions the gravest lateral motion is far graver than the gravest longitudinal or torsional motion, and consequently it will generally happen that the principal tone of either of the latter kinds agrees more or less perfectly in pitch with some overtone of the former kind. Under such circumstances the regular modes of vibrations become unstable, and a small irregularity may produce a great effect. The difficulty of exciting purely longitudinal vibrations in a bar is similar to that of getting a string to vibrate in one plane.

With this explanation we may proceed to consider the three classes of vibrations independently, commencing with longitudinal vibrations, which will in fact raise no mathematical questions beyond those already disposed of in the previous chapters.

150. When a rod is stretched by a force parallel to its length, the stretching is in general accompanied by lateral contraction in such a manner that the augmentation of volume is less than if the displacement of every particle were parallel to the axis. In the case of a short rod and of a particle situated near the cylindrical boundary, this lateral motion would be comparable in magnitude with the longitudinal motion, and could not be overlooked without risk of considerable error. But where a rod, whose length is great in proportion to the linear dimensions of its section, is subject to a stretching of one sign throughout, the longitudinal motion accumulates, and thus in the case of ordinary rods vibrating longitudinally in the graver modes, the inertia of the lateral motion may be neglected. Moreover we shall see later how a correction may be introduced, if necessary.

Let x be the distance of the layer of particles composing any section from the equilibrium position of one end, when the rod is unstretched, either by permanent tension or as the result of vibrations, and let ξ be the displacement, so that the actual position is given by $x + \xi$. The equilibrium and actual position of a neighbouring layer being $x + \delta x$, $x + \delta x + \xi + \dfrac{d\xi}{dx} \delta x$ respectively, the *elongation* is $d\xi/dx$, and thus, if T be the tension per unit area acting across the section,

$$T = q \frac{d\xi}{dx}. \quad\dotfill(1).$$

Consider now the forces acting on the slice bounded by x and $x + \delta x$. If the area of the section be ω, the tension at x is by (1) $q\omega \, d\xi/dx$, acting in the negative direction, and at $x + \delta x$ the tension is

$$q\omega \left(\frac{d\xi}{dx} + \frac{d^2\xi}{dx^2} \delta x \right),$$

acting in the positive direction; and thus the force on the slice due to the action of the adjoining parts is on the whole

$$q\omega \frac{d^2\xi}{dx^2} \delta x.$$

The mass of the element is $\rho\omega \, \delta x$, if ρ be the original density, and therefore if X be the accelerating force acting on it, the equation of equilibrium is

$$X + \frac{q}{\rho} \frac{d^2\xi}{dx^2} = 0 \quad\dotfill(2).$$

In what follows we shall not require to consider the operation of an impressed force. To find the equation of motion we have only to replace X by the reaction against acceleration $-\ddot{\xi}$, and thus if $q : \rho = a^2$, we have

$$\frac{d^2\xi}{dt^2} = a^2 \frac{d^2\xi}{dx^2} \dots\dots\dots\dots\dots\dots\dots(3).$$

This equation is of the same form as that applicable to the transverse displacements of a stretched string, and indicates the undisturbed propagation of waves of any type in the positive and negative directions. The velocity a is relative to the *unstretched* condition of the bar; the apparent velocity with which a disturbance is propagated in space will be greater in the ratio of the stretched and unstretched lengths of any portion of the bar. The distinction is material only in the case of permanent tension.

151. For the actual magnitude of the velocity of propagation, we have

$$a^2 = q : \rho = q\omega : \rho\omega,$$

which is the ratio of the whole tension necessary (according to Hooke's law) to double the length of the bar and the longitudinal density. If the same bar were stretched with total tension T, and were flexible, the velocity of propagation of waves along it would be $\sqrt{(T : \rho\omega)}$. In order then that the velocity might be the same in the two cases, T must be $q\omega$, or, in other words, the tension would have to be that theoretically necessary in order to double the length. The tones of longitudinally vibrating rods are thus very high in comparison with those obtainable from strings of comparable length.

In the case of steel the value of q is about 22×10^8 grammes weight per square centimetre. To express this in absolute units of force on the C. G. S.[1] system, we must multiply by 980. In the same system the density of steel (identical with its specific gravity referred to water) is 7·8. Hence for steel

$$a = \sqrt{\frac{980 \times 22 \times 10^8}{7\cdot8}} = 530{,}000$$

approximately, which shews that the velocity of sound in steel is about 530,000 centimetres per second, or about 16 times greater

---

[1] Centimetre, Gramme, Second. This system is recommended by a Committee of the British Association. *Brit. Ass. Report*, 1873.

than the velocity of sound in air. In glass the velocity is about the same as in steel.

It ought to be mentioned that in strictness the value of q determined by statical experiments is not that which ought to be used here. As in the case of gases, which will be treated in a subsequent chapter, the rapid alterations of state concerned in the propagation of sound are attended with thermal effects, one result of which is to increase the effective value of q beyond that obtained from observations on extension conducted at a constant temperature. But the data are not precise enough to make this correction of any consequence in the case of solids.

152. The solution of the general equation for the longitudinal vibrations of an unlimited bar, namely

$$\xi = f(x - at) + F(x + at),$$

being the same as that applicable to a string, need not be further considered here.

When both ends of a bar are free, there is of course no permanent tension, and at the ends themselves there is no temporary tension. The condition for a free end is therefore

$$\frac{d\xi}{dx} = 0 \ \dots\dots\dots\dots\dots\dots\dots\dots\dots\dots(1).$$

To determine the normal modes of vibration, we must assume that ξ varies as a harmonic function of the time—cos nat. Then as a function of x, ξ must satisfy

$$\frac{d^2\xi}{dx^2} + n^2\xi = 0 \dots\dots\dots\dots\dots\dots\dots\dots(2),$$

of which the complete integral i.

$$\xi = A \cos nx + B \sin nx \dots\dots\dots\dots\dots(3),$$

where A and B are independent of x.

Now since $d\xi/dx$ vanishes always when $x = 0$, we get $B = 0$; and again since $d\xi/dx$ vanishes when $x = l$—the natural length of the bar, sin $nl = 0$, which shews that n is of the form

$$n = \frac{i\pi}{l} \ \dots\dots\dots\dots\dots\dots\dots\dots(4),$$

i being integral.

Accordingly, the normal modes are given by equations of the form

$$\xi = A \cos \frac{i\pi x}{l} \cos \frac{i\pi a t}{l} \ldots\ldots\ldots\ldots\ldots (5),$$

in which of course an arbitrary constant may be added to t, if desired.

The complete solution for a bar with both ends free is therefore expressed by

$$\xi = \sum_{i=0}^{i=\infty} \cos \frac{i\pi x}{l} \left\{ A_i \cos \frac{i\pi a t}{l} + B_i \sin \frac{i\pi a t}{l} \right\} \ldots\ldots(6),$$

where A_i and B_i are arbitrary constants, which may be determined in the usual manner, when the initial values of ξ and $\dot{\xi}$ are given.

A zero value of i is admissible; it gives a term representing a displacement ξ constant with respect both to space and time, and amounting in fact only to an alteration of the origin.

The period of the gravest component in (6) corresponding to $i = 1$, is $2l/a$, which is the time occupied by a disturbance in travelling twice the length of the rod. The other tones found by ascribing integral values to i form a complete harmonic scale; so that according to this theory the note given by a rod in longitudinal vibration would be in all cases musical.

In the gravest mode the centre of the rod, where $x = \frac{1}{2}l$, is a place of no motion, or node; but the periodic elongation or compression $d\xi/dx$ is there a maximum.

153. The case of a bar with one end free and the other fixed may be deduced from the general solution for a bar with both ends free, and of twice the length. For whatever may be the initial state of the bar free at $x = 0$ and fixed at $x = l$, such displacements and velocities may always be ascribed to the sections of a bar extending from 0 to $2l$ and free at both ends as shall make the motions of the parts from 0 to l identical in the two cases. It is only necessary to suppose that from l to $2l$ the displacements and velocities are initially equal and opposite to those found in the portion from 0 to l at an equal distance from the centre $x = l$. Under these circumstances the centre must by the symmetry remain at rest throughout the motion, and then the portion from 0 to l satisfies all the required conditions. We conclude that the vibrations of a bar free at one end and fixed at the other are identical with those of one half of a bar of twice the length of which both ends are free, the latter vibrating only in the uneven modes, obtained by making i in succession all *odd* integers. The tones of the bar still belong to a harmonic scale, but the even tones (octave, &c. of the fundamental) are wanting.

The period of the gravest tone is the time occupied by a pulse in travelling *four* times the length of the bar.

154. When both ends of a bar are fixed, the conditions to be satisfied at the ends are that the value of ξ is to be invariable. At $x = 0$, we may suppose that $\xi = 0$. At $x = l$, ξ is a small constant α, which is zero if there be no permanent tension. Independently of the vibrations we have evidently $\xi = x\,\alpha \div l$, and we should obtain our result most simply by assuming this term at once. But it may be instructive to proceed by the general method.

Assuming that as a function of the time ξ varies as

$$A \cos nat + B \sin nat,$$

we see that as a function of x it must satisfy

$$\frac{d^2\xi}{dx^2} + n^2\xi = 0,$$

of which the general solution is

$$\xi = C \cos nx + D \sin nx \dots\dots\dots\dots(1).$$

But since ξ vanishes with x for all values of t, $C = 0$, and thus we may write

$$\xi = \Sigma \sin nx \, \{A \cos nat + B \sin nat\}.$$

The condition at $x = l$ now gives

$$\Sigma \sin nl \, \{A \cos nat + B \sin nat\} = \alpha,$$

from which it follows that for every finite admissible value of n

$$\cdot \sin nl = 0, \quad \text{or} \quad n = \frac{i\pi}{l}.$$

But for the zero value of n, we get

$$A_0 \sin nl = \alpha,$$

and the corresponding term in ξ is

$$\xi = A_0 \sin nx = \alpha \frac{\sin nx}{\sin nl} = \alpha \frac{x}{l}.$$

The complete value of ξ is accordingly

$$\xi = \alpha \frac{x}{l} + \Sigma_{i=1}^{i=\infty} \sin \frac{i\pi x}{l} \left\{ A_i \cos \frac{i\pi at}{l} + B_i \sin \frac{i\pi at}{l} \right\} \ldots(2).$$

The series of tones form a complete harmonic scale (from which however any of the members may be missing in any actual case of vibration), and the period of the gravest component is the time taken by a pulse to travel twice the length of the rod, the same therefore as if both ends were free. It must be observed that we have here to do with the *unstretched* length of the rod, and that the period for a given natural length is independent of the permanent tension.

The solution of the problem of the doubly fixed bar in the case of no permanent tension might also be derived from that of a doubly free bar by mere differentiation with respect to x. For in the latter problem $d\xi/dx$ satisfies the necessary differential equation, viz.

$$\frac{d^2}{dt^2}\left(\frac{d\xi}{dx}\right) = a^2 \frac{d^2}{dx^2}\left(\frac{d\xi}{dx}\right),$$

inasmuch as ξ satisfies

$$\frac{d^2\xi}{dt^2} = a^2 \frac{d^2\xi}{dx^2};$$

and at both ends $d\xi/dx$ vanishes. Accordingly $d\xi/dx$ in this problem satisfies all the conditions prescribed for ξ in the case when both ends are fixed. The two series of tones are thus identical.

155. The effect of a small load M attached to any point of the rod is readily calculated approximately, as it is sufficient to assume the type of vibration to be unaltered (§ 88). We will take the case of a rod fixed at $x = 0$, and free at $x = l$. The kinetic energy is proportional to

$$\tfrac{1}{2}\int_0^l \rho\omega \sin^2 \frac{i\pi x}{2l}\, dx + \tfrac{1}{2}M \sin^2 \frac{i\pi x}{2l},$$

or to

$$\frac{\rho\omega l}{4}\left(1 + \frac{2M}{\rho\omega l} \sin^2 \frac{i\pi x}{2l}\right).$$

Since the potential energy is unaltered, we see by the principles of Chapter IV., that the effect of the small load M at a distance x from the fixed end is to increase the period of the component tones in the ratio

$$1 : 1 + \frac{M}{\rho \omega l} \sin^2 \frac{i \pi x}{2l}.$$

The small quantity $M : \rho \omega l$ is the ratio of the load to the whole mass of the rod.

If the load be attached at the free end, $\sin^2(i \pi x / 2l) = 1$, and the effect is to depress the pitch of every tone by the same small interval. It will be remembered that i is here an *uneven* integer.

If the point of attachment of M be a node of any component, the pitch of that component remains unaltered by the addition.

156   Another problem worth notice occurs when the load at the free end is great in comparison with the mass of the rod. In this case we may assume as the type of vibration, a condition of uniform extension along the length of the rod.

If ξ be the displacement of the load M, the kinetic energy is

$$T = \tfrac{1}{2} M \dot{\xi}^2 + \tfrac{1}{2} \dot{\xi}^2 \int_0^l \rho \omega \frac{x^2}{l^2} dx = \tfrac{1}{2} \dot{\xi}^2 (M + \tfrac{1}{3} \rho \omega l) \dots\dots\dots (1).$$

The tension corresponding to the displacement ξ is $q \omega \xi / l$, and thus the potential energy of the displacement is

$$V = \frac{q \omega \xi^2}{2l} \dots\dots\dots\dots\dots\dots\dots\dots\dots (2).$$

The equation of motion is

$$(M + \tfrac{1}{3} \rho \omega l) \ddot{\xi} + \frac{q \omega}{l} \xi = 0,$$

and if $\xi \propto \cos pt$

$$p^2 = \frac{q \omega}{l} \div (M + \tfrac{1}{3} \rho \omega l) \dots\dots\dots\dots\dots\dots (3).$$

The correction due to the inertia of the rod is thus equivalent to the addition to M of one-third of the mass of the rod.

156 a.   So long as a rod or a wire is uniform, waves of longitudinal vibration are propagated along it without change of type, but any interruption, or alteration of mechanical properties, will in general give rise to reflection. If two uniform wires be joined, the problem of determining the reflection at the junction may be conducted as in § 148 *a*. The conditions to be satisfied at the junction are (i) the continuity of ξ, and (ii) the continuity of $q\omega \, d\xi/dx$, measuring the tension. If ρ_1, ρ_2, ω_1, ω_2, a_1, a_2 denote the volume densities, the sections, and the velocities in the two wires, the ratio of the reflected to the incident amplitude is given by

$$\frac{K}{H} = \frac{\rho_1 \omega_1 a_1 - \rho_2 \omega_2 a_2}{\rho_1 \omega_1 a_1 + \rho_2 \omega_2 a_2} \quad\dots\dots\dots\dots\dots\dots (1).$$

The reflection vanishes, or the incident wave is propagated through the junction without loss, if

$$\rho_1 \omega_1 a_1 = \rho_2 \omega_2 a_2 \dots\dots\dots\dots\dots\dots\dots (2).$$

This result illustrates the difficulty which is met with in obtaining effective transmission of sound from air to metal, or from metal to air, in the mechanical telephone. Thus the value of ρa is about 100,000 times greater in the case of steel than in the case of air.

157. Our mathematical discussion of longitudinal vibrations may close with an estimate of the error involved in neglecting the inertia of the lateral motion of the parts of the rod not situated on the axis. If the ratio of lateral contraction to longitudinal extension be denoted by μ, the lateral displacement of a particle distant r from the axis will be $\mu r \epsilon$ in the case of equilibrium, where ϵ is the extension. Although in strictness this relation will be modified by the inertia of the lateral motion, yet for the present purpose it may be supposed to hold good, § 88.

The constant μ is a numerical quantity, lying between 0 and $\frac{1}{2}$. If μ were negative, a longitudinal tension would produce a lateral swelling, and if μ were greater than $\frac{1}{2}$, the lateral contraction would be great enough to overbalance the elongation, and cause a diminution of volume on the whole. The latter state of things would be inconsistent with stability, and the former can scarcely be possible in ordinary solids. At one time it was supposed that μ was necessarily equal to $\frac{1}{4}$, so that there was only one independent elastic constant, but experiments have since shewn that μ is variable. For glass and brass Wertheim found experimentally $\mu = \frac{1}{3}$.

If η denote the lateral displacement of the particle distant r

from the axis, and if the section be circular, the kinetic energy due to the lateral motion is

$$\delta T = \pi \rho \int_0^l \int_0^r \dot{\eta}^2 dx \, . \, r dr = \frac{\rho \omega \mu^2 r^2}{4} \, . \int_0^l \left(\frac{d\dot{\xi}}{dx}\right)^2 dx.$$

Thus the whole kinetic energy is

$$T + \delta T = \frac{\rho \omega}{2} \int_0^l \dot{\xi}^2 dx + \frac{\rho \omega \mu^2 r^2}{4} \int_0^l \left(\frac{d\dot{\xi}}{dx}\right)^2 dx.$$

In the case of a bar free at both ends, we have

$$\xi \propto \cos \frac{i\pi x}{l}, \quad \frac{d\xi}{dx} \propto -\frac{i\pi}{l} \sin \frac{i\pi x}{l},$$

and thus

$$T + \delta T : T = 1 + \frac{i^2 \mu^2 \pi^2}{2} \frac{r^2}{l^2}.$$

The effect of the inertia of the lateral motion is therefore to increase the period in the ratio

$$1 : 1 + \frac{i^2 \mu^2 \pi^2}{4} \frac{r^2}{l^2}.$$

This correction will be nearly insensible for the graver modes of bars of ordinary proportions of length to thickness.

[A more complete solution of the problem of the present section has been given by Pochhammer[1], who applies the general equations for an elastic solid to the case of an infinitely extended cylinder of circular section. The result for longitudinal vibrations, so far as the term in r^2/l^2, is in agreement with that above determined. A similar investigation has also been published by Chree[2], who has also treated the more general question[3] in which the cylindrical section is not restricted to be circular.]

158. Experiments on longitudinal vibrations may be made with rods of deal or of glass. The vibrations are excited by friction § 138, with a wet cloth in the case of glass; but for metal or wooden rods it is necessary to use leather charged with powdered rosin. "The longitudinal vibrations of a pianoforte string may be excited by gently rubbing it longitudinally with a piece of india rubber, and those of a violin string by placing the bow obliquely across the string, and moving it along the string longitudinally, keeping the same point of the bow upon the string. The note is unpleasantly shrill in both cases."

[1] *Crelle*, Bd. 81, 1876.   [2] *Quart. Math. Journ.*, Vol. 21, p. 287, 1886.
[3] *Ibid*, Vol. 23, p. 317, 1889.

"If the peg of the violin be turned so as to alter the pitch of the lateral vibrations very considerably, it will be found that the pitch of the longitudinal vibrations has altered very slightly. The reason of this is that in the case of the lateral vibrations the change of velocity of wave-transmission depends chiefly on the change of tension, which is considerable. But in the case of the longitudinal vibrations, the change of velocity of wave-transmission depends upon the change of extension, which is comparatively slight[1]."

In Savart's experiments on longitudinal vibrations, a peculiar sound, called by him a "son rauque," was occasionally observed, whose pitch was an octave below that of the longitudinal vibration. According to Terquem[2] the cause of this sound is a transverse vibration, whose appearance is due to an approximate agreement between its own period and that of the sub-octave of the longitudinal vibration § 68 b. If this view be correct, the phenomenon would be one of the second order, probably referable to the fact that longitudinal compression of a bar tends to produce curvature.

159. The second class of vibrations, called torsional, which depend on the resistance opposed to twisting, is of very small importance. A solid or hollow cylindrical rod of circular section may be twisted by suitable forces, applied at the ends, in such a manner that each transverse section remains in its own plane. But if the section be not circular, the effect of a twist is of a more complicated character, the twist being necessarily attended by a warping of the layers of matter originally composing the normal sections. Although the effects of the warping might probably be determined in any particular case if it were worth while, we shall confine ourselves here to the case of a circular section, when there is no motion parallel to the axis of the rod.

The force with which twisting is resisted depends upon an elastic constant different from q, called the rigidity. If we denote it by n, the relation between q, n, and μ may be written

$$n = \frac{q}{2(\mu+1)} \dots\dots\dots\dots\dots\dots(1)[3],$$

[1] Donkin's *Acoustics*, p. 154.

[2] *Ann. de Chimie*, LVII. 129—190.

[3] Thomson and Tait, § 683. This, it should be remarked, applies to isotropic material only.

shewing that n lies between $\frac{1}{2}q$ and $\frac{1}{3}q$. In the case of $\mu = \frac{1}{3}$, $n = \frac{3}{8}q$.

Let us now suppose that we have to do with a rod in the form of a thin tube of radius r and thickness dr, and let θ denote the angular displacement of any section, distant x from the origin. The rate of twist at x is represented by $d\theta/dx$, and the shear of the material composing the pipe by $r\,d\theta/dx$. The opposing force per unit of area is $nr\,d\theta/dx$; and since the area is $2\pi r\,dr$, the moment round the axis is

$$2n\pi r^3\,dr\,\frac{d\theta}{dx}.$$

Thus the force of restitution acting on the slice dx has the moment

$$2n\pi r^3\,dr\,dx\,\frac{d^2\theta}{dx^2}.$$

Now the moment of inertia of the slice under consideration is $2\pi r\,dr \cdot dx \cdot \rho \cdot r^2$, and therefore the equation of motion assumes the form

$$\rho\,\frac{d^2\theta}{dt^2} = n\,\frac{d^2\theta}{dx^2}\dots\dots\dots\dots\dots\dots\dots(2).$$

Since this is independent of r, the same equation applies to a cylinder of finite thickness or to one solid throughout.

The velocity of wave propagation is $\sqrt{(n/\rho)}$, and the whole theory is precisely similar to that of longitudinal vibrations, the condition for a free end being $d\theta/dx = 0$, and for a fixed end $\theta = 0$, or, if a permanent twist be contemplated, $\theta = $ constant.

The velocity of longitudinal vibrations is to that of torsional vibrations in the ratio $\sqrt{q} : \sqrt{n}$ or $\sqrt{(2+2\mu)} : 1$. The same ratio applies to the frequencies of vibration for bars of equal length vibrating in corresponding modes under corresponding terminal conditions. If $\mu = \frac{1}{3}$, the ratio of frequencies would be

$$\sqrt{q} : \sqrt{n} = \sqrt{8} : \sqrt{3} = 1\cdot 63,$$

corresponding to an interval rather greater than a fifth.

In any case the ratio of frequencies must lie between

$$\sqrt{2} : 1 = 1\cdot 414, \quad \text{and} \quad \sqrt{3} : 1 = 1\cdot 732.$$

Longitudinal and torsional vibrations were first investigated by Chladni.

CHAPTER VIII.

160. In the present chapter we shall consider the lateral vibrations of thin elastic rods, which in their natural condition are straight. Next to those of strings, this class of vibrations is perhaps the most amenable to theoretical and experimental treatment. There is difficulty sufficient to bring into prominence some important points connected with the general theory, which the familiarity of the reader with circular functions may lead him to pass over too lightly in the application to strings; while at the same time the difficulties of analysis are not such as to engross attention which should be devoted to general mathematical and physical principles.

Daniel Bernoulli[1] seems to have been the first who attacked the problem. Euler, Riccati, Poisson, Cauchy, and more recently Strehlke[2], Lissajous[3], and A. Seebeck[4] are foremost among those who have advanced our knowledge of it.

161. The problem divides itself into two parts, according to the presence, or absence, of a permanent longitudinal tension. The consideration of permanent tension entails additional complication, and is of interest only in its application to stretched strings, whose stiffness, though small, cannot be neglected altogether. Our attention will therefore be given principally to the two extreme cases, (1) when there is no permanent tension, (2) when the tension is the chief agent in the vibration.

---

[1] *Comment. Acad. Petrop.* t. XIII. [2] Pogg. *Ann.* Bd. XXVII. p. 505, 1833.

[3] *Ann. d. Chimie* (3), xxx. 385, 1850.

[4] *Abhandlungen d. Math. Phys. Classe d. K. Sächs. Gesellschaft d. Wissenschaften.* Leipzig, 1852.

With respect to the section of the rod, we shall suppose that one principal axis lies in the plane of vibration, so that the bending at every part takes place in a direction of maximum or minimum (or stationary) flexural rigidity. For example, the surface of the rod may be one of revolution, each section being circular, though not necessarily of constant radius. Under these circumstances the potential energy of the bending for each element of length is proportional to the square of the curvature multiplied by a quantity depending on the material of the rod, and on the moment of inertia of the transverse section about an axis through its centre of inertia perpendicular to the plane of bending. If ω be the area of the section, $\kappa^2\omega$ its moment of inertia, q Young's modulus, ds the element of length, and dV the corresponding potential energy for a curvature $1 \div R$ of the axis of the rod,

$$dV = \tfrac{1}{2} q \kappa^2 \omega \frac{ds}{R^2} \quad\quad\quad\quad\quad\quad(1).$$

This result is readily obtained by considering the extension of the various filaments of which the bar may be supposed to be made up. Let η be the distance from the axis of the projection on the plane of bending of a filament of section $d\omega$. Then the length of the filament is altered by the bending in the ratio

$$1 : 1 + \frac{\eta}{R},$$

R being the radius of curvature. Thus on the side of the axis for which η is positive, viz. on the *outward* side, a filament is extended, while on the other side of the axis there is compression. The force necessary to produce the extension η/R is $q\,\eta/R\,.\,d\omega$ by the definition of Young's modulus; and thus the whole couple by which the bending is resisted amounts to

$$\int q\,\frac{\eta}{R}\,.\,\eta\,.\,d\omega = \frac{q}{R}\,\kappa^2\omega,$$

if ω be the area of the section and κ its radius of gyration about a line through the axis, and perpendicular to the plane of bending. The angle of bending corresponding to a length of axis ds is $ds \div R$, and thus the work required to bend ds to curvature $1 \div R$ is

$$\tfrac{1}{2} q\,\kappa^2 \omega \frac{ds}{R^2},$$

since the *mean* is half the *final* value of the couple.

[For a more complete discussion of the legitimacy of the

foregoing method of calculation the reader must be referred to works upon the Theory of Elasticity. The question of lateral vibrations has been specially treated by Pochhammer[1] on the basis of the general equations.]

For a circular section κ is one-half the radius.

That the potential energy of the bending would be proportional, *cœteris paribus*, to the square of the curvature, is evident before-hand. If we call the coefficient B, we may take

$$V = \tfrac{1}{2} \int B \frac{ds}{R^2},$$

or, in view of the approximate straightness,

$$V = \tfrac{1}{2} \int B \left(\frac{d^2 y}{dx^2}\right)^2 dx \dots \dots \dots (2),$$

in which y is the lateral displacement of that point on the axis of the rod whose abscissa, measured parallel to the undisturbed position, is x. In the case of a rod whose sections are similar and similarly situated B is a constant, and may be removed from under the integral sign.

The kinetic energy of the moving rod is derived partly from the motion of translation, parallel to y, of the elements composing it, and partly from the rotation of the same elements about axes through their centres of inertia perpendicular to the plane of vibration. The former part is expressed by

$$\tfrac{1}{2} \int \rho \omega \, \dot{y}^2 dx \dots \dots \dots (3),$$

if ρ denote the volume-density. To express the latter part, we have only to observe that the angular displacement of the element dx is dy/dx, and therefore its angular velocity $d^2 y/dt\,dx$. The square of this quantity must be multiplied by half the moment of inertia of the element, that is, by $\tfrac{1}{2}\kappa^2 \rho \omega \, dx$. We thus obtain

$$T = \tfrac{1}{2} \int \rho \omega \, \dot{y}^2 dx + \tfrac{1}{2} \int \kappa^2 \rho \omega \left(\frac{d}{dt}\frac{dy}{dx}\right)^2 dx \dots \dots \dots (4).$$

[1] *Crelle*, Bd. 81, 1876.

162. In order to form the equation of motion we may avail ourselves of the principle of virtual velocities. If for simplicity we confine ourselves to the case of uniform section, we have

$$\delta V = B \int \frac{d^2 y}{dx^2} \frac{d^2 \delta y}{dx^2} dx$$

$$= B \frac{d^2 y}{dx^2} \frac{d\delta y}{dx} - B \frac{d^3 y}{dx^3} \delta y + B \int \frac{d^4 y}{dx^4} \delta y \, dx \ldots\ldots(1),$$

where the terms free from the integral sign are to be taken between the limits. This expression includes only the internal forces due to the bending. In what follows we shall suppose that there are no forces acting from without, or rather none that do work upon the system. A force of constraint, such as that necessary to hold any point of the bar at rest, need not be regarded, as it does no work and therefore cannot appear in the equation of virtual velocities.

The virtual moment of the accelerations is

$$\int \rho \omega \frac{d^2 y}{dt^2} \delta y \, dx + \int \rho \omega \kappa^2 \frac{d^2}{dt^2} \left(\frac{dy}{dx}\right) \delta \left(\frac{dy}{dx}\right) dx$$

$$= \int \rho \omega \left(\frac{d^2 y}{dt^2} - \kappa^2 \frac{d^4 y}{dx^2 dt^2}\right) \delta y \, dx + \rho \omega \kappa^2 \delta y \frac{d^3 y}{dt^2 dx} \ldots\ldots (2).$$

Thus the variational equation of motion is

$$\int \left\{ B \frac{d^4 y}{dx^4} + \rho \omega \left(\frac{d^2 y}{dt^2} - \kappa^2 \frac{d^4 y}{dx^2 dt^2}\right)\right\} \delta y \, dx$$

$$+ B \frac{d^2 y}{dx^2} \delta \left(\frac{dy}{dx}\right) + \left\{\rho \omega \kappa^2 \frac{d^3 y}{dt^2 dx} - B \frac{d^3 y}{dx^3}\right\} \delta y = 0 \ldots\ldots .(3),$$

in which the terms free from the integral sign are to be taken between the limits. From this we derive as the equation to be satisfied at all points of the length of the bar

$$B \frac{d^4 y}{dx^4} + \rho \omega \left(\frac{d^2 y}{dt^2} - \kappa^2 \frac{d^4 y}{dx^2 dt^2}\right) = 0,$$

while at each end

$$B \frac{d^2 y}{dx^2} \delta \left(\frac{dy}{dx}\right) + \left\{\rho \omega \kappa^2 \frac{d^3 y}{dt^2 dx} - B \frac{d^3 y}{dx^3}\right\} \delta y = 0 :$$

or, if we introduce the value of B viz. $q\kappa^2\omega$, and write $q/\rho = b^2$,

$$\frac{d^2 y}{dt^2} + b^2 \kappa^2 \frac{d^4 y}{dx^4} - \kappa^2 \frac{d^4 y}{dx^2 dt^2} = 0 \ldots\ldots\ldots\ldots (4),$$

and for each end

$$b^2 \frac{d^2 y}{dx^2} \delta \left(\frac{dy}{dx} \right) + \left\{ \frac{d^3 y}{dt^2 dx} - b^2 \frac{d^3 y}{dx^3} \right\} \delta y = 0 \quad \ldots\ldots\ldots(5).$$

In these equations b expresses the velocity of transmission of longitudinal waves.

The condition (5) to be satisfied at the ends assumes different forms according to the circumstances of the case. It is possible to conceive a constraint of such a nature that the ratio $\delta(dy/dx) : \delta y$ has a prescribed finite value. The second boundary condition is then obtained from (5) by introduction of this ratio. But in all the cases that we shall have to consider, there is either no constraint or the constraint is such that either $\delta(dy/dx)$ or δy vanishes, and then the boundary conditions take the form

$$\frac{d^2 y}{dx^2} \delta \left(\frac{dy}{dx} \right) = 0, \qquad \left\{ \frac{d^3 y}{dt^2 dx} - b^2 \frac{d^3 y}{dx^3} \right\} \delta y = 0 \ldots\ldots\ldots\ldots(6).$$

We must now distinguish the special cases that may arise. If an end be free, δy and $\delta(dy/dx)$ are both arbitrary, and the conditions become

$$\frac{d^2 y}{dx^2} = 0, \qquad \frac{d^3 y}{dt^2 dx} - b^2 \frac{d^3 y}{dx^3} = 0 \ldots\ldots\ldots\ldots\ldots(7),$$

the first of which may be regarded as expressing that no couple acts at the free end, and the second that no force acts.

If the direction at the end be free, but the end itself be constrained to remain at rest by the action of an applied force of the necessary magnitude, in which case for want of a better word the rod is said to be *supported*, the conditions are

$$\frac{d^2 y}{dx^2} = 0, \qquad \delta y = 0 \quad \ldots\ldots\ldots\ldots\ldots\ldots\ldots(8),$$

by which (5) is satisfied.

A third case arises when an extremity is constrained to maintain its direction by an applied couple of the necessary magnitude, but is free to take any position. We have then

$$\delta \left(\frac{dy}{dx} \right) = 0, \qquad \frac{d^3 y}{dt^2 dx} - b^2 \frac{d^3 y}{dx^3} = 0 \quad \ldots\ldots\ldots\ldots (9).$$

Fourthly, the extremity may be constrained both as to position and direction, in which case the rod is said to be *clamped*. The conditions are plainly

$$\delta\left(\frac{dy}{dx}\right) = 0, \qquad \delta y = 0 \quad \dots\dots\dots\dots \text{(10)}.$$

Of these four cases the first and the last are the more important; the third we shall omit to consider, as there are no experimental means by which the contemplated constraint could be realized. Even with this simplification a considerable variety of problems remain for discussion, as either end of the bar may be free, clamped or supported, but the complication thence arising is not so great as might have been expected. We shall find that different cases may be treated together and that the solution for one case may sometimes be derived immediately from that of another.

In experimenting on the vibrations of bars, the condition for a clamped end may be realized with the aid of a vice of massive construction. In the case of a free end there is of course no difficulty so far as the end itself is concerned; but, when both ends are free, a question arises as to how the weight of the bar is to be supported. In order to interfere with the vibration as little as possible, the supports must be confined to the neighbourhood of the nodal points. It is sometimes sufficient merely to lay the bar on bridges, or to pass a loop of string round the bar and draw it tight by screws attached to its ends. For more exact purposes it would perhaps be preferable to carry the weight of the bar on a pin traversing a hole drilled through the middle of the thickness in the plane of vibration.

When an end is to be 'supported,' it may be pressed into contact with a fixed plate whose plane is perpendicular to the length of the bar.

163. Before proceeding further we shall introduce a supposition, which will greatly simplify the analysis, without seriously interfering with the value of the solution. We shall assume that the terms depending on the angular motion of the sections of the bar may be neglected, which amounts to supposing the *inertia* of each section concentrated at its centre. We shall afterwards (§ 186) investigate a correction for the rotatory inertia, and shall prove that under ordinary circumstances it is small. The equation of motion now becomes

$$\frac{d^2y}{dt^2} + \kappa^2 b^2 \frac{d^4y}{dx^4} = 0 \dots\dots\dots\dots\dots(1),$$

and the boundary conditions for a free end

$$\frac{d^2y}{dx^2} = 0, \qquad \frac{d^3y}{dx^3} = 0 \dots\dots\dots\dots\dots(2).$$

The next step in conformity with the general plan will be the assumption of the harmonic form of y. We may conveniently take

$$y = u \cos\left(\frac{\kappa b}{l^2} m^2 t\right) \dots\dots\dots\dots\dots(3),$$

where l is the length of the bar, and m is an abstract number, whose value has to be determined. Substituting in (1), we obtain

$$\frac{d^4u}{dx^4} = \frac{m^4}{l^4} u \dots\dots\dots\dots\dots(4).$$

If $u = e^{p\,mx/l}$ be a solution, we see that p is one of the fourth roots of unity, viz. $+1$, -1, $+i$, $-i$; so that the complete solution is

$$u = A \cos m\frac{x}{l} + B \sin m\frac{x}{l} + C\,e^{mx/l} + D\,e^{-mx/l}\dots\dots\dots(4a),$$

containing four arbitrary constants.

[The simplest case occurs when the motion is strictly periodic with respect to x, C and D vanishing. If λ be the wave-length and τ the period of the vibration, we have

$$\frac{2\pi}{\lambda} = \frac{m}{l}, \qquad \frac{2\pi}{\tau} = \kappa b \frac{m^2}{l^2},$$

so that
$$\tau = \frac{\lambda^2}{2\pi\kappa b}\dots\dots\dots\dots\dots(4b).]$$

In the case of a finite rod we have still to satisfy the four boundary conditions,—two for each end. These determine the ratios $A : B : C : D$, and furnish besides an equation which m must satisfy. Thus a series of particular values of m are alone admissible, and for each m the corresponding u is determined in everything except a constant multiplier. We shall distinguish the different functions u belonging to the same system by suffixes.

The value of y at any time may be expanded in a series of the functions u (§§ 92, 93). If ϕ_1, ϕ_2, &c. be the normal co-ordinates, we have

$$y = \phi_1 u_1 + \phi_2 u_2 + \ldots \quad \ldots\ldots\ldots\ldots\ldots(5),$$

and
$$T = \tfrac{1}{2}\rho\omega \int (\dot{\phi}_1 u_1 + \dot{\phi}_2 u_2 + \ldots)^2 dx$$

$$= \tfrac{1}{2}\rho\omega \left\{ \dot{\phi}_1^2 \int u_1^2 dx + \dot{\phi}_2^2 \int u_2^2 dx + \ldots \right\} \quad \ldots\ldots\ldots (6).$$

We are fully justified in asserting at this stage that each integrated product of the functions vanishes, and therefore the process of the following section need not be regarded as more than a *verification*. It is however required in order to determine the value of the integrated squares.

164. Let u_m, $u_{m'}$ denote two of the normal functions corresponding respectively to m and m'. Then

$$\frac{d^4 u_m}{dx^4} = \frac{m^4}{l^4} u_m, \qquad \frac{d^4 u_{m'}}{dx^4} = \frac{m'^4}{l^4} u_{m'} \quad \ldots\ldots\ldots\ldots(1);$$

or, if dashes indicate differentiation with respect to $(m x/l)$, $(m' x/l)$,

$$u_m'''' = u_m, \qquad u_{m'}'''' = u_{m'} \quad \ldots\ldots\ldots\ldots\ldots (2).$$

If we subtract equations (1) after multiplying them by $u_{m'}$, u_m respectively, and then integrate over the length of the bar, we have

$$\frac{m'^4 - m^4}{l^4} \int u_m u_m \, dx = \int \left(u_m \frac{d^4 u_{m'}}{dx^4} - u_{m'} \frac{d^4 u_m}{dx^4} \right) dx$$

$$= u_m \frac{d^3 u_m}{dx^3} - u_{m'} \frac{d^3 u_m}{dx^3} + \frac{du_{m'}}{dx} \frac{d^2 u_m}{dx^2} - \frac{du_m}{dx} \frac{d^2 u_m}{dx^2} \ldots \ldots (3),$$

the integrated terms being taken between the limits.

Now whether the end in question be clamped, supported, or free[1], each term vanishes on account of one or other of its

---

[1] The reader should observe that the cases here specified are particular, and that the right-hand member of (3) vanishes, provided that

$$u_m : \frac{d^3 u_m}{dx^3} = u_m : \frac{d^3 u_{m'}}{dx^3},$$

and
$$\frac{du_m}{dx} : \frac{d^2 u_m}{dx^2} = \frac{du_{m'}}{dx} : \frac{d^2 u_{m'}}{dx^2}.$$

These conditions include, for instance, the case of a rod whose end is urged towards its position of equilibrium by a force proportional to the displacement, as by a spring without inertia.

factors. We may therefore conclude that, if u_m, $u_{m'}$ refer to two modes of vibration (corresponding of course to the same terminal conditions) of which a rod is capable, then

$$\int u_m u_{m'} dx = 0 \quad \dots\dots\dots\dots\dots\dots(4),$$

provided m and m' be different.

The attentive reader will perceive that in the process just followed, we have in fact retraced the steps by which the fundamental differential equation was itself proved in § 162. It is the original *variational* equation that has the most immediate connection with the conjugate property. If we denote y by u and δy by v,

$$\delta V = B \int \frac{d^2 u}{dx^2} \frac{d^2 v}{dx^2} dx,$$

and the equation in question is

$$B \int \frac{d^2 u}{dx^2} \frac{d^2 v}{dx^2} dx + \rho\omega \int \ddot{u} v dx = 0 \quad \dots\dots\dots\dots(5).$$

Suppose now that u relates to a normal component vibration, so that $\ddot{u} + n^2 u = 0$, where n is some constant; then

$$n^2 \rho\omega \int uv dx = B \int \frac{d^2 u}{dx^2} \frac{d^2 v}{dx^2} dx.$$

By similar reasoning, if v be a normal function, and u represent any displacement possible to the system,

$$n'^2 \rho\omega \int uv dx = B \int \frac{d^2 u}{dx^2} \frac{d^2 v}{dx^2} dx.$$

We conclude that if u and v be both normal functions, *which have different periods*,

$$\int uv dx = 0 \dots\dots\dots\dots\dots\dots(6);$$

and this proof is evidently as direct and general as could be desired.

The reader may investigate the formula corresponding to (6), when the term representing the rotatory inertia is retained.

By means of (6) we may verify that the admissible values of n are real. For if n^2 were complex, and $u = \alpha + i\beta$ were a normal function, then $\alpha - i\beta$, the conjugate of u, would be a normal function also, corresponding to the conjugate of n^2, and then the product of the two functions, being a sum of squares, would not vanish, when integrated[1].

If in (3) m and m' be the same, the equation becomes identically true, and we cannot at once infer the value of $\int u_m^2 dx$. We must take m' equal to $m + \delta m$, and trace the limiting form of the equation as δm tends to vanish. [It should be observed that the function $u_{m+\delta m}$ is not a normal function of the system; it is supposed to be derived from u_m by variation of m in $(4a)$ § 163, the coefficients A, B, C, D being retained constant.] In this way we find

$$\frac{4m^3}{l^4} \int u_m^2 dx = u \frac{d}{dm} \frac{d^3u}{dx^3} - \frac{du}{dm} \frac{d^3u}{dx^3} + \frac{d^2u}{dx^2} \frac{d}{dm} \frac{du}{dx} - \frac{du}{dx} \frac{d}{dm} \frac{d^2u}{dx^2},$$

the right-hand side being taken between the limits.

Now $\dfrac{du}{dx} = \dfrac{m}{l} u'$, &c., $\dfrac{du}{dm} = \dfrac{x}{l} u'$, &c.,

and thus

$$\frac{4m^3}{l^4} \int u_m^2 dx = \frac{3m^2}{l^3} u u''' + \frac{m^3 x}{l^4} u u'''' - \frac{m^3 x}{l^4} u' u'''$$

$$+ \frac{m^2}{l^3} u' u'' + \frac{m^3 x}{l^4} (u'')^2 - \frac{2m^2}{l^3} u' u'' - \frac{m^3 x}{l^4} u' u''',$$

in which $u'''' = u$, so that

$$\frac{4m}{l} \int u_m^2 dx = 3u u''' + \frac{mx}{l} u^2 - \frac{2mx}{l} u' u''' - u' u'' + \frac{mx}{l} (u'')^2 \ldots (7),$$

between the limits.

Now whether an end be clamped, supported, or free,
$$u u''' = 0, \qquad u' u'' = 0,$$

and thus, if we take the origin of x at one end of the rod,

$$\int_0^l u^2 dx = \left\{ \frac{x}{4} (u^2 - 2u' u''' + u''^2) \right\}_0^l$$

$$= \tfrac{1}{4} l (u^2 - 2u' u''' + u''^2)_{x=l} \ldots \ldots \ldots \ldots (8).$$

The form of our integral is independent of the terminal condition at $x = 0$. If the end $x = l$ be free, u'' and u''' vanish, and accordingly

$$\int_0^l u^2 dx = \tfrac{1}{4} l \, u^2(l) \ldots \ldots \ldots \ldots \ldots (9),$$

[1] This method is, I believe, due to Poisson.

that is to say, for a rod with one end free the mean value of u^2 is one-fourth of the terminal value, and that whether the other end be clamped, supported, or free.

Again, if we suppose that the rod is clamped at $x = l$, u and u' vanish, and (8) gives

$$\int_0^l u^2 dx = \tfrac{1}{4} l \left[u''(l) \right]^2.$$

Since this must hold good whatever be the terminal condition at the other end, we see that for a rod, one end of which is fixed and the other free,

$$\int_0^l u^2 dx = \tfrac{1}{4} l u^2 \text{(free end)} = \tfrac{1}{4} l u''^2 \text{(fixed end)},$$

shewing that in this case u^2 at the free end is the same as u''^2 at the clamped end.

The annexed table gives the values of four times the mean of u^2 in the different cases.

clamped, free.........	u^2 (free end), or u''^2 (clamped end)
free, free	u^2 (free end)
clamped, clamped ...	u''^2 (clamped end)
supported, supported	$-2u'u'''$ (supported end) $= 2u'^2$
supported, free	u^2 (free end), or $-2u'u'''$ (supported end)
supported, clamped	u''^2 (clamped end), or $-2u'u'''$ (supported end)

By the introduction of these values the expression for T assumes a simpler form. In the case, for example, of a clamped-free or a free-free rod,

$$T = \frac{\rho l \omega}{8} \{ \dot{\phi_1}^2 u_1^2 (l) + \dot{\phi_2}^2 u_2^2 (l) + \dots \} \dots\dots\dots(10),$$

where the end $x = l$ is supposed to be free.

165. A similar method may be applied to investigate the values of $\int u'^2 dx$, and $\int u''^2 dx$. In the derivation of equation (7) of the preceding section nothing was assumed beyond the truth of the equation $u'''' = u$, and since this equation is equally true of any of the derived functions, we are at liberty to replace u by u' or u''. Thus

$$\frac{4m}{l}\int_0^l u'^2 dx = 3u'u + \frac{mx}{l}u^2 - 2\frac{mx}{l}u''u - u''u''' + \frac{mx}{l}u'''^2$$

$$= 3uu' + \frac{mu}{l}u'^2 - u''u''' + \frac{mx}{l}u'''^2,$$

taken between the limits, since the term $u\,u''$ vanishes in all three cases.

For a free-free rod

$$\frac{4m}{l}\int_0^l u'^2 dx = 3\,(uu')_l - 3\,(uu')_0 + m\,(u'^2)_l$$

$$= 6\,(uu')_l + m\,(u'^2)_l\ldots\ldots\ldots\ldots\ldots\ldots\ldots(1),$$

for, as we shall see, the values of $u\,u'$ must be equal and opposite at the two ends. Whether u be positive or negative at $x = l$, $u\,u'$ is positive.

For a rod which is clamped at $x = 0$ and free at $x = l$

$$\frac{4m}{l}\int_0^l u'^2 dx = 3\,(uu')_l + mu_l'^2 + (u''u''')_0$$

[We have already seen that $u_0'' = \pm\,u_l$; and it may be proved from the formulæ of § 173 that

$$-\frac{u_0'''}{u_l} = \frac{u_0''}{u_l} = \frac{\cos m + \cosh m}{\sin m \sinh m},$$

so that $\dfrac{(u''u''')_0}{(u'u)_l} = -\dfrac{(\cos m + \cosh m)^2}{\sin^2 m \sinh^2 m} = -1.$]

Thus $\dfrac{4m}{l}\displaystyle\int_0^l u'^2 dx = 2\,(uu')_l + mu_l'^2\ldots\ldots\ldots\ldots\ldots(2),$

a result that we shall have occasion to use later.

By applying the same equation to the evaluation of $\int u''^2 dx$, we find

$$\frac{4m}{l}\int u''^2 dx = 3u''u' + \frac{mx}{l}u''^2 - 2\frac{mx}{l}u'''u' - u'''u + \frac{mx}{l}u^2$$

$$= m\,(u''^2 - 2u'u''' + u^2)_l,$$

since $u'u''$ and $u\,u'''$ vanish.

Comparing this with (8) § 164, we see that

$$\int u''^2 dx = \int u^2 dx \dots\dots\dots\dots\dots\dots (3),$$

whatever the terminal conditions may be.

The same result may be arrived at more directly by integrating by parts the equation

$$\frac{m^4}{l^4} u^2 = u \frac{d^4 u}{dx^4}.$$

166. We may now form the expression for V in terms of the normal co-ordinates.

$$V = \frac{b^2 \kappa^2 \rho \omega}{2} \int \left\{ \phi_1 \frac{d^2 u_1}{dx^2} + \phi_2 \frac{d^2 u_2}{dx^2} + \dots \right\}^2 dx$$

$$= \frac{b^2 \kappa^2 \rho \omega}{2} \left\{ \phi_1^2 \int \left(\frac{d^2 u_1}{dx^2}\right)^2 dx + \phi_2^2 \int \left(\frac{d^2 u_2}{dx^2}\right)^2 dx + \dots \right\}$$

$$= \frac{b^2 \kappa^2 \rho \omega}{2 l^4} \left\{ m_1^4 \phi_1^2 \int u_1^2 dx + m_2^4 \phi_2^2 \int u_2^2 dx + \dots \right\} \dots\dots\dots(1).$$

If the functions u be those proper to a rod free at $x = l$, this expression reduces to

$$V = \frac{b^2 \kappa^2 \rho \omega}{8 l^3} \left\{ m_1^4 [u_1(l)]^2 \phi_1^2 + m_2^4 [u_2(l)]^2 \phi_2^2 + \dots \right\} \dots\dots\dots(2).$$

In any case the equations of motion are of the form

$$\rho \omega \int u_1^2 dx \; \ddot{\phi_1} + \frac{b^2 \kappa^2 \rho \omega}{l^4} m_1^4 \int u_1^2 dx \; \phi_1 = \Phi_1 \dots\dots\dots(3),$$

and, since $\Phi_1 \delta \phi_1$ is by definition the work done by the impressed forces during the displacement $\delta \phi_1$,

$$\Phi_1 = \int Y u_1 \rho \omega \, dx \dots\dots\dots\dots (4),$$

if $Y \rho \omega \, dx$ be the lateral force acting on the element of mass $\rho \omega \, dx$. If there be no impressed forces, the equation reduces to

$$\ddot{\phi_1} + \frac{b^2 \kappa^2 m_1^4}{l^4} \phi_1 = 0 \dots\dots\dots\dots\dots (5),$$

as we know it ought to do.

167. The significance of the reduction of the integrals $\int u^2 dx$ to dependence on the terminal values of the function and its derivatives may be placed in a clearer light by the following line of argument. To fix the ideas, consider the case of a rod clamped at $x = 0$, and free at $x = l$, vibrating in the normal mode expressed by u. If a small addition Δl be made to the rod at the free end, the form of u (considered as a function of x) is changed, but, in accordance with the general principle established in Chapter IV. (§ 88), we may calculate the period under the altered circumstances without allowance for the change of type, if we are content to neglect the square of the change. In consequence of the straightness of the rod at the place where the addition is made, there is no alteration in the potential energy, and therefore the alteration of period depends entirely on the variation of T. This quantity is increased in the ratio

$$\int_0^l u^2 dx \;:\; \int_0^{l+\Delta l} u^2 dx,$$

or

$$1 \;:\; 1 + \frac{u_l^2 \Delta l}{\int_0^l u^2 dx},$$

which is also the ratio in which the square of the period is augmented. Now, as we shall see presently, the actual period varies as l^2, and therefore the change in the square of the period is in the ratio

$$1 \;:\; 1 + 4\Delta l/l.$$

A comparison of the two ratios shews that

$$u_l^2 \;:\; \int u^2 dx = 4 \;:\; l.$$

The above reasoning is not insisted upon as a demonstration, but it serves at least to explain the reduction of which the integral is susceptible. Other cases in which such integrals occur may be treated in a similar manner, but it would often require care to predict with certainty what amount of discontinuity in the varied type might be admitted without passing out of the range of the principle on which the argument depends. The reader may, if he pleases, examine the case of a string in the middle of which a small piece is interpolated.

168. In treating problems relating to vibrations the usual course has been to determine in the first place the forms of the normal functions, viz. the functions representing the normal types, and afterwards to investigate the integral formulæ by means of which the particular solutions may be combined to suit arbitrary initial circumstances. I have preferred to follow a different order, the better to bring out the generality of the method, *which does not depend upon a knowledge of the normal functions.* In pursuance of the same plan, I shall now investigate the connection of the arbitrary constants with the initial circumstances, and solve one or two problems analogous to those treated under the head of Strings.

The general value of y may be written

$$y = \left(A_1 \cos \frac{\kappa b}{l^2} m_1^2 t + B_1 \sin \frac{\kappa b}{l^2} m_1^2 t \right) u_1$$

$$+ \left(A_2 \cos \frac{\kappa b}{l^2} m_2^2 t + B_2 \sin \frac{\kappa b}{l^2} m_2^2 t \right) u_2$$

$$+ \dots \dots \dots \dots \dots \dots \dots \dots \dots \dots \dots \dots \dots \dots \dots (1),$$

so that initially

$$y_0 = A_1 u_1 + A_2 u_2 + \dots \quad \dots \dots \dots \dots \dots \dots (2),$$

$$\dot{y}_0 = \frac{\kappa b}{l^2} \{ m_1^2 B_1 u_1 + m_2^2 B_2 u_2 + \dots \} \quad \dots \dots \dots \dots (3).$$

If we multiply (2) by u_r and integrate over the length of the rod, we get

$$\int y_0 u_r dx = A_r \int u_r^2 dx \dots \dots \dots \dots \dots (4),$$

and similarly from (3)

$$\frac{l^2}{\kappa b} \int \dot{y}_0 u_r dx = m_r^2 B_r \int u_r^2 dx \quad \dots \dots \dots \dots (5),$$

formulæ which determine the arbitrary constants A_r, B_r.

It must be observed that we do not need to prove analytically the possibility of the expansion expressed by (1). If *all* the particular solutions are included, (1) necessarily represents the most general vibration possible, and may therefore be adapted to represent any admissible initial state.

Let us now suppose that the rod is originally at rest, in its position of equilibrium, and is set in motion by a blow which imparts velocity to a small portion of it. Initially, that is, at the moment when the rod becomes free, $y_0 = 0$, and $\dot{y}_0$ differs from zero only in the neighbourhood of one point ($x = c$).

From (4) it appears that the coefficients A vanish, and from (5) that

$$m_r^2 B_r \int u_r^2 dx = \frac{l^2}{\kappa b} u_r(c) \int \dot{y}_0 dx.$$

Calling $\int \dot{y}_0 \rho \omega dx$, the whole momentum of the blow, Y, we have

$$B_r = \frac{l^2 Y}{\kappa b \rho \omega} \frac{u_r(c)}{m_r^2 \int u_r^2 dx} \dots\dots\dots\dots\dots(6),$$

and for the final solution

$$y = \frac{l^2 Y}{\kappa b \rho \omega} \left\{ \frac{u_1(c) u_1(x)}{m_1^2 \int u_1^2 dx} \sin\left(\frac{\kappa b}{l^2} m_1^2 t\right) + \dots \right.$$
$$\left. + \frac{u_r(c) u_r(x)}{m_r^2 \int u_r^2 dx} \sin\left(\frac{\kappa b}{l^2} m_r^2 t\right) + \dots\dots \right\} \dots\dots(7).$$

In adapting this result to the case of a rod free at $x = l$, we may replace

$$\int u_r^2 dx \quad \text{by} \quad \tfrac{1}{4} l [u_r(l)]^2.$$

If the blow be applied at a node of one of the normal components, that component is missing in the resulting motion. The present calculation is but a particular case of the investigation of § 101.

169. As another example we may take the case of a bar, which is initially at rest but deflected from its natural position by a lateral force acting at $x = c$. Under these circumstances the coefficients B vanish, and the others are given by (4), § 168.

Now

$$\int_0^l y_0 u_r dx = \frac{l^4}{m_r^4} \int_0^l y_0 \frac{d^4 u_r}{dx^4} dx,$$

and on integrating by parts

$$\int_0^l y_0 \frac{d^4 u_r}{dx^4} dx = y_0 \frac{d^3 u_r}{dx^3} - \frac{dy_0}{dx} \frac{d^2 u_r}{dx^2}$$
$$+ \frac{d^2 y_0}{dx^2} \frac{du_r}{dx} - \frac{d^3 y_0}{dx^3} u_r + \int_0^l \frac{d^4 y_0}{dx^4} u_r dx,$$

in which the terms free from the integral sign are to be taken between the limits; by the nature of the case y_0 satisfies the same terminal conditions as does u_r, and thus all these terms vanish at both limits. If the external force initially applied to the element dx be Ydx, the equation of equilibrium of the bar gives

$$\rho\omega\,\kappa^2 b^2 \frac{d^4 y_0}{dx^4} = Y \dotfill (1)$$

and accordingly

$$\int_0^l y_0 u_r\,dx = \frac{l^4}{\rho\omega\,\kappa^2 b^2 m_r^4} \int_0^l Y u_r(x)\,dx.$$

If we now suppose that the initial displacement is due to a force applied in the immediate neighbourhood of the point $x = c$, we have

$$\int_0^l y_0 u_r\,dx = \frac{l^4 u_r(c)}{\rho\omega\,\kappa^2 b^2 m_r^4} \int Y dx,$$

and for the complete value of y at time t,

$$y = \Sigma \left\{ \frac{l^4 u_r(c)\,u_r(x)}{m_r^4 \kappa^2 b^2 \int \rho\omega u_r^2\,dx} \cos \frac{\kappa b}{l^2} m_r^2 t \right\} \int Y dx \dotfill (2).$$

In deriving the above expression we have not hitherto made any special assumptions as to the conditions at the ends, but if we now confine ourselves to the case of a bar which is clamped at $x = 0$ and free at $x = l$, we may replace

$$\int u_r^2\,dx \quad \text{by} \quad \tfrac{1}{4} l\,[u_r(l)]^2.$$

If we suppose further that the force to which the initial deflection is due acts at the end, so that $c = l$, we get

$$y = 4 \Sigma \left\{ \frac{l^3 u_r(x)}{m_r^4 \kappa^2 b^2 \rho\omega u_r(l)} \cos \frac{\kappa b}{l^2} m_r^2 t \right\} \int Y dx \dotfill (3).$$

When $t = 0$, this equation must represent the initial displacement. In cases of this kind a difficulty may present itself as to how it is possible for the series, every term of which satisfies the condition $y''' = 0$, to represent an initial displacement in which this condition is violated. The fact is, that after triple differentiation with respect to x, the series no longer converges for $x = l$, and accordingly the value of y''' is not to be arrived at by making the differentiations first and summing the terms afterwards. The truth of this statement will be apparent if we consider a point distant dl from the end, and replace

$$u'''(l - dl) \quad \text{by} \quad u'''(l) - u^{\text{IV}}(l)\,dl,$$

in which $u^{\text{IV}}(l)$ is equal to

$$\frac{m^4}{l^4} u\,(l).$$

For the solution of the present problem by normal co-ordinates the reader is referred to § 101.

170. The forms of the normal functions in the various particular cases are to be obtained by determining the ratios of the four constants in the general solution of

$$\frac{d^4 u}{dx^4} = \frac{m^4}{l^4} u.$$

If for the sake of brevity x' be written for (mx/l), the solution may be put into the form

$$u = A\,(\cos x' + \cosh x') + B\,(\cos x' - \cosh x')$$
$$+ \, C\,(\sin x' + \sinh x') + D\,(\sin x' - \sinh x') \dots \dots (1),$$

where $\cosh x$ and $\sinh x$ are the hyperbolic cosine and sine of x, defined by the equations

$$\cosh x = \tfrac{1}{2}(e^x + e^{-x}), \quad \sinh x = \tfrac{1}{2}(e^x - e^{-x})\dots\dots\dots\dots(2).$$

I have followed the usual notation, though the introduction of a special symbol might very well be dispensed with, since

$$\cosh x = \cos ix, \quad \sinh x = -\,i \sin ix \dots\dots\dots\dots(3),$$

where $i = \sqrt{(-1)}$; and then the connection between the formulæ of circular and hyperbolic trigonometry would be more apparent. The rules for differentiation are expressed in the equations

$$\frac{d}{dx}\cosh x = \sinh x, \quad \frac{d}{dx}\sinh x = \cosh x$$

$$\frac{d^2}{dx^2}\cosh x = \cosh x, \quad \frac{d^2}{dx^2}\sinh x = \sinh x.$$

In differentiating (1) any number of times, the same four compound functions as there occur are continually reproduced. The only one of them which does not vanish with x' is $\cos x' + \cosh x'$, whose value is then 2.

Let us take first the case in which both ends are free. Since d^2u/dx^2 and d^3u/dx^3 vanish with x, it follows that $B = 0$, $D = 0$, so that

$$u = A\,(\cos x' + \cosh x') + C\,(\sin x' + \sinh x')\dots\dots\dots(4).$$

We have still to satisfy the necessary conditions when $x = l$, or $x' = m$. These give

$$\left. \begin{array}{l} A\left(-\cos m + \cosh m\right) + C\left(-\sin m + \sinh m\right) = 0 \\ A\left(\sin m + \sinh m\right) + C\left(-\cos m + \cosh m\right) = 0 \end{array} \right\} \dots \dots (5),$$

equations whose compatibility requires that

$$(\cosh m - \cos m)^2 = \sinh^2 m - \sin^2 m,$$

or in virtue of the relation

$$\cosh^2 m - \sinh^2 m = 1 \dots \dots \dots \dots \dots (6),$$

$$\cos m \, \cosh m = 1 \dots \dots \dots \dots \dots (7).$$

This is the equation whose roots are the admissible values of m. If (7) be satisfied, the two ratios of $A : C$ given in (5) are equal, and either of them may be substituted in (4). The constant multiplier being omitted, we have for the normal function

$$u = (\sin m - \sinh m)\left\{\cos \frac{mx}{l} + \cosh \frac{mx}{l}\right\}$$

$$- (\cos m - \cosh m)\left\{\sin \frac{mx}{l} + \sinh \frac{mx}{l}\right\} \dots \dots (8),$$

or, if we prefer it

$$u = (\cos m - \cosh m)\left\{\cos \frac{mx}{l} + \cosh \frac{mx}{l}\right\}$$

$$+ (\sin m + \sinh m)\left\{\sin \frac{mx}{l} + \sinh \frac{mx}{l}\right\} \dots \dots (9);$$

and the simple harmonic component of this type is expressed by

$$y = Pu \cos\left(\frac{\kappa b}{l^2} m^2 t + \epsilon\right) \dots \dots \dots (10).$$

171. The frequency of the vibration is $\dfrac{\kappa b}{2\pi l^2} m^2$, in which b is a velocity depending only on the material of which the bar is formed, and m is an abstract number. Hence for a given material and mode of vibration the frequency varies directly as κ—the radius of gyration of the section about an axis perpendicular to the plane of bending—and inversely as the *square* of the length. These results might have been anticipated by the argument from dimensions, if it were considered that the frequency is necessarily determined by the value of l, together with that of κb—the only quantity depending on space, time and mass, which occurs in the differential equation. If everything concerning a bar be given, except its absolute magnitude, the frequency varies inversely as the linear dimension.

These laws find an important application in the case of tuning-forks, whose prongs vibrate as rods, fixed at the ends where they join the stalk, and free at the other ends. Thus the period of vibration of forks of the same material and shape varies as the linear dimension. The period will be approximately independent of the thickness perpendicular to the plane of bending, but will vary inversely with the thickness in the plane of bending. When the thickness is given, the period is as the square of the length.

In order to lower the pitch of a fork we may, for temporary purposes, load the ends of the prongs with soft wax, or file away the metal near the base, thereby weakening the spring. To raise the pitch, the ends of the prongs, which act by inertia, may be filed.

The value of b attains its maximum in the case of steel, for which it amounts to about 5237 metres per second. For brass the velocity would be less in about the ratio $1\cdot5 : 1$, so that a tuning-fork made of brass would be about a fifth lower in pitch than if the material were steel.

[For the design of steel vibrators and for rough determinations of frequency, especially when below the limit of hearing, the theoretical formula is often convenient. If the section of the bar be rectangular and of thickness t in the plane of vibration, $k^2 = \frac{1}{12}t^2$; and then with the above value of b, and the values of m given later, we get as applicable to the gravest mode

$$\text{(clamped-free) frequency} = 84590\ t/l^2,$$
$$\text{(free-free) frequency}\ \ \ = 538400\ t/l^2.$$

l and t being expressed in centimetres.

The first of these may be used to calculate the pitch of steel tuning-forks.

The lateral vibrations of a bar may be excited by a blow, as when a tuning-fork is struck against a pad. This method is also employed for the harmonicon, in which strips of metal or glass are supported at the nodes, in such a manner that the free vibrations are but little impeded. A frictional maintenance may be obtained with a bow, or by the action of the wetted fingers upon a slender rod of glass suitably attached. The electro-magnetic maintenance of forks has been already considered, § 64. It may be applied with equal facility to the case of metal bars, or even to that of wooden planks carrying iron armatures, free at both ends and supported at the nodes. The maintenance by a stream of wind of the vibrations of harmonium and organ reeds may also be referred to

The sound of a bar vibrating laterally may be reinforced by a suitably tuned resonator, which may be placed under the middle portion or under one end. On this principle dinner gongs have been constructed, embracing one octave or more of the diatonic scale.]

172. The solution for the case when both ends are clamped may be immediately derived from the preceding by a double differentiation. Since y satisfies at both ends the terminal conditions

$$\frac{d^2y}{dx^2} = 0, \qquad \frac{d^3y}{dx^3} = 0,$$

it is clear that y'' satisfies

$$y'' = 0, \qquad \frac{dy'}{dx} = 0,$$

which are the conditions for a clamped end. Moreover the general differential equation is also satisfied by y''. Thus we may take, omitting a constant multiplier, as before,

$$u = (\sin m - \sinh m) \{\cos x' - \cosh x'\}$$
$$- (\cos m - \cosh m) \{\sin x' - \sinh x'\} \dots\dots\dots(1),$$

while m is given by the same equation as before, namely,

$$\cos m \cosh m = 1 \dots\dots\dots\dots\dots(2).$$

We conclude that the component tones have the same pitch in the two cases.

In each case there are four systems of points determined by the evanescence of y and its derivatives. Where y vanishes, there is a node; where y' vanishes, a loop, or place of maximum displacement; where y'' vanishes, a point of inflection; and where y''' vanishes, a place of maximum curvature. Where there are in the first case (free-free) points of inflection and of maximum curvature, there are in the second (clamped-clamped) nodes and loops respectively; and *vice versâ*, points of inflection and of maximum curvature for a doubly-clamped rod correspond to nodes and loops of a rod whose ends are free.

173. We will now consider the vibrations of a rod clamped at $x = 0$, and free at $x = l$. Reverting to the general integral (1) § 170, we see that A and C vanish in virtue of the conditions at $x = 0$, so that

$$u = B(\cos x' - \cosh x') + D(\sin x' - \sinh x')\ldots\ldots\ldots\ldots(1).$$

The remaining conditions at $x = l$ give

$$\left.\begin{array}{l} B(\ \ \cos m + \cosh m) + D(\sin m + \sinh m) = 0 \\ B(-\sin m + \sinh m) + D(\cos m + \cosh m) = 0 \end{array}\right\},$$

whence, omitting the constant multiplier,

$$u = (\sin m + \sinh m)\left\{\cos\frac{mx}{l} - \cosh\frac{mx}{l}\right\}$$

$$- (\cos m + \cosh m)\left\{\sin\frac{mx}{l} - \sinh\frac{mx}{l}\right\}\ldots\ldots(2),$$

or

$$u = (\cos m + \cosh m)\left\{\cos\frac{mx}{l} - \cosh\frac{mx}{l}\right\}$$

$$+ (\sin m - \sinh m)\left\{\sin\frac{mx}{l} - \sinh\frac{mx}{l}\right\}\ldots\ldots(3),$$

where m must be a root of

$$\cos m \cosh m + 1 = 0 \ldots\ldots\ldots\ldots\ldots\ldots(4).$$

The periods of the component tones in the present problem are thus different from, though, as we shall see presently, nearly related to, those of a rod both whose ends are clamped, or free.

If the value of u in (2) or (3) be differentiated twice, the result (u'') satisfies of course the fundamental differential equation. At $x = 0$, d^2u''/dx^2, d^3u''/dx^3 vanish, but at $x = l$ u'' and du''/dx vanish. The function u'' is therefore applicable to a rod clamped at l and free at 0, proving that the points of inflection and of maximum curvature in the original curve are at the same distances from the clamped end, as the nodes and loops respectively are from the free end.

174. In default of tables of the hyperbolic cosine or its logarithm, the admissible values of m may be calculated as follows. Taking first the equation

$$\cos m \; \cosh m = 1 \dots\dots\dots\dots\dots\dots(1),$$

we see that m, when large, must approximate in value to $\frac{1}{2}(2i + 1)\pi$, i being an integer. If we assume

$$m = \tfrac{1}{2}(2i + 1)\pi - (-1)^i \beta \dots\dots\dots\dots\dots(2),$$

β will be positive and comparatively small in magnitude.

Substituting in (1), we find

$$\cot \tfrac{1}{2}\beta = e^m = e^{\frac{1}{2}(2i+1)\pi} \, e^{-(-1)^i \beta} \; ;$$

or, if $e^{\frac{1}{2}(2i+1)\pi}$ be called a,

$$a \tan \tfrac{1}{2}\beta = e^{(-1)^i \beta} \dots\dots\dots\dots\dots\dots(3),$$

an equation which may be solved by successive approximation after expanding $\tan \tfrac{1}{2}\beta$ and $e^{(-1)^i \beta}$ in ascending powers of the small quantity β. The result is

$$\beta_i = \frac{2}{a} + (-1)^i \frac{4}{a^2} + \frac{34}{3a^3} + (-1)^i \frac{112}{3a^4} + \dots\dots\dots(4)[1],$$

which is sufficiently accurate, even when $i = 1$.

By calculation

$$\beta_1 = \text{·}0179666 - \text{·}0003228 + \text{·}0000082 - \text{·}0000002 = \text{·}0176518.$$

β_2, β_3, β_4, β_5 are found still more easily. After β_3 the first term of the series gives β correctly as far as six significant figures. The table contains the value of β, the angle whose circular measure is β, and the value of $\sin \tfrac{1}{2}\beta$, which will be required further on.

Free-Free Bar.

	β.	β expressed in degrees, minutes, and seconds.	$\sin \dfrac{\beta}{2}$.
1	$10^{-1} \times \text{·}176518$	$1° \; 0' \; 40''\text{·}94$	$10^{-2} \times \text{·}88258$
2	$10^{-3} \times \text{·}777010$	$2' \; 40''\text{·}2699$	$10^{-3} \times \text{·}38850$
3	$10^{-4} \times \text{·}335505$	$6''\text{·}92029$	$10^{-4} \times \text{·}16775$
4	$10^{-5} \times \text{·}144989$	$\text{·}299062$	$10^{-6} \times \text{·}72494$
5	$10^{-7} \times \text{·}626556$	$\text{·}0129237$	$10^{-7} \times \text{·}31328$

[1] This process is somewhat similar to that adopted by Strehlke.

The values of m which satisfy (1) are

$$m_1 = \ \ 4\cdot7123890 + \beta_1 = \ \ 4\cdot7300408$$
$$m_2 = \ \ 7\cdot8539816 - \beta_2 = \ \ 7\cdot8532046$$
$$m_3 = 10\cdot9955743 + \beta_3 = 10\cdot9956078$$
$$m_4 = 14\cdot1371669 - \beta_4 = 14\cdot1371655$$
$$m_5 = 17\cdot2787596 + \beta_5 = 17\cdot2787596$$

after which $m = \frac{1}{2}(2i+1)\pi$ to seven decimal places.

We will now consider the roots of the equation

$$\cos m \cosh m = -1 \quad\dots\dots\dots\dots\dots(5)^1.$$

[Assuming

$$m_i = \tfrac{1}{2}(2i-1)\pi - (-1)^i \alpha_i \dots\dots\dots\dots (6),$$

we have

$$e^{m_i} = \cot \tfrac{1}{2}\alpha_i = e^{\frac{1}{2}(2i-1)\pi} \cdot e^{-(-1)^i \alpha_i},$$

or

$$a \tan \tfrac{1}{2}\alpha_{i+1} = e^{-(-1)^i \alpha_{i+1}} \dots\dots\dots\dots\dots\dots(7),$$

a having the value previously defined.

Thus, as in (4),

$$\alpha_{i+1} = \frac{2}{a} - (-1)^i \frac{4}{a^2} + \frac{34}{3a^3} - (-1)^i \frac{112}{3a^4} + \dots\dots(8),$$

α_{i+1} being *approximately* equal to β_i.

The values calculated from (8) are

$$\alpha_2 = 10^{-1} \times \ 182979 \quad \alpha_4 = 10^{-4} \times \ \cdot335527,$$
$$\alpha_3 = 10^{-3} \times \ \cdot775804, \quad \alpha_5 = 10^{-5} \times \ \cdot144989,$$

after which the difference between α_{i+1} and β_i does not appear.]

The value of α_1 may be obtained by trial and error from the equation

$$\log_{10} \cot \tfrac{1}{2}\alpha_1 - \cdot6821882 - \cdot43429448 \, \alpha_1 = 0,$$

and will be found to be

$$\alpha_1 = \cdot3043077.$$

Another method by which m_1 may be obtained directly will be given presently.

The values of m, which satisfy (5), are

$$m_1 = \ \ 1\cdot5707963 + \alpha_1 = \ \ 1\cdot875104$$
$$m_2 = \ \ 4\cdot7123890 - \alpha_2 = \ \ 4\cdot694091$$
$$m_3 = \ \ 7\cdot8539816 + \alpha_3 = \ \ 7\cdot854757$$
$$m_4 = 10\cdot9955743 - \alpha_4 = 10\cdot995541$$
$$m_5 = 14\cdot1371669 + \alpha_5 = 14\cdot137168$$
$$m_6 = 17\cdot2787596 - \alpha_6 = 17\cdot278759,$$

---

[1] The calculation of the roots of (5) given in the first edition was affected by an error, which has been pointed out by Greenhill (*Math. Mess.*, Dec. 1886).

after which $m = \frac{1}{2}(2i-1)\pi$ sensibly. The frequencies are proportional to m^2, and are therefore for the higher tones nearly in the ratio of the squares of the odd numbers. However, in the case of overtones of very high order, the pitch may be slightly disturbed by the rotatory inertia, whose effect is here neglected.

175. Since the component vibrations of a system, not subject to dissipation, are necessarily of the harmonic type, all the values of m^2, which satisfy

$$\cos m \, \cosh m = \pm 1 \dots\dots\dots\dots\dots\dots(1),$$

must be real. We see further that, if m be a root, so are also $-m$, $m\sqrt{(-1)}$, $-m\sqrt{(-1)}$. Hence, taking first the lower sign, we have

$$\frac{1}{2}(\cos m \cosh m + 1) = 1 - \frac{m^4}{12} + \frac{m^8}{12^2.35} - \dots\dots$$

$$= \left(1 - \frac{m^4}{m_1^4}\right)\left(1 - \frac{m^4}{m_2^4}\right) \&c\dots\dots\dots\dots\dots(2).$$

If we take the logarithms of both sides, expand, and equate coefficients, we get

$$\Sigma \frac{1}{m^4} = \frac{1}{12}; \quad \Sigma \frac{1}{m^8} = \frac{1}{12^2}.\frac{33}{35}; \quad \&c\dots\dots\dots\dots(3).$$

This is for a clamped-free rod.

From the known value of Σm^{-8}, the value of m_1 may be derived with the aid of approximate values of m_2, m_3,...... We find

$$\Sigma m^{-8} = \cdot006547621,$$

and

$$m_2^{-8} = \cdot000004242$$

$$m_3^{-8} = \cdot000000069$$

$$m_4^{-8} = \cdot000000005,$$

whence

$$m_1^{-8} = \cdot006543305$$

giving

$$m_1 \;\; = \cdot1875104, \quad \text{as before.}$$

In like manner, if both ends of the bar be clamped or free,

$$1 - \frac{m^4}{12.35} + \dots = \left(1 - \frac{m^4}{m_1^4}\right)\left(1 - \frac{m^4}{m_2^4}\right) \&c. \dots\dots\dots (4),$$

whence $\Sigma \dfrac{1}{m^4} = \dfrac{1}{12.35}$ &c., where of course the summation is exclusive of the zero value of m.

176. The frequencies of the series of tones are proportional to m^2. The interval between any tone and the gravest of the series may conveniently be expressed in octaves and fractions of an octave. This is effected by dividing the difference of the logarithms of m^2 by the logarithm of 2. The results are as follows:

1·4629	2·6478
2·4358	4·1332
3·1590	5·1036
3·7382, &c	5·8288, &c.

where the first column relates to the tones of a rod both whose ends are clamped, or free; and the second column to the case of a rod clamped at one end but free at the other. Thus from the second column we find that the first overtone is 2·6478 octaves higher than the gravest tone. The fractional part may be reduced to mean semitones by multiplication by 12. The interval is then two octaves + 7·7736 mean semitones. It will be seen that the rise of pitch is much more rapid than in the case of strings.

If a rod be clamped at one end and free at the other, the pitch of the gravest tone is 2 (log 4·7300 − log 1·8751) ÷ log 2 or 2·6698 octaves lower than if both ends were clamped, or both free.

177. In order to examine more closely the curve in which the rod vibrates, we will transform the expression for u into a form more convenient for numerical calculation, taking first the case when both ends are free. Since $m = \frac{1}{2}(2i+1)\pi - (-1)^i\beta$, $\cos m = \sin\beta$, $\sin m = \cos i\pi \times \cos\beta$; and therefore, m being a root of $\cos m \cosh m = 1$, $\cosh m = \operatorname{cosec}\beta$.

Also
$$\sinh^2 m = \cosh^2 m - 1 = \tan^2 m = \cot^2\beta,$$

or, since $\cot\beta$ is positive,
$$\sinh m = \cot\beta.$$

Thus
$$\frac{\sin m - \sinh m}{\cos m - \cosh m} = \frac{1 - \cos i\pi \sin\beta}{\cos\beta}$$

$$= \frac{(\cos\frac{1}{2}\beta - \cos i\pi \sin\frac{1}{2}\beta)^2}{(\cos\frac{1}{2}\beta - \cos i\pi \sin\frac{1}{2}\beta)(\cos\frac{1}{2}\beta + \cos i\pi \sin\frac{1}{2}\beta)}$$

$$= \frac{\cos\frac{1}{2}\beta \cos i\pi - \sin\frac{1}{2}\beta}{\cos\frac{1}{2}\beta \cos i\pi + \sin\frac{1}{2}\beta}.$$

We may therefore take, omitting the constant multiplier,

$$u = (\cos \tfrac{1}{2}\beta \cos i\pi + \sin \tfrac{1}{2}\beta) \left\{ \sin \frac{mx}{l} + \sinh \frac{mx}{l} \right\}$$

$$- (\cos \tfrac{1}{2}\beta \cos i\pi - \sin \tfrac{1}{2}\beta) \left\{ \cos \frac{mx}{l} + \cosh \frac{mx}{l} \right\}$$

$$= \sqrt{2} \cos i\pi \, \sin \left\{ \frac{mx}{l} - \frac{\pi}{4} + (-1)^i \frac{\beta}{2} \right\}$$

$$+ \sin \tfrac{1}{2}\beta \, e^{mx/l} - \cos i\pi \, \cos \tfrac{1}{2}\beta \, e^{-mx/l} \dots\dots\dots\dots\dots (1).$$

If we further throw out the factor $\sqrt{2}$, and put $l = 1$, we may take

$$u = F_1 + F_2 + F_3,$$

where

$$\left. \begin{aligned} F_1 &= \cos i\pi \sin \{ mx - \tfrac{1}{4}\pi + \tfrac{1}{2}(-1)^i \beta \} \\ \log F_2 &= \quad mx \log e + \log \sin \tfrac{1}{2}\beta - \log \sqrt{2} \\ \log \pm F_3 &= - mx \log e + \log \cos \tfrac{1}{2}\beta - \log \sqrt{2} \end{aligned} \right\} \dots\dots\dots\dots (2),$$

from which u may be calculated for different values of i and x.

At the centre of the bar, $x = \tfrac{1}{2}$, and F_2, F_3 are numerically equal in virtue of $e^m = \cot \tfrac{1}{2}\beta$. When i is *even*, these terms cancel. For F_1, we have $F_1 = (-1)^i \sin \tfrac{1}{2} i\pi$, which is equal to zero when i is even, and to ± 1 when i is odd. When i is even, therefore, the sum of the three terms vanishes, and there is accordingly a node in the middle.

When $x = 0$, u reduces to $-2(-1)^i \sin \{\tfrac{1}{4}\pi - \tfrac{1}{2}(-1)^i \beta\}$, which (since β is always small) shews that for no value of i is there a node at the end. If a long bar of steel (held, for example, at the centre) be gently tapped with a hammer while varying points of its length are damped with the fingers, an unusual deadness in the sound will be noticed. as the end is closely approached.

178. We will now take some particular cases.

Vibration with two nodes. $i = 1$.

If $i = 1$, the vibration is the gravest of which the rod is capable. Our formulæ become

$$F_1 = - \sin \{ x (270^\circ + 1^\circ \, 0' \, 40'' \cdot 94) - 45^\circ - 30' \, 20'' \cdot 47 \}$$

$$\log F_2 = \quad 2 \cdot 054231 \, x + 3 \cdot 7952391$$

$$\log F_3 = - 2 \cdot 054231 \, x + \overline{1} \cdot 8494681,$$

from which is calculated the following table, giving the values of u for x equal to ·00, ·05, ·10, &c.

The values of $u : u(\cdot5)$ for the intermediate values of x (in the last column) were found by interpolation formulæ. If o, p, q, r, s, t be six consecutive terms, that intermediate between q and r is·

$$\frac{q+r}{2} + \frac{q+r-(p+s)}{4^2} + \frac{3}{4^4}\left\{2\left[q+r-(p+s)\right]-(p+s)+o+t\right\}.$$

x	F_1	F_2	F_3	u	$u : u(\cdot5)$
·000	+ ·7133200	+ ·0062408	+ ·7070793	+ 1·4266401	+ 1·645219
·025	...	...	...	...	1·454176
·050	·5292548	·0079059	·5581572	1·0953179	1·263134
·075	...	...	.	...	1·072162
·100	·3157243	·0100153	·4406005	·7663401	·8837528
·125	...	...	...	...	·6969004
·150	+ ·0846166	·0126874	·3478031	·4451071	·5133028
·175	...	...	...	...	·3341625
·200	− ·1512020	·0160726	·2745503	+ ·1394209	+ ·1607819
·225	...	...	...	...	− ·0054711
·250	·3786027	·0203609	·2167256	− ·1415162	·1631982
·275	...	...	...	...	·3109982
·300	·5849255	·0257934	·1710798	·3880523	·4475066
·325	...	...	...	...	·5714137
·350	·7586838	·0326753	·1350477	·5909608	·6815032
·375	...	...	...	...	·7766629
·400	·8902038	·0413934	·1066045	·7422059	·8559210
·425	...	...	...	...	·9184491
·450	·9721635	·0524376	·0841519	·8355740	·9635940
·475	...	...	...	...	·9908730
·500	− 1·000000	+ ·0664285	·0664282	− ·8671433	− 1·0000000

Since the vibration curve is symmetrical with respect to the middle of the rod, it is unnecessary to continue the table beyond $x = \cdot5$. The curve itself is shewn in fig. 28.

Fig. 28.

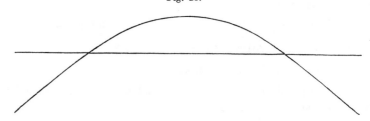

To find the position of the node, we have by interpolation

$$x = \cdot200 + \frac{\cdot1607819}{\cdot1662530} \times \cdot025 = \cdot22418,$$

which is the fraction of the whole length by which the node is distant from the nearer end.

Vibration with three nodes. $i = 2.$

$$F_1 = \sin\{(450^0 - 2'\,40''\,\cdot27)\,x - 45^0 + 1'\,20''\,\cdot135\}$$

$$\log F_2 = \quad 3\cdot410604\,x + \bar{4}\cdot4388816$$

$$\log(-F_3) = -\,3\cdot410604\,x + \bar{1}\cdot8494850.$$

x	$u : -u(0)$	x	$u : -u(0)$
·000	− 1·0000	·250	+ ·5847
·025	·8040	·275	·6374
·050	·6079	·300	·6620
·075	·4147	·325	·6569
·100	·2274	·350	·6245
·125	− ·0487	·375	·5652
·150	+ ·1175	·400	·4830
·175	·2672	·425	·3805
·200	·3972	·450	·2627
·225	·5037	·475	·1340
		·500	·0000

In this table, as in the preceding, the values of u were calculated directly for $x = \cdot000$, $\cdot050$, $\cdot100$ &c., and interpolated for the intermediate values. For the position of the node the table gives by ordinary interpolation $x = \cdot132$. Calculating from the above formulæ, we find

$$u(\cdot1321) = -\,\cdot000076,$$

$$u(\cdot1322) = +\,\cdot000881,$$

whence $x = \cdot132108$, agreeing with the result obtained by Strehlke. The place of maximum excursion may be found from the derived function. We get

$$u'(\cdot3083) = +\,\cdot0006077, \quad u'(\cdot3084) = -\,\cdot0002227,$$

whence $u'(\cdot308373) = 0.$

Hence u is a maximum, when $x = \cdot308373$; it then attains the value $\cdot6636$, which, it should be observed, is much less than the excursion at the end.

The curve is shewn in fig. 29.

Fig. 29.

Vibration with four nodes. $i = 3$.

$$F_1 = - \sin \{ (630^\circ + 6''\cdot92)\, x - 45^\circ - 3''\cdot46 \},$$

$$\log F_2 = \quad 4\cdot775332\, x + \bar{5}\cdot0741527,$$

$$\log F_3 = - 4\cdot775332\, x + \bar{1}\cdot8494850.$$

From this $u(0) = 1\cdot41424$, $u(\tfrac{1}{2}) = 1\cdot00579$. The positions of the nodes are readily found by trial and error. Thus

$$u(\cdot3558) = - \cdot000037 \qquad u(\cdot3559) = + \cdot001047,$$

whence $u(\cdot355803) = 0$. The value of x for the node near the end is $\cdot0944$, (Seebeck).

The position of the loop is best found from the derived function. It appears that $u' = 0$, when $x = \cdot2200$, and then $u = - \cdot9349$. There is also a loop at the centre, where however the excursion is not so great as at the two others.

Fig. 30.

We saw that at the centre of the bar F_2 and F_3 are numerically equal. In the neighbourhood of the middle, F_3 is evidently very small, if i be moderately great, and thus the equation for the nodes reduces approximately to

$$\frac{mx}{l} - \frac{\pi}{4} + (-1)^i \frac{\beta}{2} = \pm n\pi,$$

n being an integer. If we transform the origin to the centre of the rod, and replace m by its approximate value $\tfrac{1}{2}(2i+1)\pi$, we find

$$\frac{x}{l} = \frac{\pm 2n - i}{2i + 1},$$

shewing that near the middle of the bar the nodes are uniformly spaced, the interval between consecutive nodes being $2l \div (2i + 1)$. This theoretical result has been verified by the measurements of Strehlke and Lissajous.

For methods of approximation applicable to the nodes near the ends, when i is greater than 3, the reader is referred to the memoir by Seebeck already mentioned § 160, and to Donkin's *Acoustics* (p. 194).

179. The calculations are very similar for the case of a bar clamped at one end and free at the other. If $u \propto F$, and $F = F_1 + F_2 + F_3$, we have in general

$$F_1 = \cos \{mx + \tfrac{1}{4}\pi + \tfrac{1}{2}(-1)^i \alpha\},$$

$$F_2 = \frac{(-1)^i}{\sqrt{2}} \sin \tfrac{1}{2}\alpha \, e^{mx}; \quad F_3 = -\frac{1}{\sqrt{2}} \cos \tfrac{1}{2}\alpha \, e^{-mx}.$$

If $i = 1$, we obtain for the calculation of the gravest vibration-curve

$$F_1 = \cos \left\{ \frac{180}{\pi} mx^0 + 45^0 - 8^0 \, 43' \cdot 0665 \right\},$$

$$\log(-F_2) = \quad mx \log e + \overline{1} \cdot 0300909.$$

$$\log(-F_3) = -mx \log e + \overline{1} \cdot 8444383.$$

These give on calculation

$$F(0) = \cdot 000000, \qquad F(\cdot 6) = \quad \cdot 743452,$$

$$F(\cdot 2) = \cdot 102974, \qquad F(\cdot 8) = 1 \cdot 169632,$$

$$F(\cdot 4) = \cdot 370625, \qquad F(1 \cdot 0) = 1 \cdot 612224,$$

from which fig. 31 was constructed.

Fig. 31.

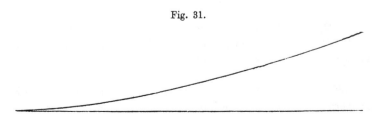

The distances of the nodes from the free end in the case of a rod clamped at the other end are given by Seebeck and by Donkin

2nd tone ·2261.

3rd tone ·1321, ·4999.

4th tone ·0944, ·3558, ·6439.

i^{th} tone $\dfrac{1\cdot3222}{4i-2}$, $\dfrac{4\cdot9820}{4i-2}$, $\dfrac{9\cdot0007}{4i-2}$, $\dfrac{4j-3}{4i-2}$, $\dfrac{4i-10\cdot9993}{4i-2}$, $\dfrac{4i-7\cdot0175}{4i-2}$.

"The last row in this table must be understood as meaning that $\dfrac{4j-3}{4i-2}$ may be taken as the distance of the jth node from the free end, except for the first three and the last two nodes."

When both ends are free, the distances of the nodes from the nearer end are

1st tone ·2242.

2nd tone ·1321   ·5.

3rd tone ·0944   ·3558.

i^{th} tone $\dfrac{1\cdot3222}{4i+2}$ $\dfrac{4\cdot9820}{4i+2}$ $\dfrac{9\cdot0007}{4i+2}$ $\dfrac{4j-3}{4i+2}$.

The points of inflection for a free-free rod (corresponding to the nodes of a clamped-clamped rod) are also given by Seebeck ;—

	1st point.	2nd point.	κ^{th} point.
1st tone	No inflection point.		
2nd tone......	·5000		
3rd tone......	·3593		
i^{th} tone	$\dfrac{5\cdot0175}{4i+2}$	$\dfrac{8\cdot9993}{4i+2}$	$\dfrac{4\kappa+1}{4i+2}$

Except in the case of the extreme nodes (which have no corresponding inflection-point), the nodes and inflection-points always occur in close proximity.

180. The case where one end of a rod is free and the other *supported* does not need an independent investigation, as it may be referred to that of a rod with both ends free *vibrating in an even mode*, that is, with a node in the middle. For at the central node y and y'' vanish, which are precisely the conditions for a supported end. In like manner the vibrations of a clamped-supported rod are the same as those of one-half of a rod both whose ends are clamped, vibrating with a central node.

181. The last of the six combinations of terminal conditions occurs when both ends are supported. Referring to (1) § 170, we see that the conditions at $x = 0$, give $A = 0$, $B = 0$; so that

$$u = (C + D) \sin x' + (C - D) \sinh x'.$$

Since u and u'' vanish when $x' = m$, $C - D = 0$, and $\sin m = 0$.

Hence the solution is

$$y = \sin \frac{i\pi x}{l} \cos \frac{i^2 \pi^2 \kappa b}{l^2} t \ \dots\dots\dots\dots (1),$$

where i is an integer. An arbitrary constant multiplier may of course be prefixed, and a constant may be added to t.

It appears that the normal curves are the same as in the case of a string stretched between two fixed points, but the sequence of tone is altogether different, the frequency varying as the *square* of i. The nodes and inflection-points coincide, and the loops (which are also the points of maximum curvature) bisect the distances between the nodes.

182. The theory of a vibrating rod may be applied to illustrate the general principle that the natural periods of a system fulfil the maximum-minimum condition, and that the greatest of the natural periods exceeds any that can be obtained by a variation of type. Suppose that the vibration curve of a clamped-free rod is that in which the rod would dispose itself if deflected by a force applied at its free extremity. The equation of the curve may be taken to be

$$v = -3lx^2 + x^3,$$

which satisfies $d^4y/dx^4 = 0$ throughout, and makes y and y' vanish at 0, and y'' at l. Thus, if the configuration of the rod at time t be

$$y = (-3lx^2 + x^3) \cos pt \ \dots\dots\dots\dots\dots (1),$$

the potential energy is by (1) § 161, $6 q\kappa^2 \omega l^3 \cos^2 pt$, while the kinetic energy is $\dfrac{33}{70}\,\rho\omega\,l^7\,p^2\sin^2 pt$; and thus $p^2 = \dfrac{140}{11}\,\dfrac{\kappa^2 b^2}{l^4}$

Now p_1 (the true value of p for the gravest tone) is equal to

$$\frac{\kappa b}{l^2} \times (1\!\cdot\!8751)^2\,;$$

so that

$$p_1 : p = (1\!\cdot\!8751)^2 \sqrt{\frac{11}{140}} = \cdot 98556,$$

shewing that the real pitch of the gravest tone is rather (but comparatively little) lower than that calculated from the hypothetical type. It is to be observed that the hypothetical type in question violates the terminal condition $y''' = 0$. This circumstance, however, does not interfere with the application of the principle, for the assumed type may be any which would be admissible as an initial configuration; but it tends to prevent a very close agreement of periods.

We may expect a better approximation, if we found our calculation on the curve in which the rod would be deflected by a force acting at some little distance from the free end, between which and the point of action of the force ($x = c$) the rod would be straight, and therefore without potential energy. Thus

$$\text{potential energy} = 6\,q\kappa^2\omega c^3 \cos^2 pt.$$

The kinetic energy can be readily found by integration from the value of y.

From 0 to c $y = -3cx^2 + x^3$;

and from c to l $y = c^2(c - 3x)$,

as may be seen from the consideration that y and y' must not suddenly change at $x = c$. The result is

$$\text{kinetic energy} = \rho\omega\,p^2\sin^2 pt\left[\frac{33}{70}c^7 + \tfrac{1}{2}c^4(l-c)(c^2 + 3l^2)\right],$$

whence

$$\frac{1}{p^2} = \frac{1}{6\kappa^2 b^2}\left[\frac{33}{70}c^4 + \frac{c}{2}(l-c)(c^2 + 3l^2)\right]\dots\dots\dots\dots(2).$$

The maximum value of $1/p^2$ will occur when the point of application of the force is in the neighbourhood of the node of the second normal component vibration. If we take $c = \tfrac{3}{4}l$, we obtain a result which is too high in the musical scale by the interval expressed by the ratio 1 : ·9977, and is accordingly extremely near the truth. This example may give an idea how nearly the period of a vibrating system may be calculated by simple means without the solution of differential or transcendental equations.

The type of vibration just considered would be that actually assumed by a bar which is itself devoid of inertia, but carries a load M at its free end, provided that the rotatory inertia of M could be neglected. We should have, in fact,

$$V = 6q\kappa^2\omega l^3 \cos^2 pt, \qquad T = 2Ml^6 p^2 \sin^2 pt,$$

so that

$$p^2 = \frac{3q\kappa^2\omega}{Ml^3} \quad\dots\dots\dots\dots\dots\dots (3).$$

Even if the inertia of the bar be not altogether negligible in comparison with M, we may still take the same type as the basis of an approximate calculation :

$$V = 6q\kappa^2\omega l^3 \cos^2 pt,$$

$$T = \left(2Ml^6 + \frac{33}{70}\rho\omega l^7\right) p^2 \sin^2 pt,$$

whence

$$\frac{1}{p^2} = \frac{l^3}{3q\kappa^2\omega}\left(M + \frac{33}{140}\rho\omega l\right) \quad\dots\dots\dots\dots\dots (4),$$

that is, M is to be increased by about one quarter of the mass of the rod. Since this result is accurate when M is infinite, and does not differ much from the truth, even when $M = 0$, it may be regarded as generally applicable as an approximation. The error will always be on the side of estimating the pitch too high.

183. But the neglect of the rotatory inertia of M could not be justified under the ordinary conditions of experiment. It is as easy to imagine, though not to construct, a case in which the inertia of translation should be negligible in comparison with the inertia of rotation, as the opposite extreme which has just been considered. If both kinds of inertia in the mass M be included, even though that of the bar be neglected altogether, the system possesses two distinct and independent periods of vibration.

Let z and θ denote the values of y and dy/dx at $x = l$. Then the equation of the curve of the bar is

$$y = \frac{3z - l\theta}{l^2} x^2 + \frac{l\theta - 2z}{l^3} x^3,$$

and

$$V = \frac{2q\kappa^2\omega}{l^3}\left\{3z^2 - 3zl\theta + l^2\theta^2\right\} \dots\dots\dots\dots (1);$$

while for the kinetic energy

$$T = \tfrac{1}{2}M\dot{z}^2 + \tfrac{1}{2}M\kappa'^2\dot{\theta}^2 \dots\dots\dots\dots\dots (2),$$

if κ' be the radius of gyration of M about an axis perpendicular to the plane of vibration.

The equations of motion are therefore

$$\left.\begin{aligned}
M\ddot{z} + \frac{2q\kappa^2\omega}{l^3}\ (6z - 3l\theta) &= 0 \\
M\kappa'^2\ddot{\theta} + \frac{2q\kappa^2\omega}{l^3}(-3lz + 2l^2\theta) &= 0
\end{aligned}\right\} \dots\dots\dots (3);$$

whence, if z and θ vary as $\cos pt$, we find

$$p^2 = \frac{2q\kappa^2\omega}{Ml\kappa'^2}\left\{1 + \frac{3\kappa'^2}{l^2} \pm \sqrt{1 + \frac{3\kappa'^2}{l^2} + \frac{9\kappa'^4}{l^4}}\right\} \dots\dots (4),$$

corresponding to the two periods, which are always different.

If we neglect the rotatory inertia by putting $\kappa' = 0$, we fall back on our previous result

$$p^2 = \frac{3q\kappa^2\omega}{Ml^3}.$$

The other value of p^2 is then infinite.

If $\kappa' : l$ be merely small, so that its higher powers may be neglected,

$$\left.\begin{aligned}
p^2 &= \frac{4q\kappa^2\omega}{Ml\kappa'^2}\left(1 + \frac{9}{4}\frac{\kappa'^2}{l^2}\right) \\
p^2 &= \frac{3q\kappa^2\omega}{Ml^3}\left(1 - \frac{9}{4}\frac{\kappa'^2}{l^2}\right)
\end{aligned}\right\} \dots\dots\dots\dots (5).$$

If on the other hand κ'^2 be very great, so that rotation is prevented,

$$p^2 = \frac{12\,q\kappa^2\omega}{Ml^3}\quad\text{or}\quad\frac{q\kappa^2\omega}{Ml\kappa'^2} \dots\dots\ \dots\dots (6),$$

the latter of which is very small. It appears that when rotation is prevented, the pitch is an octave higher than if there were no rotatory inertia at all. These conclusions might also be derived directly from the differential equations; for if $\kappa' = \infty$, $\theta = 0$, and then

$$M\ddot{z} + \frac{12\,q\kappa^2\omega}{l^3}\,z = 0\,;$$

but if $\kappa' = 0$, $\theta = 3z/2l$, by the second of equations (3), and in that case

$$M\ddot{z} + \frac{3q\kappa^2\omega}{l^3}\,z = 0.$$

184. If any addition to a bar be made at the end, the period of vibration is prolonged. If the end in question be free, suppose first that the piece added is without inertia. Since there would be no alteration in either the potential or kinetic energies, the pitch would be unchanged; but in proportion as the additional part acquires inertia, the pitch falls (§ 88).

In the same way a small continuation of a bar beyond a clamped end would be without effect, as it would acquire no motion. No change will ensue if the new end be also clamped; but as the first clamping is relaxed, the pitch falls, in consequence of the diminution in the potential energy of a given deformation.

The case of a 'supported' end is not quite so simple. Let the original end of the rod be A, and let the added piece which is at first supposed to have no inertia, be AB. Initially the end A is fixed, or held, if we like so to regard it, by a spring of infinite stiffness. Suppose that this spring, which has no inertia, is gradually relaxed. During this process the motion of the new end B diminishes, and at a certain point of relaxation, B comes to rest. During this process the pitch falls. B, being now at rest, may be supposed to become fixed, and the abolition of the spring at A entails another fall of pitch, to be further increased as AB acquires inertia.

185. The case of a rod which is not quite uniform may be treated by the general method of § 90. We have in the notation there adopted

$$c_r = \int B_0 \left(\frac{d^2u_r}{dx^2}\right)^2 dx, \quad \delta c_r = \int \delta B \left(\frac{d^2u_r}{dx^2}\right)^2 dx$$

$$a_r = \int \rho\omega_0 u_r{}^2 dx, \quad \delta a_r = \int \delta\overline{\rho\omega}\, u_r{}^2 dx,$$

whence, P_r being the uncorrected value of p_r,

$$p_r{}^2 = P_r{}^2 \left\{ 1 + \frac{\int \delta B \left(\frac{d^2 u_r}{dx^2}\right)^2 dx}{\int B_0 \left(\frac{d^2 u_r}{dx^2}\right)^2 dx} - \frac{\int \delta \overline{\rho \omega}\, u_r{}^2 dx}{\int \overline{\rho \omega_0}\, u_r{}^2 dx} \right\}$$

$$= P_r{}^2 \left\{ 1 + \frac{\int \delta B\, u_r''^2 dx}{B_0 \int u_r{}^2 dx} - \frac{\int \delta \rho \omega\, u_r\, dx}{\rho \omega_0 \int u_r{}^2 dx} \right\} \dots \dots \dots (1).$$

[If the motion be strictly periodic with respect to x, u_r'' is proportional to u_r, and both quantities vanish at a node. Accordingly an irregularity situated at a node of this kind of motion has no effect upon the period. A similar conclusion will hold good approximately for the interior nodes of a bar vibrating with numerous subdivisions, even though, as when the terminals are clamped or free, the mode of motion be not strictly periodic with respect to x.]

If the rod be clamped at 0 and free at l,

$$p_r{}^2 = \frac{B_0 m^4}{\rho \omega_0 l^4} \left\{ 1 + \frac{4}{l u_l{}^2} \int_0^l \frac{\delta B}{B_0} u''^2 dx - \frac{4}{l u_l{}^2} \int_0^l \frac{\delta \rho \omega}{\rho \omega_0} u^2 dx \right\}.$$

The same formula applies to a doubly free bar.

The effect of a small load dM is thus given by

$$p^2 = \frac{B_0 m^4}{\rho \omega_0 l^4} \left\{ 1 - 4 \frac{u^2 dM}{u_l{}^2 M} \right\} \dots \dots \dots \dots (2),$$

where M denotes the mass of the whole bar. If the load be at the end, its effect is the same as a lengthening of the bar in the ratio $M : M + dM$. (Compare § 167.)

[In (2) dM is supposed to act by inertia only; but a similar formula may conveniently be employed when an irregularity of mass dM depends upon a variation of section, without a change of mechanical properties. Since $B = q\kappa^2 \omega$,

$$\delta B / B_0 = \delta (\kappa^2 \omega)/(\kappa^2 \omega)_0 ;$$

so that the effect of a local excrescence is given by

$$p^2/P^2 = 1 + \frac{4u''^2}{l u_l{}^2} \int \frac{\delta (\kappa^2 \omega)}{(\kappa^2 \omega)_0}\, dx - \frac{4u^2}{l u_l{}^2} \int \frac{\delta \omega}{\omega_0}\, dx \dots \dots \dots \dots (3).$$

If the thickness in the plane of bending be constant, $\delta \kappa^2 = 0$, and $\quad\quad\quad\quad \delta (\kappa^2 \omega)/(\kappa^2 \omega)_0 = \delta \omega / \omega_0.$

Further,
$$\int \frac{\delta\omega\, dx}{l\omega_0} = \frac{dM}{M};$$

and thus
$$p^2/P^2 = 1 + 4\, \frac{dM}{M}\, \frac{u''^2 - u^2}{u_l^2} \dots\dots\dots\dots(4).$$

If, however, the thickness in the plane perpendicular to that of bending be constant, and in the plane of bending variable (2γ),

then
$$\delta\,(\kappa^2\omega)/(\kappa^2\omega)_0 = \delta\gamma^3/\gamma_0^3 = 3\,\delta\gamma/\gamma_0 = 3\,\delta\omega/\omega_0;$$

and in place of (4)
$$p^2/P^2 = 1 + 4\, \frac{dM}{M}\, \frac{3u''^2 - u^2}{u_l^2} \dots\dots\dots\dots (5).$$

If a tuning-fork be filed (dM negative) near the stalk (clamped end), the pitch is lowered; and if it be filed near the free end, the pitch is raised. Since $u_0''^2 = u_l^2$, the effects of a given stroke of the file are equal and opposite in the circumstances of (4), but in the circumstances of (5) the effect at the stalk is three times as great as at the free end.]

186. The same principle may be applied to estimate the correction due to the rotatory inertia of a uniform rod. We have only to find what addition to make to the kinetic energy, supposing that the bar vibrates according to the same law as would obtain, were there no rotatory inertia.

Let us take, for example, the case of a bar clamped at 0 and free at l, and assume that the vibration is of the type,

$$y = u \cos pt,$$

where u is one of the functions investigated in § 179. The kinetic energy of the rotation is

$$\tfrac{1}{2} \int \rho\omega\kappa^2 \left(\frac{d^2y}{dx\,dt}\right)^2 dx = \frac{\rho\omega\kappa^2 m^2 p^2}{2l^2} \sin^2 pt \int_0^l u'^2\, dx$$

$$= \frac{\rho\omega\kappa^2 m p^2}{8l} \sin^2 pt\, (2uu' + mu'^2)_l,$$

by (2) § 165.

To this must be added

$$\frac{\rho\omega}{2}\, p^2 \sin^2 pt \int_0^l u^2\, dx, \quad \text{or} \quad \frac{\rho\omega l}{8}\, p^2 \sin^2 pt\, u_l^2;$$

so that the kinetic energy is increased in the ratio

$$1 \;:\; 1 + \frac{m\kappa^2}{l^2} \left(2\frac{u'}{u} + m\frac{u'^2}{u^2}\right)_l.$$

The altered frequency bears to that calculated without allowance for rotatory inertia a ratio which is the square root of the reciprocal of the preceding. Thus

$$p : P = 1 - \tfrac{1}{2}\frac{m\kappa^2}{l^2}\left(2\frac{u'}{u} + m\frac{u'^2}{u^2}\right)_l \dots\dots\dots (1).$$

By use of the relations $\cosh m = -\sec m$, $\sinh m = \cos i\pi . \tan m$, we may express $u' : u$ when $x = l$ in the form

$$\frac{u'}{u} = \frac{-\sin m}{\cos i\pi + \cos m} = \frac{\cos\alpha}{1 - \cos i\pi \sin\alpha},$$

if we substitute for m from

$$m = \tfrac{1}{2}(2i - 1)\pi - (-1)^i\alpha.$$

In the case of the gravest tone, $\alpha = \cdot3043$, or, in degrees and minutes, $\alpha = 17^{\circ}\ 26'$, whence

$$\frac{u'}{u} = \cdot73413, \qquad 2\frac{u'}{u} + m\frac{u'^2}{u^2} = 2\cdot4789.$$

Thus

$$p : P = 1 - 2\cdot3241\,\frac{\kappa^2}{l^2} \dots\dots\dots\dots (2),$$

which gives the correction for rotatory inertia in the case of the gravest tone.

When the order of the tone is moderate, α is very small, and then

$$u' : u = 1 \quad \text{sensibly}.$$

and

$$p : P = 1 - \left(1 + \frac{m}{2}\right)\frac{m\kappa^2}{l^2} \dots\dots\dots\dots (3),$$

shewing that the correction increases in importance with the order of the component.

In all ordinary bars $\kappa : l$ is very small, and the term depending on its square may be neglected without sensible error.

187. When the rigidity and density of a bar are variable from point to point along it, the normal functions cannot in general be expressed analytically, but their nature may be investigated by the methods of Sturm and Liouville explained in § 142.

If, as in § 162, B denote the variable flexural rigidity at any point of the bar, and $\rho\omega\,dx$ the mass of the element, whose length is dx, we find as the general differential equation

$$\frac{d^2}{dx^2}\left(B\frac{d^2y}{dx^2}\right) + \rho\omega\,\frac{d^2y}{dt^2} = 0 \quad\dots\dots\dots\dots\dots (1),$$

the effects of rotatory inertia being omitted. If we assume that $y \propto \cos\nu t$, we obtain as the equation to determine the form of the normal functions

$$\frac{d^2}{dx^2}\left(B\frac{d^2y}{dx^2}\right) = \nu^2\rho\omega y \quad\dots\dots\dots\dots\dots (2),$$

in which ν^2 is limited by the terminal conditions to be one of an infinite series of definite quantities $\nu_1^2, \nu_2^2, \nu_3^2\dots\dots$

Let us suppose, for example, that the bar is clamped at both ends, so that the terminal values of y and dy/dx vanish. The first normal function, for which ν^2 has its lowest value ν_1^2, has no internal root, so that the vibration-curve lies entirely on one side of the equilibrium-position. The second normal function has one internal root, the third function has two internal roots, and, generally, the rth function has $r-1$ internal roots.

Any two different normal functions are conjugate, that is to say, their product will vanish when multiplied by $\rho\omega\,dx$, and integrated over the length of the bar.

Let us examine the number of roots of a function $f(x)$ of the form

$$f(x) = \phi_m u_m(x) + \phi_{m+1}u_{m+1}(x) + \dots + \phi_n u_n(x)\dots\dots(3),$$

compounded of a finite number of normal functions, of which the function of lowest order is $u_m(x)$ and that of highest order is $u_n(x)$. If the number of internal roots of $f(x)$ be μ, so that there are $\mu+4$ roots in all, the derived function $f'(x)$ cannot have less than $\mu+1$ internal roots besides two roots at the extremities, and the second derived function cannot have less than $\mu+2$ roots. No roots can be lost when the latter function is multiplied by B, and another double differentiation with respect to x will leave at least μ internal roots. Hence by (2) and (3) we conclude that

$$\nu_m^2\phi_m u_m(x) + \nu_{m+1}^2\phi_{m+1}u_{m+1}(x) + \dots + \nu_n^2\phi_n u_n(x)\dots(4)$$

has at least as many roots as $f(x)$. Since (4) is a function of the same form as $f(x)$, the same argument may be repeated, and a series of functions obtained every member of which has at least as many roots as $f(x)$ has. When the operation by which (4) was derived from (3) has been repeated sufficiently often, a function is arrived at whose form differs as little as we please from that of the component normal function of highest order $u_n(x)$; and we conclude that $f(x)$ cannot have more than $n-1$ internal roots. In like manner we may prove that $f(x)$ cannot have less than $m-1$ internal roots.

The application of this theorem to demonstrate the possibility of expanding an arbitrary function in an infinite series of normal functions would proceed exactly as in § 142.

[An analytical investigation of certain cases where the section of a rod is supposed to be variable, will be found in a memoir by Kirchhoff[1]].

188. When the bar, whose lateral vibrations are to be considered, is subject to longitudinal tension, the potential energy of any configuration is composed of two parts, the first depending on the stiffness by which the bending is directly opposed, and the second on the reaction against the extension, which is a necessary accompaniment of the bending, when the ends are nodes. The second part is similar to the potential energy of a deflected string; the first is of the same nature as that with which we have been occupied hitherto in this Chapter, though it is not entirely independent of the permanent tension.

Consider the extension of a filament of the bar of section $d\omega$, whose distance from the axis projected on the plane of vibration is η. Since the sections, which were normal to the axis originally, remain normal during the bending, the length of the filament bears to the corresponding element of the axis the ratio $R+\eta:R$, R being the radius of curvature. Now the axis itself is extended in the ratio $q:q+T$, reckoning from the unstretched state, if $T\omega$ denote the whole tension to which the bar is subjected. Hence the actual tension on the filament is $\{T+\eta(T+q)/R\}\,d\omega$, from which we find for the moment of the couple acting across the section

$$\int \left\{ T + \frac{\eta}{R}(T+q) \right\} \eta\, d\omega = \frac{q+T}{R}\, \kappa^2 \omega,$$

[1] *Berlin Monatsber.*, 1879; *Collected Works*, p. 339. See also Todhunter and Pearson's *History of the Theory of Elasticity*, Vol. II., Part ii., § 1302.

and for the whole potential energy due to stiffness

$$\tfrac{1}{2}(q+T)\kappa^2\omega \int \left(\frac{d^2y}{dx^2}\right)^2 dx \dots\dots\dots\dots\dots (1),$$

an expression differing from that previously used (§ 162) by the substitution of $(q+T)$ for q.

Since q is the tension required to stretch a bar of unit area to twice its natural length, it is evident that in most practical cases T would be negligible in comparison with q.

The expression (1) denotes the work that would be gained during the straightening of the bar, if the length of each element of the axis were preserved constant during the process. But when a stretched bar or string is allowed to pass from a displaced to the natural position, the length of the axis is decreased. The amount of the decrease is $\tfrac{1}{2}\int (dy/dx)^2\,dx$, and the corresponding gain of work is

$$\tfrac{1}{2}T\omega \int \left(\frac{dy}{dx}\right)^2 dx.$$

Thus

$$V = \tfrac{1}{2}(q+T)\kappa^2\omega \int \left(\frac{d^2y}{dx^2}\right)^2 dx + \tfrac{1}{2}T\omega \int \left(\frac{dy}{dx}\right)^2 dx \dots\dots(2).$$

The variation of the first part due to a hypothetical displacement is given in § 162. For the second part, we have

$$\tfrac{1}{2}\delta \int \left(\frac{dy}{dx}\right)^2 dx = \int \frac{dy}{dx}\frac{d\delta y}{dx}\,dx = \left\{ \frac{dy}{dx}\delta y \right\} - \int \frac{d^2y}{dx^2}\delta y\,dx \dots\dots(3).$$

In all the cases that we have to consider, δy vanishes at the limits. The general differential equation is accordingly

$$\kappa^2(q+T)\frac{d^4y}{dx^4} - T\frac{d^2y}{dx^2} + \rho\frac{d^2y}{dt^2} - \kappa^2\rho\frac{d^4y}{dx^2\,dt^2} = 0,$$

or, if we put $q+T=b^2\rho, \quad T=a^2\rho,$

$$\kappa^2\left(b^2\frac{d^4y}{dx^4} - \frac{d^4y}{dx^2\,dt^2}\right) - a^2\frac{d^2y}{dx^2} + \frac{d^2y}{dt^2} = 0 \dots\dots\dots(4).$$

For a more detailed investigation of this equation the reader is referred to the writings of Clebsch[1] and Donkin.

189. If the ends of the rod, or wire, be clamped, $dy/dx = 0$, and the terminal conditions are satisfied. If the nature of the support be such that, while the extremity is constrained to be a node, there

---

[1] *Theorie der Elasticität fester Körper.* Leipzig, 1862.

is no couple acting on the bar, d^2y/dx^2 must vanish, that is to say, the end must be straight. This supposition is usually taken to represent the case of a string stretched over bridges, as in many musical instruments; but it is evident that the part beyond the bridge must partake of the vibration, and that therefore its length cannot be altogether a matter of indifference.

If in the general differential equation we take y proportional to $\cos nt$, we get

$$\kappa^2 \left(b^2 \frac{d^4y}{dx^4} + n^2 \frac{d^2y}{dx^2} \right) - a^2 \frac{d^2y}{dx^2} - n^2y = 0 \dots\dots\dots(1),$$

which is evidently satisfied by

$$y = \sin i \frac{\pi x}{l} \cos nt \dots\dots\dots\dots (2),$$

if n be suitably determined. The same solution also makes y and y'' vanish at the extremities. By substitution we obtain for n,

$$n^2 = \frac{i^2 \pi^2}{l^2} \frac{a^2 l^2 + i^2 \pi^2 \kappa^2 b^2}{l^2 + i^2 \pi^2 \kappa^2} \dots\dots\dots\dots(3),$$

which determines the frequency.

If we suppose the wire infinitely thin, $n^2 = i^2 \pi^2 a^2 \div l^2$, the same as was found in Chapter VI., by starting from the supposition of perfect flexibility. If we treat $\kappa : l$ as a very small quantity, the approximate value of n is

$$n = \frac{i \pi a}{l} \left\{ 1 + i^2 \frac{\pi^2 \kappa^2}{2l^2} \left(\frac{b^2}{a^2} - 1 \right) \right\}.$$

For a wire of circular section of radius r, $\kappa^2 = \frac{1}{4} r^2$, and if we replace b and a by their values in terms of q, T, and ρ,

$$n = \frac{i \pi a}{l} \left\{ 1 + \frac{i^2 \pi^2}{8} \frac{r^2}{l^2} \frac{q}{T} \right\} \dots\dots\dots\dots(4),$$

which gives the correction for rigidity[1]. Since the expression within brackets involves i, it appears that the harmonic relation of the component tones is disturbed by the stiffness.

190. The investigation of the correction for stiffness when the ends of the wire are clamped is not so simple, in consequence of the change of type which occurs near the ends. In order to pass from the case of the preceding section to that now under con-

---

[1] Donkin's *Acoustics*, Art. 18?.

sideration an additional constraint must be introduced, with the effect of still further raising the pitch. The following is, in the main, the investigation of Seebeck and Donkin.

If the rotatory inertia be neglected, the differential equation becomes

$$\left(D^4 - \frac{a^2}{\kappa^2 b^2} D^2 - \frac{n^2}{b^2 \kappa^2}\right) y = 0 \dots \dots \dots \dots (1),$$

where D stands for $\frac{d}{dx}$. In the equation

$$D^4 - \frac{a^2}{\kappa^2 b^2} D^2 - \frac{n^2}{b^2 \kappa^2} = 0,$$

one of the values of D^2 must be positive, and the other negative. We may therefore take

$$D^4 - \frac{a^2}{\kappa^2 b^2} D^2 - \frac{n^2}{b^2 \kappa^2} = (D^4 - \alpha^2)(D^2 + \beta^2) \dots \dots \dots (2),$$

and for the complete integral of (1)

$$y = A \cosh \alpha x + B \sinh \alpha x + C \cos \beta x + D \sin \beta x \dots \dots (3),$$

where α and β are functions of n determined by (2).

The solution must now be made to satisfy the four boundary conditions, which, as there are only three disposable ratios, lead to an equation connecting α, β, l. This may be put into the form

$$\frac{\sinh \alpha l \ \sin \beta l}{1 - \cosh \alpha l \ \cos \beta l} + \frac{2\alpha\beta}{\alpha^2 - \beta^2} = 0 \dots \dots \dots \dots (4).$$

The value of $\frac{2\alpha\beta}{\alpha^2 - \beta^2}$, determined by (2), is $\frac{2nb\kappa}{a^2}$, so that

$$\frac{\sinh \alpha l \ \sin \beta l}{1 - \cosh \alpha l \ \cos \beta l} + \frac{2nb\kappa}{a^2} = 0 \dots \dots \dots \dots (5).$$

From (2) we find also that

$$\left. \begin{aligned} \alpha^2 &= \frac{a^2}{2b^2\kappa^2}\left\{\sqrt{1 + 4\frac{n^2 b^2 \kappa^2}{a^4}} + 1\right\} \\ \beta^2 &= \frac{a^2}{2b^2\kappa^2}\left\{\sqrt{1 + 4\frac{n^2 b^2 \kappa^2}{a^4}} - 1\right\} \end{aligned} \right\} \dots \dots \dots (6).$$

Thus far our equations are rigorous, or rather as rigorous as the differential equation on which they are founded; but we shall now introduce the supposition that the vibration considered is but slightly affected by the existence of rigidity. This being the case, the approximate expression for y is

$$y = \sin \frac{i\pi x}{l} \cos \left(\frac{i\pi}{l} at \right),$$

and therefore

$$\beta = i\pi/l, \qquad n = i\pi a/l \quad \dots\dots\dots\dots\dots\dots\dots\dots(7),$$

nearly.

The introduction of these values into the second of equations (6) proves that $n^2 b^2 \kappa^2/a^4$ or $b^2\kappa^2/a^2 l^2$ is a small quantity under the circumstances contemplated, and therefore that $\alpha^2 l^2$ is a large quantity. Since $\cosh \alpha l$, $\sinh \alpha l$ are both large, equation (5) reduces to

$$\tan \beta l = \frac{2nb\kappa}{a^2},$$

or, on substitution of the approximate value for β derived from (6),

$$\tan \frac{nl}{a} = 2\frac{nb\kappa}{a^2}.$$

The approximate value of nl/a is $i\pi$. If we take $nl/a = i\pi + \theta$ we get

$$\tan (i\pi + \theta) = \tan \theta = \theta = 2\frac{nb\kappa}{a^2} = 2i\pi \frac{b}{a} \frac{\kappa}{l},$$

so that

$$n = i\frac{\pi a}{l} \left(1 + 2\frac{b}{a}\frac{\kappa}{l} \right) \quad \dots\dots\dots\dots\dots(8).$$

According to this equation the component tones are all raised in pitch by the same small interval, and therefore the harmonic relation is not disturbed by the rigidity. It would probably be otherwise if terms involving $\kappa^2 : l^2$ were retained; it does not therefore follow that the harmonic relation is better preserved in spite of rigidity when the ends are clamped than when they are free, but only that there is no additional disturbance in the former case, though the absolute alteration of pitch is much greater. It should be remarked that $b : a$ or $\sqrt{(q+T)} : \sqrt{T}$, is a large quantity, and that, if our result is to be correct, $\kappa : l$ must be small enough to bear multiplication by $b : a$ and yet remain small.

The theoretical result embodied in (8) has been compared with experiment by Seebeck, who found a satisfactory agreement. The constant of stiffness was deduced from observations of the rapidity of the vibrations of a small piece of the wire, when one end was clamped in a vice.

[As the result of a second approximation Seebeck gives (*loc. cit.*)

$$n^2 = n_0{}^2 \left\{ 1 + 4\frac{b\kappa}{al} + (12 + i^2\pi^2)\frac{b^2\kappa^2}{a^2l^2} \right\} \dots\dots\dots(9)].$$

191. It has been shewn in this chapter that the theory of bars, even when simplified to the utmost by the omission of unimportant quantities, is decidedly more complicated than that of perfectly flexible strings. The reason of the extreme simplicity of the vibrations of strings is to be found in the fact that waves of the harmonic type are propagated with a velocity independent of the wave length, so that an arbitrary wave is allowed to travel without decomposition. But when we pass from strings to bars, the constant in the differential equation, viz. $d^2y/dt^2 + \kappa^2b^2\,d^4y/dx^4 = 0$, is no longer expressible as a velocity, and therefore the velocity of transmission of a train of harmonic waves cannot depend on the differential equation alone, but must vary with the wave length. Indeed, if it be admitted that the train of harmonic waves can be propagated at all, this consideration is sufficient by itself to prove that the velocity must vary inversely as the wave length. The same thing may be seen from the solution applicable to waves propagated in one direction, viz. $y = \cos\dfrac{2\pi}{\lambda}(Vt - x)$, which satisfies the differential equation if

$$V = \frac{2\pi\kappa b}{\lambda} \quad \dots\dots\dots\dots\dots\dots\dots(1).$$

Let us suppose that there are two trains of waves of equal amplitudes, but of different wave lengths, travelling in the same direction. Thus

$$y = \cos 2\pi\left(\frac{t}{\tau} - \frac{x}{\lambda}\right) + \cos 2\pi\left(\frac{t}{\tau'} - \frac{x}{\lambda'}\right)$$

$$= 2\cos\pi\left\{t\left(\frac{1}{\tau} - \frac{1}{\tau'}\right) - x\left(\frac{1}{\lambda} - \frac{1}{\lambda'}\right)\right\}\cos\pi\left\{t\left(\frac{1}{\tau} + \frac{1}{\tau'}\right) - x\left(\frac{1}{\lambda} + \frac{1}{\lambda'}\right)\right\}\dots(2).$$

If $\tau' - \tau$, $\lambda' - \lambda$ be small, we have a train of waves, whose amplitude slowly varies from one point to another between the values 0 and 2, forming a series of groups separated from one another by regions comparatively free from disturbance. In the case of a string or of a column of air, λ varies as τ, and then the groups move forward with the same velocity as the component trains, and there is no change of type. It is otherwise when, as in the case of a bar vibrating transversely, the velocity of propagation is a function of the wave length. The position at time t of the middle of the group which was initially at the origin is given by

$$t\left(\frac{1}{\tau}-\frac{1}{\tau'}\right)-x\left(\frac{1}{\lambda}-\frac{1}{\lambda'}\right)=0,$$

which shews that the velocity of the group is

$$\left(\frac{1}{\tau}-\frac{1}{\tau'}\right)\div\left(\frac{1}{\lambda}-\frac{1}{\lambda'}\right)=\delta\left(\frac{1}{\tau}\right)\div\delta\left(\frac{1}{\lambda}\right).$$

If we suppose that the velocity V of a train of waves varies as λ^n, we find

$$\frac{d\,(1/\tau)}{d\,(1/\lambda)}=\frac{d\,(V/\lambda)}{d\,(1/\lambda)}=-(n-1)\,V\ \dots\dots\dots\dots(3).$$

In the present case $n=-1$, and accordingly the velocity of the groups is *twice* that of the component waves[1].

192. On account of the dependence of the velocity of propagation on the wave length, the condition of an infinite bar at any time subsequent to an initial disturbance confined to a limited portion, will have none of the simplicity which characterises the corresponding problem for a string; but nevertheless Fourier's investigation of this problem may properly find a place here.

It is required to determine a function of x and t, so as to satisfy

$$\frac{d^2y}{dt^2}+\frac{d^4y}{dx^4}=0\ \dots\dots\dots\dots\dots(1),$$

and make initially $y=\phi(x)$, $\dot{y}=\psi(x)$.

A solution of (1) is

$$y=\cos q^2t\ \cos q\,(x-\alpha)\dots\dots\dots\dots\dots(2),$$

where q and α are constants, from which we conclude that

$$y=\int_{-\infty}^{+\infty}d\alpha\,F(\alpha)\int_{-\infty}^{+\infty}dq\cos q^2t\ \cos q\,(x-\alpha)$$

[1] In the corresponding problem for waves on the surface of deep water, the velocity of propagation varies directly as the square root of the wave length, so that $n=\frac{1}{2}$. The velocity of a group of such waves is therefore *one half* of that of the component trains. [See note on Progressive Waves, appended to this volume.]

is also a solution, where $F(\alpha)$ is an arbitrary function of α. If now we put $t = 0$,

$$y_0 = \int_{-\infty}^{+\infty} d\alpha F(\alpha) \int_{-\infty}^{+\infty} dq \cos q(x - \alpha),$$

which shews that $F(\alpha)$ must be taken to be $\dfrac{1}{2\pi} \phi(\alpha)$, for then by Fourier's double integral theorem $y_0 = \phi(x)$. Moreover, $\dot{y} = 0$; hence

$$y = \frac{1}{2\pi} \int_{-\infty}^{+\infty} d\alpha \, \phi(\alpha) \int_{-\infty}^{+\infty} dq \cos q^2 t \, \cos q(x - \alpha) \ldots \ldots (3)$$

satisfies the differential equation, and makes initially

$$y = \phi(x), \qquad \dot{y} = 0.$$

By Stokes' theorem (§ 95), or independently, we may now supply the remaining part of the solution, which has to satisfy the differential equation while it makes initially $y = 0$, $\dot{y} = \psi(x)$; it is

$$y = \frac{1}{2\pi} \int_{-\infty}^{+\infty} d\alpha \, \psi(\alpha) \int_{-\infty}^{+\infty} dq \frac{1}{q^2} \sin q^2 t \, \cos q(x - \alpha) \ldots \ldots (4).$$

The final result is obtained by adding the right-hand members of (3) and (4).

In (3) the integration with respect to q may be effected by means of the formula

$$\int_{-\infty}^{+\infty} dq \cos q^2 t \cos qz = \sqrt{\frac{\pi}{t}} \sin\left(\frac{\pi}{4} + \frac{z^2}{4t}\right) \ldots \ldots \ldots \ldots (5),$$

which may be proved as follows. If in the well-known integral formula

$$\int_{-\infty}^{+\infty} e^{-a^2 x^2} \, dx = \frac{\sqrt{\pi}}{a},$$

we put $x + b$ for x, we get

$$\int_{-\infty}^{+\infty} e^{-a^2 (x^2 + 2bx)} \, dx = \frac{\sqrt{\pi}}{a} e^{a^2 b^2}.$$

Now suppose that $a^2 = i = e^{\frac{1}{2}i\pi}$, where $i = \sqrt{(-1)}$, and retain only the real part of the equation. Thus

$$\int_{-\infty}^{+\infty} \cos(x^2 + 2bx) \, dx = \sqrt{\pi} \sin(b^2 + \tfrac{1}{4}\pi),$$

whence

$$\int_{-\infty}^{+\infty} \cos x^2 \cos 2bx \, dx = \sqrt{\pi} \sin (b^2 + \tfrac{1}{4}\pi),$$

from which (5) follows by a simple change of variable. Thus equation (3) may be written

$$y = \frac{1}{2\sqrt{\pi t}} \int_{-\infty}^{+\infty} d\alpha \, \phi(\alpha) \sin \left\{ \frac{\pi}{4} + \frac{(x-\alpha)^2}{4t} \right\},$$

or, if $\dfrac{\alpha - x}{2\sqrt{t}} = \mu$,

$$y = \frac{1}{\sqrt{2\pi}} \int_{-\infty}^{+\infty} d\mu \, (\cos \mu^2 + \sin \mu^2) \, \phi(x + 2\mu\sqrt{t}) \, \ldots\ldots (6).$$

192 *a.* If the axis of the rod be curved instead of straight, we obtain problems which may be regarded as extensions of those of the present and of the last chapters. The most important case under this head is that of a circular ring, whose section we will regard as also circular, and of radius (c) small in comparison with the radius (a) of the circular axis.

The investigation of the flexural modes of vibration, executed in the plane of the ring, is analogous to the case of a cylinder (see § 233), and was first effected by Hoppe[1]. If s be the number of periods in the circumference, the coefficient p of the time in the expression for the vibrations is given by.

$$p^2 = \frac{1}{4} \frac{s^2 (s^2-1)^2}{1 + s^2} \frac{q}{\rho} \frac{c^2}{a^4} \quad \ldots\ldots\ldots\ldots\ldots(1),$$

where q is Young's modulus and ρ the density of the material. This may be compared with equation (9) § 233. To fall back upon the case of a straight axis we have only to suppose s and a to be infinite in such a manner that $2\pi a/s$ is equal to the proposed linear period. The vibrations in question are then purely transverse.

In the class of vibrations considered above the circular axis remains unextended, and (§ 232) the periods are comparatively long. For the other class of vibrations in the plane of the ring, Hoppe found

$$p^2 = (1 + s^2) \frac{q}{\rho} \frac{1}{a^2} \quad \ldots\ldots\ldots\ldots\ldots\ldots(2).$$

[1] *Crelle*, Bd. 63, p. 158, 1871.

The frequencies are here independent of c, and the vibrations are analogous to the longitudinal vibrations of straight rods.

If $s = 0$ in (2), we have the solution for vibrations which are purely radial.

For flexural vibrations perpendicular to the plane of the ring, the result[1] corresponding to (1) is

$$p^2 = \frac{1}{4} \frac{s^2 (s^2 - 1)^2}{1 + \mu + s^2} \frac{q}{\rho} \frac{c^2}{a^4} \quad \dots\dots\dots\dots\dots(3),$$

the difference consisting only in the occurrence of Poisson's ratio (μ) in the denominator.

Our limits will not allow of our dwelling further upon the problem of this section. A complete investigation will be found in Love's *Treatise on Elasticity*, Chapter XVIII. The effect of a small curvature upon the lateral vibrations of a limited bar has been especially considered by Lamb[2].

[1] Michell, *Messenger of Mathematics*, XIX., 1889.

[2] *Proc. Lond. Math. Soc.*, XIX., p. 365, 1888.

CHAPTER IX.

193. THE theoretical membrane is a perfectly flexible and infinitely thin lamina of solid matter, of uniform material and thickness, which is stretched in all directions by a tension so great as to remain sensibly unaltered during the vibrations and displacements contemplated. If an imaginary line be drawn across the membrane in any direction, the mutual action between the two portions separated by an element of the line is proportional to the length of the element and perpendicular to its direction. If the force in question be $T_1 ds$, T_1 may be called the *tension of the membrane;* it is a quantity of one dimension in mass and -2 in time.

The principal problem in connection with this subject is the investigation of the transverse vibrations of membranes of different shapes, whose boundaries are fixed. Other questions indeed may be proposed, but they are of comparatively little interest; and, moreover, the methods proper for solving them will be sufficiently illustrated in other parts of this work. We may therefore proceed at once to the consideration of a membrane stretched over the area included within a fixed, closed, plane boundary.

194. Taking the plane of the boundary as that of xy, let w denote the small displacement therefrom of any point P of the membrane. Round P take a small area S, and consider the forces acting upon it parallel to z. The resolved part of the tension is expressed by

$$T_1 \int \frac{dw}{dn} \, ds,$$

where ds denotes an element of the boundary of S, and dn an element of the normal to the curve drawn outwards. This is balanced by the reaction against acceleration measured by $\rho S \ddot{w}$,

ρ being a symbol of one dimension in mass and -2 in length denoting the superficial density. Now by Green's theorem, if $\nabla\cdot = d^2/dx^2 + d^2/dy^2$,

$$\int \frac{dw}{dn}\, ds = \iint \nabla^2 w\, dS = \nabla^2 w \,.\, S \quad \text{ultimately,}$$

and thus the equation of motion is

$$\frac{d^2w}{dt^2} = \frac{T_1}{\rho}\left(\frac{d^2w}{dx^2} + \frac{d^2w}{dy^2}\right) \quad \dots\dots\dots\dots\dots(1).$$

The condition to be satisfied at the boundary is of course $w = 0$.

The differential equation may also be investigated from the expression for the potential energy, which is found by multiplying the tension by the superficial stretching. The altered area is

$$\iint \sqrt{1 + \left(\frac{dw}{dx}\right)^2 + \left(\frac{dw}{dy}\right)^2}\, dx\, dy \,;$$

and thus

$$V = \tfrac{1}{2}T_1 \iint \left\{\left(\frac{dw}{dx}\right)^2 + \left(\frac{dw}{dy}\right)^2\right\} dx\, dy \dots\dots\dots\dots(2),$$

from which δV is easily found by an integration by parts.

If we write $T_1 \div \rho = c^2$, then c is of the nature of a velocity, and the differential equation is

$$\frac{d^2w}{dt^2} = c^2 \left(\frac{d^2w}{dx^2} + \frac{d^2w}{dy^2}\right) \dots\dots\dots\dots\dots\dots (3).$$

195.   We shall now suppose that the boundary of the membrane is the rectangle formed by the coordinate axes and the lines $x = a$, $y = b$. For every point within the area (3) § 194 is satisfied, and for every point on the boundary $w = 0$.

A particular integral is evidently

$$w = \sin\frac{m\pi x}{a} \sin\frac{n\pi y}{b} \cos pt \dots\dots\dots\dots\dots(1),$$

where

$$p^2 = c^2\pi^2\left(\frac{m^2}{a^2} + \frac{n^2}{b^2}\right) \dots\dots\dots\dots\dots\dots\dots (2),$$

and m and n are integers; and from this the general solution may be derived. Thus

$$w = \sum_{m=1}^{m=\infty} \sum_{n=1}^{n=\infty} \sin\frac{m\pi x}{a} \sin\frac{n\pi y}{b}\{A_{mn}\cos pt + B_{mn}\sin pt\} \dots\dots (3).$$

That this result is really general may be proved *a posteriori*, by shewing that it may be adapted to express arbitrary initial circumstances.

Whatever function of the co-ordinates w may be, it can be expressed for all values of x between the limits 0 and a by the series

$$Y_1 \sin \frac{\pi x}{a} + Y_2 \sin \frac{2\pi x}{a} + \ldots\ldots,$$

where the coefficients Y_1, Y_2, &c. are independent of x. Again whatever function of y any one of the coefficients Y may be, it can be expanded between 0 and b in the series

$$C_1 \sin \frac{\pi y}{b} + C_2 \sin \frac{2\pi y}{b} + \ldots\ldots,$$

where C_1 &c. are constants. From this we conclude that any function of x and y can be expressed within the limits of the rectangle by the double series

$$\sum_{m=1}^{m=\infty} \sum_{n=1}^{n=\infty} A_{mn} \sin \frac{m\pi x}{a} \sin \frac{n\pi y}{b};$$

and therefore that the expression for w in (3) can be adapted to arbitrary initial values of w and $\dot{w}$. In fact

$$\left. \begin{aligned} A_{mn} &= \frac{4}{ab} \int_0^a \int_0^b w_0 \sin \frac{m\pi x}{a} \sin \frac{n\pi y}{b} \, dx \, dy, \\ B_{mn} &= \frac{4}{abp} \int_0^a \int_0^b \dot{w}_0 \sin \frac{m\pi x}{a} \sin \frac{n\pi y}{b} \, dx \, dy, \end{aligned} \right\} \ldots\ldots(4).$$

The character of the normal functions of a given rectangle,

$$\sin \frac{m\pi x}{a} \sin \frac{n\pi y}{b},$$

as depending on m and n, is easily understood. If m and n be both unity, w retains the same sign over the whole of the rectangle, vanishing at the edge only; but in any other case there are nodal lines running parallel to the axes of coordinates. The number of the nodal lines parallel to x is $n-1$, their equations being

$$y = \frac{b}{n}, \frac{2b}{n}, \ldots\ldots\frac{(n-1)b}{n}.$$

In the same way the equations of the nodal lines parallel to y are

$$x = \frac{a}{m}, \; \frac{2a}{m}, \; \ldots\ldots \frac{(m-1)\,a}{m},$$

being $m-1$ in number. The nodal system divides the rectangle into mn equal parts, in each of which the numerical value of w is repeated.

196. The expression for w in terms of the normal functions is

$$w = \Sigma\Sigma\, \phi_{mn} \sin\frac{m\pi x}{a} \sin\frac{n\pi y}{b} \; \ldots\ldots\ldots\ldots (1),$$

where ϕ_{mn} &c. are the normal coordinates. We proceed to form the expression for V in terms of ϕ_{mn}. We have

$$\left(\frac{dw}{dx}\right)^2 = \pi^2 \left\{\Sigma\Sigma\, \phi_{mn}\frac{m}{a}\cos\frac{m\pi x}{a}\sin\frac{n\pi y}{b}\right\}^2,$$

$$\left(\frac{dw}{dy}\right)^2 = \pi^2 \left\{\Sigma\Sigma\, \phi_{mn}\frac{n}{b}\sin\frac{m\pi x}{a}\cos\frac{n\pi y}{b}\right\}^2.$$

In integrating these expressions over the area of the rectangle the products of the normal coordinates disappear, and we find

$$V = \frac{T_1}{2}\iint\left\{\left(\frac{dw}{dx}\right)^2 + \left(\frac{dw}{dy}\right)^2\right\} dx\,dy$$

$$= \frac{T_1}{2}\frac{ab\pi^2}{4}\Sigma\Sigma\left(\frac{m^2}{a^2}+\frac{n^2}{b^2}\right)\phi_{mn}^2 \; \ldots\ldots\ldots\ldots (2),$$

the summation being extended to all integral values of m and n.

The expression for the kinetic energy is proved in the same way to be

$$T = \frac{\rho}{2}\frac{ab}{4}\Sigma\Sigma\,\dot{\phi}_{mn}^2 \ldots\ldots\ldots\ldots\ldots\ldots(3),$$

from which we deduce as the normal equation of motion

$$\ddot{\phi}_{mn} + c^2\pi^2\left(\frac{m^2}{a^2}+\frac{n^2}{b^2}\right)\phi_{mn} = \frac{4}{ab\rho}\,\Phi_{mn}\ldots\ldots\ldots\ldots(4).$$

In this equation

$$\Phi_{mn} = \int_0^a\int_0^b Z\sin\frac{m\pi x}{a}\sin\frac{n\pi y}{b}\,dx\,dy\ldots\ldots\ldots(5),$$

if $Z\,dx\,dy$ denote the transverse force acting on the element $dx\,dy$.

Let us suppose that the initial condition is one of rest under the operation of a constant force Z, such as may be supposed to arise from gaseous pressure. At the time $t = 0$, the impressed force is removed, and the membrane left to itself. Initially the equation of equilibrium is

$$c^2 \pi^2 \left(\frac{m^2}{a^2} + \frac{n^2}{b^2} \right)(\phi_{mn})_0 = \frac{4}{ab\rho} \, \Phi_{mn} \dots\dots\dots\dots(6),$$

whence $(\phi_{mn})_0$ is to be found. The position of the system at time t is then given by

$$\phi_{mn} = (\phi_{mn})_0 \cos \left(\sqrt{\frac{m^2}{a^2} + \frac{n^2}{b^2}} \cdot c\pi t \right) \dots\dots\dots (7),$$

in conjunction with (1).

In order to express Φ_{mn}, we have merely to substitute for Z its value in (5), or in this case simply to remove Z from under the integral sign. Thus

$$\Phi_{mn} = Z \int_0^a \int_0^b \sin \frac{m\pi x}{a} \sin \frac{n\pi y}{b} dx\, dy,$$

$$= Z \frac{ab}{mn\pi^2} (1 - \cos m\pi)(1 - \cos n\pi).$$

We conclude that Φ_{mn} vanishes, unless m and n are *both* odd, and that then

$$\Phi_{mn} = \frac{4ab}{mn\pi^2} Z.$$

Accordingly, m and n being both odd,

$$\phi_{mn} = \frac{16Z}{\pi^2 \rho} \frac{\cos pt}{mnp^2} \dots\dots\dots\dots\dots\dots (8),$$

where

$$p^2 = c^2 \pi^2 \left(\frac{m^2}{a^2} + \frac{n^2}{b^2} \right) \dots\dots\dots\dots\dots (9).$$

This is an example of (8), § 101.

If the membrane, previously at rest in its position of equilibrium, be set in motion by a blow applied at the point (α, β), the solution is

$$\phi_{mn} = \frac{4}{ab p} \sin \frac{m\pi \alpha}{a} \sin \frac{n\pi \beta}{b} \iint \dot{w}_0 dx\, dy \cdot \sin pt \dots (10).$$

[As an example of forced vibrations, suppose that a harmonic force acts at the centre. Unless m and n are both odd, $\Phi_{mn} = 0$, and in the case reserved

$$\Phi_{mn} = \pm Z_1 \cos qt \dots\dots\dots\dots\dots\dots(11),$$

where Z_1 is the whole force acting at time t, and $\pm$ represents $\sin \frac{1}{2} m\pi \sin \frac{1}{2} n\pi$. From (4) and (9) we have

$$\phi_{mn} = \frac{\pm 4 Z_1 \cos qt}{ab\rho \left(p_{mn}^2 - q^2 \right)} \quad \dots \dots \dots \dots (12)$$

and w is then given by (1).

In the case of a *square* membrane, p is a symmetrical function of m and n. When m and n are unequal, the terms occur in pairs, such as

$$\frac{\pm 4 Z_1 \cos qt}{a^2 \rho \left(p_{mn}^2 - q^2 \right)} \left\{ \sin \frac{m\pi x}{a} \sin \frac{n\pi y}{a} + \sin \frac{n\pi x}{a} \sin \frac{m\pi y}{a} \right\} \dots (13),$$

a combination symmetrical as between x and y. The vibration is of course similarly related as well to the four sides as to the four corners of the square.

In the neighbourhood of the centre, where the force is applied, the series loses its convergency, and the displacement w tends to become (logarithmically) infinite.]

197. The frequency of the natural vibrations is found by ascribing different integral values to m and n in the expression

$$\frac{p}{2\pi} = \frac{c}{2} \sqrt{\frac{m^2}{a^2} + \frac{n^2}{b^2}} \quad \dots \dots \dots \dots \dots (1).$$

For a given mode of vibration the pitch falls when either side of the rectangle is increased. In the case of the gravest mode, when $m = 1$, $n = 1$, additions to the shorter side are the more effective; and when the form is very elongated, additions to the longer side are almost without effect.

When a^2 and b^2 are incommensurable, no two pairs of values of m and n can give the same frequency, and each fundamental mode of vibration has its own characteristic period. But when a^2 and b^2 are commensurable, two or more fundamental modes may have the same periodic time, and may then coexist in any proportions, while the motion still retains its simple harmonic character. In such cases the specification of the period does not completely determine the type. The full consideration of the problem now presenting itself requires the aid of the theory of numbers; but it will be sufficient for the purposes of this work to consider a few of the simpler cases, which arise when the membrane is square. The reader will find fuller information in Riemann's lectures on partial differential equations.

If $a = b$,

$$\frac{p}{2\pi} = \frac{c}{2a} \sqrt{m^2 + n^2} \dotfill (2).$$

The lowest tone is found by putting m and n equal to unity, which gives only one fundamental mode:—

$$w = \sin \frac{\pi x}{a} \sin \frac{\pi y}{a} \cos pt \dotfill (3).$$

Next suppose that one of the numbers m, n is equal to 2, and the other to unity. In this way two distinct types of vibration are obtained, whose periods are the same. If the two vibrations be synchronous in phase, the whole motion is expressed by

$$w = \left\{ C \sin \frac{2\pi x}{a} \sin \frac{\pi y}{a} + D \sin \frac{\pi x}{a} \sin \frac{2\pi y}{a} \right\} \cos pt \dots (4);$$

so that, although every part vibrates synchronously with a harmonic motion, the type of vibration is to some extent arbitrary. Four particular cases may be especially noted. First, if $D = 0$,

$$w = C \sin \frac{2\pi x}{a} \sin \frac{\pi y}{a} \cos pt \dotfill (5),$$

which indicates a vibration with one node along the line $x = \frac{1}{2}a$. Similarly if $C = 0$, we have a node parallel to the other pair of edges. Next, however, suppose that C and D are finite and equal. Then w is proportional to

$$\sin \frac{2\pi x}{a} \sin \frac{\pi y}{a} + \sin \frac{\pi x}{a} \sin \frac{2\pi y}{a},$$

which may be put into the form

$$2 \sin \frac{\pi x}{a} \sin \frac{\pi y}{a} \left(\cos \frac{\pi x}{a} + \cos \frac{\pi y}{a} \right).$$

This expression vanishes, when

$$\sin \pi x / a = 0, \quad \text{or} \quad \sin \pi y / a = 0$$

or again, when

$$\cos \pi x / a + \cos \pi y / a = 0.$$

The first two equations give the edges, which were originally assumed to be nodal; while the third gives $y + x = a$, representing one diagonal of the square.

In the fourth case, when $C = -D$, we obtain for the nodal lines, the edges of the square together with the diagonal $y = x$. The figures represent the four cases.

Fig. 32.

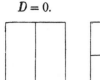

$D = 0.$ $C = 0.$ $C - D = 0.$ $C + D = 0.$

[Frequency (referred to gravest) = 1·58.]

For other relative values of C and D the interior nodal line is curved, but is always analytically expressed by

$$C \cos \frac{\pi x}{a} + D \cos \frac{\pi y}{a} = 0 \quad \dots\dots\dots\dots\dots (6),$$

and may be easily constructed with the help of a table of logarithmic cosines.

The next case in order of pitch occurs when $m = 2$, $n = 2$. The values of m and n being equal, no alteration is caused by their interchange, while no other pair of values gives the same frequency of vibration. The only type to be considered is accordingly

$$w = \sin \frac{2\pi x}{a} \cdot \sin \frac{2\pi y}{a} \cos pt,$$

whose nodes, determined by the equation

Fig. 33.

$$\sin \frac{\pi x}{a} \cdot \sin \frac{\pi y}{a} \cos \frac{\pi x}{a} \cos \frac{\pi y}{a} = 0,$$

are (in addition to the edges) the straight lines Fig. (33)

$$x = \tfrac{1}{2}a \qquad y = \tfrac{1}{2}a.$$

[Frequency = 2·00.]

The next case which we shall consider is obtained by ascribing to m, n the values 3, 1, and 1, 3 successively. We have

$$w = \left\{ C \sin \frac{3\pi x}{a} \sin \frac{\pi y}{a} + D \sin \frac{\pi x}{a} \sin \frac{3\pi y}{a} \right\} \cos pt.$$

The nodes are given by

$$\sin \frac{\pi x}{a} \sin \frac{\pi y}{a} \left\{ C \left(4 \cos^2 \frac{\pi x}{a} - 1 \right) + D \left(4 \cos^2 \frac{\pi y}{a} - 1 \right) \right\} = 0,$$

or, if we reject the first two factors, which correspond to the edges,

$$C \left(4 \cos^2 \frac{\pi x}{a} - 1 \right) + D \left(4 \cos^2 \frac{\pi y}{a} - 1 \right) = 0 \dots\dots (7).$$

If $C = 0$, we have $y = \frac{1}{3} a$, $y = \frac{2}{3} a$.

If $D = 0$, $x = \frac{1}{3} a$, $x = \frac{2}{3} a$.

If $C = -D$, $\cos \dfrac{\pi x}{a} = \pm \cos \dfrac{\pi y}{a}$,

whence, $y = x$, $y = a - x$,

which represent the two diagonals.

Lastly, if $C = D$, the equation of the node is

$$\cos^2 \frac{\pi x}{a} + \cos^2 \frac{\pi y}{a} = \frac{1}{2},$$

or $$1 + \cos \frac{2\pi x}{a} + \cos \frac{2\pi y}{a} = 0 \dots\dots\dots\dots\dots (8),$$

Fig. 34.

$C = 0.$ $D = 0.$ $C + D = 0.$ $C - D = 0.$

[Frequency = 2·24.]

In case (4) when $x = \frac{1}{2} a$, $y = \frac{1}{4} a$, or $\frac{3}{4} a$; and similarly when $y = \frac{1}{2} a$, $x = \frac{1}{4} x$, or $\frac{3}{4} a$. Thus one half of each of the lines joining the middle points of opposite edges is intercepted by the curve.

[The diameters of the nodal curve parallel to the sides of the square are thus equal to $\frac{1}{2} a$. Those measured along the diagonals are sensibly smaller, equal to $\frac{1}{3} \sqrt{2} \cdot a$, or ·471 a.]

It should be noticed that in whatever ratio to one another C and D may be taken, the nodal curve always passes through the four points of intersection of the nodal lines of the first two cases, $C = 0$, $D = 0$. If the vibrations of these cases be compounded with corresponding phases, it is evident that in the shaded compartments of Fig. (35) the directions of displacement are the same, and that therefore no part of the nodal curve is to be found there; whatever the ratio of amplitudes, the curve must be drawn through the unshaded portions. When on the other hand the phases are opposed, the nodal curve will pass exclusively through the shaded portions.

Fig. 35.

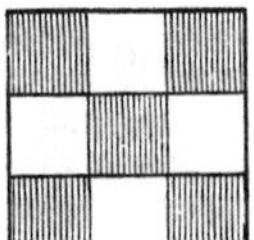

When $m = 3$, $n = 3$, the nodes are the straight lines parallel to the edges shewn in Fig. (36).

The last case [Frequency = 2·55] which we shall consider is obtained by putting

Fig. 36.

$$m = 3, \quad n = 2, \quad \text{or} \quad m = 2, \quad n = 3.$$

The nodal system is

[Frequency = 3·00.]

$$C \sin \frac{3\pi x}{a} \sin \frac{2\pi y}{a} + D \sin \frac{2\pi x}{a} \sin \frac{3\pi y}{a} = 0,$$

or, if the factors corresponding to the edges be rejected,

$$C \left(4 \cos^2 \frac{\pi x}{a} - 1 \right) \cos \frac{\pi y}{a} + D \cos \frac{\pi x}{a} \left(4 \cos^2 \frac{\pi y}{a} - 1 \right) = 0......(9).$$

If C or D vanish, we fall back on the nodal systems of the component vibrations, consisting of straight lines parallel to the edges. If $C = D$, our equation may be written

$$\left(\cos \frac{\pi x}{a} + \cos \frac{\pi y}{a} \right) \left(4 \cos \frac{\pi x}{a} \cos \frac{\pi y}{a} - 1 \right) = 0......(10),$$

of which the first factor represents the diagonal $y + x = a$, and the second a hyperbolic curve.

If $C = -D$, we obtain the same figure relatively to the other diagonal[1].

198. The pitch of the natural modes of a square membrane, which is nearly, but not quite uniform, may be investigated by the general method of § 90.

We will suppose in the first place that m and n are equal. In this case, when the pitch of a uniform membrane is given, the mode of its vibration is completely determined. If we now conceive a variation of density to ensue, the natural type of vibration is in general modified, but the period may be calculated approximately without allowance for the change of type.

We have

$$T = \tfrac{1}{2} \iint (\rho_0 + \delta\rho) \, \dot{\phi}_{mm}{}^2 \sin^2 \frac{m\pi x}{a} \sin^2 \frac{m\pi y}{a} \, dx dy$$

$$= \tfrac{1}{2} \dot{\phi}_{mm}{}^2 \left\{ \rho_0 \frac{a^2}{4} + \iint \delta\rho \sin^2 \frac{m\pi x}{a} \sin^2 \frac{m\pi y}{a} \, dx dy \right\},$$

[1] Lamé, *Leçons sur l'élasticité*, p. 129.

of which the second term is the increment of T due to $\delta\rho$. Hence if $w \propto \cos pt$, and P denote the value of p previously to variation, we have

$$p_{mm}{}^2 : P_{mm}{}^2 = 1 - \frac{4}{a^2} \int_0^a \int_0^a \frac{\delta\rho}{\rho_0} \sin^2 \frac{m\pi x}{a} \sin^2 \frac{m\pi y}{a} \, dx dy \dots\dots (1),$$

where $\qquad P_{mm}{}^2 = \dfrac{2\,c^2\pi^2 m^2}{a^2}$, and $c^2 = T_1 \div \rho_0$.

For example, if there be a small load M attached to the middle of the square,

$$p_{mm}{}^2 : P_{mm}{}^2 = 1 - \frac{4M}{a^2\rho_0} \sin^4 m \frac{\pi}{2} \dots\dots\dots\dots (2),$$

in which $\sin^4 \frac{1}{2}m\pi$ vanishes, if m be even, and is equal to unity, if m be odd. In the former case the centre is on the nodal line of the unloaded membrane, and thus the addition of the load produces no result.

When, however, m and n are unequal, the problem, though remaining subject to the same general principles, presents a peculiarity different from anything we have hitherto met with. The natural type for the unloaded membrane corresponding to a specified period is now to some extent arbitrary; but the introduction of the load will in general remove the indeterminate element. In attempting to calculate the period on the assumption of the undisturbed type, the question will arise how the selection of the undisturbed type is to be made, seeing that there are an indefinite number, which in the uniform condition of the membrane give identical periods. The answer is that those types must be chosen which differ infinitely little from the actual types assumed under the operation of the load, and such a type will be known by the criterion of its making the period calculated from it a maximum or minimum.

As a simple example, let us suppose that a small load M is attached to the membrane at a point lying on the line $x = \frac{1}{2}a$, and that we wish to know what periods are to be substituted for the two equal periods of the unloaded membrane, found by making

$$m = 2, \ n = 1, \quad \text{or} \quad m = 1, \ n = 2.$$

It is clear that the normal types to be chosen, are those whose nodes are represented in the first two cases of Fig. (32). In the first case the increase in the period due to the load is zero, which is the least that it can be; and in the second case the increase is the greatest possible. If β be the ordinate of M, the kinetic energy is altered in the ratio

$$\frac{\rho}{2}\frac{a^2}{4} : \frac{\rho}{2}\frac{a^2}{4} + \frac{M}{2}\sin^2\frac{2\pi\beta}{a} ;$$

and thus

$$p_{12}{}^2 : P_{12}{}^2 = 1 - \frac{4M}{a^2\rho}\sin^2\frac{2\pi\beta}{a}\dots\dots\dots\dots\dots(3)$$

while

$$p_{21}{}^2 = P_{21}{}^2 = P_{12}{}^2.$$

The ratio characteristic of the interval between the two natural tones of the loaded membrane is thus approximately

$$1 + \frac{2M}{a^2\rho}\sin^2\frac{2\pi\beta}{a}\dots\dots\dots\dots\dots(4).$$

If $\beta = \frac{1}{2}a$, neither period is affected by the load.

As another example, the case where the values of m and n are 3 and 1, considered in § 197, may be referred to. With a load in the middle, the two normal types to be selected are those corresponding to the last two cases of Fig. (34), in the former of which the load has no effect on the period.

The problem of determining the vibration of a square membrane which carries a relatively heavy load is more difficult, and we shall not attempt its solution. But it may be worth while to recall to memory the fact that the actual period is greater than any that can be calculated from a hypothetical type, which differs from the actual one.

199. The preceding theory of square membranes includes a good deal more than was at first intended. Whenever in a vibrating system certain parts remain at rest, they may be supposed to be absolutely fixed, and we thus obtain solutions of other questions than those originally proposed. For example, in the present case, wherever a diagonal of the square is nodal, we obtain a solution applicable to a membrane whose fixed boundary is an· isosceles right-angled triangle. Moreover, any mode of vibration possible to the triangle corresponds to some natural mode of the square, as may be seen by supposing two triangles put together, the vibrations being equal and opposite at points which are the images of each other in the common hypothenuse. Under these circumstances it is evident that the hypothenuse would remain at rest without constraint, and therefore the vibration in question is included among those of which a complete square is capable.

The frequency of the gravest tone of the triangle is found by putting $m = 1$, $n = 2$ in the formula

$$\frac{p}{2\pi} = \frac{c}{2a} \sqrt{(m^2 + n^2)}\dots\dots\dots\dots\dots(1),$$

and is therefore equal to $c\sqrt{5}/2a$.

The next tone occurs, when $m = 3$, $n = 1$. In this case

$$\frac{p}{2\pi} = \frac{c\sqrt{10}}{2a}\dots\dots\dots\dots\dots(2),$$

as might also be seen by noticing that the triangle divides itself into two, Fig. (37), whose sides are less than those of the whole triangle in the ratio $\sqrt{2} : 1$.

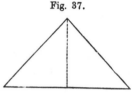

Fig. 37.

For the theory of the vibrations of a membrane whose boundary is in the form of an equilateral triangle, the reader is referred to Lamé's *Leçons sur l'élasticité*. It is proved that the frequency of the gravest tone is $c \div h$, where h is the height of the triangle, which is the same as the frequency of the gravest tone of a square whose diagonal is h.

200. When the fixed boundary of the membrane is circular, the first step towards a solution of the problem is the expression of the general differential equation in polar co-ordinates. This may be effected analytically; but it is simpler to form the polar equation *de novo* by considering the forces which act on the polar element of area $r\, d\theta\, dr$. As in § 194 the force of restitution acting on a small area of the membrane is

$$- T_1 \int \frac{dw}{dn}\, ds = - T_1 \left\{ \frac{d}{dr}\left(\frac{dw}{dr}\, r\, d\theta \right) dr + \frac{d}{d\theta}\left(\frac{dw}{r\, d\theta}\, dr \right) d\theta \right\}$$

$$= - T_1 \cdot r\, d\theta\, dr \left\{ \frac{d^2 w}{dr^2} + \frac{1}{r}\frac{dw}{dr} + \frac{1}{r^2}\frac{d^2 w}{d\theta^2} \right\};$$

and thus, if $T_1/\rho = c^2$ as before, the equation of motion is

$$\frac{d^2 w}{dt^2} = c^2 \left\{ \frac{d^2 w}{dr^2} + \frac{1}{r}\frac{dw}{dr} + \frac{1}{r^2}\frac{d^2 w}{d\theta^2} \right\}\dots\dots\dots\dots(1).$$

The subsidiary condition to be satisfied at the boundary is that $w = 0$, when $r = a$.

In order to investigate the normal component vibrations we have now to assume that w is a harmonic function of the time.

Thus, if $w \propto \cos(pt - \epsilon)$, and for the sake of brevity we write $p/c = k$, the differential equation appears in the form

$$\frac{d^2w}{dr^2} + \frac{1}{r}\frac{dw}{dr} + \frac{1}{r^2}\frac{d^2w}{d\theta^2} + k^2 w = 0 \dots\dots\dots\dots(2),$$

in which k is the reciprocal of a linear quantity.

Now whatever may be the nature of w as a function of r and θ, it can be expanded in Fourier's series

$$w = w_0 + w_1 \cos(\theta + \alpha_1) + w_2 \cos 2(\theta + \alpha_2) + \dots\dots(3),$$

in which w_0, w_1, &c. are functions of r, but not of θ. The result of substituting from (3) in (2) may be written

$$\Sigma \left\{ \frac{d^2w_n}{dr^2} + \frac{1}{r}\frac{dw_n}{dr} + \left(k^2 - \frac{n^2}{r^2}\right)w_n \right\} \cos n(\theta + \alpha_n) = 0,$$

the summation extending to all integral values of n. If we multiply this equation by $\cos n(\theta + \alpha_n)$, and integrate with respect to θ between the limits 0 and 2π, we see that each term must vanish separately, and we thus obtain to determine w_n as a function of r

$$\frac{d^2w_n}{dr^2} + \frac{1}{r}\frac{dw_n}{dr} + \left(k^2 - \frac{n^2}{r^2}\right)w_n = 0 \dots\dots\dots\dots(4),$$

in which it is a matter of indifference whether the factor $\cos n(\theta + \alpha_n)$ be supposed to be included in w_n or not.

The solution of (4) involves two distinct functions of r, each multiplied by an arbitrary constant. But one of these functions becomes infinite when r vanishes, and the corresponding particular solution must be excluded as not satisfying the prescribed conditions at the origin of co-ordinates. This point may be illustrated by a reference to the simpler equation derived from (4) by making k and n vanish, when the solution in question reduces to $w = \log r$, which, however, does not at the origin satisfy $\nabla^2 w = 0$, as may be seen from the value of $\int (dw/dn)\,ds$, integrated round a small circle with the origin for centre. In like manner the complete integral of (4) is too general for our present purpose, since it covers the case in which the centre of the membrane is subjected to an external force.

The other function of r, which satisfies (4), is the Bessel's function of the nth order, denoted by $J_n(kr)$, and may be expressed in several ways. The ascending series (obtained immediately from the differential equation) is

$$J_n(z) = \frac{z^n}{2^n \Gamma(n+1)} \left\{ 1 - \frac{z^2}{2 \cdot 2n+2} + \frac{z^4}{2 \cdot 4 \cdot 2n+2 \cdot 2n+4} \right.$$
$$\left. - \frac{z^6}{2 \cdot 4 \cdot 6 \cdot 2n+2 \cdot 2n+4 \cdot 2n+6} + \ldots \right\} \ldots \ldots (5),$$

from which the following relations between functions of consecutive orders may readily be deduced:

$$J_0'(z) = -J_1(z) \ldots\ldots\ldots\ldots\ldots\ldots\ldots(6),$$
$$2J_n'(z) = J_{n-1}(z) - J_{n+1}(z) \ldots\ldots\ldots\ldots(7),$$
$$\frac{2n}{z} J_n(z) = J_{n-1}(z) + J_{n+1}(z) \ldots\ldots\ldots\ldots(8).$$

When n is an integer, $J_n(z)$ may be expressed by the definite integral

$$J_n(z) = \frac{1}{\pi} \int_0^\pi \cos(z \sin \omega - n\omega) \, d\omega \ldots\ldots\ldots\ldots(9),$$

which is Bessel's original form. From this expression it is evident that J_n and its differential coefficients with respect to z are always less than unity.

The ascending series (5), though infinite, is convergent for all values of n and z; but, when z is great, the convergence does not begin for a long time, and then the series becomes useless as a basis for numerical calculation. In such cases another series proceeding by descending powers of z may be substituted with advantage. This series is

$$J_n(z) = \sqrt{\frac{2}{\pi z}} \left\{ 1 - \frac{(1^2 - 4n^2)(3^2 - 4n^2)}{1 \cdot 2 \cdot (8z)^2} + \ldots \right\} \cos\left(z - \frac{\pi}{4} - n\frac{\pi}{2}\right)$$
$$+ \sqrt{\frac{2}{\pi z}} \left\{ \frac{1^2 - 4n^2}{1 \cdot 8z} - \frac{(1^2 - 4n^2)(3^2 - 4n^2)(5^2 - 4n^2)}{1 \cdot 2 \cdot 3 \cdot (8z)^3} + \ldots \right\}$$
$$\times \sin\left(z - \frac{\pi}{4} - n\frac{\pi}{2}\right) \ldots\ldots\ldots\ldots\ldots\ldots(10);$$

it terminates, if $2n$ be equal to an odd integer, but otherwise, it runs on to infinity, and becomes ultimately divergent. Nevertheless when z is great, the convergent part may be employed in calculation; for it can be proved that the sum of any number of terms differs from the true value of the function by less than the last term included. We shall have occasion later, in connection with another problem, to consider the derivation of this descending series.

As Bessel's functions are of considerable importance in theoretical acoustics, I have thought it advisable to give a table for the functions J_0 and J_1, extracted from Lommel's [1] work, and due

---

[1] Lommel, *Studien über die Bessel'schen Functionen.* Leipzig, 1868.

originally to Hansen. The functions J_0 and J_1 are connected by the relation $J_0' = -J_1$.

z	$J_0(z)$	$J_1(z)$	z	$J_0(z)$	$J_1(z)$	z	$J_0(z)$	$J_1(z)$
0·0	1·0000	0·0000	4·5	·3205	·2311	9·0	·0903	·2453
0·1	·9975	·0499	4·6	·2961	·2566	9·1	·1142	·2324
0·2	·9900	·0995	4·7	·2693	·2791	9·2	·1367	·2174
0·3	·9776	·1483	4·8	·2404	·2985	9·3	·1577	·2004
0·4	·9604	·1960	4·9	·2097	·3147	9·4	·1768	·1816
0·5	·9385	·2423	5·0	·1776	·3276	9·5	·1939	·1613
0·6	·9120	·2867	5·1	·1443	·3371	9·6	·2090	·1395
0·7	·8812	·3290	5·2	·1103	·3432	9·7	·2218	·1166
0·8	·8463	·3688	5·3	·0758	·3460	9·8	·2323	·0928
0·9	·8075	·4060	5·4	·0412	·3453	9·9	·2403	·0684
1·0	·7652	·4401	5·5	− ·0068	·3414	10·0	·2459	·0435
1·1	·7196	·4709	5·6	+ ·0270	·3343	10·1	·2490	+ ·0184
1·2	·6711	·4983	5·7	·0599	·3241	10·2	·2496	− ·0066
1·3	·6201	·5220	5·8	·0917	·3110	10·3	·2477	·0313
1·4	·5669	·5419	5·9	·1220	·2951	10·4	·2434	·0555
1·5	·5118	·5579	6·0	·1506	·2767	10·5	·2366	·0789
1·6	·4554	·5699	6·1	·1773	·2559	10·6	·2276	·1012
1·7	·3980	·5778	6·2	·2017	·2329	10·7	·2164	·1224
1·8	·3400	·5815	6·3	·2238	·2081	10·8	·2032	·1422
1·9	·2818	·5812	6·4	·2433	·1816	10·9	·1881	·1604
2·0	·2239	·5767	6·5	·2601	·1538	11·0	·1712	·1768
2·1	·1666	·5683	6·6	·2740	·1250	11·1	·1528	·1913
2·2	·1104	·5560	6·7	·2851	·0953	11·2	·1330	·2039
2·3	·0555	·5399	6·8	·2931	·0652	11·3	·1121	·2143
2·4	+ ·0025	·5202	6·9	·2981	·0349	11·4	·0902	·2225
2·5	− ·0484	·4971	7·0	·3001	− ·0047	11·5	·0677	·2284
2·6	·0968	·4708	7·1	·2991	+ ·0252	11·6	·0446	·2320
2·7	·1424	·4416	7·2	·2951	·0543	11·7	− ·0213	·2333
2·8	·1850	·4097	7·3	·2882	·0826	11·8	+ ·0020	·2323
2·9	·2243	·3754	7·4	·2786	·1096	11·9	·0250	·2290
3·0	·2601	·3391	7·5	·2663	·1352	12·0	·0477	·2234
3·1	·2921	·3009	7·6	·2516	·1592	12·1	·0697	·2157
3·2	·3202	·2613	7·7	·2346	·1813	12·2	·0908	·2060
3·3	·3443	·2207	7·8	·2154	·2014	12·3	·1108	·1943
3·4	·3643	·1792	7·9	·1944	·2192	12·4	·1296	·1807
3·5	·3801	·1374	8·0	·1717	·2346	12·5	·1469	·1655
3·6	·3918	·0955	8·1	·1475	·2476	12·6	·1626	·1487
3·7	·3992	·0538	8·2	·1222	·2580	12·7	·1766	·1307
3·8	·4026	+ ·0128	8·3	·0960	·2657	12·8	·1887	·1114
3·9	·4018	− ·0272	8·4	·0692	·2708	12·9	·1988	·0912
4·0	·3972	·0660	8·5	·0419	·2731	13·0	·2069	·0703
4·1	·3887	·1033	8·6	+ ·0146	·2728	13·1	·2129	·0489
4·2	·3766	·1386	8·7	− ·0125	·2697	13·2	·2167	·0271
4·3	·3610	·1719	8·8	·0392	·2641	13·3	·2183	− ·0052
4·4	·3423	·2028	8·9	·0653	·2559	13·4	·2177	+ ·0166

201. In accordance with the notation for Bessel's functions the expression for a normal component vibration may therefore be written

$$w = P J_n(kr) \cos n(\theta + \alpha) \cos(pt + \epsilon)\ldots\ldots\ldots(1);$$

and the boundary condition requires that

$$J_n(ka) = 0\ldots\ldots\ldots\ldots\ldots\ldots(2),$$

an equation whose roots give the admissible values of k, and therefore of p.

The complete expression for w is obtained by combining the particular solutions embodied in (1) with all admissible values of k and n, and is necessarily general enough to cover any initial circumstances that may be imagined. We conclude that any function of r and θ may be expanded within the limits of the circle $r = a$ in the series

$$w = \Sigma\Sigma J_n(kr) \{\phi \cos n\theta + \psi \sin n\theta\} \ldots\ldots\ldots(3).$$

For every integral value of n there are a series of values of k, given by (2); and for each of these the constants ϕ and ψ are arbitrary.

The determination of the constants is effected in the usual way. Since the energy of the motion is equal to

$$\tfrac{1}{2}\rho \int_0^a \int_0^{2\pi} \dot{w}^2 r\, d\theta\, dr \ldots\ldots\ldots\ldots\ldots(4),$$

and when expressed by means of the normal co-ordinates can only involve their squares, it follows that the product of any two of the terms in (3) vanishes, when integrated over the area of the circle. Thus, if we multiply (3) by $J_n(kr) \cos n\theta$, and integrate, we find

$$\int_0^a \int_0^{2\pi} w J_n(kr) \cos n\theta\, r dr\, d\theta$$

$$= \phi \int\int [J_n(kr)]^2 \cos^2 n\theta\, r dr\, d\theta = \phi . \pi \int_0^a [J_n(kr)]^2 r dr\ldots\ldots(5),$$

by which ϕ is determined. The corresponding formula for ψ is obtained by writing $\sin n\theta$ for $\cos n\theta$. A method of evaluating the integral on the right will be given presently. Since ϕ and ψ each contain two terms, one varying as $\cos pt$ and the other as $\sin pt$, it is now evident how the solution may be adapted so as to agree with arbitrary initial values of w and $\dot{w}$.

202. Let us now examine more particularly the character of the fundamental vibrations. If $n = 0$, w is a function of r only, that is to say, the motion is symmetrical with respect to the centre of the membrane. The nodes, if any, are the concentric circles, whose equation is

$$J_0(kr) = 0 \quad\ldots\ldots\ldots\ldots\ldots\ldots\ldots (1).$$

When n has an integral value different from zero, w is a function of θ as well as of r, and the equation of the nodal system takes the form

$$J_n(kr)\, \cos n\,(\theta - \alpha) = 0 \ldots\ldots\ldots\ldots\ldots\ldots (2).$$

The nodal system is thus divisible into two parts, the first consisting of the concentric circles represented by

$$J_n(kr) = 0 \quad\ldots\ldots\ldots\ldots\ldots\ldots\ldots (3),$$

and the second of the diameters

$$\theta = \alpha + (2m + 1)\,\pi/2n \ldots\ldots\ldots\ldots\ldots\ldots (4),$$

where m is an integer. These diameters are n in number, and are ranged uniformly round the centre; in other respects their position is arbitrary. The radii of the circular nodes will be investigated further on.

203. The important integral formula

$$\int_0^a J_n(kr)\, J_n(k'r)\, r\, dr = 0 \quad\ldots\ldots\ldots\ldots (1),$$

where k and k' are different roots of

$$J_n(ka) = 0 \quad\ldots\ldots\ldots\ldots\ldots\ldots\ldots (2),$$

may be verified analytically by means of the differential equations satisfied by $J_n(kr)$, $J_n(k'r)$; but it is both simpler and more instructive to begin with the more general problem, where the boundary of the membrane is not restricted to be circular.

The variational equation of motion is

$$\delta V + \rho \iint \ddot{w}\, \delta w\, dx\, dy = 0 \quad\ldots\ldots\ldots\ldots (3)$$

where

$$V = \tfrac{1}{2} T_1 \iint \left\{ \left(\frac{dw}{dx}\right)^2 + \left(\frac{dw}{dy}\right)^2 \right\} dx\, dy \quad\ldots\ldots\ldots (4),$$

and therefore

$$\delta V = T_1 \iint \left\{ \frac{dw}{dx}\frac{d\delta w}{dx} + \frac{dw}{dy}\frac{d\delta w}{dy} \right\} dx\, dy \ldots\ldots\ldots (5).$$

In these equations w refers to the actual motion, and δw to a hypothetical displacement consistent with the conditions to which the system is subjected. Let us now suppose that the system is executing one of its normal component vibrations, so that $w = u$, and

$$\ddot{u} + p^2 u = 0 \dots\dots\dots\dots\dots(6),$$

while δw is proportional to another normal function v.

Since $k = p/c$, we get from (3)

$$k^2 \iint u\,v\,dx\,dy = \iint \left\{ \frac{du}{dx}\frac{dv}{dx} + \frac{du}{dy}\frac{dv}{dy} \right\} dx\,dy \dots\dots (7).$$

The integral on the right is symmetrical with respect to u and v and thus

$$(k'^2 - k^2) \iint u\,v\,dx\,dy = 0 \dots\dots\dots\dots (8),$$

where k'^2 bears the same relation to v that k^2 bears to u.

Accordingly, if the normal vibrations represented by u and v have different periods,

$$\iint u\,v\,dx\,dy = 0 \dots\dots\dots\dots\dots (9).$$

In obtaining this result, we have made no assumption as to the boundary conditions beyond what is implied in the absence of reactions against acceleration, which, if they existed, would appear in the fundamental equation (3).

If in (8) we suppose $k' = k$, the equation is satisfied identically, and we cannot infer the value of $\iint u^2 dx\,dy$. In order to evaluate this integral we must follow a rather different course.

If u and v be functions satisfying within a certain contour the equations $\nabla^2 u + k^2 u = 0$, $\nabla^2 v + k'^2 v = 0$, we have

$$(k'^2 - k^2) \iint u\,v\,dx\,dy = \iint (v\,\nabla^2 u - u\,\nabla^2 v)\,dx\,dy$$

$$= \int \left(v\frac{du}{dn} - u\frac{dv}{dn} \right) ds \dots\dots\dots(10),$$

by Green's theorem. Let us now suppose that v is derived from u by slightly varying k, so that

$$v = u + \frac{du}{dk}\delta k, \quad k' = k + \delta k;$$

substituting in (10), we find

$$2k \iint u^2 dx\,dy = \int \left(\frac{du}{dk\,.dn} - u\frac{d^2 u}{dn\,dk} \right) ds \dots\dots\dots\dots(11);$$

or, if u vanish on the boundary,

$$2k \iint u^2 dx\, dy = \int \frac{du}{dk} \frac{du}{dn}\, ds \dots\dots\dots\dots\dots (12).$$

For the application to a circular area of radius r, we have

$$\left. \begin{aligned} u &= \cos n\theta\ J_n(kr) \\ v &= \cos n\theta\ J_n(k'r) \end{aligned} \right\} \dots\dots\dots\dots\dots(13),$$

and thus from (10) on substitution of polar co-ordinates and integra tion with respect to θ,

$$(k'^2 - k^2) \int_0^r J_n(kr)\, J_n(k'r)\, r dr$$

$$= r J_n(k'r) \frac{d}{dr} J_n(kr) - r J_n(kr) \frac{d}{dr} J_n(k'r) \dots\dots\dots\dots(14).$$

Accordingly, if

$$\frac{d}{dr} J_n(k'r)\ :\ J_n(k'r) = \frac{d}{dr} J_n(kr)\ :\ J_n(kr),$$

and k and k' be different,

$$\int_0^r J_n(kr)\, J_n(k'r)\, r dr = 0 \dots\dots\dots\dots\dots(15),$$

an equation first proved by Fourier for the case when

$$J_n(kr) = J_n(k'r) = 0.$$

Again from (11)

$$2k \int_0^r J_n^2(kr)\, r dr = r \frac{dJ}{dk} \frac{dJ}{dr} - r J \frac{d^2 J}{dr\, dk}$$

$$= kr^2 J'^2 - kr^2 J \left(J'' + \frac{1}{kr} J' \right),$$

dashes denoting differentiation with respect to kr.   Now

$$J'' + \frac{1}{kr} J' + \left(1 - \frac{n^2}{k^2 r^2} \right) J = 0,$$

and thus

$$2 \int_0^r J_n^2(kr)\, r dr = r^2 J_n'^2(kr) + r^2 \left(1 - \frac{n^2}{k^2 r^2} \right) J_n^2(kr) \dots\dots (16).$$

This result is general; but if, as in the application to membranes with fixed boundaries, $J_n(kr) = 0$,

then $$2 \int_0^r J_n^2(kr)\, r dr = r^2 J_n'^2(kr) \dots\dots\dots\dots(17).$$

204. We may use the result just arrived at to simplify the expressions for T and V. From

$$w = \Sigma\Sigma \left\{ \phi_{mn} J_n(k_{mn}r) \cos n\theta + \psi_{mn} J_n(k_{mn}r) \sin n\theta \right\} \ldots\ldots\ldots(1),$$

we find

$$T = \tfrac{1}{4} \rho\pi a^2 \, \Sigma\Sigma \, J_n'^2(k_{mn}a) \left\{ \dot{\phi}_{mn}^2 + \dot{\psi}_{mn}^2 \right\} \ldots\ldots\ldots\ldots(2),$$

$$V = \tfrac{1}{4} \rho\pi a^2 \, \Sigma\Sigma \, p_{mn}^2 J_n'^2(k_{mn}a) \left\{ \phi_{mn}^2 + \psi_{mn}^2 \right\} \ldots\ldots(3);$$

whence is derived the normal equation of motion

$$\ddot{\phi}_{mn} + p_{mn}^2 \phi_{mn} = \frac{2\,\Phi_{mn}}{\rho\pi a^2 J_n'^2(k_{mn}a)}\ldots\ldots\ldots\ldots\ldots(4),$$

and a similar equation for ψ_{mn}. The value of Φ_{mn} is to be found from the consideration that $\Phi_{mn}\delta\phi_{mn}$ denotes the work done by the impressed forces during a hypothetical displacement $\delta\phi_{mn}$; so that if Z be the impressed force, reckoned per unit of area,

$$\Phi_{mn} = \iint Z J_n(k_{mn}r) \cos n\theta \, r dr \, d\theta \ldots\ldots\ldots (5).$$

These expressions and equations do not apply to the case $n = 0$, when ϕ and ψ are amalgamated. We then have

$$\left. \begin{array}{l} T = \tfrac{1}{2} \rho\pi a^2 J_0'^2(k_{m0}a)\,\dot{\phi}_{m0}^2 \\ V = \tfrac{1}{2} \rho\pi a^2 p_{m0}^2 J_0'^2(k_{m0}a)\,\phi_{m0}^2 \end{array} \right\} \ldots\ldots\ldots\ldots(6),$$

$$\ddot{\phi}_{m0} + p_{m0}^2 \phi_{m0} = \frac{\Phi_{m0}}{\rho\pi a^2 J_0'^2(k_{m0}a)}\ldots\ldots\ldots\ldots\ldots(7).$$

As an example, let us suppose that the initial velocities are zero, and the initial configuration that assumed under the influence of a constant pressure Z; thus

$$\Phi_{m0} = Z \, . \, 2\pi \int_0^a J_0(k_{m0}r)\,rdr.$$

Now by the differential equation,

$$rJ_0(kr) = - \left\{ rJ_0''(kr) + \frac{1}{k} J_0'(kr) \right\},$$

and thus

$$\int_0^a J_0(kr)\,rdr = - \frac{a}{k} J_0'(ka) \ldots\ldots\ldots\ldots\ldots(8);$$

so that

$$\Phi_{m0} = - \frac{2\pi a}{k_{m0}} Z J_0'(k_{m0}a).$$

Substituting this in (7), we see that the initial value of ϕ_{m0} is

$$(\phi_{m0})_{t=0} = \frac{-2Z}{k_{m0}\,p_{m0}^2\rho a \, J_0'(k_{m0}a)} \ldots\ldots\ldots\ldots (9)$$

For values of n other than zero, Φ and the initial value of ϕ_{mn} vanish. The state of the system at time t is expressed by

$$\phi_{m0} = (\phi_{m0})_{t=0} \cdot \cos p_{m0} t \quad \dots\dots\dots\dots(10),$$

$$w = \Sigma \phi_{m0} J_0(k_{m0} r) \dots\dots\dots\dots\dots\dots(11),$$

the summation extending to all the admissible values of k_{m0}.

As an example of *forced* vibrations, we may suppose that Z, still constant with respect to space, varies as a harmonic function of the time. This may be taken to represent roughly the circumstances of a small membrane set in vibration by a train of aerial waves. If $Z = \cos qt$, we find, nearly as before,

$$w = \frac{2}{\rho \mu} \cos qt \, \Sigma \frac{J_0(k_{m0} r)}{k_{m0}(q^2 - p_{m0}^2) J_0'(k_{m0} a)} \quad \dots\dots\dots(12).$$

The forced vibration is of course independent of θ. It will be seen that, while none of the symmetrical normal components are missing, their relative importance may vary greatly, especially if there be a near approach in value between q and one of the series of quantities p_{m0}. If the approach be very close, the effect of dissipative forces must be included.

[Again, suppose that the force is applied locally at the centre. By (5)

$$\Phi_{m0} = Z_1 \cos qt \dots\dots\dots\dots\dots\dots(13),$$

if $Z_1 \cos qt$ denote the whole force at time t. From (7)

$$\phi_{m0} = \frac{Z_1 \cos qt}{\rho \pi a^2 (p_{m0}^2 - q^2) J_0'^2 (k_{m0} a)} \quad \dots\dots\dots(14),$$

and w is then given by (11). The series is convergent, unless $r = 0$.

But this problem would be more naturally attacked by including in the solutions of (4) § 200 the second Bessel's function § 341. In this method $k = q/c$; and the ratio of constants by which the two functions of r are multiplied is determined by the boundary condition. When q coincides with one of the values of p, the second function disappears from the solution.]

205. The pitches of the various simple tones and the radii of the nodal circles depend on the roots of the equation

$$J_n(ka) = J_n(z) = 0.$$

If these (exclusive of zero) taken in order of magnitude be called $z_n{}^{(1)}$, $z_n{}^{(2)}$, $z_n{}^{(3)}$ ...... $z_n{}^{(s)}$ ......, then the admissible values of p are to be found by multiplying the quantities $z_n{}^{(s)}$ by c/a. The particular solution may then be written

$$w = J_n\left(z_n{}^{(s)}\frac{r}{a}\right)\{A_n{}^{(s)}\cos n\theta + B_n{}^{(s)}\sin n\theta\}\cos\left\{\frac{c}{a}z_n{}^{(s)}t - \epsilon_n{}^{(s)}\right\}...(1).$$

The lowest tone of the group n corresponds to $z_n{}^{(1)}$; and since in this case $J_n(z_n{}^{(1)}r/a)$ does not vanish for any value of r less than a, there is no interior nodal circle. If we put $s = 2$, J_n will vanish, when

$$z_n{}^{(2)}\frac{r}{a} = z_n{}^{(1)},$$

that is, when

$$r = a\frac{z_n{}^{(1)}}{z_n{}^{(2)}},$$

which is the radius of the one interior nodal circle. Similarly if we take the root $z_n{}^{(s)}$, we obtain a vibration with $s-1$ nodal circles (exclusive of the boundary) whose radii are

$$a\frac{z_n{}^{(1)}}{z_n{}^{(s)}},\ \ a\frac{z_n{}^{(2)}}{z_n{}^{(s)}},\ \ ......\ a\frac{z_n{}^{(s-1)}}{z_n{}^{(s)}}.$$

All the roots of the equation $J_n(ka) = 0$ are *real*. For, if possible, let $ka = \lambda + i\mu$ be a root; then $k'a = \lambda - i\mu$ is also a root, and thus by (14) § 203,

$$4i\lambda\mu\int_0^a J_n(kr)\,J_n(k'r)\,r\,dr = 0.$$

Now $J_n(kr)$, $J_n(k'r)$ are conjugate complex quantities, whose product is necessarily positive; so that the above equation requires that either λ or μ vanish. That λ cannot vanish appears from the consideration that if ka were a pure imaginary, each term of the ascending series for J_n would be positive, and therefore the sum of the series incapable of vanishing. We conclude that $\mu = 0$, or that k is real[1]. The same result might be arrived at from the consideration that only circular functions of the time can enter into the analytical expression for a normal component vibration.

The equation $J_n(z) = 0$ has no equal roots (except zero). From equations (7) and (8) § 200 we get

$$J_n' = \frac{n}{z}J_n - J_{n+1},$$

[1] Riemann, *Partielle Differentialgleichungen*, Braunschweig, 1869, p. 260.

whence we see that if J_n, J_n' vanished for the same value of z, J_{n+1} would also vanish for that value. But in virtue of (8) § 200 this would require that *all* the functions J_n vanish for the value of z in question [1].

206. The actual values of z_n may be found by interpolation from Hansen's tables so far as these extend; or formulæ may be calculated from the descending series by the method of successive approximation, expressing the roots directly. For the important case of the symmetrical vibrations $(n = 0)$, the values of z_0 may be found from the following, given by Stokes [2]:

$$\frac{z_0^{(s)}}{\pi} = s - \cdot 25 + \frac{\cdot 050661}{4s - 1} - \frac{\cdot 053041}{(4s - 1)^3} + \frac{\cdot 262051}{(4s - 1)^5} \ldots \ldots (1).$$

For $n = 1$, the formula is

$$\frac{z_1^{(s)}}{\pi} = s + \cdot 25 - \frac{\cdot 151982}{4s + 1} + \frac{\cdot 015399}{(4s + 1)^3} - \frac{\cdot 245270}{(4s + 1)^5} \ldots \ldots (2).$$

The latter series is convergent enough, even for the first root, corresponding to $s = 1$. The series (1) will suffice for values of s greater than unity; but the first root must be calculated independently. The accompanying table (A) is taken from Stokes' paper, with a slight difference of notation.

It will be seen either from the formulæ, or the table, that the difference of successive roots of high order is approximately π. This is true for all values of n, as is evident from the descending series (10) § 200.

[The general formula, analogous to (1) and (2), for the roots of $J_n(z)$ has been investigated by Prof. McMahon. If $m = 4n^2$, and

$$a = \tfrac{1}{4} \pi (2n - 1 + 4s) \ldots \ldots \ldots \ldots \ldots (3),$$

we have $$z_n^{(s)} = a - \frac{m - 1}{8a} - \frac{4 (m - 1) (7m - 31)}{3 (8a)^3}$$

$$- \frac{32 (m - 1) (83m^2 - 982m + 3779)}{15 (8a)^5} + \ldots \ldots \ldots (4).$$

---

[1] Bourget, "Mémoire sur le mouvement vibratoire des membranes circulaires," *Ann. de l'école normale*, t. III., 1866. In one passage M. Bourget implies that he has proved that no two Bessel's functions of integral order can have the same root, but I cannot find that he has done so. The theorem, however, is probably true; in the case of functions, whose orders differ by 1 or 2, it may be easily proved from the formulæ of § 200.

[2] *Camb. Phil. Trans.* Vol. IX. "On the numerical calculation of a class of definite integrals and infinite series." [In accordance with the calculation of Prof. McMahon the numerator of the last term in (2) has been altered from $\cdot 245835$ to $\cdot 245270$.]

This formula may be applied not only to integral values of n as in (1) and (2), but also when n is fractional. The cases of $n = \frac{1}{2}$, and $n = \frac{3}{2}$ are considered in § 207.]

M. Bourget has given in his memoir very elaborate tables of the frequencies of the different simple tones and of the radii of the nodal circles. Table B includes the values of z, which satisfy $J_n(z)$, for $n = 0, 1, \ldots 5$, $s = 1, 2, \ldots 9$.

TABLE A.

s	$\dfrac{z}{\pi}$ for $J_0(z) = 0$.	Diff.	$\dfrac{z}{\pi}$ for $J_1(z) = 0$.	Diff.
1	·7655	·9916	1·2197	1·0133
2	1·7571	·9975	2·2330	1·0053
3	2·7546	·9988	3·2383	1·0028
4	3·7534	·9993	4·2411	1·0017
5	4·7527	·9995	5·2428	1·0011
6	5·7522	·9997	6·2439	1·0009
7	6·7519	9997	7·2448	1·0006
8	7·7516	9998	8·2454	1·0005
9	8·7514	·9999	9·2459	1·0004
10	9·7513	·9999	10·2463	1·0003
11	10·7512	·9999	11·2466	1·0003
12	11·7511		12·2469	

When n is considerable the calculation of the earlier roots becomes troublesome. For very high values of n, $z_n^{(1)}/n$ approximates to a ratio of equality, as may be seen from the consideration that the pitch of the gravest tone of a very acute sector must tend to coincide with that of a long parallel strip, whose width is equal to the greatest width of the sector.

TABLE B.

s	$n = 0$	$n = 1$	$n = 2$	$n = 3$	$n = 4$	$n = 5$
1	2·404	3·832	5·135	6·379	7·586	8·780
2	5·520	7·016	8·417	9·760	11·064	12·339
3	8·654	10·173	11·620	13·017	14·373	15·700
4	11·792	13·323	14·796	16·224	17·616	18·982
5	14·931	16·470	17·960	19·410	20·827	22·220
6	18·071	19·616	21·117	22·583	24·018	25·431
7	21·212	22·760	24·270	25·749	27·200	28·628
8	24·353	25·903	27·421	28·909	30·371	31·813
9	27·494	29·047	30·571	32·050	33·512	34·983

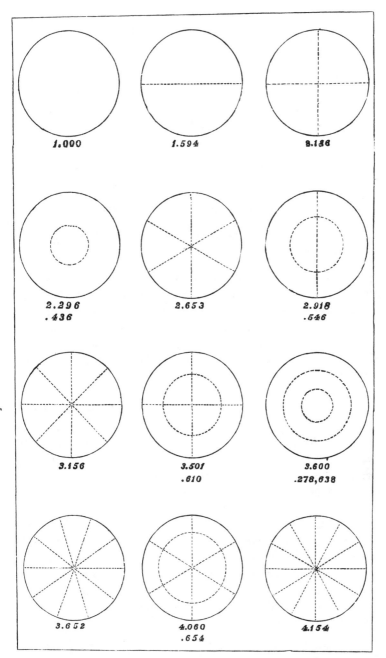

The figures represent the more important normal modes of vibration, and the numbers affixed give the frequency referred to the gravest as unity, together with the radii of the circular nodes expressed as fractions of the radius of the membrane. In the case of six nodal diameters the frequency stated is the result of a rough calculation by myself.

The tones corresponding to the various fundamental modes of the circular membrane do not belong to a harmonic scale, but there are one or two approximately harmonic relations which may be worth notice. Thus

$$\tfrac{4}{3} \times 1{\cdot}594 = 2{\cdot}125 = 2{\cdot}136 \text{ nearly,}$$

$$\tfrac{5}{3} \times 1{\cdot}594 = 2{\cdot}657 = 2{\cdot}653 \text{ nearly,}$$

$$2 \times 1{\cdot}594 = 3{\cdot}188 = 3{\cdot}156 \text{ nearly;}$$

so that the four gravest modes with nodal diameters only would give a consonant chord.

The area of the membrane is divided into segments by the nodal system in such a manner that the sign of the vibration changes whenever a node is crossed. In those modes of vibration which have nodal diameters there is evidently no displacement of the centre of inertia of the membrane. In the case of symmetrical vibrations the displacement of the centre of inertia is proportional to

$$\int_0^a J_0(kr)\, r\, dr = \div \int_0^a \left\{ J_0''(kr) + \frac{1}{kr} J_0'(kr) \right\} r\, dr = -\frac{a}{k} J_0'(ka),$$

an expression which does not vanish for any of the admissible values of k, since $J_0'(z)$ and $J_0(z)$ cannot vanish simultaneously. In all the symmetrical modes there is therefore a displacement of the centre of inertia of the membrane.

207. Hitherto we have supposed the circular area of the membrane to be complete, and the circumference only to be fixed; but it is evident that our theory virtually includes the solution of other problems, for example—some cases of a membrane bounded by two concentric circles. The *complete* theory for a membrane in the form of a ring requires the second Bessel's function.

The problem of the membrane in the form of a semi-circle may be regarded as already solved, since any mode of vibration of which the semi-circle is capable must be applicable to the complete circle also. In order to see this, it is only necessary to attribute to any point in the complementary semi-circle the opposite motion to that which obtains at its optical image in the bounding diameter. This line will then require no constraint to keep it nodal. Similar considerations apply to any sector whose angle is an aliquot part of two right angles.

When the opening of the sector is arbitrary, the problem may be solved in terms of Bessel's functions of fractional order. If the fixed radii are $\theta = 0$, $\theta = \beta$, the particular solution is

$$w = P J_{\nu\pi/\beta}(kr) \, \sin\frac{\nu\pi\theta}{\beta} \cos(pt - \epsilon) \, \ldots\ldots\ldots (1),$$

where ν is an integer. We see that if β be an aliquot part of π, $\nu\pi/\beta$ is integral, and the solution is included among those already used for the complete circle.

An interesting case is when $\beta = 2\pi$, which corresponds to the problem of a complete circle, of which the radius $\theta = 0$ is constrained to be nodal.

Fig. 38.

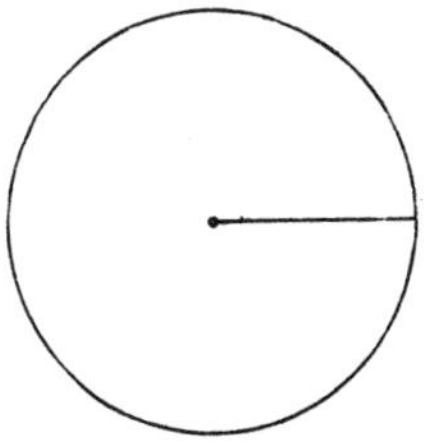

We have

$$w = P J_{\frac{1}{2}\nu}(kr) \, \sin \tfrac{1}{2}\nu\theta \cos(pt - \epsilon).$$

When ν is even, this gives, as might be expected, modes of vibration possible without the constraint; but, when ν is odd, new modes make their appearance. In fact, in the latter case the descending series for J terminates, so that the solution is expressible in finite terms. Thus, when $\nu = 1$,

$$w = P\frac{\sin kr}{\sqrt{(kr)}} \sin \tfrac{1}{2}\theta \, \cos(pt - \epsilon) \, \ldots\ldots\ldots\ldots(2).$$

The values of k are given by

$$\sin ka = 0, \quad \text{or} \quad ka = s\pi.$$

Thus the circular nodes divide the fixed radius into equal parts, and the series of tones form a harmonic scale. In the case of the gravest mode, the whole of the membrane is at any moment deflected on the same side of its equilibrium position. It is remarkable that the application of the constraint to the radius $\theta = 0$ makes the problem easier than before.

Fig. 39.

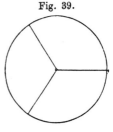

If we take $\nu = 3$, the solution is

$$w = P \frac{1}{\sqrt{(kr)}} \left(\frac{\sin kr}{kr} - \cos kr \right) \sin \tfrac{3}{2}\theta \cos (pt - \epsilon) \dots\dots (3).$$

In this case the nodal radii are Fig. (39)

$$\theta = \tfrac{2}{3}\pi, \quad \theta = \tfrac{4}{3}\pi ;$$

and the possible tones are given by the equation

$$\tan ka = ka \dots\dots\dots\dots (4).$$

To calculate the roots of $\tan x = x$ we may assume

$$x = (s + \tfrac{1}{2})\pi - y = X - y,$$

where y is a positive quantity, which is small when x is large.

Substituting this, we find $\cot y = X - y$, whence

$$y = \frac{1}{X}\left(1 + \frac{y}{X} + \frac{y^2}{X^2} + \dots\right) - \frac{y^3}{3} - \frac{2y^5}{15} - \frac{17y^7}{315} - \dots$$

This equation is to be solved by successive approximation. It will readily be found that

$$y = X^{-1} + \frac{2}{3} X^{-3} + \frac{13}{15} X^{-5} + \frac{146}{105} X^{-7} + \dots,$$

so that the roots of $\tan x = x$ are given by

$$x = X - X^{-1} - \frac{2}{3} X^{-3} - \frac{13}{15} X^{-5} - \frac{146}{105} X^{-7} - \dots\dots (5),$$

where $\qquad X = (s + \tfrac{1}{2})\pi.$

In the first quadrant there is no root after zero since $\tan x > x$, and in the second quadrant there is none because the signs of x and $\tan x$ are opposite. The first root after zero is thus in the third quadrant, corresponding to $s = 1$. Even in this case the series converges sufficiently to give the value of the root with considerable accuracy, while for higher values of s it is all that could be desired. The actual values of x/π are 1·4303, 2·4590, 3·4709, 4·4774, 5·4818, 6·4844, &c.

208. The effect on the periods of a slight inequality in the density of the circular membrane may be investigated by the general method § 90, of which several examples have already been given. It will be sufficient here to consider the case of a small load M attached to the membrane at a point whose radius vector is r'.

We will take first the symmetrical types $(n = 0)$, which may still be supposed to apply notwithstanding the presence of M. The kinetic energy T is (6) § 204 altered from

$$\tfrac{1}{2} \rho \pi a^2 J_0'^2 (k_{m0} a) \, \dot{\phi}_{m0}^2 \text{ to } \tfrac{1}{2} \rho \pi a^2 J_0'^2 (k_{m0} a) \, \dot{\phi}_{m0}^2 + \tfrac{1}{2} M \dot{\phi}_{m0}^2 J_0^2 (k_{m0} r'),$$

and therefore

$$p_{m0}^2 : P_{m0}^2 = 1 - \frac{M}{\rho \pi a^2} \frac{J_0^2 (k_{m0} r')}{J_0'^2 (k_{m0} a)} \dots\dots\dots\dots(1),$$

where P_{m0}^2 denotes the value of p_{m0}^2, when there is no load.

The unsymmetrical normal types are not fully determinate for the unloaded membrane; but for the present purpose they must be taken so as to make the resulting periods a maximum or minimum, that is to say, so that the effect of the load is the greatest and least possible. Now, since a load can never raise the pitch, it is clear that the influence of the load is the least possible, viz. zero, when the type is such that a nodal diameter (it is indifferent which) passes through the point at which the load is attached. The unloaded membrane must be supposed to have two coincident periods, of which *one* is unaltered by the addition of the load. The other type is to be chosen, so that the alteration of period is as great as possible, which will evidently be the case when the radius vector r' bisects the angle between two adjacent nodal diameters. Thus, if r' correspond to $\theta = 0$, we are to take

$$w = \phi_{mn} J_n (k_{mn} r) \cos n\theta;$$

so that (2) § 204

$$T = \tfrac{1}{4} \rho \pi a^2 \dot{\phi}_{mn}^2 J_n'^2 (k_{mn} a) + \tfrac{1}{2} M \dot{\phi}_{mn}^2 J_n^2 (k_{mn} r').$$

The *altered* p_{mn}^2 is therefore given by

$$p_{mn}^2 : P_{mn}^2 = 1 - \frac{2M}{\rho \pi a^2} \frac{J_n^2 (k_{mn} r')}{J_n'^2 (k_{mn} a)} \dots\dots\dots\dots(2).$$

Of course, if r' be such that the load lies on one of the nodal circles, neither period is affected.

For example, let M be at the centre of the membrane. $J_n (0)$ vanishes, except when $n = 0$; and $J_0 (0) = 1$. It is only the symmetrical vibrations whose pitch is influenced by a central load, and for them by (1)

$$p_{m0}^2 : P_{m0}^2 = 1 - \frac{M}{J_0'^2 (k_{m0} a) \, \rho \pi a^2} \dots\dots\dots\dots(3).$$

By (6) § 200 $J_0'(z) = - J_1(z)$,

so that the application of the formula requires only a knowledge of the values of $J_1(z)$, when $J_0(z)$ vanishes, § 200. For the gravest mode the value of $J_0'(k_{m0}a)$ is ·51903[1]. When $k_{m0}a$ is considerable.

$$J_1^2(k_{m0}a) = 2 \div \pi k_{m0}a$$

approximately; so that for the higher components the influence of the load in altering the pitch increases.

The influence of a small irregularity in disturbing the nodal system may be calculated from the formulæ of § 90. The most obvious effect is the breaking up of nodal diameters into curves of hyperbolic form due to the introduction of subsidiary symmetrical vibrations. In many cases the disturbance is favoured by close agreement between some of the natural periods.

209. We will next investigate how the natural vibrations of a uniform membrane are affected by a slight departure from the exact circular form.

Whatever may be the nature of the boundary, w satisfies the equation

$$\frac{d^2w}{dr^2} + \frac{1}{r}\frac{dw}{dr} + \frac{1}{r^2}\frac{d^2w}{d\theta^2} + k^2w = 0 \dots\dots\dots\dots(1),$$

where k is a constant to be determined. By Fourier's theorem w may be expanded in the series

$$w = w_0 + w_1 \cos(\theta + \alpha_1) + w_2 \cos 2(\theta + \alpha_2) + \dots\dots$$
$$+ w_n \cos n(\theta + \alpha_n) + \dots\dots,$$

where w_0, w_1, &c. are functions of r only. Substituting in (1), we see that w_n must satisfy

$$\frac{d^2w_n}{dr^2} + \frac{1}{r}\frac{dw_n}{dr} + \left(k^2 - \frac{n^2}{r^2}\right)w_n = 0,$$

of which the solution is

$$w_n \propto J_n(kr);$$

for, as in § 200, the other function of r cannot appear.

The general expression for w may thus be written

$$w = A_0 J_0(kr) + J_1(kr)(A_1 \cos\theta + B_1 \sin\theta)$$
$$+ \dots + J_n(kr)(A_n \cos n\theta + B_n \sin n\theta) + \dots\dots(2).$$

For all points on the boundary w is to vanish.

---

[1] The succeeding values are approximately ·341, ·271, ·232, ·206, ·187, &c.

In the case of a nearly circular membrane the radius vector is nearly constant. We may take $r = a + \delta r$, δr being a small function of θ. Hence the boundary condition is

$$0 = A_0 [J_0(ka) + k \delta r J_0'(ka)] + \ldots\ldots$$
$$+ [J_n(ka) + k \delta r J_n'(ka)] [A_n \cos n\theta + B_n \sin n\theta]$$
$$+ \ldots\ldots\ldots\ldots\ldots\ldots\ldots\ldots\ldots\ldots\ldots\ldots\ldots\ldots\ldots\ldots\ldots(3),$$

which is to hold good for all values of θ.

Let us consider first those modes of vibration which are nearly symmetrical, for which therefore approximately

$$w = A_0 J_0(kr).$$

All the remaining coefficients are small relatively to A_0, since the type of vibration can only differ a little from what it would be, were the boundary an exact circle. Hence if the squares of the small quantities be omitted, (3) becomes

$$A_0 [J_0(ka) + k \delta r J_0'(ka)] + J_1(ka) [A_1 \cos \theta + B_1 \sin \theta]$$
$$+ \ldots + J_n(ka) [A_n \cos n\theta + B_n \sin n\theta] + \ldots = 0 \ldots\ldots(4).$$

If we integrate this equation with respect to θ between the limits 0 and 2π, we obtain

$$2\pi J_0(ka) + J_0'(ka) \int_0^{2\pi} k \delta r \, d\theta = 0,$$

or

$$J_0 \left\{ ka + k \int_0^{2\pi} \delta r \, \frac{d\theta}{2\pi} \right\} = 0 \ldots\ldots\ldots\ldots\ldots(5),$$

which shews that the pitch of the vibration is approximately the same as if the radius vector had uniformly its *mean value*.

This result allows us to form a rough estimate of the pitch of any membrane whose boundary is not extravagantly elongated. If σ denote the area, so that $\rho\sigma$ is the mass of the whole membrane, the frequency of the gravest tone is approximately

$$(2\pi)^{-1} \times 2 \cdot 404 \times \sqrt{\frac{\pi T_1}{\sigma \rho}} \ldots\ldots\ldots\ldots\ldots(6).[1]$$

In order to investigate the altered type of vibration, we may

---

[1] [A numerical error is here corrected.]

multiply (4) by $\cos n\theta$, or $\sin n\theta$, and then integrate as before. Thus

$$A_0 J_0'(ka) \int_0^{2\pi} k\,\delta r \cos n\theta\, d\theta + \pi A_n J_n(ka) = 0$$
$$A_0 J_0'(ka) \int_0^{2\pi} k\,\delta r \sin n\theta\, d\theta + \pi B_n J_n(ka) = 0$$
$$\Bigg\}\quad\dots\dots(7),$$

which determine the ratios $A_n : A_0$ and $B_n : A_0$.

If
$$\delta r = \delta r_0 + \delta r_1 + \dots + \delta r_n + \dots$$

be Fourier's expansion, the final expression for w may be written,

$$w : A_0 = J_0(kr)$$
$$- k J_0'(ka)\left\{\frac{J_1(kr)\,\delta r_1}{J_1(ka)} + \dots + \frac{J_n(kr)\,\delta r_n}{J_n(ka)} + \dots\right\}\dots\dots(8).$$

When the vibration is not approximately symmetrical, the question becomes more complicated. The normal modes for the truly circular membrane are to some extent indeterminate, but the irregularity in the boundary will, in general, remove the indeterminateness. The position of the nodal diameters must be taken, so that the resulting periods may have maximum or minimum values. Let us, however, suppose that the approximate type is

$$w = A_\nu J_\nu(kr)\cos\nu\theta\dots\dots\dots\dots(9),$$

and afterwards investigate how the initial line must be taken in order that this form may hold good.

All the remaining coefficients being treated as small in comparison with A_ν, we get from (4)

$$A_0 J_0(ka) + \dots + A_\nu[J_\nu(ka) + k\,\delta r J_\nu'(ka)]\cos\nu\theta$$
$$+ B_\nu J_\nu(ka)\sin\nu\theta + \dots\dots$$
$$+ J_n(ka)[A_n\cos n\theta + B_n\cos n\theta] + \dots = 0 \dots\dots (10).$$

Multiplying by $\cos\nu\theta$ and integrating,

$$\pi J_\nu(ka) + k J_\nu'(ka)\int_0^{2\pi}\delta r\cos^2\nu\theta\,d\theta = 0,$$

or

$$J_\nu\left[ka + k\int_0^{2\pi}\delta r\cos^2\nu\theta\,\frac{d\theta}{\pi}\right] = 0,$$

which shews that the effective radius of the membrane is

$$a + \int_0^{2\pi}\delta r\cos^2\nu\theta\,\frac{d\theta}{\pi}\dots\dots\dots\dots(11).$$

The ratios of A_n and B_n to A_ν may be found as before by integrating equation (10) after multiplication by $\cos n\theta$, $\sin n\theta$.

But the point of greatest interest is the pitch. The initial line is to be so taken as to make the expression (11) a maximum or minimum. If we refer to a line fixed in space by putting $\theta - \alpha$ instead of θ, we have to consider the dependence on α of the quantity

$$\int_0^{2\pi} \delta r \cos^2 \nu (\theta - \alpha) \, d\theta,$$

which may also be written

$$\cos^2 \nu\alpha \int_0^{2\pi} \delta r \cos^2 \nu\theta \, d\theta + 2 \cos \nu\alpha \sin \nu\alpha \int_0^{2\pi} \delta r \cos \nu\theta \sin \nu\theta \, d\theta$$

$$+ \sin^2 \nu\alpha \int_0^{2\pi} \delta r \sin^2 \nu\theta \, d\theta \dots\dots\dots\dots(12),$$

and is of the form

$$A \cos^2 \nu\alpha + 2B \cos \nu\alpha \sin \nu\alpha + C \sin^2 \nu\alpha,$$

A, B, C being independent of α. There are accordingly two admissible positions for the nodal diameters, one of which makes the period a maximum, and the other a minimum. The diameters of one set bisect the angles between the diameters of the other set.

There are, however, cases where the normal modes remain indeterminate, which happens when the expression (12) is independent of α. This is the case when δr is constant, or when δr is proportional to $\cos \nu\theta$. For example, if δr were proportional to $\cos 2\theta$, or in other words the boundary were slightly elliptical, the nodal system corresponding to $n = 2$ (that consisting of a pair of perpendicular diameters) would be arbitrary in position, at least to this order of approximation. But the single diameter, corresponding to $n = 1$, must coincide with one of the principal axes of the ellipse, and the periods will be different for the two axes

210. We have seen that the gravest tone of a membrane, whose boundary is approximately circular, is nearly the same as that of a mechanically similar membrane in the form of a circle of the same mean radius or area. If the area of a membrane be given, there must evidently be some form of boundary for which the pitch (of the principal tone) is the gravest possible, and this form can be no other than the circle. In the case of approximate circularity an analytical demonstration may be given, of which the following is an outline.

The general value of w being

$$w = A_0 J_0(kr) + \ldots + J_n(kr)(A_n \cos n\theta + B \sin n\theta) + \ldots \ldots (1),$$

in which for the present purpose the coefficients $A_1, B_1, \ldots$ are small relatively to A_0, we find from the condition that w vanishes when $r = a + \delta r$,

$$A_0 J_0(ka) + kA_0 J_0'(ka) \delta r + \tfrac{1}{2} k^2 A_0 J_0''(ka).(\delta r)^2 + \ldots \ldots$$
$$+ \Sigma \left[\{J_n(ka) + kJ_n'(ka)\delta r + \ldots\}\{A_n \cos n\theta + B_n \sin n\theta\}\right] = 0 \ldots (2).$$

Hence, if

$$\delta r = \alpha_1 \cos\theta + \beta_1 \sin\theta + \ldots + \alpha_n \cos n\theta + \beta_n \sin n\theta + \ldots \ldots (3),$$

we obtain on integration with respect to θ from 0 to 2π,

$$2A_0 J_0 + \tfrac{1}{2} k^2 A_0 J_0'' \sum_{n=1}^{n=\infty} (\alpha_n^2 + \beta_n^2)$$
$$+ k \sum_{n=1}^{n=\infty} \left[(\alpha_n A_n + \beta_n B_n) J_n'\right] = 0 \ldots \ldots \ldots (4),$$

from which we see, as before, that if the squares of the small quantities be neglected, $J_0(ka) = 0$, or that to this order of approximation the mean radius is also the effective radius. In order to obtain a closer approximation we first determine $A_n : A_0$ and $B_n : A_0$ by multiplying (2) by $\cos n\theta$, $\sin n\theta$, and then integrating between the limits 0 and 2π. Thus

$$A_n J_n = -k\alpha_n A_0 J_0', \quad B_n J_n = -k\beta_n A_0 J_0' \ldots \ldots (5).$$

Substituting these values in (4), we get

$$J_0(ka) = \tfrac{1}{2} k^2 \sum_{n=1}^{n=\infty} \left[(\alpha_n^2 + \beta_n^2)\left\{\frac{J_n' J_0'}{J_n} - \tfrac{1}{2} J_0''\right\}\right] \ldots (6).$$

Since J_0 satisfies the fundamental equation

$$J_0'' + \frac{1}{ka} J_0' + J_0 = 0 \ldots \ldots \ldots \ldots (7),$$

and in the present case $J_0 = 0$ approximately, we may replace J_0'' by $-\frac{1}{ka} J_0'$. Equation (6) then becomes

$$J_0(ka) = \tfrac{1}{2} k^2 J_0' \sum_{n=1}^{n=\infty} \left[(\alpha_n^2 + \beta_n^2)\left\{\frac{J_n'}{J_n} + \frac{1}{2ka}\right\}\right] \ldots (8).$$

Let us now suppose that $a + da$ is the equivalent radius of the membrane, so that

$$J_0\left[k\left(a + da\right)\right] = J_0\left(ka\right) + J_0'\left(ka\right)kda = 0.$$

Then by (8) we find

$$da = -\tfrac{1}{2}k\sum_{n=1}^{n=\infty}\left[\left(\alpha_n^2 + \beta_n^2\right)\left\{\frac{J_n'}{J_n} + \frac{1}{2ka}\right\}\right]\ \ldots\ldots\ldots\ (9).$$

Again, if $a + da'$ be the radius of the truly circular membrane of equal area,

$$da' = \frac{1}{4a}\sum_{n=1}^{n=\infty}\left(\alpha_n^2 + \beta_n^2\right)\ldots\ldots\ldots\ldots\ldots(10);$$

so that

$$da' - da = \frac{1}{2a}\sum_{n=1}^{n=\infty}\left[\left(\alpha_n^2 + \beta_n^2\right)\left\{1 + ka\,\frac{J_n'\left(ka\right)}{J_n\left(ka\right)}\right\}\right]\ldots\ldots(11).$$

The question is now as to the sign of the right-hand member. If $n = 1$, and z be written for ka,

$$1 + z\frac{J_n'\left(z\right)}{J_n\left(z\right)}$$

vanishes approximately by (7), since in general $J_1 = -J_0'$, and in the present case $J_0\left(z\right) = 0$ nearly. Thus $da' - da = 0$, as should evidently be the case, since the term in question represents merely a displacement of the circle without an alteration in the form of the boundary. When $n = 2$, (8) § 200,

$$J_2 = \frac{2}{z}J_1 - J_0,$$

from which and (7) we find that, when $J_0 = 0$,

$$\frac{J_2'}{J_2} = \frac{z^2 - 4}{2z}\ \ldots\ldots\ldots\ldots\ldots\ldots\ldots\ (12),$$

whence

$$da' - da = \frac{1}{2a}\left(\alpha_2^2 + \beta_2^2\right)\left(\frac{z^2}{2} - 1\right)\ \ldots\ldots\ldots\ldots\ (13),$$

which is positive, since $z = 2\cdot404$.

We have still to prove that

$$1 + z\frac{J_n'\left(z\right)}{J_n\left(z\right)}$$

is positive for integral values of n greater than 2, when $z = 2\cdot404$.

For this purpose we may avail ourselves of a theorem given in Riemann's *Partielle Differentialgleichungen*, to the effect that neither J_n nor J_n' has a root (other than zero) less than n. The differential equation for J_n may be put into the form

$$\frac{d^2 J_n(z)}{d(\log z)^2} + (z^2 - n^2) J_n(z) = 0 \, ;$$

while initially J_n and J_n' (as well as $dJ_n/d\log z$) are positive. Accordingly $dJ_n/d\log z$ begins by increasing and does not cease to do so before $z = n$, from which it is clear that within the range $z = 0$ to $z = n$, neither J_n nor J_n' can vanish. And since J_n and J_n' are both positive until $z = n$, it follows that, when n is an integer greater than 2·404, $da' - da$ is positive. We conclude that, unless α_2, β_2, α_3, ... all vanish, da' is greater than da, which shews that in the case of any membrane of approximately circular outline, the circle of equal area exceeds the circle of equal pitch.

We have seen that a good estimate of the pitch of an approximately circular membrane may be obtained from its area alone, but by means of equation (9) a still closer approximation may be effected. We will apply this method to the case of an ellipse, whose semi-axis major is R and eccentricity e.

The polar equation of the boundary is

$$r = R \left\{ 1 - \tfrac{1}{4} e^2 - \tfrac{7}{64} e^4 + \ldots\ldots + \tfrac{1}{4} e^2 \cos 2\theta + \ldots\ldots \right\} \ldots\ldots (14) \, ;$$

so that in the notation of this section

$$a = R \left(1 - \tfrac{1}{4} e^2 - \tfrac{7}{64} e^4 \right), \quad \alpha_2 = \tfrac{1}{4} e^2 R.$$

Accordingly by (9)

$$da = -\frac{e^4 R}{32} \cdot kR \cdot \left\{ \frac{J_2'(z)}{J_2(z)} + \frac{1}{2z} \right\},$$

or by (12), since $kR = z = 2\cdot404$,

$$da = -\frac{2\cdot779}{64} e^4 R.$$

Thus the radius of the circle of equal pitch is

$$a + da = R \left\{ 1 - \frac{1}{4} e^2 - \frac{9\cdot779 \, e^4}{64} \right\} \ldots\ldots\ldots\ldots (15)$$

in which the term containing e^4 should be correct.

The result may also be expressed in terms of e and the area σ. We have

$$R = \sqrt{\frac{\sigma}{\pi}}\left(1 + \frac{1}{4}e^2 + \frac{5}{32}e^4\right),$$

and thus

$$a + da = \sqrt{\frac{\sigma}{\pi}}\left(1 - \frac{3\cdot779}{64}e^4\right)\dots\dots\dots\dots(16),$$

from which we see how small is the influence of a moderate eccentricity, when the area is given.

211. When the fixed boundary of a membrane is neither straight nor circular, the problem of determining its vibrations presents difficulties which in general could not be overcome without the introduction of functions not hitherto discussed or tabulated. A partial exception must be made in favour of an elliptic boundary; but for the purposes of this treatise the importance of the problem is scarcely sufficient to warrant the introduction of complicated analysis. The reader is therefore referred to the original investigation of M. Mathieu[1].

[The method depends upon the use of conjugate functions. If

$$x + iy = e\cos(\xi + i\eta)\dots\dots\dots\dots\dots(1),$$

then the curves $\eta = \text{const.}$ are confocal ellipses, and $\xi = \text{const.}$ are confocal hyperbolas. In terms of ξ, η the fundamental equation $(\nabla^2 + k^2)u = 0$ becomes

$$\frac{d^2u}{d\xi^2} + \frac{d^2u}{d\eta^2} + k'^2(\cosh^2\eta - \cos^2\xi)u = 0\dots\dots\dots(2),$$

where $k' = ke$.

The solution of (2) may be found in the form

$$u = \Xi(\xi)\,.\,H(\eta)\dots\dots\dots\dots\dots(3),$$

in which Ξ is a function of ξ only, and H a function of η only, provided

$$\frac{d^2\Xi}{d\xi^2} - (k'^2\cos^2\xi - a)\Xi = 0\dots\dots\dots\dots(4),$$

$$\frac{d^2H}{d\eta^2} + (k'^2\cosh^2\eta - a)H = 0\dots\dots\dots\dots(5),$$

a being an arbitrary constant[2].

---

[1] Liouville, XIII., 1868; *Cours de physique mathématique*, 1873, p. 122.
[2] Pockels, *Über die partielle Differentialgleichung* $\Delta u + k^2 u = 0$, p. 114.

Michell[1] has shewn that the elliptic transformation (1) is the only one which yields an equation capable of satisfaction in the form (3).]

Soluble cases may be invented by means of the general solution

$$w = A_0 J_0(kr) + \ldots + (A_n \cos n\theta + B_n \sin n\theta) J_n(kr) + \ldots\ldots$$

For example we might take

$$w = J_0(kr) - \lambda J_1(kr) \cos \theta,$$

and attaching different values to λ, trace the various forms of boundary to which the solution will then apply.

Useful information may sometimes be obtained from the theorem of § 88, which allows us to prove that any contraction of the fixed boundary of a vibrating membrane must cause an elevation of pitch, because the new state of things may be conceived to differ from the old merely by the introduction of an additional constraint. Springs, without inertia, are supposed to urge the line of the proposed boundary towards its equilibrium position, and gradually to become stiffer. At each step the vibrations become more rapid, until they approach a limit, corresponding to infinite stiffness of the springs and absolute fixity of their points of application. It is not necessary that the part cut off should have the same density as the rest, or even any density at all.

For instance, the pitch of a regular polygon is intermediate between those of the inscribed and circumscribed circles. Closer limits would however be obtained by substituting for the circumscribed circle that of equal area according to the result of § 210. In the case of the hexagon, the ratio of the radius of the circle of equal area to that of the circle inscribed is 1·050, so that the mean of the two limits cannot differ from the truth by so much as $2\frac{1}{2}$ per cent. In the same way we might conclude that the sector of a circle of 60° is a graver form than the equilateral triangle obtained by substituting the chord for the arc of the circle.

The following table giving the relative frequency in certain calculable cases for the gravest tone of membranes under similar mechanical conditions and of *equal area* (σ), shews the effect of a greater or less departure from the circular form.

[1] *Messenger of Mathematics*, vol. XIX. p. 86, 1890.

Circle..	$2\cdot404 . \sqrt{\pi} = 4\cdot261.$
Square...	$\sqrt{2} . \pi = 4\cdot443.$
Quadrant of a circle.......................	$\dfrac{5\cdot135}{2} . \sqrt{\pi} = 4\cdot551.$
Sector of a circle 60°.......................	$6\cdot379 \sqrt{\dfrac{\pi}{6}} = 4\cdot616.$
Rectangle 3 × 2..............................	$\sqrt{\dfrac{13}{6}} . \pi = 4\cdot624.$
Equilateral triangle........................	$2\pi . \sqrt{\tan 30°} = 4\cdot774.$
Semicircle....................................	$3\cdot832 \sqrt{\dfrac{\pi}{2}} = 4\cdot803.$
Rectangle 2 × 1.............................. Right-angled isosceles triangle............	$\pi \sqrt{\dfrac{5}{2}} = 4\cdot967.$
Rectangle 3 × 1..............................	$\pi \sqrt{\dfrac{10}{3}} = 5\cdot736.$

For instance, if a square and a circle have the same area, the former is the more acute in the ratio 4·443 : 4·261, or 1·043 : 1.

For the circle the absolute frequency is

$$(2\pi)^{-1} \times 2\cdot404\, c \sqrt{\frac{\pi}{\sigma}}, \quad \text{where} \quad c = \sqrt{T_1} \div \sqrt{\rho}.$$

In the case of similar forms the frequency is inversely as the linear dimension.

[From the principle that an extension of boundary is always accompanied by a fall of pitch, we may infer that the gravest mode of a membrane of any shape, and of any variable density, is devoid of internal nodal lines.]

212. The theory of the free vibrations of a membrane was first successfully considered by Poisson[1]. His theory in the case of the rectangle left little to be desired, but his treatment of the circular membrane was restricted to the symmetrical vibrations. Kirchhoff's solution of the similar, but much more difficult, problem of the circular plate was published in 1850, and Clebsch's *Theory of Elasticity* (1862) gives the general theory of the circular membrane including the effects of stiffness and

---

[1] *Mém. de l'Académie*, t. VIII. 1829.

of rotatory inertia[1]. It will therefore be seen that there was not much left to be done in 1866; nevertheless the memoir of Bourget already referred to contains a useful discussion of the problem accompanied by very complete numerical results, the whole of which however were not new.

213. In his experimental investigations M. Bourget made use of various materials, of which paper proved to be as good as any. The paper is immersed in water, and after removal of the superfluous moisture by blotting-paper is placed upon a frame of wood whose edges have been previously coated with glue. The contraction of the paper in drying produces the necessary tension, but many failures may be met with before a satisfactory result is obtained. Even a well stretched membrane requires considerable precautions in use, being liable to great variations in pitch in consequence of the varying moisture of the atmosphere. The vibrations are excited by organ-pipes, of which it is necessary to have a series proceeding by small intervals of pitch, and they are made evident to the eye by means of a little sand scattered on the membrane. If the vibration be sufficiently vigorous, the sand accumulates on the nodal lines, whose form is thus defined with more or less precision. Any inequality in the tension shews itself by the circles becoming elliptic.

The principal results of experiment are the following :—

A circular membrane cannot vibrate in unison with every sound. It can only place itself in unison with sounds more acute than that heard when the membrane is gently tapped.

As theory indicates, these possible sounds are separated by less and less intervals, the higher they become.

The nodal lines are only formed distinctly in response to certain definite sounds. A little above or below confusion ensues, and when the pitch of the pipe is decidedly altered, the membrane remains unmoved. There is not, as Savart supposed, a continuous transition from one system of nodal lines to another.

The nodal lines are circles or diameters or combinations of circles and diameters, as theory requires. However, when the

---

[1] [The reader who wishes to pursue the subject from a mathematical point of view is referred to an excellent discussion by Pockels (Leipzig, 1891) of the differential equation $\nabla^2 u + k^2 u = 0$.]

number of diameters exceeds two, the sand tends to heap itself confusedly towards the middle of the membrane, and the nodes are not well defined.

The same general laws were verified by MM. Bernard and Bourget in the case of square membranes[1]; and these authors consider that the results of theory are decisively established in opposition to the views of Savart, who held that a membrane was capable of responding to any sound, no matter what its pitch might be. But I must here remark that the distinction between forced and free vibrations does not seem to have been sufficiently borne in mind. When a membrane is set in motion by aerial waves having their origin in an organ-pipe, the vibration is properly speaking *forced*. Theory asserts, not that the membrane is only capable of vibrating with certain defined frequencies, but that it is only capable of so vibrating *freely*. When however the period of the force is not approximately equal to one of the natural periods, the resulting vibration may be insensible.

In Savart's experiments the sound of the pipe was two or three octaves higher than the gravest tone of the membrane, and was accordingly never far from unison with one of the series of overtones. MM. Bourget and Bernard made the experiment under more favourable conditions. When they sounded a pipe somewhat lower in pitch than the gravest tone of the membrane, the sand remained at rest, but was thrown into vehement vibration as unison was approached. So soon as the pipe was decidedly higher than the membrane, the sand returned again to rest. A modification of the experiment was made by first tuning a pipe about a third higher than the membrane when in its natural condition. The membrane was then heated until its tension had increased sufficiently to bring the pitch above that of the pipe. During the process of cooling the pitch gradually fell, and the point of coincidence manifested itself by the violent motion of the sand, which at the beginning and end of the experiment was sensibly at rest.

M. Bourget found a good agreement between theory and observation with respect to the radii of the circular nodes, though the test was not very precise, in consequence of the sensible width of the bands of sand; but the relative pitch of the various simple tones deviated considerably from the theoretical estimates. The

[1] *Ann. de Chim.* LX. 449—479. 1860.

committee of the French Academy appointed to report on M. Bourget's memoir suggest as the explanation the want of perfect fixity of the boundary. It should also be remembered that the theory proceeds on the supposition of perfect flexibility—a condition of things not at all closely approached by an ordinary membrane stretched with a comparatively small force. But perhaps the most important disturbing cause is the resistance of the air, which acts with much greater force on a membrane than on a string or bar in consequence of the large surface exposed. The gravest mode of vibration, during which the displacement is at all points in the same direction, might ·be affected very differently from the higher modes, which would not require so great a transference of air from one side to the other.

[In the case of kettle-drums the matter is further complicated by the action of the shell, which limits the motion of the air upon one side of the membrane. From the fact that kettle-drums are struck, not in the centre, but at a point about midway between the centre and edge, we may infer that the vibrations which it is desired to excite are not of the symmetrical class. The sound is indeed but little affected when the central point is touched with the finger. Under these circumstances the principal vibration (1) is that with one nodal diameter and no nodal circle, and to this corresponds the greater part of the sound obtained in the normal use of the instrument. Other tones, however, are audible, which correspond with vibrations characterized (2) by two nodal diameters and no nodal circle, (3) by three nodal diameters and no nodal circles, (4) by one nodal diameter and one nodal circle. By observation with resonators upon a large kettle-drum of 25 inches diameter the pitch of (2) was found to be about a fifth above (1), that of (3) about a major seventh above (1), and that of (4) a little higher again, forming an imperfect octave with the principal tone. For the corresponding modes of a uniform perfectly flexible membrane vibrating *in vacuo*, the theoretical intervals are those represented by the ratios 1·34, 1·66, 1·83 respectively[1].

The vibrations of soap films have been investigated by Melde[2]. The frequencies for surfaces of equal area in the form of the circle, the square and the equilateral triangle, were found to be as

---

[1] *Phil. Mag.*, vol. vii., p. 160, 1879.

[2] *Pogg. Ann.*, 159, p. 275, 1876. *Akustik*, p. 131, 1883.

1·000 : 1·049 : 1·175. In membranes of this kind the tension is due to capillarity, and is independent of the thickness of the film.]

213 *a.* The forced vibrations of square and circular membranes have been further experimentally studied by Elsas[1], who has confirmed the conclusions of Savart as to the responsiveness of a membrane to sounds of arbitrary pitch. In these experiments the vibrations of a fork were communicated to the membrane by means of a light thread, attached normally at the centre ; and the position of the nodal curves and of the maxima of disturbance was traced in the usual manner by sand and lycopodium. A series of figures accompanies the memoir, shewing the effect of sounds of progressively rising pitch.

In many instances the curves found do not exhibit the symmetries demanded by the supposed conditions. Thus in the case of the square membrane all the curves should be similarly related to the four corners, and in the case of the circular membrane all the curves should be circles. The explanation is probably to be sought in the difficulty of attaining equality of tension. If there be any irregularity, the effect will be to introduce modes of vibration which should not appear, as having nodes at the point of excitation, and this especially when there is a near agreement of periods. Or again, an irregularity may operate to disturb the balance between two modes of theoretically identical pitch, which should be excited to the same degree. Indeed the passage through such a point of isochronism may be expected to be highly unstable in the absence of moderate dissipative forces.

The theoretical solution of these questions has already (§§ 196, 204) been given, but would need much further development for an accurate determination of the nodal curves relating to periods not included among the natural periods. But the general course of the phenomenon can be traced without difficulty.

If the imposed frequency be less than the lowest natural frequency, the vibration is devoid of (internal) nodes. For a nodal line, if it existed, being of necessity either endless or terminated at the boundary[2], would divide the membrane into two parts. Of

---

[1] *Nova Acta der Ksl. Leop. Carol. Deutschen Akademie*, Bd. xlv. Nr. 1. Halle, 1882.

[2] Otherwise the extremity would have to remain at rest under the action of component tensions from the surrounding parts which are all in one direction.

these one part would be vibrating freely with a frequency less than the lowest natural to the whole membrane, an impossible condition of things (§ 211). The absence of nodal curves under the above-mentioned conditions is one of the conclusions drawn by Elsas from his observations.

As the frequency of the imposed vibration rises through the lowest natural frequency, a nodal curve manifests itself round the point of excitation, and gradually extends. The course of things is most easily followed in the case of the circular membrane, excited at the centre. The nodal curves are then of necessity also circles, and it is evident that the first appearance of a nodal circle can take place only at the centre. Otherwise there would be a circular annulus of finite internal diameter, vibrating freely with a frequency only infinitesimally higher than that of the entire circle. At first sight indeed it might appear that even an infinitely small nodal circle would entail a finite elevation of pitch, but a consideration of the solution (§ 204) as expressed by a combination of Bessel's functions of the first and second kinds, shews that this is not the case. At the point of isochronism the second function disappears, and immediately afterwards re-enters with an infinitely small coefficient. But inasmuch as this function is itself infinite when $r = 0$, a nodal circle of vanishing radius is possible. Accordingly the fixation of the centre of a vibrating circular membrane does not alter the pitch, a conclusion which may be extended to the fixation of any number of detached points of a membrane of any shape.

The effect of gradually increasing frequency upon the nodal system of a circular membrane may be thus summarized. Below the first proper tone there is no internal node. As this point is reached, the mode of vibration identifies itself with the corresponding free mode, and then an infinitely small nodal circle manifests itself. As the frequency further increases, this circle expands, until when the second proper tone is reached, it coincides with the nodal circle of the free vibration of this frequency. Another infinitely small circle now appears, and it, as well as the first, continually expands, until they coincide with the nodal system of a free vibration in the third proper tone. This process continues as the pitch rises, every circle moving continually outwards. At each coincidence with a natural frequency the nodal system identifies itself with that of the free vibration, and a new circle begins to form itself at the centre.

The behaviour of a square membrane is of course more difficult
to follow in detail. The transition from Fig. (34) case (4), corresponding to $m = 3$, $n = 1$, and $m = 1$, $n = 3$, to Fig. (36) where $m = 3$,
$n = 3$, can be traced in Elsas's curves through such forms as

Fig. 39 a.

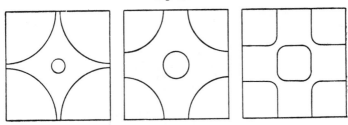

CHAPTER X.

VIBRATIONS OF PLATES[1].

214. In order to form according to Green's method the equations of equilibrium and motion for a thin solid plate of uniform isotropic material and constant thickness, we require the expression for the potential energy of bending. It is easy to see that for each unit of area the potential energy V is a positive homogeneous symmetrical quadratic function of the two principal curvatures. Thus, if ρ_1, ρ_2 be the principal radii of curvature, the expression for V will be

$$A \left(\frac{1}{\rho_1^2} + \frac{1}{\rho_2^2} + \frac{2\mu}{\rho_1 \rho_2} \right) \dots\dots\dots\dots\dots (1),$$

where A and μ are constants, of which A must be positive, and μ must be numerically less than unity. Moreover if the material be of such a character that it undergoes no lateral contraction when a bar is pulled out, the constant μ must vanish. This amount of information is almost all that is required for our purpose, and we may therefore content ourselves with a mere statement of the relations of the constants in (1) with those by means of which the elastic properties of bodies are usually defined.

From Thomson and Tait's *Natural Philosophy*, §§ 639, 642, 720, it appears that, if $2h$ be the thickness, q Young's modulus,

---

[1] [This Chapter deals only with *flexural* vibrations. The extensional vibrations of an infinite plane plate are briefly considered in Chapter X. A, as a particular case of those of an infinite cylindrical shell. They are not of much acoustical importance.]

and μ the ratio of lateral contraction to longitudinal elongation when a bar is pulled out, the expression for V is

$$V = \frac{qh^3}{3(1-\mu^2)} \left\{ \frac{1}{\rho_1^2} + \frac{1}{\rho_2^2} + \frac{2\mu}{\rho_1 \rho_2} \right\}$$

$$= \frac{qh^3}{3(1-\mu^2)} \left\{ \left(\frac{1}{\rho_1} + \frac{1}{\rho_2} \right)^2 - \frac{2(1-\mu)}{\rho_1 \rho_2} \right\} \dots\dots\dots (2)^1.$$

[Equation (2) gives the interpretation of the constants of (1) in its application to a homogeneous plate of isotropic material; but the expression (1) itself is of far wider scope. The material composing the plate may vary from layer to layer, and the elastic character of any layer need not be isotropic, but only symmetrical with respect to the normal. As a particular case, the middle layer, or indeed any other layer, may be supposed to be *physically* inextensible.

Similar remarks apply to the investigations of the following chapter relating to curved shells.]

If w be the small displacement perpendicular to the plane of the plate at the point whose rectangular coordinates in the plane of the plate are x, y,

$$\frac{1}{\rho_1} + \frac{1}{\rho_2} = \nabla^2 w, \qquad \frac{1}{\rho_1 \rho_2} = \frac{d^2w}{dx^2} \frac{d^2w}{dy^2} - \left(\frac{d^2w}{dx\,dy} \right)^2,$$

and thus for a unit of area, we have

$$V = \frac{qh^3}{3(1-\mu^2)} \left[(\nabla^2 w)^2 - 2(1-\mu) \left\{ \frac{d^2w}{dx^2} \frac{d^2w}{dy^2} - \left(\frac{d^2w}{dx\,dy} \right)^2 \right\} \right] \quad (3),$$

which quantity has to be integrated over the surface (S) of the plate.

---

[1] The following comparison of the notations used by the principal writers may save trouble to those who wish to consult the original memoirs.

Rigidity $= n$ (Thomson) $= \mu$ (Lamé).

Young's modulus $= E$ (Clebsch) $= M$ (Thomson) $= \dfrac{9nk}{3k+n}$ (Thomson)

$= \dfrac{n(3m-n)}{m}$ (Thomson) $= q$ (Kirchhoff and Donkin) $= 2K \dfrac{1+3\theta}{1+\theta}$ (Kirchhoff).

Ratio of lateral contraction to longitudinal elongation $= \mu$ (Clebsch and Donkin)

$= \sigma$ (Thomson) $= \dfrac{m-n}{2m}$ (Thomson) $= \dfrac{\theta}{1+2\theta}$ (Kirchhoff) $= \dfrac{\lambda}{2\lambda+2\mu}$ (Lamé).

Poisson assumed this ratio to be $\frac{1}{4}$, and Wertheim $\frac{1}{3}$.

215. We proceed to find the variation of V, but it should be previously noticed that the second term in V, namely $\iint \dfrac{dS}{\rho_1 \rho_2}$, represents the *total curvature* of the plate, and is therefore dependent only on the state of things at the edge.

$$\delta V = \frac{2qh^3}{3(1-\mu^2)} \iint \left\{ \nabla^2 w \cdot \nabla^2 \delta w - (1-\mu)\, \delta \frac{1}{\rho_1 \rho_2} \right\} dS \ \ldots\ldots (1);$$

so that we have to consider the two variations

$$\iint \nabla^2 w \cdot \nabla^2 \delta w \cdot dS \quad \text{and} \quad \iint \delta (\rho_1 \rho_2)^{-1}\, dS.$$

Now by Green's theorem

$$\iint \nabla^2 w \cdot \nabla^2 \delta w \cdot dS = \iint \nabla^4 w \cdot \delta w \cdot dS$$

$$- \int \frac{d\nabla^2 w}{dn} \cdot \delta w \cdot ds + \int \nabla^2 w \frac{d\delta w}{dn}\, ds \ \ldots\ldots\ldots (2),$$

in which ds denotes an element of the boundary, and d/dn denotes differentiation with respect to the normal of the boundary drawn outwards.

The transformation of the second part is more difficult. We have

$$\delta \iint \frac{dS}{\rho_1 \rho_2} = \iint \left\{ \frac{d^2 w}{dx^2} \frac{d^2 \delta w}{dy^2} + \frac{d^2 w}{dy^2} \frac{d^2 \delta w}{dx^2} - 2 \frac{d^2 w}{dx\,dy} \frac{d^2 \delta w}{dx\,dy} \right\} dS.$$

The quantity under the sign of integration may be put into the form

$$\frac{d}{dy} \left(\frac{d\delta w}{dy} \frac{d^2 w}{dx^2} - \frac{d\delta w}{dx} \frac{d^2 w}{dx\,dy} \right) + \frac{d}{dx} \left(\frac{d\delta w}{dx} \frac{d^2 w}{dy^2} - \frac{d\delta w}{dy} \frac{d^2 w}{dx\,dy} \right).$$

Now, if F be any function of x and y,

$$\left. \begin{aligned} \iint \frac{dF}{dy}\, dx\,dy &= \int F \sin \theta\, ds \\ \iint \frac{dF}{dx}\, dx\,dy &= \int F \cos \theta\, ds \end{aligned} \right\} \ \ldots\ldots\ldots\ldots (3),$$

where θ is the angle between x and the normal drawn outwards, and the integration on the right-hand side extends round the boundary. Using these, we find

$$\delta \iint \frac{dS}{\rho_1 \rho_2} = \int ds \sin \theta \left\{ \frac{d\delta w}{dy} \frac{d^2 w}{dx^2} - \frac{d\delta w}{dx} \frac{d^2 w}{dx\,dy} \right\}$$

$$+ \int ds \cos \theta \left\{ \frac{d\delta w}{dx} \frac{d^2 w}{dy^2} - \frac{d\delta w}{dy} \frac{d^2 w}{dx\,dy} \right\}$$

If we substitute for $d\delta w/dx$, $d\delta w/dy$ their values in terms $d\delta w/dn$, $d\delta w/ds$, from the equations (see Fig. 40)

$$\left.\begin{aligned}
\frac{d\delta w}{dx} &= \frac{d\delta w}{dn}\cos\theta - \frac{d\delta w}{ds}\sin\theta \\
\frac{d\delta w}{dy} &= \frac{d\delta w}{dn}\sin\theta + \frac{d\delta w}{ds}\cos\theta
\end{aligned}\right\}\dots\dots\dots\dots(4).$$

Fig. 40.

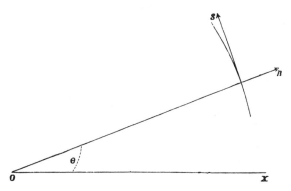

we obtain

$$\delta\iint\frac{dS}{\rho_1\rho_2} = \int ds\,\frac{d\delta w}{dn}\left\{\sin^2\theta\,\frac{d^2w}{dx^2} + \cos^2\theta\,\frac{d^2w}{dy^2} - 2\sin\theta\cos\theta\,\frac{d^2w}{dx\,dy}\right\}$$

$$+ \int ds\,\frac{d\delta w}{ds}\left\{\cos\theta\sin\theta\left(\frac{d^2w}{dx^2} - \frac{d^2w}{dy^2}\right) + (\sin^2\theta - \cos^2\theta)\,\frac{d^2w}{dx\,dy}\right\}\dots(5).$$

The second integral by a partial integration with respect to s may be put into the form

$$\int\delta w\,\frac{d}{ds}\left\{\cos\theta\sin\theta\left(\frac{d^2w}{dy^2} - \frac{d^2w}{dx^2}\right) + (\cos^2\theta - \sin^2\theta)\,\frac{d^2w}{dx\,dy}\right\}ds.$$

Collecting and rearranging our results, we find

$$\delta V = \frac{2qh^3}{3(1-\mu^2)}\left[\iint\nabla^4 w\,\delta w\,dS\right.$$

$$- \int\delta w\,ds\left\{\frac{d\nabla^2 w}{dn} + (1-\mu)\frac{d}{ds}\left(\cos\theta\sin\theta\left(\frac{d^2w}{dy^2} - \frac{d^2w}{dx^2}\right)\right.\right.$$

$$\left.\left. + (\cos^2\theta - \sin^2\theta)\,\frac{d^2w}{dx\,dy}\right)\right\}$$

$$+ \int\frac{d\delta w}{dn}\,ds\left\{\mu\nabla^2 w + (1-\mu)\left(\cos^2\theta\,\frac{d^2w}{dx^2} + \sin^2\theta\,\frac{d^2w}{dy^2}\right.\right.$$

$$\left.\left.\left. + 2\cos\theta\sin\theta\,\frac{d^2w}{dx\,dy}\right)\right\}\right]\dots(6).$$

There will now be no difficulty in forming the equations of motion. If ρ be the volume density, and $Z.\rho.2h.dS$ the transverse force acting on the element dS,

$$\delta V - \iint 2Z\rho h\, \delta w\, dS + \iint 2\rho h \ddot{w}\, \delta w\, dS = 0 \dots\dots(7)^1$$

is the general variational equation, which must be true whatever function (consistent with the constitution of the system) δw may be supposed to be. Hence by the principles of the Calculus of Variations

$$\frac{qh^2}{3\rho(1-\mu^2)} \nabla^4 w - Z + \ddot{w} = 0 \dots\dots\dots\dots(8),$$

at every point of the plate.

If the edges of the plate be free, there is no restriction on the hypothetical boundary values of δw and $d\delta w/dn$, and therefore the coefficients of these quantities in the expression for δV must vanish. The conditions to be satisfied at a free edge are thus

$$\left. \begin{aligned} \frac{d\nabla^2 w}{dn} + (1-\mu)\frac{d}{ds}&\left\{\cos\theta\sin\theta\left(\frac{d^2 w}{dy^2}-\frac{d^2 w}{dx^2}\right)\right. \\ &\left. + (\cos^2\theta - \sin^2\theta)\frac{d^2 w}{dx\,dy}\right\} = 0 \\ \mu\nabla^2 w + (1-\mu)&\left\{\cos^2\theta\frac{d^2 w}{dx^2} + \sin^2\theta\frac{d^2 w}{dy^2}\right. \\ &\left. + 2\cos\theta\sin\theta\frac{d^2 w}{dx\,dy}\right\} = 0 \end{aligned} \right\} \dots\dots(9).$$

If the whole circumference of the plate be clamped, $\delta w = 0$, $d\delta w/dn = 0$, and the satisfaction of the boundary conditions is already secured. If the edge be 'supported'^{2}, $\delta w = 0$, but $d\delta w/dn$ is arbitrary. The second of the equations (9) must in this case be satisfied by w.

216. The boundary equations may be simplified by getting rid of the extrinsic element involved in the use of Cartesian coordinates. Taking the axis of x parallel to the normal of the bounding curve, we see that we may write

$$\cos^2\theta\frac{d^2 w}{dx^2} + \sin^2\theta\frac{d^2 w}{dy^2} + 2\cos\theta\sin\theta\frac{d^2 w}{dx\,dy} = \frac{d^2 w}{dn^2}.$$

Also
$$\nabla^2 w = \frac{d^2 w}{dn^2} + \frac{d^2 w}{d\sigma^2} \dots\dots\dots\dots\dots(1),$$

---

^{1} The rotatory inertia is here neglected. ^{2} Compare § 162.

where σ is a fixed axis coinciding with the tangent at the point under consideration. In general $d^2w/d\sigma^2$ differs from d^2w/ds^2. To obtain the relation between them, we may proceed thus. Expand w by Maclaurin's theorem in ascending powers of the small quantities n and σ, and substitute for n and σ their values in terms of s, the arc of the curve.

Thus in general

$$w = w_0 + \frac{dw}{dn_0}n + \frac{dw}{d\sigma_0}\sigma + \tfrac{1}{2}\frac{d^2w}{dn_0^2}n^2 + \frac{d^2w}{dn_0\,d\sigma_0}n\sigma + \tfrac{1}{2}\frac{d^2w}{d\sigma_0^2}\sigma^2 + \dots,$$

while on the curve $\sigma = s + \text{cubes}$, $n = -\tfrac{1}{2}s^2/\rho + \dots$, where ρ is the radius of curvature. Accordingly for points on the curve,

$$w = w_0 - \tfrac{1}{2}\frac{dw}{dn_0}\frac{s^2}{\rho} + \frac{dw}{d\sigma_0}s + \tfrac{1}{2}\frac{d^2w}{d\sigma_0^2}s^2 + \text{cubes of } s,$$

and therefore
$$\frac{d^2w}{ds^2} = \frac{d^2w}{d\sigma^2} - \frac{1}{\rho}\frac{dw}{dn} \quad\dots\dots\dots\dots\dots\dots(2);$$

whence from (1)
$$\nabla^2 w = \frac{d^2w}{dn^2} + \frac{1}{\rho}\frac{dw}{dn} + \frac{d^2w}{ds^2} \quad\dots\dots\dots\dots(3).$$

We conclude that the second boundary condition in (9) § 215 may be put into the form

$$\frac{d^2w}{dn^2} + \mu\left(\frac{1}{\rho}\frac{dw}{dn} + \frac{d^2w}{ds^2}\right) = 0 \quad\dots\dots\dots\dots(4).$$

In the same way by putting $\theta = 0$, we see that

$$\cos\theta\sin\theta\left(\frac{d^2w}{dy^2} - \frac{d^2w}{dx^2}\right) + (\cos^2\theta - \sin^2\theta)\frac{d^2w}{dx\,dy}$$

is equivalent to $d^2w/dn\,d\sigma$, where it is to be understood that the axes of n and σ are fixed. The first boundary condition now becomes

$$\frac{d}{dn}\nabla^2 w + (1-\mu)\frac{d}{ds}\left(\frac{d^2w}{dn\,d\sigma}\right) = 0 \quad\dots\dots\dots\dots(5).$$

If we apply these equations to the rectangle whose sides are parallel to the coordinate axes, we obtain as the conditions to be satisfied along the edges parallel to y,

$$\left.\begin{aligned}\frac{d}{dx}\left\{\frac{d^2w}{dx^2} + (2-\mu)\frac{d^2w}{dy^2}\right\} &= 0 \\ \frac{d^2w}{dx^2} + \mu\frac{d^2w}{dy^2} &= 0\end{aligned}\right\} \quad\dots\dots\dots\dots(6).$$

In this case the distinction between σ and s disappears, and ρ, the radius of curvature, is infinitely great. The conditions for the other pair of edges are found by interchanging x and y. These results may be obtained equally well from (9) § 215 directly, without the preliminary transformation.

217.   If we suppose $Z = 0$, and write

$$\frac{qh^2}{3\rho(1-\mu^2)} = c^4 \quad \ldots\ldots\ldots\ldots\ldots\ldots(1),$$

the general equation becomes

$$\ddot{w} + c^4 \nabla^4 w = 0 \quad \ldots\ldots\ldots\ldots\ldots\ldots(2),$$

or, if $w \propto \cos(pt - \epsilon)$,

$$\nabla^4 w = k^4 w \quad \ldots\ldots\ldots\ldots\ldots\ldots(3),$$

where

$$k^4 = p^2/c^4 \ldots\ldots\ldots\ldots\ldots\ldots\ldots(4).$$

Any two values of w, u and v, corresponding to the same boundary conditions, are *conjugate*, that is to say

$$\iint uv \, dS = 0 \ldots\ldots\ldots\ldots\ldots\ldots(5),$$

provided that the periods be different. In order to prove this from the ordinary differential equation (3), we should have to retrace the steps by which (3) was obtained. This is the method adopted by Kirchhoff for the circular disc, but it is much simpler and more direct to use the variational equation

$$\delta V + 2\rho h \iint \ddot{w} \, \delta w \, dS = 0 \ldots\ldots\ldots\ldots\ldots(6),$$

in which w refers to the actual motion, and δw to an arbitrary displacement consistent with the nature of the system. δV is a symmetrical function of w and δw, as may be seen from § 215, or from the general character of V (§ 94).

If we now suppose in the first place that $w = u$, $\delta w = v$, we have

$$\delta V = 2\rho h p^2 \iint uv \, dS;$$

and in like manner if we put $w = v$, $\delta w = u$, which we are equally entitled to do,

$$\delta V = 2\rho h p'^2 \iint uv \, dS,$$

whence $$(p^2 - p'^2)\iint uv\,dS = 0 \quad\dots\dots\dots\dots\dots\dots(7).$$

This demonstration is valid whatever may be the form of the boundary, and whether the edge be clamped, supported, or free, in whole or in part.

As for the case of membranes in the last Chapter, equation (7) may be employed to prove that the admissible values of p^2 are real; but this is evident from physical considerations.

218. For the application to a circular disc, it is necessary to express the equations by means of polar coordinates. Taking the centre of the disc as pole, we have for the general equation to be satisfied at all points of the area

$$(\nabla^4 - k^4)\,w = 0 \quad\dots\dots\dots\dots\dots\dots(1),$$

where (§ 200) $\qquad \nabla^2 = \dfrac{d^2}{dr^2} + \dfrac{1}{r}\dfrac{d}{dr} + \dfrac{1}{r^2}\dfrac{d^2}{d\theta^2}.$

To express the boundary condition (§ 216) for a free edge $(r = a)$, we have

$$\frac{d}{dn}\nabla^2 w = \frac{d}{dr}\nabla^2 w, \quad \frac{d}{ds}\left(\frac{d^2 w}{dn\,d\sigma}\right) = \frac{d}{a\,d\theta}\frac{d}{dr}\left(\frac{dw}{r\,d\theta}\right), \quad \frac{d^2 w}{ds^2} = \frac{d^2 w}{a^2\,d\theta^2},$$

ρ = radius of curvature = a; and thus

$$\left. \begin{aligned} \frac{d}{dr}\left(\frac{d^2 w}{dr^2} + \frac{1}{r}\frac{dw}{dr}\right) + \frac{d^2}{d\theta^2}\left(\frac{2 - \mu}{a^2}\frac{dw}{dr} - \frac{3 - \mu}{a^3}w\right) = 0 \\ \frac{d^2 w}{dr^2} + \mu\left(\frac{1}{a}\frac{dw}{dr} + \frac{1}{a^2}\frac{d^2 w}{d\theta^2}\right) = 0 \end{aligned} \right\} \dots\dots\dots(2).$$

After the differentiations are performed, r is to be made equal to a.

If w be expanded in Fourier's series

$$w = w_0 + w_1 + \dots + w_n + \dots,$$

each term separately must satisfy (2), and thus, since

$$w_n \propto \cos(n\theta - \alpha),$$

$$\left. \begin{aligned} \frac{d}{dr}\left(\frac{d^2 w_n}{dr^2} + \frac{1}{r}\frac{dw_w}{dr}\right) - n^2\left(\frac{2 - \mu}{a^2}\frac{dw_n}{dr} - \frac{3 - \mu}{a^3}w_n\right) = 0 \\ \frac{d^2 w_n}{dr^2} + \mu\left(\frac{1}{a}\frac{dw_n}{dr} - \frac{n^2}{a^2}w_n\right) = 0 \end{aligned} \right\} \dots\dots(3).$$

The superficial differential equation may be written

$$(\nabla^2 + k^2)(\nabla^2 - k^2) w = 0,$$

which becomes for the general term of the Fourier expansion

$$\left(\frac{d^2}{dr^2} + \frac{1}{r}\frac{d}{dr} - \frac{n^2}{r^2} + k^2\right)\left(\frac{d^2}{dr^2} + \frac{1}{r}\frac{d}{dr} - \frac{n^2}{r^2} - k^2\right) w_n = 0,$$

shewing that the complete value of w_n will be obtained by adding together, with arbitrary constants prefixed, the general solutions of

$$\left(\frac{d^2}{dr^2} + \frac{1}{r}\frac{d}{dr} - \frac{n^2}{r^2} \pm k^2\right) w_n = 0 \quad \dots\dots\dots\dots\dots(4).$$

The equation with the upper sign is the same as that which obtains in the case of the vibrations of circular membranes, and as in the last Chapter we conclude that the solution applicable to the problem in hand is $w_n \propto J_n(kr)$, the second function of r being here inadmissible.

In the same way the solution of the equation with the lower sign is $w_n \propto J_n(ikr)$, where $i = \sqrt{(-1)}$ as usual. [See § 221 a.]

The simple vibration is thus

$$w_n = \cos n\theta \left\{\alpha J_n(kr) + \beta J_n(ikr)\right\} + \sin n\theta \left\{\gamma J_n(kr) + \delta J_n(ikr)\right\}.$$

The two boundary equations will determine the admissible values of k and the values which must be given to the ratios $\alpha : \beta$ and $\gamma : \delta$. From the form of these equations it is evident that we must have $\qquad \alpha : \beta = \gamma : \delta,$

and thus w_n may be expressed in the form

$$w_n = P\cos(n\theta - \alpha)\left\{J_n(kr) + \lambda J_n(ikr)\right\}\cos(pt - \epsilon)\dots\dots(5).$$

As in the case of a membrane the nodal system is composed of the n diameters symmetrically distributed round the centre, but otherwise arbitrary, denoted by

$$\cos(n\theta - \alpha) = 0 \dots\dots\dots\dots\dots\dots\dots(6),$$

together with the concentric circles, whose equation is

$$J_n(kr) + \lambda J_n(ikr) = 0\dots\dots\dots\dots\dots\dots(7).$$

219. In order to determine λ and k we must introduce the boundary conditions. When the edge is free, we obtain from (3) § 218

$$\left.\begin{array}{l}-\lambda = \dfrac{n^2(\mu-1)\left\{ka\,J_n{}'(ka) - J_n(ka)\right\} - k^3a^3J_n{}'(ka)}{n^2(\mu-1)\left\{ika J_n{}'(ika) - J_n(ika)\right\} + ik^3a^3J_n{}'(ika)} \\[3mm] -\lambda = \dfrac{(\mu-1)\left\{ka J_n{}'(ka) - n^2 J_n(ka)\right\} - k^2a^2 J_n(ka)}{(\mu-1)\left\{ika J_n{}'(ika) - n^2 J_n(ika)\right\} + k^2a^2 J_n(ika)}\end{array}\right\} \dots(1),$$

in which use has been made of the differential equations satisfied by $J_n(kr)$, $J_n(ikr)$. In each of the fractions on the right the denominator may be derived from the numerator by writing ik in place of k. By elimination of λ the equation is obtained whose roots give the admissible values of k.

When $n = 0$, the result assumes a simple form, viz.

$$2(1-\mu) + ika\,\frac{J_0(ika)}{J_0'(ika)} + ka\,\frac{J_0(ka)}{J_0'(ka)} = 0 \ldots\ldots\ldots\ldots(2).$$

This, of course, could have been more easily obtained by neglecting n from the beginning.

The calculation of the lowest root for each value of n is troublesome, and in the absence of appropriate tables must be effected by means of the ascending series for the functions $J_n(kr)$, $J_n(ikr)$. In the case of the higher roots recourse may be had to the semiconvergent descending series for the same functions. Kirchhoff finds

$$\tan(ka - \tfrac{1}{2}n\pi) = \frac{B/(8ka) + C/(8ka)^2 - D/(8ka)^3 + \ldots}{A + B/(8ka) + D/(8ka)^3 + \ldots} \ \ldots\ldots(3),$$

where

$$A = \gamma = (1-\mu)^{-1},$$

$$B = \gamma(1-4n^2) - 8,$$

$$C = \gamma(1-4n^2)(9-4n^2) + 48(1+4n^2),$$

$$D = -\gamma\tfrac{1}{3}\{(1-4n^2)(9-4n^2)(13-4n^2)\} + 8(9+136n^2+80n^4).$$

When ka is great,

$$\tan(ka - \tfrac{1}{2}n\pi) = 0 \quad \text{approx.};$$

whence

$$ka = \tfrac{1}{2}\pi(n+2h) \ \ldots\ldots\ldots\ldots\ldots\ldots\ldots(4),$$

where h is an integer.

It appears by a numerical comparison that h is identical with the number of circular nodes, and (4) expresses a law discovered by Chladni, that the frequencies corresponding to figures with a given number of nodal diameters are, with the exception of the lowest, approximately proportional to the squares of consecutive even or uneven numbers, according as the number of the diameters is itself even or odd. Within the limits of application of (4), we see also that the pitch is approximately unaltered, when any number is subtracted from h, provided twice that number be added to n. This law, of which traces appear in the following table, may be expressed by saying that towards raising the pitch nodal circles have twice the effect of nodal diameters. It is probable, however, that, strictly speaking, no two normal components have exactly the same pitch.

h	$n = 0$			$n = 1$		
	CH.	P.	W.	CH.	P.	W.
0	...	...	...	...	...	...
1	Gis	Gis +	A +	b	h −	c −
2	gis' +	b' −	b' +	e'' +	f'' +	fis'' +

h	$n = 2$			$n = 3$		
	CH.	P.	W.	CH.	P.	W.
0	C	C	C	d	dis −	dis −
1	g'	gis' +	a' −	d''.dis''	dis'' +	e'' −

The table, extracted from Kirchhoff's memoir, gives the pitch of the more important overtones of a free circular plate, the gravest being assumed to be C^1. The three columns under the heads Ch, P, W refer respectively to the results as observed by Chladni and as calculated from theory with Poisson's and Wertheim's values of μ. A *plus* sign denotes that the actual pitch is a little higher, and a *minus* sign that it is a little lower, than that written. The discrepancies between theory and observation are considerable, but perhaps not greater than may be attributed to irregularity in the plate.

220. The radii of the nodal circles in the symmetrical case ($n = 0$) were calculated by Poisson, and compared by him with results obtained experimentally by Savart. The following numbers are taken from a paper by Strehlke[2], who made some careful measurements. The radius of the disc is taken as unity.

	Observation.	Calculation.
One circle ...	0·67815	0·68062.
Two circles...	0·39133	0·39151.
	0·84149	0·84200.
Three circles	0·25631	0·25679.
	0·59107	0·59147.
	0·89360	0·89381.

[1] Gis corresponds to G♯ of the English notation, and h to b natural.
[2] Pogg. *Ann.* xcv. p. 577. 1855.

The calculated results appear to refer to Poisson's value of μ, but would vary very little if Wertheim's value were substituted.

The following table gives a comparison of Kirchhoff's theory (n not zero) with measurements by Strehlke made on less accurate discs.

Radii of Circular Nodes.

	Observation.				Calculation.	
					$\mu = \frac{1}{4}$ (P.).	$\mu = \frac{1}{3}$ (W.).
$n=1,\ h=1$	0·781	0·783	0·781	0·783	0·78136	0·78088
$n=2,\ h=1$	0·79	0·81	0·82		0·82194	0·82274
$n=3,\ h=1$	0·838	0·842			0·84523	0·84681
$n=1,\ h=2$	0·488	0·492			0·49774	0·49715
	0·869	0·869			0·87057	0·87015

The most general motion of the uniform circular plate is expressed by the superposition, with arbitrary amplitudes and phases, of the normal components already investigated. The determination of the amplitude and phase to correspond to arbitrary initial displacements and velocities is effected precisely as in the corresponding problem for the membrane by the aid of the characteristic property of the normal functions proved in § 217.

221. When the plate is truly symmetrical, whether uniform or not, theory indicates, and experiment verifies, that the position of the nodal diameters is arbitrary, or rather dependent only on the manner in which the plate is supported, and excited. By varying the place of support, any desired diameter may be made nodal. It is generally otherwise when there is any sensible departure from exact symmetry. The two modes of vibration, which originally, in consequence of the equality of periods, could be combined in any proportion without ceasing to be simple harmonic, are now separated and affected with different periods. At the same time the position of the nodal diameters becomes determinate, or rather limited to two alternatives. The one set is derived from the other by rotation through half the angle included between two adjacent diameters of the same set. This supposes that the deviation from uniformity is small; otherwise the nodal system will no longer be composed of approximate circles and diameters at all. The cause of the deviation may be an irregularity either in the material or in the thickness or in the form of the boundary. The effect of a small load at any point may be investigated as in the parallel problem of the membrane § 208. If the place at which the load is attached does not lie on a nodal circle, the normal types are made determinate. The diametral system corresponding to one of the types passes through the place in question, and for this type the period is unaltered. The period of the other type is increased.

[The divergence of free periods, which is due to slight inequalities, would seem to afford an explanation of some curious observations by Savart[1]. When a circular plate, vibrating with nodal diameters, is under the influence of the bow applied at any part of the circumference, the nodal diameters indicated by sand are so situated that the bow lies in the middle of a vibrating segment. If, however, the bow be suddenly withdrawn, the nodal system oscillates, or even revolves, during the subsidence of the motion. It is evident that no such displacement could be expected, were the plate absolutely symmetrical. The same would be true, even in the case of asymmetry, if the bow were so applied as to excite one only of the two determinate vibrations then possible. But in general the effect of the bow must be to excite both kinds of vibrations, and then the matter is more complicated. It would seem that so long as the constraining action of the bow lasts, both vibrations are forced to keep the same time, and the effect is much the same as in the case of symmetry. But on withdrawal of the bow the free vibrations which then ensue take place each in its proper frequency, and a phase difference soon arises by which the effects are modified.

Let us suppose that the origin of θ is so chosen in relation to the irregularities that the types of vibration are represented by $\cos n\theta$, $\sin n\theta$. Then in general the free vibrations, resulting from the action of the bow at an arbitrary point of the circumference, may be taken to be

$$\cos n\alpha \sin n\theta \cos pt - \sin n\alpha \cos n\theta \cos (pt + \epsilon)\ldots\ldots\ldots(1),$$

where ϵ is the difference of phase which has accumulated since the commencement of the free vibrations. In the case of symmetry $\epsilon = 0$, and (1) becomes

$$\sin n(\theta - \alpha) \cos pt\ldots\ldots\ldots\ldots\ldots\ldots(2),$$

[1] *Ann. Chim.*, vol. 36, p. 257, 1827.

which represents a fixed nodal system

$$\theta = \alpha + m\,(\pi/n).\ldots\ldots\ldots\ldots\ldots\ldots\ldots(3),$$

in any arbitrary position depending upon the point of application of the bow. A similar fixity of the nodal system occurs, in spite of the variable ϵ, when α is so chosen that $\cos n\alpha = 0$ or $\sin n\alpha = 0$. But in general there is no fixed nodal system. When ϵ is a multiple of 2π, that is when the two vibrations are restored to the same phase, there is a nodal system represented by (3). And when ϵ is an odd multiple of π, so that the two vibrations are in opposite phases, we have in place of (2)

$$\sin n(\theta + \alpha)\cos pt.\ldots\ldots\ldots\ldots\ldots\ldots(4),$$

with a nodal system

$$\theta = -\alpha + m\,(\pi/n).\ldots\ldots\ldots\ldots\ldots\ldots(5).$$

In these cases there is a nodal system, and in a sense the system may be said to oscillate between the positions given by (3) and (5); but it must not be overlooked that at intermediate times there is no true nodal system at all. Thus, when $\epsilon = \tfrac{1}{2}\pi$, (1) becomes

$$\cos n\alpha \sin n\theta \cos pt + \sin n\alpha \cos n\theta \sin pt.$$

The squared amplitude of this motion is

$$\cos^2 n\alpha \sin^2 n\theta + \sin^2 n\alpha \cos^2 n\theta,$$

a quantity which does not vanish for any value of θ. In general the squared amplitude is

$$\cos^2 n\alpha \sin^2 n\theta + \sin^2 n\alpha \cos^2 n\theta - 2\cos n\alpha \sin n\alpha \cos n\theta \sin n\theta \cos \epsilon,$$

or, as it may also be written,

$$\tfrac{1}{2} - \tfrac{1}{2}\cos 2n\alpha \cos 2n\theta - \tfrac{1}{2}\sin 2n\alpha \sin 2n\theta \cos \epsilon\ldots\ldots\ldots\ldots(6).$$

This quantity is a maximum or a minimum when

$$\tan 2n\theta = \cos \epsilon \tan 2n\alpha.\ldots\ldots\ldots\ldots\ldots(7).$$

The minimum of motion thus oscillates backwards and forwards between $\theta = +\alpha$ and $\theta = -\alpha$; but as we have seen, it is only in these extreme positions that the minimum is zero.

A like phenomenon occurs during the free vibrations of a circular membrane, or in fact of any system of revolution such that the position of nodal lines is arbitrary so long as the symmetry is complete.]

The two other cases of a circular plate in which the edge is either clamped or *supported* would be easier than the preceding in their theoretical treatment, but are of less practical interest on account of the difficulty of experimentally realising the conditions assumed. The general result that the nodal system is composed of concentric circles, and diameters symmetrically distributed, is applicable to all the three cases.

221 a. The use in the telephone of a thin circular plate clamped at the edge lends a certain interest to the calculation of the periods and modes of vibration of such a plate. It will suffice to consider the symmetrical modes.

By (5) § 218 we may take as representing the motion in this case

$$w = J_0(kr) + \lambda J_0(ikr) = J_0(kr) + \lambda I_0(kr)\ldots\ldots(1),$$

from which

$$\frac{dw}{kdr} = J_0'(kr) + i\lambda J_0'(ikr) = -J_1(kr) + \lambda I_1(kr)\ldots\ldots(2);$$

where we write

$$I_0(z) = J_0(iz) = 1 + \frac{z^2}{2^2} + \frac{z^4}{2^2 \cdot 4^2} + \cdots \quad \ldots\ldots(3),$$

$$I_1(z) = iJ_0'(iz) = \frac{z}{2} + \frac{z^3}{2^2 \cdot 4} + \frac{z^5}{2^2 \cdot 4^2 \cdot 5} + \cdots \quad \ldots(4).$$

Since the plate is clamped at $r = a$, both w and dw/dr must there vanish. Hence, writing $ka = z$, we get as the frequency equation

$$\frac{J_1(z)}{J_0(z)} + \frac{I_1(z)}{I_0(z)} = 0\ldots\ldots\ldots\ldots(5).$$

In (5) I_1 and I_0 are both positive, so that the signs of J_1 and J_0 must be opposite. Hence by Table B § 206 the first root must lie between 2·4 and 3·8, the second between 5·5 and 7·0, and so on. The values of the earlier roots might be obtained without much difficulty from the series for I_0 and I_1 by using the table § 200 for J_0 and J_1; but it will be convenient for the present and further purposes to give a short table[1] of the functions I_0 and I_1 themselves. For large values of the argument descending series, analogous to (10) § 200, may be employed.

[1] Calculated by A. Lodge, *Brit. Ass. Rep.*, 1889.

z	$I_0(z)$	$I_1(z)$	z	$I_0(z)$	$I_1(z)$
0·0	1·0000	0·0000	3·0	4·8808	3·9534
·2	1·0100	·1005	3·2	5·7472	4·7343
·4	1·0404	·2040	3·4	6·7848	5·6701
·6	1·0920	·3137	3·6	8·0277	6·7927
·8	1·1665	·4329	3·8	9·5169	8·1404
1·0	1·2661	·5652	4·0	11·3019	9·7595
1·2	1·3937	·7147	4·2	13·4425	11·7056
1·4	1·5534	·8861	4·4	16·0104	14·0462
1·6	1·7500	1·0848	4·6	19·0926	16·8626
1·8	1·9896	1·3172	4·8	22·7937	20·2528
2·0	2·2796	1·5906	5·0	27·2399	24·3356
2·2	2·6291	1·9141	5·2	32·5836	29·2543
2·4	3·0493	2·2981	5·4	39·0088	35·1821
2·6	3·5533	2·7554	5·6	46·7376	42·3283
2·8	4·1573	3·3011	5·8	56·0381	50·9462
			6·0	67·2344	61·3419

The first root of (5) is $z = 3\cdot20$. This then is the value of ka for the gravest symmetrical vibration. The next value of z is about 6·3. Since the frequency varies as k^2 (§ 217), the interval between the tones is nearly two octaves.

Returning to the first root, we have for the frequency (n) § 217,

$$n = \frac{p}{2\pi} = \frac{(3\cdot2)^2 c^2}{2\pi a^2} = \frac{(3\cdot2)^2 \sqrt{q} \cdot h}{2\pi a^2 \sqrt{3\rho(1-\mu^2)}} \quad \ldots\ldots (6).$$

This is the general formula. For rough calculations μ^2 in the denominator may be omitted. If for the case of iron we take

$$\rho = 7\cdot7, \quad q = 2\cdot0 \times 10^{12},$$

we find
$$n = \frac{2\cdot4 \times 10^5 \cdot 2h}{a^2} \ldots\ldots\ldots\ldots\ldots (7),$$

$2h$ and a being expressed in centimetres.

A telephone plate measured by the author gave

$$a = 2\cdot2, \quad 2h = \cdot020.$$

According to these values

$$n = 991 \text{ vibrations per second.}$$

222. We have seen that in general Chladni's figures as traced by sand agree very closely with the circles and diameters of theory; but in certain cases deviations occur, which are usually attributed to irregularities in the plate. It must however be remembered that the vibrations excited by a bow are not strictly speaking free, and that their periods are therefore liable to a certain modification. It may be that under the action of the bow two or more normal component vibrations coexist. The whole motion may be simple harmonic in virtue of the external force, although the natural periods would be a little different. Such an explanation is suggested by the regular character of the figures obtained in certain cases.

Another cause of deviation may perhaps be found in the manner in which the plates are supported. The requirements of theory are often difficult to meet in actual experiment. When this is so, we may have to be content with an imperfect comparison; but we must remember that a discrepancy may be the fault of the experiment as well as of the theory.

[In the ordinary use of sand to investigate the vibrations of flat plates and membranes the movement to the nodes is irregular in its character. If a grain be situated elsewhere than at a node, it is made to jump by a sufficiently vigorous transverse vibration. The result may be a movement either towards or from a node; but after a succession of such jumps the grain ultimately finds its way to a node as the only place where it can remain undisturbed. Grains which have already arrived at a node remain there, while others are constantly shifting their position.

It was found by Savart that very fine powder, such as lycopodium, behaves differently from sand. Instead of collecting at the nodes, it heaps itself up at the places of greatest motion. This effect was traced by Faraday[1] to the influence of currents of air, themselves the result of the vibration. In a vacuum all powders move to the nodes.

In some cases the movement of sand to the nodes, or to some of them, takes place in a more direct manner as the result of friction. Thus, in his investigation of the longitudinal vibrations of thin narrow strips of glass, held horizontally, Savart[2] observed the delineation of nodes apparently dependent upon an accompaniment of vibrations of a transverse character. The special peculiarity of this phenomenon was the non-correspondence of the lines traced by sand upon the two faces of the glass when tested

[1] On a Peculiar Class of Acoustical Figures, *Phil. Trans.*, 1831, p. 299.
[2] *Ann. d. Chim.*, vol. 14, p. 113, 1820.

in succession, a fact sufficient to shew that the transverse motion was connected with a failure of uniformity. In consequence of this there are developed transverse vibrations of the same (high) pitch as that of the principal longitudinal motion, and therefore attended with many nodes. These nodes are of course the same whichever face of the glass is uppermost, and it might be supposed that they would all be indicated by the sand, as would happen if the transverse vibrations existed alone. But the combination of the two kinds of motion causes a creeping of the sand towards the *alternate* nodes, the movements of the sand at corresponding points on the two sides of the plate being always in opposite directions. On the one side an inwards longitudinal motion (for example) is attended by an upwards transverse motion, but when the plate is reversed the same inwards longitudinal motion is associated with a transverse motion directed downwards. If there were no transverse motion, the longitudinal force upon any particle resulting from friction would vanish in the long run, but in consequence of the transverse motion this balance is upset, and in a manner different upon the two sides of the plate. The above considerations appear to afford sufficient ground for an explanation of the remarkable phenomenon observed by Savart, but an attempt to follow the matter further into detail would lead us too far[1].]

223. The first attempt to solve the problem with which we have just been occupied is due to Sophie Germain, who succeeded in obtaining the correct differential equation, but was led to erroneous boundary conditions. For a free plate the latter part of the problem is indeed of considerable difficulty. In Poisson's memoir 'Sur l'équilibre et le mouvement des corps élastiques[2],' that eminent mathematician gave *three* equations as necessary to be satisfied at all points of a free edge, but Kirchhoff has proved that in general it would be impossible to satisfy them all. It happens, however, that an exception occurs in the case of the symmetrical vibrations of a circular plate, when one of the equations is true identically. Owing to this peculiarity, Poisson's theory of the symmetrical vibrations is correct, notwithstanding the error in his view as to the boundary conditions. In 1850 the subject was

---

[1] See Terquem, *C. R.*, XLVI., p. 775, 1858.
[2] *Mém. de l'Acad. d. Sc. à Par.* 1829.

resumed by Kirchhoff[1], who first gave the *two* equations appropriate to a free edge, and completed the theory of the vibrations of a circular disc.

224. The correctness of Kirchhoff's boundary equations has been disputed by Mathieu[2], who, without explaining where he considers Kirchhoff's error to lie, has substituted a different set of equations. He proves that if u and u' be two normal functions, so that $w = u \cos pt$, $w = u' \cos p't$ are possible vibrations, then

$$(p^2 - p'^2) \iint uu'\, dx\, dy$$

$$= c^4 \int ds \left\{ u' \frac{d\nabla^2 u}{dn} - \nabla^2 u \frac{du'}{dn} - \nabla^2 u' \frac{du}{dn} + u \frac{d\nabla^2 u'}{dn} \right\}\ldots\ldots(1).$$

This follows, if it be admitted that u, u' satisfy respectively the equations

$$c^4 \nabla^4 u = p^2 u, \qquad c^4 \nabla^4 u' = p'^2 u'.$$

Since the left-hand member is zero, the same must be true of the right-hand member; and this, according to Mathieu, cannot be the case, unless at all points of the boundary both u and u' satisfy one of the four following pairs of equations:

$$\left. \begin{array}{c} u = 0 \\ \dfrac{du}{dn} = 0 \end{array} \right\},\quad \left. \begin{array}{c} \nabla^2 u = 0 \\ \dfrac{d\nabla^2 u}{dn} = 0 \end{array} \right\},\quad \left. \begin{array}{c} u = 0 \\ \nabla^2 u = 0 \end{array} \right\},\quad \left. \begin{array}{c} \dfrac{du}{dn} = 0 \\ \dfrac{d\nabla^2 u}{dn} = 0 \end{array} \right\}$$

The second pair would seem the most likely for a free edge, but it is found to lead to an impossibility. Since the first and third pairs are obviously inadmissible, Mathieu concludes that the fourth pair of equations must be those which really express the condition of a free edge. In his belief in this result he is not shaken by the fact that the corresponding conditions for the free end of a bar would be $du/dx = 0$, $d^3u/dx^3 = 0$, the first of which is contradicted by the roughest observation of the vibration of a large tuning-fork.

---

[1] *Crelle*, t. XL. p. 51. Ueber das Gleichgewicht und die Bewegung einer elastischen Scheibe.

[2] *Liouville*, t. XIV. 1869.

The fact is that although any of the four pairs of equations would secure the evanescence of the boundary integral in (1), it does not follow conversely that the integral can be made to vanish in no other way; and such a conclusion is negatived by Kirchhoff's investigation. There are besides innumerable other cases in which the integral in question would vanish, all that is really necessary being that the boundary appliances should be either at rest, or devoid of inertia.

225. The vibrations of a rectangular plate, whose edge is *supported*, may be easily investigated theoretically, the normal functions being identical with those applicable to a membrane of the same shape, whose boundary is fixed. If we assume

$$w = \sin\frac{m\pi x}{a} \sin\frac{n\pi y}{b} \cos pt \dots\dots\dots\dots(1),$$

we see that at all points of the boundary,

$$w = 0, \quad d^2w/dx^2 = 0, \quad d^2w/dy^2 = 0,$$

which secure the fulfilment of the necessary conditions (§ 215). The value of p, found by substitution in $c^4 \nabla^4 w = p^2 w$,

is
$$p = c^2\pi^2 \left(\frac{m^2}{a^2} + \frac{n^2}{b^2}\right)\dots\dots\dots\dots\dots\dots(2),$$

shewing that the analogy to the membrane does not extend to the sequence of tones.

It is not necessary to repeat here the discussion of the primary and derived nodal systems given in Chapter IX. It is enough to observe that if two of the fundamental modes (1) have the same period in the case of the membrane, they must also have the same period in the case of the plate. The derived nodal systems are accordingly identical in the two cases.

The generality of the value of w obtained by compounding with arbitrary amplitudes and phases all possible particular solutions of the form (1) requires no fresh discussion.

Unless the contrary assertion had been made, it would have seemed unnecessary to say that the nodes of a *supported* plate have nothing to do with the ordinary Chladni's figures, which belong to a plate whose edges are free.

The realization of the conditions for a supported edge is scarcely attainable in practice. Appliances are required capable of holding the boundary of the plate at rest, and of such a nature that they give rise to no couples about tangential axes. We may conceive the plate to be held in its place by friction against the walls of a cylinder circumscribed closely round it.

226. The problem of a rectangular plate, whose edges are free, is one of great difficulty, and has for the most part resisted attack[1]. If we suppose that the displacement w is independent of y, the general differential equation becomes identical with that with which we were concerned in Chapter VIII. If we take the solution corresponding to the case of a bar whose ends are free, and therefore satisfying $d^2w/dx^2 = 0$, $d^3w/dx^3 = 0$, when $x = 0$ and when $x = a$, we obtain a value of w which satisfies the general fferential equation, as well as the pair of boundary equations

$$\left. \begin{array}{c} \dfrac{d}{dx}\left\{ \dfrac{d^2w}{dx^2} + (2-\mu)\,\dfrac{d^2w}{dy^2} \right\} = 0 \\[2mm] \dfrac{d^2w}{dx^2} + \mu\,\dfrac{d^2w}{dy^2} = 0 \end{array} \right\} \dots\dots\dots\dots(1),$$

which are applicable to the edges parallel to y; but the second boundary condition for the other pair of edges, namely

$$\frac{d^2w}{dy^2} + \mu\,\frac{d^2w}{dx^2} = 0 \ \dots\dots\dots\dots\dots\dots(2),$$

will be violated, unless $\mu = 0$. This shews that, except in the case reserved, it is not possible for a free rectangular plate to vibrate after the manner of a bar; unless indeed as an approximation, when the length parallel to one pair of edges is so great that the conditions to be satisfied at the second pair of edges may be left out of account.

Although the constant μ (which expresses the ratio of lateral contraction to longitudinal extension when a bar is drawn out) is positive for every known substance, in the case of a few substances—cork, for example—it is comparatively very small. There is, so far as we know, nothing absurd in the idea of a substance

---

[1] [The case where two opposite edges are free while the other two edges are supported, has been discussed by Voigt (*Göttingen Nachrichten*, 1893) p. 225.]

for which μ vanishes. The investigation of the problem under this condition is therefore not devoid of interest, though the results will not be strictly applicable to ordinary glass or metal plates, for which the value of μ is about $\frac{1}{3}$.[1]

If u_1, u_2, &c. denote the normal functions for a free bar investigated in Chapter VIII., corresponding to 2, 3, ... nodes, the vibrations of a rectangular plate will be expressed by

$$w = u_1(x/a), \quad w = u_2(x/a), \text{ &c.,}$$

or
$$w = u_1(y/b), \quad w = u_2(y/b), \text{ &c.}$$

In each of these primitive modes the nodal system is composed of straight lines parallel to one or other of the edges of the rectangle. When $b = a$, the rectangle becomes a square, and the vibrations

$$w = u_n(x/a), \quad w = u_n(y/a),$$

having necessarily the same period, may be combined in any proportion, while the whole motion still remains simple harmonic. Whatever the proportion may be, the resulting nodal curve will of necessity pass through the points determined by

$$u_n(x/a) = 0, \quad u_n(y/a) = 0.$$

Now let us consider more particularly the case of $n = 1$. The nodal system of the primitive mode, $w = u_1(x/a)$, consists of a pair of straight lines parallel to y, whose distance from the nearest edge is $\cdot2242a$. The points in which these lines are met by the corresponding pair for $w = u_1(y/a)$, are those through which the nodal curve of the compound vibration must in all cases pass. It is evident that they are symmetrically disposed on the diagonals of the square. If the two primitive vibrations be taken equal, but in opposite phases (or, algebraically, with equal and opposite amplitudes), we have

$$w = u_1(x/a) - u_1(y/a) \dots\dots\dots\dots\dots(3),$$

---

[1] In order to make a plate of material, for which μ is not zero, vibrate in the manner of a bar, it would be necessary to apply constraining couples to the edges parallel to the plane of bending to prevent the assumption of a contrary curvature. The effect of these couples would be to raise the pitch, and therefore the calculation founded on the type proper to $\mu=0$ would give a result somewhat higher in pitch than the truth.

from which it is evident that w vanishes when $x = y$, that is along
the diagonal which passes through the origin.   Fig. 41.
That w will also vanish along the other diagonal
follows from the symmetry of the functions, and
we conclude that the nodal system of (3) comprises
both the diagonals (Fig. 41). This is a well-known
mode of vibration of a square plate.

Fig. 41.

A second notable case is when the amplitudes are equal and
their phases the same, so that

$$w = u_1(x/a) + u_1(y/a)... \ ....(4).$$

The most convenient method of constructing graphically
the curves, for which $w = $ const., is that employed by Maxwell
in similar cases. The two systems of curves (in this instance
straight lines) represented by $u_1(x/a) = $ const., $u_1(y/a) = $ const., are
first laid down, the values of the constants forming an arithmetical progression with the same common difference in the two
cases. In this way a network is obtained which the required
curves cross diagonally. The execution of the proposed plan
requires an inversion of the table given in Chapter VIII., § 178,
expressing the march of the function u_1, of which the result is as
follows :—

u_1	$x : a$	u_1	$x : a$
+ 1·00	·5000	− ·25	·1871
·75	·3680	·50	·1518
·50	·3106	·75	·1179
·25	·2647	1·00	·0846
·00	·2242	1·25	·0517
		− 1·50	·0190

The system of lines represented by the above values of x (completed symmetrically on the further side of the central line) and
the corresponding system for y are laid down in Fig. 42. From
these the curves of equal displacement are deduced. At the
centre of the square we have w a maximum and equal to 2 on the
scale adopted. The first curve proceeding outwards is the locus of
points at which $w = 1$. The next is the nodal line, separating the
regions of opposite displacement. The remaining curves taken in

order give the displacements $-1, -2, -3$. The numerically greatest negative displacement occurs at the corners of the square, where it amounts to $2 \times 1\cdot645 = 3\cdot290$.[1]

The nodal curve thus constructed agrees pretty closely with the observations of Strehlke[2]. His results, which refer to three carefully worked plates of glass, are embodied in the following polar equations:

$$r = \left. \begin{matrix} \cdot40143 & \cdot0171 \\ \cdot40143 + \cdot0172 \\ \cdot4019 & \cdot0168 \end{matrix} \right\} \cos 4t + \left. \begin{matrix} \cdot00127 \\ \cdot00127 \\ \cdot0013 \end{matrix} \right\} \cos 8t,$$

Fig. 42.

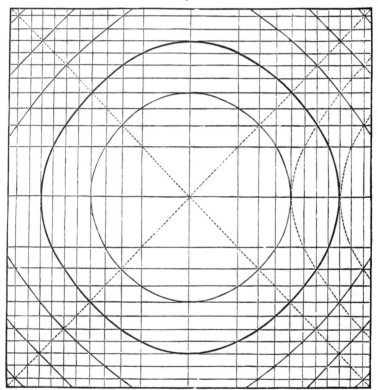

the centre of the square being pole. From these we obtain for the radius vector parallel to the sides of the square $(t = 0)$ $\cdot41980$,

---

[1] On the nodal lines of a square plate. *Phil. Mag.* August, 1873.

[2] Pogg. *Ann.* Vol. cxlvi. p. 319. 1872.

·41981, ·4200, while the calculated result is ·4154. The radius vector measured along a diagonal is ·3856, ·3855, ·3864, and by calculation ·3900.

By crossing the network in the other direction we obtain the locus of points for which $u_1(x/a) - u_1(y/a)$ is constant, which are the curves of constant displacement for that mode in which the diagonals are nodal. The *pitch* of the vibration is (according to theory) the same in both cases.

Fig. 43.

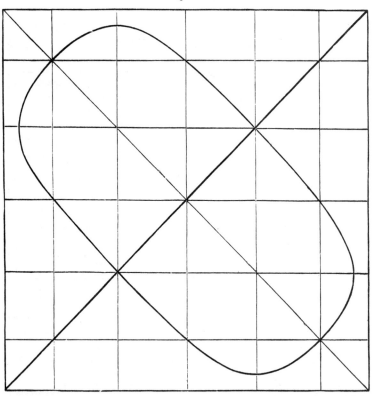

The primitive modes represented by $w = u_2(x/a)$ or $w = u_2(y/a)$ may be combined in like manner. Fig. 43 shews the nodal curve for the vibration

$$w = u_2(x/a) \pm u_2(y/a) \dots\dots\dots\dots\dots(5).$$

The form of the curve is the same relatively to the other diagonal, if the sign of the ambiguity be altered.

227. The method of superposition does not depend for its application on any particular form of normal function. Whatever the form may be, the mode of vibration, which when $\mu = 0$ passes into that just discussed, must have the same period, whether the approximately straight nodal lines are parallel to x or to y. If the two synchronous vibrations be superposed, the resultant has still the same period, and the general course of its nodal system may be traced by means of the consideration that no point of the plate can be nodal at which the primitive vibrations have the same sign. To determine exactly the line of compensation, a complete knowledge of the primitive normal functions, and not merely of the points at which they vanish, would in general be necessary. Doctor Young and the brothers Weber appear to have had the idea of superposition as capable of giving rise to new varieties of vibration, but it is to Sir Charles Wheatstone[1] that we owe the first systematic application of it to the explanation of Chladni's figures. The results actually obtained by Wheatstone are however only very roughly applicable to a plate, in consequence of the form of normal function implicitly assumed. In place of Fig. 42 (itself, be it remembered, only an approximation) Wheatstone finds for the node of the compound vibration the inscribed square shewn in Fig. 44. This form is really applicable, not to a plate vibrating in virtue of rigidity, but to a stretched membrane, so supported that every point of the circumference is free to move along lines perpendicular to the plane of the membrane. The boundary condition applicable under these circumstances is $\frac{dw}{dn} = 0$; and it is easy to shew that the normal functions which involve only one co-ordinate are

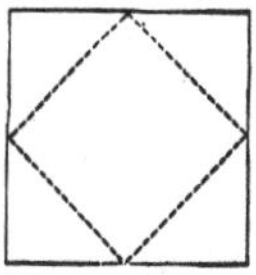

Fig. 44.

$$w = \cos (m\pi x/a), \text{ or } w = \cos (m\pi y/a),$$

the origin being at a corner of the square. Thus the vibration

$$w = \cos \frac{2\pi x}{a} + \cos \frac{2\pi y}{a} \dots\dots\dots\dots\dots(1)$$

has its nodes determined by

$$\cos \frac{\pi (x + y)}{a} \cos \frac{\pi (x - y)}{a} = 0,$$

[1] *Phil. Trans.* 1833.

whence $x + y = \frac{1}{2}a$ or $\frac{3}{2}a$, or $x - y = \pm \frac{1}{2}a$, equations which represent the inscribed square.

If
$$w = \cos \frac{2\pi x}{a} - \cos \frac{2\pi y}{a} \quad \dots\dots\dots\dots\dots (2),$$

the nodal system is composed of the two diagonals. This result, which depends only on the symmetry of the normal functions, is strictly applicable to a square plate.

When $m = 3$,
$$w = \cos \frac{3\pi x}{a} + \cos \frac{3\pi y}{a} \quad \dots\dots\dots\dots\dots(3),$$

and the equations of the nodal lines are

Fig. 45.

$$x + y = \frac{a}{3}, \ a, \ \frac{5a}{3}; \qquad x - y = \pm \frac{a}{3},$$

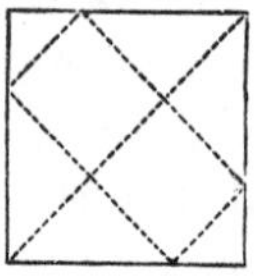

shewn in Fig. 45. If the other sign be taken, we obtain a similar figure with reference to the other diagonal.

When $m = 4$,

Fig. 46.

$$w = \cos \frac{4\pi x}{a} + \cos \frac{4\pi y}{a} \dots\dots\dots\dots (4),$$

giving the nodal lines

$$x + y = \frac{a}{4}, \ \frac{3a}{4}, \ \frac{5a}{4}, \ \frac{7a}{4}, \quad x - y = \pm \frac{a}{4}, \ \pm \frac{3a}{4} \ (\text{Fig. 46}).$$

With the other sign
$$w = \cos \frac{4\pi x}{a} - \cos \frac{4\pi y}{a} \quad \dots\dots\dots\dots (5),$$

we obtain

Fig. 47.

$$x + y = \frac{a}{2}, \ a, \ \frac{3a}{2}, \qquad x - y = 0, \ \pm \frac{a}{2} \ (\text{Fig. 47}),$$

representing a system composed of the diagonals, together with the inscribed square.

These forms, which are strictly applicable to the membrane, resemble the figures obtained by means of sand on a square plate more closely than might have been expected. The sequence of tones is however quite different. From § 176 we see that, if μ were zero, the interval between the form (43) derived from three primitive nodes, and (41) or (42) derived from two, would be

1·4629 octaves; and the interval between (41) or (42) and (46) or (47) would be 2·4358 octaves. Whatever may be the value of μ the forms (41) and (42) should have exactly the same pitch, and the same should be true of (46) and (47). With respect to the first-mentioned pair this result is not in agreement with Chladni's observations, who found a difference of more than a whole tone, (42) giving the higher pitch. If however (42) be left out of account, the comparison is more satisfactory. According to theory ($\mu = 0$), if (41) gave d, (43) should give $g'-$, and (46), (47) should give $g''+$. Chladni found for (43) $g'\sharp +$, and for (46), (47) $g''\sharp$ and $g''\sharp +$ respectively.

228. The gravest mode of a square plate has yet to be considered. The nodes in this case are the two lines drawn through the middle points of opposite sides. That there must be such a mode will be shewn presently from considerations of symmetry, but neither the form of the normal function, nor the pitch, has yet been determined, even for the particular case of $\mu = 0$. A rough calculation however may be founded on an assumed type of vibration.

If we take the nodal lines for axes, the form $w = xy$ satisfies $\nabla^4 w = 0$, as well as the boundary conditions proper for a free edge at all points of the perimeter except the actual corners. This is in fact the form which the plate would assume if held at rest by four forces numerically equal, acting at the corners perpendicularly to the plane of the plate, those at the ends of one diagonal being in one direction, and those at the ends of the other diagonal in the opposite direction. From this it follows that $w = xy \cos pt$ would be a possible mode of vibration, if the mass of the plate were concentrated equally in the four corners. By (3) § 214, we see that

$$V = \frac{2qh^3 a^2}{3(1+\mu)} \cos^2 pt \dots\dots\dots\dots(1),$$

inasmuch as

$$d^2 w/dx^2 = d^2 w/dy^2 = 0, \qquad d^2 w/dx\,dy = \cos pt.$$

For the kinetic energy, if ρ be the volume density, and M the additional mass at each corner,

$$T = \tfrac{1}{2} p^2 \sin^2 pt \left\{ \int_{-\frac{1}{2}a}^{+\frac{1}{2}a} \int_{-\frac{1}{2}a}^{+\frac{1}{2}a} 2\rho h\, x^2 y^2 \, dx\, dy + \tfrac{1}{4} M a^4 \right\}$$

$$= \tfrac{1}{2} p^2 \sin^2 pt \left\{ \frac{\rho h a^6}{8 \times 9} + \frac{a^4}{4} M \right\} \dots\dots\dots\dots(2).$$

Hence

$$\frac{1}{p^2} = \frac{\rho\,(1+\mu)\,a^4}{96\,qh^2}\left(1 + 36\,\frac{M}{M'}\right) \dots\dots\dots\dots (3),$$

where M' denotes the mass of the plate without the loads. This result tends to become accurate when M is relatively great; otherwise by § 89 it is sensibly less than the truth. But even when $M = 0$, the error is probably not very great. In this case we should have

$$p^2 = \frac{96\,qh^2}{\rho\,(1+\mu)\,a^4}\dots\dots\dots\dots\dots(4),$$

giving a pitch which is somewhat too high. The gravest mode next after this is when the diagonals are nodes, of which the pitch, if $\mu = 0$, would be given by

$$p'^2 = \frac{qh^2}{\rho a^4}\,\frac{(4\cdot7300)^4}{3}\dots\dots\dots\dots\dots (5),$$

(see § 174).

We may conclude that if the material of the plate were such that $\mu = 0$, the interval between the two gravest tones would be somewhat greater than that expressed by the ratio 1·318. Chladni makes the interval a fifth.

229. That there must exist modes of vibration in which the two shortest diameters are nodes may be inferred from such considerations as the following. In Fig. (48) suppose that GH is a plate of which the edges HO, GO are *supported*, and the edges GC, CH free. This plate, since it tends to a definite position of equilibrium, must be capable of vibrating in certain fundamental modes. Fixing our attention on one of these, let us conceive a distribution of w over the three remaining quadrants, such that in any two that adjoin, the values of w are equal and opposite at points which are the images of each other in the line of separation. If the whole plate vibrate according to the law thus determined, no constraint will be required in order to keep the lines GE, FH fixed, and therefore the whole plate may be regarded as free. The same argument may be used to prove that modes exist in which the diagonals are nodes, or in which both the diagonals and the diameters just considered are together nodal.

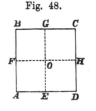

Fig. 48.

The principle of symmetry may also be applied to other forms of plate. We might thus infer the possibility of nodal diameters in a circle, or of nodal principal axes in an ellipse. When the

Fig. 49.   Fig. 50.   Fig. 51.

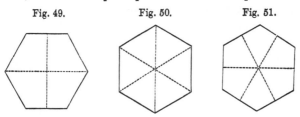

boundary is a regular hexagon, it is easy to see that Figs. (49), (50), (51) represent possible forms.

It is interesting to trace the continuity of Chladni's figures, as the form of the plate is gradually altered. In the circle, for example, when there are two perpendicular nodal diameters, it is a matter of indifference as respects the pitch and the type of vibration, in what position they be taken. As the circle develops into a square by throwing out corners, the position of these diameters becomes definite. In the two alternatives the pitch of the vibration is different, for the projecting corners have not the same efficiency in the two cases. The vibration of a square plate shewn in Fig. (42) corresponds to that of a circle when there is one circular node. The correspondence of the graver modes of a hexagon or an ellipse with those of a circle may be traced in like manner.

230. For plates of uniform material and thickness and of invariable shape, the period of the vibration in any fundamental mode varies as the square of the linear dimension, provided of course that the boundary conditions are the same in all the cases compared. When the edges are clamped, we may go further and assert that the removal of *any* external portion is attended by a rise of pitch, whether the material and the thickness be uniform, or not.

Let AB be a part of a clamped edge (it is of no consequence whether the remainder of the boundary be clamped, or not), and

Fig. 52.

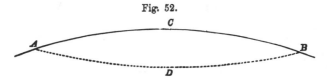

let the piece $ACBD$ be removed, the new edge ADB being also clamped. The pitch of any fundamental vibration is sharper than before the change. This is evident, since the altered vibrations might be obtained from the original system by the introduction of a constraint clamping the edge ADB. The effect of the constraint is to raise the pitch of every component, and the portion $ACBD$ being plane and at rest throughout the motion, may be removed. In order to follow the sequence of changes with greater security from error, it is best to suppose the line of clamping to advance by stages between the two positions ACB, ADB. For example, the pitch of a uniform clamped plate in the form of a regular hexagon is lower than for the inscribed circle and higher than for the circumscribed circle.

When a plate is free, it is not true that an addition to the edge always increases the period. In proof of this it may be sufficient to notice a particular case.

AB is a narrow thin plate, itself without inertia but carrying loads at A, B, C. It is clear that the addition to the breadth

Fig. 53.

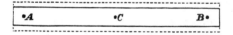

indicated by the dotted line would augment the stiffness of the bar, and therefore *lessen* the period of vibration. The same consideration shews that for a uniform free plate of given area there is no lower limit of pitch; for by a sufficient elongation the period of the gravest component may be made to exceed any assignable quantity. When the edges are clamped, the form of gravest pitch is doubtless the circle.

If all the dimensions of a plate, including the thickness, be altered in the same proportion, the period is proportional to the linear dimension, as in every case of a solid body vibrating in virtue of its own elasticity.

The period also varies inversely as the square root of Young's modulus, if μ be constant, and directly as the square root of the mass of unit of volume of the substance.

231. Experimenting with square plates of thin wood whose grain ran parallel to one pair of sides, Wheatstone[1] found that the pitch of the vibrations was different according as the approximately straight nodal lines were parallel or perpendicular to the fibre of the wood. This effect depends on a variation in the flexural rigidity in the two directions. The two sets of vibrations having different periods cannot be combined in the usual manner, and consequently it is not possible to make such a plate of wood vibrate with nodal diagonals. The inequality of periods may however be obviated by altering the ratio of the sides, and then the ordinary mode of superposition giving nodal diagonals is again possible. This was verified by Wheatstone.

A further application of the principle of superposition is due to König[2]. In order that two modes of vibration may combine, it is only necessary that the periods agree. Now it is evident that the sides of a rectangular plate may be taken in such a ratio, that (for instance) the vibration with two nodes parallel to one pair of sides may agree in pitch with the vibration having three nodes parallel to the other pair of sides. In such a case new nodal figures arise by composition of the two primary modes of vibration.

232. When the plate whose vibrations are to be considered is naturally curved, the difficulties of the question are generally much increased. But there is one case in which the complication due to curvature is more than compensated by the absence of a free edge; and this case happens to be of considerable interest, being the best representative of a bell which admits of simple analytical treatment.

A long cylindrical shell of circular section and uniform thickness is evidently capable of vibrations of a flexural character in which the axis remains at rest and the surface cylindrical, while the motion of every part is perpendicular to the generating lines. The problem may thus be treated as one of two dimensions only, and depends upon the consideration of the potential and kinetic energies of the various deformations of which the section is capable. The same analysis also applies to the corresponding vibrations of a ring, formed by the revolution of a small closed area about an external axis (§ 192 a).

[1] *Phil. Trans.* 1833.
[2] Pogg. *Ann.* 1884, cxxii. p. 238.

The cylinder, or ring, is susceptible of two classes of vibrations depending respectively on extensibility and flexural rigidity, and analogous to the longitudinal and lateral vibrations of straight bars. When, however, the cylinder is thin, the forces resisting bending become small in comparison with those by which extension is opposed; and, as in the case of straight bars, the vibrations depending on bending are graver and more important than those which have their origin in longitudinal rigidity. In the limiting case of an infinitely thin shell (or ring), the flexural vibrations become independent of any extension of the circumference as a whole, and may be calculated on the supposition that each part of the circumference retains its natural length throughout the motion.

But although the vibrations about to be considered are analogous to the transverse vibrations of straight bars in respect of depending on the resistance to flexure, we must not fall into the common mistake of supposing that they are exclusively normal. It is indeed easy to see that a motion of a cylinder or ring in which each particle is displaced in the direction of the radius would be incompatible with the condition of no extension. In order to satisfy this condition it is necessary to ascribe to each part of the circumference a tangential as well as a normal motion, whose relative magnitudes must satisfy a certain differential equation. Our first step will be the investigation of this equation.

233. The original radius of the circle being a, let the equilibrium position of any element of the circumference be defined by the vectorial angle θ. During the motion let the polar coordinates of the element become

$$r = a + \delta r, \quad \phi = \theta + \delta\theta.$$

If ds represent the arc of the deformed curve corresponding to $a\,d\theta$, we have

$$(ds)^2 = (a\,d\theta)^2 = (d\delta r)^2 + r^2\,(d\theta + d\delta\theta)^2;$$

whence we find, by neglecting the squares of the small quantities δr, $\delta\theta$,

$$\frac{\delta r}{a} + \frac{d\delta\theta}{d\theta} = 0 \dots\dots\dots\dots\dots\dots(1),$$

as the required relation.

In whatever manner the original circle may be deformed at time t, δr may be expanded by Fourier's theorem in the series

$$\delta r = a \left\{ A_1 \cos \theta + B_1 \sin \theta + A_2 \cos 2\theta + B_2 \sin 2\theta + \ldots \right.$$

$$\left. + A_s \cos s\theta + B_s \sin s\theta + \ldots \right\} \ldots\ldots\ldots\ldots(2),$$

and the corresponding tangential displacement required by the condition of no extension will be

$$\delta \theta = - A_1 \sin \theta + B_1 \cos \theta + \ldots - \frac{A_s}{s} \sin s\theta + \frac{B_s}{s} \cos s\theta - \ldots \ldots\ldots(3),$$

the constant that might be added to $\delta\theta$ being omitted.

If $\sigma a d\theta$ denote the mass of the element $a d\theta$, the kinetic energy T of the whole motion will be

$$T = \tfrac{1}{2}\sigma a \int_0^{2\pi} \left\{ \left(\frac{d\delta r}{dt}\right)^2 + a^2 \left(\frac{d\delta\theta}{dt}\right)^2 \right\} d\theta$$

$$= \tfrac{1}{2}\sigma\pi a^3 \left\{ 2\left(\dot{A}_1^2 + \dot{B}_1^2\right) + \frac{5}{4}\left(\dot{A}_2^2 + \dot{B}_2^2\right) + \ldots \right.$$

$$\left. + \left(1 + \frac{1}{s^2}\right)\left(\dot{A}_s^2 + \dot{B}_s^2\right) + \ldots \right\} \ldots\ldots\ldots(4),$$

the products of the co-ordinates A_s, B_s disappearing in the integration.

We have now to calculate the form of the potential energy V. Let ρ be the radius of curvature of any element ds; then for the corresponding element of V we may take $\tfrac{1}{2}B ds \{\delta(1/\rho)\}^2$, where B is a constant depending on the material and on the thickness. Thus

$$V = \tfrac{1}{2}Ba \int_0^{2\pi} \left(\delta\frac{1}{\rho}\right)^2 d\theta \ \ldots\ldots\ldots\ldots (5).$$

Now

$$1/\rho = u + d^2u/d\phi^2,$$

and

$$u = \frac{1}{r} = \frac{1}{a}\left\{1 - A_1 \cos \phi - B_1 \sin \phi - \ldots\right\},$$

for in the small terms the distinction between ϕ and θ may be neglected.

Hence

$$\delta\frac{1}{\rho} = \frac{1}{a}\Sigma\left\{(s^2 - 1)(A_s \cos s\phi + B_s \sin s\phi)\right\},$$

and

$$V = \frac{B}{2a} \int_0^{2\pi} \{\Sigma \, (s^2 - 1) \, (A_s \cos s\theta + B_s \sin s\theta)\}^2 \, d\theta$$

$$= \pi \frac{B}{2a} \Sigma \, (s^2 - 1)^2 \, (A_s^2 + B_s^2) \dots\dots\dots\dots\dots\dots(6),$$

in which the summation extends to all positive integral values of s.

The term for which $s = 1$ contributes nothing to the potential energy, as it corresponds to a displacement of the circle as a whole, without deformation.

We see that when the configuration of the system is defined as above by the co-ordinates A_1, B_1, &c., the expressions for T and V involve only squares; in other words, these are the *normal* co-ordinates, whose independent harmonic variation expresses the vibration of the system.

If we consider only the terms involving $\cos s\theta$, $\sin s\theta$, we have by taking the origin of θ suitably,

$$\delta r = a A_s \cos s\theta, \qquad \delta\theta = -\frac{A_s}{s} \sin s\theta \dots\dots\dots\dots(7),$$

while the equation defining the dependence of A_s upon the time is

$$\sigma a^3 \left(1 + \frac{1}{s^2}\right) \ddot{A}_s + \frac{B}{a} (s^2 - 1)^2 A_s = 0 \dots\dots\dots(8),$$

from which we conclude that, if A_s varies as $\cos(pt - \epsilon)$,

$$p^2 = \frac{B}{\sigma a^4} \cdot \frac{s^2 (s^2 - 1)^2}{s^2 + 1} \dots\dots\dots\dots\dots(9).$$

This result was given by Hoppe for a ring in a memoir published in Crelle, Bd. 63, 1871. His method, though more complete than the preceding, is less simple, in consequence of his not recognising explicitly that the motion contemplated corresponds to complete inextensibility of the circumference.

[In the application of (9) to a ring we have, § 192 a,

$$\frac{B}{\sigma} = \frac{c^2}{4} \frac{q}{\rho} \dots\dots\dots\dots\dots\dots(10),$$

where q is Young's modulus, ρ the volume density, and c the

radius of the circular section. For the cylindrical shell, (18)
§ 235 g,

$$\frac{B}{\sigma} = \frac{4mnh^2}{3(m+n)\rho} \quad \dotsc\dotsc\dotsc\dotsc\dotsc (11),$$

$2h$ denoting the thickness, and m, n the elastic constants in
Thomson and Tait's notation.]

According to Chladni the frequencies of the tones of a ring
are as

$$3^2 : 5^2 : 7^2 : 9^2 \dotsc\dotsc\dotsc$$

If we refer each tone to the gravest of the series, we find for
the ratios characteristic of the intervals

$$2\cdot778, \quad 5\cdot445, \quad 9, \quad 13\cdot44, \quad \&c.$$

The corresponding numbers obtained from the above theoretical
formula (9), by making s successively equal to 2, 3, 4, &c., are

$$2\cdot828, \quad 5\cdot423, \quad 8\cdot771, \quad 12\cdot87, \quad \&c.,$$

agreeing pretty nearly with those found experimentally.

[Observations upon the tones of thin metallic cylinders, open
at one end, have been made by Fenkner[1]. Since the pitch proved
to be very nearly independent of the height of the cylinders, the
vibrations may be regarded as approximately two-dimensional.
In accordance with (9), (11), Fenkner found the frequency proportional to the thickness directly, and to the square of the radius
inversely. As regards the sequence of tones from a given
cylinder[2], the numbers, referred to the gravest ($s = 2$) as unity,
were $2\cdot67$, $5\cdot00$, $8\cdot00$, $12\cdot00$, &c. The agreement with (9) would
be improved if these numbers were raised by about $\frac{1}{12}$ part,
equivalent to an alteration in the pitch of the gravest tone.

The influence of rotation of the shell about its axis has been
examined by Bryan[3]. It appears that the nodes are carried
round, but with an angular velocity less than that of the rotation.
If the latter be denoted by ω, the nodal angular velocity is

$$\frac{s^2 - 1}{s^2 + 1} \omega.]$$

[1] *Wied. Ann.* vol. 8, p. 185, 1879.

[2] Melde, *Akustik*, Leipzig, 1883, p. 223.

[3] *Proc. Camb. Phil. Soc,* vol. VII. p. 101, 1890.

234. When $s = 1$, the frequency is zero, as might have been anticipated. The principal mode of vibration corresponds to $s = 2$, and has four nodes, distant from each other by 90°. These so-called nodes are not, however, places of absolute rest, for the tangential motion is there a maximum. In fact the tangential vibration at these points is half the maximum normal motion. In general for the sth term the maximum tangential motion is $(1/s)$ of the maximum normal motion, and occurs at the nodes of the latter.

When a bell-shaped body is sounded by a blow, the point of application of the blow is a place of maximum normal motion of the resulting vibrations, and the same is true when the vibrations are excited by a violin-bow, as generally in lecture-room experiments. Bells of glass, such as finger-glasses, are however more easily thrown into regular vibration by friction with the wetted finger carried round the circumference. The pitch of the resulting sound is the same as of that elicited by a tap with the soft part of the finger; but inasmuch as the tangential motion of a vibrating bell has been very generally ignored, the production of sound in this manner has been felt as a difficulty. It is now scarcely necessary to point out that the effect of the friction is in the first instance to excite tangential motion, and that the point of application of the friction is the place where the tangential motion is greatest, and therefore where the normal motion vanishes.

235. The existence of tangential vibration in the case of a bell was verified in the following manner. A so-called air-pump receiver was securely fastened to a table, open end uppermost, and set into vibration with the moistened finger. A small chip in the rim, reflecting the light of a candle, gave a bright spot whose motion could be observed with a Coddington lens suitably fixed. As the finger was carried round, the line of vibration was seen to revolve with an angular velocity double that of the finger; and the amount of excursion (indicated by the length of the line of light), though variable, was finite in every position. There was, however, some difficulty in observing the correspondence between the momentary direction of vibration and the situation of the point of excitement. To effect this satisfactorily it was found necessary to apply the friction in the neighbourhood of one point. It then became evident that the spot moved tangentially when the bell was excited at points distant therefrom 0, 90, 180, or 270 degrees.; and normally when the friction was applied at the intermediate points corresponding to 45, 135, 225 and 315 degrees. Care is sometimes required in order to make the bell vibrate in its gravest mode without sensible admixture of overtones.

If there be a small load at any point of the circumference, a slight augmentation of period ensues, which is different according as the loaded point coincides with a node of the normal or of the tangential motion, being greater in the latter case than in the former. The sound produced depends therefore on the place of excitation; in general both tones are heard, and by interference give rise to *beats*, whose frequency is equal to the difference between the frequencies of the two tones. This phenomenon may often be observed in the case of large bells.

235 *a*. In determining the number of nodal meridians (2*s*) corresponding to any particular tone of a bell, advantage may be taken of beats, whether due to accidental irregularities or introduced for the purpose by special loading (compare §§ 208, 209). By tapping cautiously round a circle of *latitude* the places may be investigated where the beats disappear, owing to the absence of one or other of the component tones. But here a decision must not be made too hastily. The inaudibility of the beats may be favoured by an unsuitable position of the ear or of the mouth of the resonator used in connection with the ear. By travelling round, a situation is soon found where the observation can be made to the best advantage. In the neighbourhood of the place where the blow is being tried there is a loop of the vibration which is most excited and a (coincident) node of the vibration which is least excited. When the ear is opposite to a node of the first vibration, and therefore to a loop of the second, the original inequality is redressed, and distinct beats may be heard even though the deviation of the blow from a nodal point may be very small. The accurate determination in this way of two consecutive places where no beats are generated is all that is absolutely necessary for the purpose in view. The ratio of the entire circumference of the circle of latitude to the arc between the points in question is in fact 4*s*. Thus, if the arc between consecutive points proved to be 45°, we should infer that we were dealing with the case of *s* = 2, in which the deformation is elliptical. As a greater security against error, it is advisable in practice to determine a larger number of points where no beats occur. Unless the deviation from symmetry be considerable, these points should be uniformly distributed along the circle of latitude[1].

In the above process for determining nodes we are supposed to hear distinctly the tone corresponding to the vibration under investigation. For this purpose the beats are of assistance in directing the attention; but in dealing with the more difficult subjects, such as church bells, it is advisable to have recourse to resonators. A set of v. Helmholtz's pattern, as manufactured by König, are very convenient. The one next higher in pitch to the tone under examination is chosen and tuned by advancing the finger across the aperture. Without the security afforded by resonators, the determination of the octave is very uncertain.

The only class of bells, for which an approximate theory can be given, are those with thin walls, §§ 233, 235 c. Of such the following glass bells may be regarded as examples :—

$$\text{I.} \quad c', \quad e''\flat, \quad c'''\sharp.$$
$$\text{II.} \quad a, \quad c''\sharp, \quad b''.$$
$$\text{III.} \quad f'\sharp, \quad b''.$$

The value of s for the gravest tone was 2, for the second 3, and for the third tone 4.

Similar observations have been made upon a so-called hemispherical bell, of nearly uniform thickness, and weighing about 3 cwt. Four tones could be plainly heard,

$$e\flat, \quad f'\sharp, \quad e'', \quad b'',$$

the pitch being taken from a harmonium. The gravest tone has a long duration. When the bell is struck by a hard body, the higher tones are at first predominant, but after a time they die away, and leave $e\flat$ in possession of the field. If the striking body be soft, the original preponderance of the higher elements is less marked.

By the method described there was no difficulty in shewing that the four tones correspond respectively to $s = 2, 3, 4, 5$. Thus for the gravest tone the vibration is elliptical with 4 nodal meridians, for the next tone there are 6 nodal meridians, and so on.

---

[1] The bells, or gongs, as they are sometimes called, of striking clocks often give disagreeable beats. A remedy may be found in a suitable rotation of the bell round its axis.

Tapping along a meridian shewed that the sounds became less clear as the edge was departed from, and this in a continuous manner with no suggestion of a nodal circle of latitude. A question to which we shall recur in connection with church bells here suggests itself. Which of the various coexisting tones characterizes the pitch of the bell as a whole? It would appear to be the third in order, for the founders gave the pitch as E natural.

In church bells there is great concentration of metal at the "sound-bow" where the clapper strikes, indeed to such an extent that we can hardly expect much correspondence with what occurs in the case of thin uniform bells. But the method already described suffices to determine the number of nodal meridians for all the more important tones. From a bell of 6 cwt. by Mears and Stainbank 6 tones could be obtained, viz.:

$$e', \quad c'', \quad f''+, \quad b''\flat, \quad d''', \quad f'''.$$
$$(4) \quad (4) \quad (6) \quad (6) \quad (8)$$

The pitch of this bell as given by the makers is d'', so that it is the fifth in the above series of tones which characterizes the bell. The number of nodal meridians in the various components is indicated within the parentheses. Thus in the case of the tone e' there are 4 nodal meridians. A similar method of examination along a meridian shewed that there was no nodal circle of latitude. At the same time differences of intensity were observed. This tone is most fully developed when the blow is delivered about midway between the crown and the rim of the bell.

The next tone is c''. Observation shewed that for this vibration also there are four, and but four, nodal meridians. But now there is a well-defined nodal circle of latitude, situated about a quarter of the way up from the rim towards the crown. As heard with a resonator, this tone disappears when the blow is accurately delivered at some point of this circle, but revives with a very small displacement on either side. The nodal circle and the four meridians divide the surface into segments, over each of which the normal motion is of one sign.

To the tone f'' correspond 6 nodal meridians. There is no well-defined nodal circle. The sound is indeed very faint when the tap is much displaced from the sound-bow; it was thought to fall to a minimum when a position about half-way up was reached.

The three graver tones are heard loudly from the sound-bow. But the next in order, $b''\flat$, is there scarcely audible, unless the blow is delivered to the rim itself in a tangential direction. The maximum effect occurs about half-way up. Tapping round the circle revealed 6 nodal meridians.

The fifth tone, d''', is heard loudly from the sound-bow, but soon falls off when the locality of the blow is varied, and in the upper three-fourths of the bell it is very faint. No distinct circular node could be detected. Tapping round the circumference shewed that there were 8 nodal meridians.

The highest tone recorded, f''', was not easy of observation, and the mode of vibration could not be fixed satisfactorily.

Similar results have been obtained from a bell of 4 cwt., cast by Taylor of Loughborough for Ampton church. The nominal pitch (without regard to octave) was d, and the following were the tones observed:—

$$e'\flat - 2, \quad d'' - 6, \quad f'' + 4, \quad b''\flat - b'', \quad d''', \quad g'''.$$
$$(4) \qquad (4) \qquad (6) \qquad (6) \qquad (8)$$

In the specification of pitch the numerals following the note indicate by how much the frequency for the bell differed from that of the harmonium employed as a standard. Thus the gravest tone $e'\flat$ gave 2 beats per second, and was flat. When the number exceeds 3, it is the result of somewhat rough estimation, and cannot be trusted to be quite accurate. Moreover, as has been explained, there are in strictness two frequencies under each head, and these often differ sensibly. In the case of the 4th tone, $b''\flat - b''$ means that, as nearly as could be judged, the pitch of the bell was midway between the two specified notes of the harmonium.

Observations in the laboratory upon the above-mentioned bells having settled the modes of vibration corresponding to the five gravest tones, other bells of the church pattern could be sufficiently investigated by simple determinations of pitch. The results are collected in the following table[1], and include, besides those already given, observations upon a Belgian bell, the property of Mr Haweis, and upon the five bells of the Terling peal. As regards

---

[1] On Bells, *Phil. Mag.*, vol. 29, p. 1, 1890.

the nominal pitch of the latter bells, several observers concurred in fixing the notes of the peal as

$$f\sharp, \qquad g\sharp, \qquad a\sharp, \qquad b, \qquad c\sharp,$$

no attention being paid to the question of the octave.

Mears, 1888.	Ampton, 1888.	Belgian Bell.	Terling (5), Osborn, 1783.	Terling (4), Mears, 1810.	Terling (3), Graye, 1623.	Terling (2), Gardner, 1723.	Terling (1), Warner, 1863.
Actual Pitch by Harmonium.							
e'	$e'\flat - 2$	$d' - 4$	$g - 3$	$a + 3$	$a\sharp + 3$	$d' - 6$	$d' + 2$
c''	$d'' - 6$	$c''\sharp - d''$	$g' - 4$	$g'\sharp - 4$	$a' + 6$	$a'\sharp - 5$	$b' + 2$
$f'' +$	$f'' + 4$	$f'' + 1$	$a' + 6$	$b' + 6$	$c''\sharp + 4$	$d'' + 8$	e''
$b''\flat$	$b''\flat - b''$	$a'' - 6$	$d'' - 3$	$d''\sharp - e''$	$e'' + 6$	$g''\sharp + (10)$	$g''\sharp + 4$
$/d'''$	d'''	......	$f''\sharp - 2$	$g''\sharp - 6$	$a''\sharp$	$b'' + 2$	$c'''\sharp + 3$
f'''	g'''						
Pitch referred to fifth tone as *c*.							
d	$c\sharp - 2$		$c\sharp - 3$	$c\sharp + 3$	$c + 3$	$e\flat - 6$	$c\sharp + 2$
$b\flat$	$c - 6$		$c\sharp - 4$	$c - 4$	$b + 6$	$b - 5$	$b\flat + 2$
$e\flat +$	$e\flat + 4$		$e\flat + 6$	$e\flat + 6$	$e\flat + 4$	$e\flat + 8$	$e\flat$
$a\flat$	$a\flat - a$		$a\flat - 3$	$g - g\sharp$	$f\sharp + 6$	$a + 8$	$g + 4$
c	c		$c - 2$	$c - 6$	c	$c + 2$	$c + 3$

Examination of the table reveals the remarkable fact that in every case of the English bells it is the 5th tone in order which agrees with the nominal pitch, and that, with the exception of Terling (4), no other tone shews such agreement[1]. Moreover, as appeared most clearly in the case of the bell cast by Mears and Stainbank, the nominal pitch, as given by the makers, is *an octave below* the only corresponding tone.

The highly composite, and often discordant, character of the sounds of bells tends to explain the discrepancies sometimes manifested in estimations of pitch. Mr Simpson, who has devoted much attention to the subject, has put forward strong arguments for the opinion that the Belgian makers determine the pitch of their bells by the tone 2nd in order in the above series, so that for instance the pitch of Terling (3) would be *a* and not *a*♯. In subordination to this tone they pay attention also to the next (the 3rd in order), classifying their bells according to the character

---

[1] In this comparison the gravest tone is disregarded.

of the *third*, whether major or minor, so compounded. Thus in Terling (3) the interval, a' to c'' , is a *major* third. The comparative neglect with which the Belgians treat the 5th tone, regarded almost exclusively by English makers, may perhaps be explained by a less prominent development of this tone in Belgian bells, and by a difference in treatment. When a bell is sounded alone, or with other bells in a comparatively slow succession, attention is likely to concentrate itself upon the graver and more persistent elements of the sound rather than upon the acuter and more evanescent elements, while the contrary may be expected to occur when bells follow one another rapidly in a peal.

In any case the false octaves with which the Table abounds are simple facts of observation, and we may well believe that their correction would improve the general effect. Especially should the octave between the 2nd tone and the 5th tone be made true. Probably the lower octave of the gravest, or *hum-note*, as it is called by English founders, is of less importance. The same may be said of the *fifth*, given by the 4th tone of the series, which is much less prominent. The variations recorded in the Table would seem to shew that no insuperable obstacle stands in the way of obtaining accurate harmonic relations among the various tones.

No adequate explanation has been given of the form adopted for church bells. It appears both from experiment and from the theory of thin shells that this form is especially stiff, as regards the principal mode of deformation ($s = 2$), to forces applied normally and near the rim. Possibly the advantage of this form lies in its rendering less prominent the gravest component of the sound, or the hum-note.

CHAPTER X a.

CURVED PLATES OR SHELLS.

235 b. In the last chapter (§§ 232, 233) we have considered the comparatively simple problem of the vibration in two dimensions of a cylindrical shell, so far at least as relates to vibrations of a *flexural* character. The shell is supposed to be thin, to be composed of isotropic material, and to be bounded by infinite coaxal cylindrical surfaces. It is proposed in the present chapter to treat the problem of the cylindrical shell more generally, and further to give the theory of the flexural vibrations of spherical shells.

In considering the deformation of a thin shell the most important question which presents itself is whether the middle surface, viz. the surface which lies midway between the boundaries, does, or does not, undergo extension. In the former case the deformation may be called *extensional*, and its potential energy is proportional to the thickness of the shell, which will be denoted by $2h$. Since the inertia of the shell, and therefore the kinetic energy of a given motion, is also proportional to h, the frequencies of vibration are in this case *independent* of h, § 44.   On the other hand, when no line traced upon the middle surface undergoes extension, the potential energy of a deformation is of a higher order in the small quantity h. If the shell be conceived to be divided into laminæ, the extension in any lamina is proportional to its distance from the middle surface, and the contribution to the potential energy is proportional to the square of that distance. When the integration over the thickness is carried out, the whole potential energy is found to be proportional to h^3. Vibrations of this kind may be called inextensional, or flexural, and (§ 44) their frequencies are proportional to h, so that the sounds become graver without limit as the thickness is reduced.

Vibrations of the one class may thus be considered to depend upon the term of order h, and vibrations of the other class upon the term of order h^3, in the expression for the potential energy. In general both terms occur; and it is only in the limit that the separation into two classes becomes absolute. This is a question which has sometimes presented difficulty. That in the case of extensional vibrations the term in h^3 should be negligible in comparison with the term in h seems reasonable enough. But is it permissible in dealing with the other class of vibrations to omit the term in h while retaining the term in h^3?

The question may be illustrated by consideration of a statical problem. It is a general mechanical principle (§ 74) that, if given displacements (not sufficient by themselves to determine the configuration) be produced in a system originally in equilibrium by forces of corresponding types, the resulting deformation is determined by the condition that the potential energy shall be as small as possible. Apply this principle to the case of an elastic shell, the given displacements being such as not of themselves to involve a stretching of the middle surface. The resulting deformation will, in general, include both stretching and bending, and any expression for the energy will be of the form

$$Ah \, (\text{extension})^2 + Bh^3 \, (\text{bending})^2 \dots\dots\dots\dots\dots(1).$$

This energy is to be as small as possible. Hence, when the thickness is diminished without limit, the actual displacement will be one of pure bending, if such there be, consistent with the given conditions.

At first sight it may well appear strange that of the two terms the one proportional to the cube of the thickness is to be retained, while that proportional to the first power may be neglected. The fact, however, is that the large potential energy that would accompany any stretching of the middle surface is the very reason why such stretching does not occur. The comparative largeness of the coefficient (proportional to h) is more than neutralized by the smallness of the stretching itself, to the square of which the energy is proportional.

An example may be taken from the case of a rod, clamped at one end A, and deflected by a lateral force; it is required to trace the effect of constantly increasing stiffness of the part included between A and a neighbouring point B. In the limit we may regard the rod as clamped at B, and neglect the energy of the part AB, in spite of, or rather in consequence of, its infinite stiffness.

It would thus be a mistake to regard the omission of the term in h as especially mysterious. In any case of a constraint which is supposed to be gradually introduced (§ 92 *a*), the vibrations tend to arrange themselves into two classes, in one of which the constraint is observed, while in the other, in which the constraint is violated, the frequencies increase without limit. The analogy with the shell of gradually diminishing thickness is complete if we suppose that at the same time the elastic constants are increased in such a manner that the resistance to *bending* remains unchanged. The resistance to extension then becomes infinite, and in the limit one class of vibrations is purely inextensional, or flexural.

In the investigation which we are about to give of the vibrations of a cylindrical shell, the extensional and the inextensional classes will be considered separately. It would apparently be more direct to establish in the first instance a general expression for the potential energy complete as far as the term in h^3, from which the whole theory might be deduced. Such an expression would involve the extensions and the curvatures of the middle surface. It appears, however, that this method is difficult of application, inasmuch as the potential energy (correct to h^3) does not depend only upon the above-mentioned quantities, but also upon the manner of application of the normal forces, which are in general implied in the existence of middle surface extensions[1]

235 *c*. The first question to be considered is the expression of the conditions that the middle surface remain unextended, or if these conditions be violated, to find the values of the extensions in terms of the displacements of the various points of the surface.

[1] On the Uniform Deformation in Two Dimensions of a Cylindrical Shell, with Application to the General Theory of Deformation of Thin Shells. *Proc. Math. Soc.*, vol. xx. p. 372, 1889.

We will suppose in the first instance merely that the surface is of revolution, and that a point is determined by cylindrical co-ordinates z, r, ϕ. After deformation the co-ordinates of the above point become $z + \delta z$, $r + \delta r$, $\phi + \delta \phi$ respectively. If ds denote an element of arc traced upon the surface,

$$(ds + d\delta s)^2 = (dz + d\delta z)^2 + (r + \delta r)^2 (d\phi + d\delta \phi)^2 + (dr + d\delta r)^2,$$

so that

$$ds \, d\delta s = dz \, d\delta z + r^2 d\phi \, d\delta \phi + r \delta r \, (d\phi)^2 + dr \, d\delta r \ldots\ldots(1).$$

In this we regard z and ϕ as independent variables, so that, for example,

$$d\delta z = \frac{d\delta z}{dz} dz + \frac{d\delta z}{d\phi} d\phi \; ;$$

while

$$dr = \frac{dr}{dz} dz + \frac{dr}{d\phi} d\phi,$$

in which by hypothesis $dr/d\phi = 0$. Accordingly

$$\frac{d\delta s}{ds} = \frac{(dz)^2}{(ds)^2} \left\{ \frac{d\delta z}{dz} + \frac{dr}{dz} \frac{d\delta r}{dz} \right\} + \frac{(d\phi)^2}{(ds)^2} \left\{ r^2 \frac{d\delta \phi}{d\phi} + r \delta r \right\}$$

$$+ \frac{dz \, d\phi}{(ds)^2} \left\{ \frac{d\delta z}{d\phi} + r^2 \frac{d\delta \phi}{dz} + \frac{dr}{dz} \frac{d\delta r}{d\phi} \right\} \ldots\ldots\ldots\ldots\ldots(2),$$

in which $d\delta s/ds$ represents the *extension* of the element ds. If there be no extension of any arc traced upon the surface, (2) must vanish independently of any relations between dz and $d\phi$. Hence

$$\frac{d\delta z}{dz} + \frac{dr}{dz} \frac{d\delta r}{dz} = 0 \ldots\ldots\ldots\ldots\ldots\ldots(3),$$

$$r \frac{d\delta \phi}{d\phi} + \delta r = 0 \ldots\ldots\ldots\ldots\ldots(4),$$

$$\frac{d\delta z}{d\phi} + r^2 \frac{d\delta \phi}{dz} + \frac{dr}{dz} \frac{d\delta r}{d\phi} = 0 \ldots\ldots\ldots\ldots(5).$$

From these, by elimination of δr,

$$\frac{d\delta z}{dz} - \frac{dr}{dz} \frac{d}{dz} \left(r \frac{d\delta \phi}{d\phi} \right) = 0,$$

$$\frac{d\delta z}{d\phi} + r^2 \frac{d\delta \phi}{dz} - r \frac{dr}{dz} \frac{d^2 \delta \phi}{d\phi^2} = 0 \; ;$$

and again, by elimination of δz,

$$\frac{d}{dz} \left(r^2 \frac{d\delta \phi}{dz} \right) - r \frac{d^2 r}{dz^2} \frac{d^2 \delta \phi}{d\phi^2} = 0 \ldots\ldots\ldots\ldots(6).$$

If the distribution of thickness and the form of the boundary or boundaries be symmetrical with respect to the axis, the normal functions of the system are to be found by assuming $\delta\phi$ to be proportional to $\cos s\phi$, or $\sin s\phi$. The equation for $\delta\phi$ may then be put into the form

$$r^2 \frac{d}{dz}\left(r^2 \frac{d\delta\phi}{dz}\right) + s^2 r^3 \frac{d^2 r}{dz^2} \delta\phi = 0 \dots\dots\dots\dots (7).$$

It will be seen that the conditions of inextension go a long way towards determining the form of the normal functions.

The simplest application is to the case of a *cylinder* for which r is constant, equal say to a. Thus (3), (4), (5), (7) become simply

$$\frac{d\delta z}{dz} = 0, \qquad \delta r + a \frac{d\delta\phi}{d\phi} = 0, \qquad \frac{d\delta z}{d\phi} + a^2 \frac{d\delta\phi}{dz} = 0 \dots\dots(8),$$

$$\frac{d^2\delta\phi}{dz^2} = 0 \dots\dots\dots\dots\dots\dots (9).$$

By (9), if $\delta\phi \propto \cos s\phi$, we may take

$$a\,\delta\phi = (A_s a + B_s z)\cos s\phi \dots \dots\dots\dots\dots(10),$$

and then, by (8), $\qquad \delta r = s\,(A_s a + B_s z)\sin s\phi \dots\dots\dots\dots(11),$

$$\delta z = -s^{-1}B_s a \sin s\phi \dots\dots\dots\dots\dots(12).$$

Corresponding terms, with fresh arbitrary constants, obtained by writing $s\phi + \tfrac{1}{2}\pi$ for $s\phi$, may of course be added. If $B_s = 0$, the displacement is in two dimensions only (§ 233).

If an inextensible disc be attached to the cylinder at $z = 0$, so as to form a kind of cup, the displacements δr and $\delta\phi$ must vanish for that value of z, exception being made of the case $s = 1$. Hence $A_s = 0$, and

$$a\,\delta\phi = B_s z \cos s\phi, \quad \delta r = s B_s z \sin s\phi, \quad \delta z = -s^{-1}B_s a \sin s\phi \dots(13).$$

Again, in the case of a *cone*, for which $r = \tan\gamma . z$, the equations (3), (4), (5), (7) become

$$\left. \begin{aligned} \frac{d\delta z}{dz} + \tan\gamma\,\frac{d\delta r}{dz} &= 0, \quad z\tan\gamma\,\frac{d\delta\phi}{d\phi} + \delta r = 0 \\[2mm] \frac{d\delta z}{d\phi} + z^2 \tan^2\gamma\,\frac{d\delta\phi}{dz} + \tan\gamma\,\frac{d\delta r}{d\phi} &= 0 \end{aligned} \right\} \dots(14),$$

$$\frac{d}{dz}\left(z^2 \frac{d\delta\phi}{dz}\right) = 0 \dots\dots\dots\dots\dots(15).$$

If we take, as usual, $\delta\phi \propto \cos s\phi$, we get as the solution of (15)

$$\delta\phi = (A_s + B_s z^{-1}) \cos s\phi \dots\dots\dots(16),$$

and corresponding thereto

$$\delta r = s \tan\gamma (A_s z + B_s) \sin s\phi \dots\dots\dots(17),$$

$$\delta z = \tan^2\gamma [s^{-1} B_s - s (A_s z + B_s)] \sin s\phi \dots(18).$$

If the cone be complete up to the vertex at $z = 0$, $B_s = 0$, so that

$$\delta\phi = A_s \cos s\phi \dots\dots\dots\dots(19),$$

$$\delta r = s A_s r \sin s\phi \dots\dots\dots\dots(20),$$

$$\delta z = - s A_s \tan\gamma\, r \sin s\phi \dots\dots\dots(21).$$

For the cone and the cylinder, the second term in the general equation (7) vanishes. We shall obtain a more extensive class of soluble cases by supposing that the surface is such that

$$r^3 \frac{d^2 r}{dz^2} = \text{constant} \dots\dots\dots\dots (22),$$

an equation which is satisfied by surfaces of the second degree in general. If

$$\frac{z^2}{a^2} + \frac{r^2}{b^2} = 1 \dots\dots\dots\dots\dots(23),$$

we shall find

$$r^3 \frac{d^2 r}{dz^2} = - \frac{b^4}{a^2} \dots\dots\dots\dots(24);$$

and thus (7) takes the form

$$\frac{d^2\delta\phi}{d\alpha^2} - \frac{s^2 b^4}{a^2} \delta\phi = 0 \dots\dots\dots\dots(25),$$

if $\delta\phi \propto \cos s\phi$, and α is defined by

$$\alpha = \int r^{-2}\, dz \dots\dots\dots\dots(26),$$

or in the present case

$$\alpha = \frac{a}{2b^2} \log \frac{a + z}{a - z} \dots\dots\dots\dots(27).$$

The solution of (25) is

$$\delta\phi = \left[A \left(\frac{a+z}{a-z}\right)^{-\frac{1}{2}s} + B \left(\frac{a+z}{a-z}\right)^{+\frac{1}{2}s} \right] \cos s\phi \dots\dots(28).$$

The corresponding values of δr and δz are to be obtained from (4) and (5).

If the surface be complete through the vertex $z = a$, the term multiplied by B must disappear. Thus, omitting the constant multiplier, we may take

$$\delta\phi = \left(\frac{a-z}{a+z}\right)^{\frac{1}{2}s} \cos s\phi \quad \dots\dots\dots\dots\dots(29);$$

whence, by (4), (5),

$$\delta r = \frac{sb}{a}\frac{(a-z)^{\frac{1}{2}s+\frac{1}{2}}}{(a+z)^{\frac{1}{2}s-\frac{1}{2}}} \sin s\varphi\dots\dots\dots\dots\dots(30),$$

$$\delta z = (sz+a)\frac{b^2(a-z)^{\frac{1}{2}s}}{a^2(a+z)^{\frac{1}{2}s}} \sin s\phi\dots\dots\dots\dots(31).$$

If we measure z' from the vertex, $z' = a - z$, and we may write

$$\delta\phi = \left(\frac{z'}{r}\right)^{s} \cos s\phi\dots\dots\dots\dots\dots\dots\dots\dots\dots (32),$$

$$\delta r = sr\left(\frac{z'}{r}\right)^{s} \sin s\phi\dots\dots\dots\dots\dots\dots\dots\dots\dots(33),$$

$$\delta z = -\delta z' = \frac{b^2}{a^2}\left\{(s+1)\,a - sz'\right\}\left(\frac{z'}{r}\right)^{s} \sin s\phi \ \dots\dots(34).$$

For the parabola, a and b are infinite, while $b^2/a = 2a'$, and $r^2 = 4a'z'$. Thus we may take[1]

$$\delta\phi = r^s \cos s\phi, \quad \delta r = sr^{s+1}\sin s\phi, \quad \delta z' = -2\,(s+1)\,a'r^s \sin s\phi\dots(35).$$

We will now take into consideration the important case of the sphere, for which in (23) $b = a$. Denoting by θ the angle between the radius vector and the axis, we have $z = a\cos\theta$, $r = a\sin\theta$, and thus from (29), (30), (31)

$$\delta\phi = \cos s\phi \tan^s \tfrac{1}{2}\theta \dots\dots\dots\dots\dots\dots\dots(36),$$

$$\delta r/a = s\sin s\phi \sin\theta \tan^s \tfrac{1}{2}\theta \dots\dots\dots\dots\dots(37),$$

$$\delta z/a = (1 + s\cos\theta)\sin s\phi \tan^s \tfrac{1}{2}\theta\dots\dots\dots\dots(38).$$

The other terms of the complete solution, corresponding to (28), are to be obtained by changing the sign of s.

In the above equations the displacements are resolved parallel and perpendicular to the axis $\theta = 0$. It would usually be more convenient to resolve along the normal and the meridian. If the components in these directions be denoted by w and $a\delta\theta$, we have

$$w = \delta r \sin\theta + \delta z \cos\theta, \quad a\delta\theta = \delta r \cos\theta - \delta z \sin\theta;$$

[1] On the Infinitesimal Bending of Surfaces of Revolution. *Proc. Math. Soc.*, vol. XIII. p. 4, 1881.

so that altogether

$$\delta\phi = \cos s\phi \left[A_s \tan^s \tfrac{1}{2}\theta + B_s \cot^s \tfrac{1}{2}\theta \right] \dots \dots (39),$$

$$\delta\theta = -\sin s\phi \sin \theta \left[A_s \tan^s \tfrac{1}{2}\theta - B_s \cot^s \tfrac{1}{2}\theta \right] \dots \dots (40),$$

$$w/a = \sin s\phi \left[A_s (s + \cos \theta) \tan^s \tfrac{1}{2}\theta + B_s (s - \cos \theta) \cot^s \tfrac{1}{2}\theta \right] \dots (41).$$

To the above may be added terms derived by writing $s\phi + \tfrac{1}{2}\pi$ for $s\phi$, and changing the arbitrary constants.

235 d.   We now proceed to apply the equations of § 235 c to the principal extensions of a cylindrical surface, with a view to the formation of the expression for the potential energy.   The axial and circumferential extensions will be denoted respectively by ϵ_1, ϵ_2, and the shear by ϖ.   The first of these is given by (2) § 235 c, if we suppose that $d\phi = 0$, $dz/ds = 1$.   Since in the case of a cylinder $dr/dz = 0$, we find

$$\epsilon_1 = \frac{d\delta z}{dz} \dots \dots (1).$$

In like manner

$$\epsilon_2 = \frac{\delta r}{a} + \frac{d\delta\phi}{d\phi} \dots \dots (2).$$

The value of the shear may be arrived at by considering the difference of extensions for the two diagonals of an infinitesimal square whose sides are dz and $a\,d\phi$.   It is

$$\varpi = \frac{1}{a}\frac{d\delta z}{d\phi} + a\frac{d\delta\phi}{dz} \dots \dots (3).$$

The next part of the problem, viz. the expression of the potential energy by means of ϵ_1, ϵ_2, ϖ, appertains to the general theory of elasticity, and can only be treated here in a cursory manner.   But it may be convenient to give the leading steps of the investigation, referring for further explanations to the treatises of Thomson and Tait and of Love.   In the notation of the former (*Natural Philosophy*, § 694) the general equations in three dimensions are

$$na = S, \quad nb = T, \quad nc = U \dots \dots (4),$$

$$\left. \begin{aligned} Me &= P - \sigma (Q + R) \\ Mf &= Q - \sigma (R + P) \\ Mg &= R - \sigma (P + Q) \end{aligned} \right\} \dots \dots (5),$$

where

$$\sigma = \frac{m - n}{2m} \dots \dots (6)[1].$$

---

[1] M is Young's modulus, σ is Poisson's ratio, n is the constant of rigidity, and $(m - \tfrac{1}{3}n)$ that of compressibility.

The energy w, corresponding to unit of volume, is given by

$$2w = (m + n)(e^2 + f^2 + g^2)$$
$$+ 2(m - n)(fg + ge + ef) + n(a^2 + b^2 + c^2)\ldots\ldots(7).$$

In the application to a lamina, supposed parallel to the plane xy, we are to take $R = 0$, $S = 0$, $T = 0$, so that

$$g = -\sigma\frac{e + f}{1 - \sigma}, \quad a = 0, \quad b = 0\ldots\ldots\ldots\ldots(8).$$

Thus in terms of the extensions e, f, parallel to x, y, and of the shear c, we get

$$w = n\left\{e^2 + f^2 + \frac{m - n}{m + n}(e + f)^2 + \tfrac{1}{2}c^2\right\}\ldots\ldots\ldots(9).$$

This is the energy reckoned per unit of volume. In order to adapt the expression to our purposes, we must multiply it by the thickness ($2h$). Hence as the energy per unit area of a shell of thickness $2h$, we may take in the notation adopted at the commencement of this section,

$$2nh\left\{\epsilon_1^2 + \epsilon_2^2 + \tfrac{1}{2}\varpi^2 + \frac{m - n}{m + n}(\epsilon_1 + \epsilon_2)^2\right\}\ldots\ldots\ldots(10).$$

This expression may be applied to curved as well as to plane plates, for any modification due to curvature must involve higher powers of h. The same is true of the energy of bending.

235 e. We are now prepared for the investigation of the extensional vibrations of an infinite cylindrical shell, assumed to be periodic with respect both to z and to ϕ. It will be convenient to denote by single letters the displacements parallel to z, ϕ, r; we take

$$\delta z = u, \quad a\delta\phi = v, \quad \delta r = w\ldots\ldots\ldots\ldots(1).$$

These functions are to be assumed proportional to the sines or cosines of jz/a and $s\phi$. Various combinations may be made, of which an example[1] is

$$u = U\cos s\phi\cos jz/a, \quad v = V\sin s\phi\sin jz/a,$$
$$w = W\cos s\phi\sin jz/a\ldots\ldots(2);$$

so that (1), (2), (3), § 235 d

$$a.\epsilon_1 = -jU\cos s\phi\sin jz/a\ldots\ldots\ldots\ldots\ldots\ldots(3),$$
$$a.\epsilon_2 = (W + sV)\cos s\phi\sin jz/a\ldots\ldots\ldots\ldots(4),$$
$$a.\varpi = (-sU + jV)\sin s\phi\cos jz/a\ldots\ldots\ldots\ldots(5).$$

[1] Additions of $\tfrac{1}{2}\pi$ to $s\phi$, or to jz/a, or to both, may of course be made at pleasure.

The potential energy per unit area is thus (10) § 235 d

$$2nha^{-2}\left[\cos^2 s\phi \sin^2 jz/a\left\{j^2U^2+(W+sV)^2+\frac{m-n}{m+n}(W+sV-jU)^2\right\}\right.$$
$$\left.+\tfrac{1}{2}\sin^2 s\phi \cos^2 jz/a\,(-sU+jV)^2\right]\;\ldots\ldots(6).$$

Again, if ρ be the volume density, the kinetic energy per unit of area is

$$\rho h\left[\left(\frac{dU}{dt}\right)^2\cos^2 s\phi \cos^2 jz/a+\left(\frac{dV}{dt}\right)^2\sin^2 s\phi \sin^2 jz/a\right.$$
$$\left.+\left(\frac{dW}{dt}\right)^2\cos^2 s\phi \sin^2 jz/a\right]\;\ldots\ldots(7).$$

In the integration of (6), (7) with respect to z and ϕ, $\tfrac{1}{2}$ is the mean value of the square of each sine or cosine.[1] We may then apply Lagrange's method, regarding U, V, W as independent generalized co-ordinates. If the type of vibration be $\cos pt$, and $p^2\rho/n=k^2$, the resulting equations may be written

$$\{2(N+1)j^2+s^2-k^2a^2\}\,U-(2N+1)jsV-2NjW=0\ldots(8),$$
$$-(2N+1)jsU+\{j^2+2(N+1)s^2-k^2a^2\}\,V+2(N+1)sW=0\ldots(9),$$
$$-2NjU+2(N+1)sV+\{2(N+1)-k^2a^2\}\,W=0\ldots(10),$$

where
$$N=\frac{m-n}{m+n}\;\ldots\ldots\ldots\ldots(11).$$

The frequency equation is that expressing the evanescence of the determinant of this triad of equations. On reduction it may be written

$$[k^2a^2-j^2-s^2]\{k^2a^2[k^2a^2-2(N+1)(j^2+s^2+1)]$$
$$+4(2N+1)j^2\}+4(2N+1)j^2s^2=0\ldots\ldots(12).[2]$$

These equations include of course the theory of the extensional vibrations of a plane plate, for which $a=\infty$. In this application it is convenient to write $a\phi=y$, $s/a=\beta$, $j/a=\gamma$. The displacements are then

$$u=U\cos\beta y\cos\gamma z,\quad v=V\sin\beta y\sin\gamma z,\quad w=W\cos\beta y\sin\gamma z$$
$$\ldots(13).$$

---

[1] In the physical problem of a simple cylinder the range of integration for ϕ is from 0 to 2π; but mathematically we are not confined to one revolution. We may conceive the shell to consist of several superposed convolutions, and then s is not limited to be a whole number.

[2] Note on the Free Vibrations of an infinitely long Cylindrical Shell. *Proc. Roy. Soc.*, vol. 45, p. 446, 1889.

When a is made infinite while β, γ remain constant, the equations (10), (8), (9) ultimately assume the form $W = 0$, and

$$\{2\,(N+1)\,\gamma^2 + \beta^2 - k^2\}\,U - (2N+1)\,\gamma\beta V = 0 \dots (14),$$

$$-(2N+1)\,\gamma\beta U + \{\gamma^2 + 2\,(N+1)\,\beta^2 - k^2\}\,V = 0 \dots(15);$$

and the determinantal equation (12) becomes

$$k^2\,[k^2 - \gamma^2 - \beta^2]\,[k^2 - 2\,(N+1)\,(\gamma^2 + \beta^2)] = 0 \dots\dots(16).$$

In (16), as was to be expected, k^2 appears as a function of $(\beta^2 + \gamma^2)$. The first root $k^2 = 0$ relates to flexural vibrations, not here regarded. The second root is

$$k^2 = \beta^2 + \gamma^2 \dots\dots\dots\dots\dots\dots\dots\dots(17),$$

or

$$p^2 = \frac{n}{\rho}\,(\beta^2 + \gamma^2) \dots\dots\dots\dots\dots\dots(18).$$

At the same time (14) gives

$$\gamma U - \beta V = 0 \dots\dots\dots\dots\dots\dots\dots(19).$$

These vibrations involve only a shearing of the plate in its own plane. For example, if $\gamma = 0$, the vibration may be represented by

$$u = \cos \beta y \cos pt, \quad v = 0, \quad w = 0 \dots\dots\dots\dots(20).$$

The third root of (16)

$$k^2 = 2\,(N+1)\,(\beta^2 + \gamma^2) = \frac{4m}{m+n}\,(\beta^2 + \gamma^2) \dots\dots (21)$$

gives

$$p^2 = \frac{4mn}{m+n}\frac{\beta^2 + \gamma^2}{\rho} \dots\dots\dots\dots\dots (22).$$

The corresponding relation between U and V is

$$\beta U + \gamma V = 0 \dots\dots\dots\dots\dots\dots\dots(23).$$

A simple example of this case is given by supposing in (13), (23), $\beta = 0$. We may take

$$u = \cos \gamma z \cos pt, \quad v = 0, \quad w = 0 \dots\dots\dots\dots (24),$$

the motion being in one dimension.

Reverting to the cylinder we will consider in detail a few particular cases of importance. The first arises when $j = 0$, that is, when the vibrations are independent of z. The three equations (8), (9), (10) then reduce to

$$(s^2 - k^2 a^2)\,U = 0 \dots\dots\dots\dots\dots\dots(25),$$

$$\{2\,(N+1)\,s^2 - k^2 a^2\}\,V + 2\,(N+1)\,s W = 0 \dots\dots\dots(26),$$

$$2\,(N+1)\,s V + \{2\,(N+1) - k^2 a^2\}\,W = 0 \dots\dots\dots(27);$$

and they may be satisfied in two ways. First let $V = W = 0$; then U may be finite, provided

$$s^2 - k^2 a^2 = 0 \dots\dots\dots\dots\dots (28).$$

The corresponding type for u is

$$u = \cos s\phi \cos pt \dots\dots\dots\dots\dots (29),$$

where

$$p^2 = \frac{ns^2}{\rho a^2} \dots\dots\dots\dots\dots (30).$$

In this motion the material is sheared without dilatation of area or volume, every generating line of the cylinder moving along its own length. The frequency depends upon the circumferential wave-length, and not upon the curvature of the cylinder.

The second kind of vibrations are those for which $U = 0$, so that the motion is strictly in two dimensions. The elimination of the ratio V/W from (26), (27) gives

$$k^2 a^2 \{k^2 a^2 - 2 (N + 1)(1 + s^2)\} = 0 \dots\dots\dots (31),$$

as the frequency equation. The first root is $k^2 = 0$, indicating infinitely slow motion. The modes in question are flexural, for which, according to our present reckoning, the potential energy is evanescent. The corresponding relation between V and W is by (26)

$$sV + W = 0 \dots\dots\dots\dots\dots (32),$$

giving in (3), (4), (5),

$$\epsilon_1 = 0, \quad \epsilon_2 = 0, \quad \varpi = 0.$$

The other root of (31) is

$$k^2 a^2 = 2 (N + 1)(1 + s^2) \dots\dots\dots\dots (33),$$

or

$$p^2 = \frac{4mn}{m + n} \frac{1 + s^2}{a^2 \rho} \dots\dots\dots\dots (34);$$

while the relation between V and W is

$$V - sW = 0 \dots\dots\dots\dots\dots (35).$$

The type of the motion may be taken to be

$$u = 0, \quad v = s \sin s\phi \cos pt, \quad w = \cos s\phi \cos pt \dots\dots (36).$$

It will be observed that when s is very large, the flexural vibrations (32) tend to become exclusively radial, and the extensional vibrations (35) tend to become exclusively tangential.

Another important class of vibrations are those which are characterized by symmetry round the axis, for which accordingly $s = 0$. The general frequency equation (12) reduces in this case to

$$\{k^2 a^2 - j^2\} \{k^2 a^2 [k^2 a^2 - 2(N+1)(j^2+1)] + 4(2N+1)j^2\} = 0$$
$$\ldots (37).$$

Corresponding to the first root we have $U = 0$, $W = 0$, as is readily proved on reference to the original equations (8), (9), (10) with $s = 0$. The vibrations are the purely torsional ones represented by

$$u = 0, \quad v = \sin(jz/a)\cos pt, \quad w = 0 \ldots\ldots\ldots(38),$$

where
$$p^2 = \frac{nj^2}{\rho a^2} \ldots\ldots\ldots\ldots\ldots\ldots (39).$$

The frequency depends upon the wave-length parallel to the axis, and not upon the radius of the cylinder.

The remaining roots of (37) correspond to motions for which $V = 0$, or which take place in planes passing through the axis. The general character of these vibrations may be illustrated by the case where j is small, so that the wave-length is a large multiple of the radius of the cylinder. We find approximately from the quadratic which gives the remaining roots

$$\frac{k^2 a^2}{N+1} = 2 + \frac{2N^2 j^2}{(N+1)^2} \ldots\ldots\ldots\ldots\ldots(40),$$

or
$$k^2 a^2 = \frac{2(2N+1)j^2}{N+1} \ldots\ldots\ldots\ldots(41).$$

The vibrations of (40) are almost purely radial. If we suppose that j actually vanishes, we fall back upon

$$k^2 a^2 = 2(N+1) \ldots\ldots\ldots\ldots\ldots(42),$$

and
$$p^2 = \frac{4mn}{m+n}\frac{1}{a^2\rho} \ldots\ldots \ldots\ldots\ldots(43)[1],$$

obtainable from (33), (34) on introduction of the condition $s = 0$. The type of vibration is now

$$u = 0, \quad v = 0, \quad w = \cos pt \ldots\ldots\ldots\ldots(44).$$

---

[1] This equation was given by Love in a memoir " On the Small Free Vibrations and Deformation of a thin Elastic Shell," *Phil. Trans.*, vol. 179 (1888) p. 523, and also by Chree, Cambridge *Phil. Trans.* vol. xiv, p. 250, 1887.

On the other hand, the vibrations of (41) are ultimately purely axial. As the type we may take

$$u = \cos jz/a \cdot \cos pt, \quad v = 0, \quad w = \frac{m-n}{2m} j \sin jz/a \cdot \cos pt \dots (45),$$

where

$$p^2 = \frac{3m - n}{m} \frac{nj^2}{\rho a^2} \dots\dots\dots\dots\dots (46).$$

Now, if q denote Young's modulus, we have, § 214,

$$q = n (3m - n)/m,$$

so that

$$p^2 = \frac{qj^2}{\rho a^2} \dots\dots\dots\dots\dots\dots (47).$$

Thus u satisfies the equation

$$\frac{d^2u}{dt^2} = \frac{q}{\rho} \frac{d^2u}{dz^2},$$

which is the usual formula (§ 150) for the longitudinal vibrations of a rod, the fact that the section is here a thin annulus not influencing the result to this order of approximation.

Another particular case worthy of notice arises when $s = 1$, so that (12) assumes the form

$$k^2 a^2 (k^2 a^2 - j^2 - 1) [k^2 a^2 - 2 (N + 1)(j^2 + 2)]$$
$$+ 4j^2 (k^2 a^2 - j^2)(2N + 1) = 0 \dots (48).$$

As we have already seen. if j be zero, one of the values of k^2 vanishes. If j be small, the corresponding value of k^2 is of the order j^4. Equation (48) gives in this case

$$k^2 a^2 = \frac{2N + 1}{N + 1} j^4 \dots\dots\dots\dots (49);$$

or in terms of p and q,

$$p^2 = \frac{qj^4}{2\rho a^2} \dots\dots\dots\dots\dots (50).$$

The type of vibration is

$$\left. \begin{array}{l} u = 0 \\ v = \sin \phi \sin jz/a \cdot \cos pt \\ w = - \cos \phi \sin jz/a \cdot \cos pt \end{array} \right\} \dots\dots\dots (51),$$

and corresponds to the flexural vibrations of a rod (§ 163). In (51) v satisfies the equation

$$\frac{d^2v}{dt^2} + \frac{qa^2}{2\rho} \frac{d^4v}{dz^4} = 0,$$

in which $\frac{1}{2}a^2$ represents the square of the radius of gyration of the section of the cylindrical shell about a diameter.

This discussion of particular cases may suffice. It is scarcely necessary to add, in conclusion, that the most general deformation of the middle surface can be expressed by means of a series of such as are periodic with respect to z and ϕ, so that the problem considered is really the most general small motion of an infinite cylindrical shell.

The extensional vibrations of a cylinder of finite length have been considered by Love in his Theory of Elasticity[1] (1893), where will also be found a full investigation of the general equations of extensional deformation.

235 *f*. When a shell is deformed in such a manner that no line traced upon the middle surface changes in length, the term of order h disappears from the expression for the potential energy, and unless we are content to regard this function as zero, a further approximation is necessary. In proceeding to this the first remark that occurs is that the quality of inextension attaches only to the central lamina. Consider, for example, a portion of a cylindrical shell, which is bent so that the original curvature is increased. It is evident that while the middle lamina remains unextended, those laminæ which lie externally must be stretched, and those that lie internally must be contracted. The amount of these stretchings and contractions is proportional in the first place to the distance from the middle surface, and in the second place to the change of curvature which the middle surface undergoes. The potential energy of bending is thus a question of the *curvatures* of the middle surface. Displacements of translation or rotation, such as a rigid body is capable of, may be disregarded.

In order to take the question in its simplest form, let us refer the original surface to the normal and principal tangents at any point P as axes of co-ordinates, and let us suppose that after deformation the lines in the sheet originally coincident with the principal tangents are brought back (if necessary) so as to occupy the same positions as at first. The possibility of this will be apparent when it is remembered that, in virtue of the inextension of the sheet, the angles of intersections of all lines traced

[1] Also *Phil. Trans.* vol. 179 A, 1888.

upon it remain unaltered. The equation of the original surface in the neighbourhood of the point being

$$z = \tfrac{1}{2}\left(\frac{x^2}{\rho_1} + \frac{y^2}{\rho_2}\right)\dots\dots\dots\dots\dots(1),$$

that of the deformed surface may be written

$$z = \tfrac{1}{2}\left\{\frac{x^2}{\rho_1 + \delta\rho_1} + \frac{y^2}{\rho_2 + \delta\rho_2} + 2\tau xy\right\}\dots\dots\dots(2).$$

In strictness $(\rho_1 + \delta\rho_1)^{-1}$, $(\rho_2 + \delta\rho_2)^{-1}$ are the curvatures of the sections made by the planes x, y; but since the principal curvatures are a maximum and a minimum, they represent in general with sufficient accuracy the new principal curvatures, although these are to be found in slightly different planes. The condition of inextension shews that points which have the same x, y in (1) and (2) are corresponding points; and by Gauss's theorem it is further necessary that

$$\frac{\delta\rho_1}{\rho_1} + \frac{\delta\rho_2}{\rho_2} = 0\dots\dots\dots\dots\dots(3).$$

It thus appears that the energy of bending will depend in general upon two quantities, one giving the alterations of principal curvature, and the other τ depending upon the shift (in the material) of the principal planes.

The case of a spherical surface is in some respects exceptional. Previously to the bending there are no planes marked out as principal planes, and thus the position of these planes after bending is of no consequence. The energy depends only upon the alterations of principal curvature, and these by Gauss's theorem are equal and opposite, so that, if a denote the radius of the sphere, the new principal radii are $a + \delta\rho$, $a - \delta\rho$. If the equation of the deformed surface be

$$2z = Ax^2 + 2Bxy + Cy^2\dots\dots\dots\dots(4),$$

we have $(a + \delta\rho)^{-1} + (a - \delta\rho)^{-1} = A + C$,

$$(a + \delta\rho)^{-1} \cdot (a - \delta\rho)^{-1} = AC - B^2;$$

so that $\left(\delta\dfrac{1}{\rho}\right)^2 = \tfrac{1}{4}(A - C)^2 + B^2\dots\dots\dots\dots(5).$

We have now to express the elongations of the various laminæ of a shell when bent, and we will begin with the case where $\tau = 0$, that is, when the principal planes of curvature remain unchanged. It is evident that in this case the shear c vanishes, and we have to deal only with the elongations e and f parallel to the axes, § 235·d. In the section by the plane of zx, let s, s' denote corresponding infinitely small arcs of the middle surface and of a lamina distant h from it. If ψ be the angle between the terminal normals, $s = \rho_1 \psi$, $s' = (\rho_1 + h) \psi$, $s' - s = h \psi$. In the bending, which leaves s unchanged,

$$\delta s' = h \, \delta \psi = h s \, \delta (1/\rho_1).$$

Hence $e = \delta s'/s' = h \delta (1/\rho_1),$

and in like manner $f = h \delta (1/\rho_2)$. Thus for the energy U per unit *area* we have

$$dU = nh^2 dh \left\{ \left(\delta \frac{1}{\rho_1} \right)^2 + \left(\delta \frac{1}{\rho_2} \right)^2 + \frac{m-n}{m+n} \left(\delta \frac{1}{\rho_1} + \delta \frac{1}{\rho_2} \right)^2 \right\},$$

and on integration over the whole thickness of the shell ($2h$)

$$U = \frac{2nh^3}{3} \left\{ \left(\delta \frac{1}{\rho_1} \right)^2 + \left(\delta \frac{1}{\rho_2} \right)^2 + \frac{m-n}{m+n} \left(\delta \frac{1}{\rho_1} + \delta \frac{1}{\rho_2} \right)^2 \right\} \dots \dots (6).$$

This conclusion may be applied at once, so as to give the result applicable to a spherical shell; for, since the original principal planes are arbitrary, they can be taken so as to coincide with the principal planes after bending. Thus $\tau = 0$; and by Gauss's theorem

$$\delta (1/\rho_1) + \delta (1/\rho_2) = 0,$$

so that $U = \dfrac{4nh^3}{3} \left(\delta \dfrac{1}{\rho} \right)^2 \dots \dots \dots \dots \dots \dots (7),$

where $\delta (1/\rho)$ denotes the change of principal curvature. Since $e = -f$, $g = 0$, the various laminæ are simply sheared, and that in proportion to their distance from the middle surface. The energy is thus a function of the constant of rigidity only.

The result (6) is applicable directly to the plane plate; but this case is peculiar in that, on account of the infinitude of ρ_1, ρ_2 (3) is satisfied without any relation between $\delta \rho_1$ and $\delta \rho_2$. Thus for a plane plate

$$U = \frac{2nh^3}{3} \left\{ \frac{1}{\rho_1^2} + \frac{1}{\rho_2^2} + \frac{m-n}{m+n} \left(\frac{1}{\rho_1} + \frac{1}{\rho_2} \right)^2 \right\} \dots \dots \dots (8),$$

where $1/\rho_1$, $1/\rho_2$, are the two independent principal curvatures after bending[1].

[1] This will be found to agree with the value (2) § 214, expressed in a different notation.

We have thus far considered τ to vanish; and it remains to investigate the effect of the deformations expressed by

$$\delta z = \tau x y = \tfrac{1}{2}\tau\,(\xi^2 - \eta^2)\ldots\ldots\ldots\ldots\ldots(9),$$

where ξ, η relate to new axes inclined at 45° to those of x, y. The curvatures defined by (9) are in the planes of ξ, η, and are equal in numerical value and opposite in sign. The elongations in these directions for any lamina within the thickness of the shell are $h\tau$, $-h\tau$, and the corresponding energy (as in the case of the sphere just considered) takes the form

$$U' = \frac{4nh^3\tau^2}{3}\ldots\ldots\ldots\ldots\ldots\ldots(10).$$

This energy is to be added[1] to that already found in (6); and we get finally

$$U = \frac{2nh^3}{3}\left\{\left(\delta\frac{1}{\rho_1}\right)^2 + \left(\delta\frac{1}{\rho_2}\right)^2 + \frac{m-n}{m+n}\left(\delta\frac{1}{\rho_1} + \delta\frac{1}{\rho_2}\right)^2 + 2\tau^2\right\}\ldots(11),$$

as the complete expression of the energy, when the deformation is such that the middle surface is unextended. We may interpret τ by means of the angle χ, through which the principal planes are shifted; thus

$$\tau = \chi\left(\frac{1}{\rho_2} - \frac{1}{\rho_1}\right)\ldots\ldots\ldots\ldots\ldots\ldots(12).$$

235 g. We will now proceed with the calculation of the potential energy involved in the bending of a cylindrical shell. The problem before us is the expression of the changes of principal curvature and the shifts of the principal planes at any point $P(z, \phi)$ of the cylinder in terms of the displacements u, v, w. As in § 235 f, take as fixed co-ordinate axes the principal tangents and normal to the undisturbed cylinder at the point P, the axis of x being parallel to that of the cylinder, that of y tangential to the circular section, and that of ζ normal, measured inwards. If, as it will be convenient to do, we measure z and ϕ from the point P, we may express the undisturbed co-ordinates of a material point Q in the neighbourhood of P, by

$$x = z, \qquad y = a\phi, \qquad \zeta = \tfrac{1}{2}a\phi^2\ldots\ldots\ldots\ldots(1).$$

---

[1] There are clearly no terms involving the products of τ with the changes of principal curvature $\delta(\rho_1^{-1})$, $\delta(\rho_2^{-1})$; for a change in the sign of τ can have no influence upon the energy of the deformation defined by (2).

During the displacement the co-ordinates of Q will receive the increments

$$u, \quad w\sin\phi + v\cos\phi, \quad -w\cos\phi + v\sin\phi\,;$$

so that after displacement

$$x = z + u, \qquad y = a\phi + w\phi + v(1 - \tfrac{1}{2}\phi^2),$$
$$\zeta = \tfrac{1}{2}a\phi^2 - w(1 - \tfrac{1}{2}\phi^2) + v\phi\,;$$

or, if u, v, w be expanded in powers of the small quantities z, ϕ,

$$x = z + u_0 + \frac{du}{dz_0}z + \frac{du}{d\phi_0}\phi + \ \ldots\ldots\ldots\ldots\ldots \quad (2),$$

$$y = a\phi + w_0\phi + v_0 + \frac{dv}{dz_0}z + \frac{dv}{d\phi_0}\phi + \ \ldots\ldots\ldots \quad (3),$$

$$\zeta = \tfrac{1}{2}a\phi^2 - w_0 - \frac{dw}{dz_0}z - \frac{dw}{d\phi_0}\phi + v_0\phi$$
$$+ \tfrac{1}{2}w_0\phi^2 - \tfrac{1}{2}\frac{d^2w}{dz_0{}^2}z^2 - \frac{d^2w}{dz_0 d\phi_0}z\phi - \tfrac{1}{2}\frac{d^2w}{d\phi_0{}^2}\phi^2$$
$$+ \frac{dv}{dz_0}z\phi + \frac{dv}{d\phi_0}\phi^2 \ldots\ldots\ldots\ldots\ldots\ldots\ldots(4),$$

$u_0, v_0, \ldots$ being the values of u, v at the point P.

These equations give the co-ordinates of the various points of the deformed sheet. We have now to suppose the sheet moved as a rigid body so as to restore the position (as far as the first power of small quantities is concerned) of points infinitely near P. A purely translatory motion by which the displaced P is brought back to its original position will be expressed by the simple omission in (2), (3), (4) of the terms u_0, v_0, w_0 respectively, which are independent of z, ϕ. The effect of an arbitrary rotation is represented by the additions to x, y, ζ respectively of $y\omega_3 - \zeta\omega_2$, $\zeta\omega_1 - x\omega_3$, $x\omega_2 - y\omega_1$; where for the present purpose $\omega_1, \omega_2, \omega_3$ are small quantities of the order of the deformation, the square of which is to be neglected throughout. If we make these additions to (2), &c., substituting for x, y, ζ in the terms containing θ their approximate values, we find so far as the first powers of z, ϕ

$$x = z + \frac{du}{dz_0}z + \frac{du}{d\phi_0}\phi + a\phi\omega_3,$$

$$y = a\phi + w_0\phi + \frac{dv}{dz_0}z + \frac{dv}{d\phi_0}\phi - z\omega_3,$$

$$\zeta = -\frac{dw}{dz_0}z - \frac{dw}{d\phi_0}\phi + v_0\phi + z\omega_2 - a\phi\omega_1.$$

Now, since the sheet is assumed to be unextended, it must be possible so to determine ω_1, ω_2, ω_3 that to this order $x = z$, $y = a\phi$, $\zeta = 0$.   Hence

$$\frac{du}{dz_0} = 0, \qquad\qquad \frac{du}{d\phi_0} + a\omega_3 = 0,$$

$$\frac{dv}{dz_0} - \omega_3 = 0, \qquad\qquad w_0 + \frac{dv}{d\phi_0} = 0,$$

$$-\frac{dw}{dz_0} + \omega_2 = 0, \qquad\qquad \frac{dw}{d\phi_0} - v_0 + a\omega_1 = 0.$$

The conditions of inextension are thus (if we drop the suffices as no longer required)

$$\frac{du}{dz} = 0, \qquad w + \frac{dv}{d\phi} = 0, \qquad \frac{du}{d\phi} + a\frac{dv}{dz} = 0 \ldots\ldots\ldots(5),$$

which agree with (8) § 235 c.

Returning to (2), &c., as modified by the addition of the translatory and rotatory terms, we get

$$x = z + \text{terms of 2nd order in } z, \ \phi,$$

$$y = a\phi + \qquad\quad ,, \qquad\qquad ,,$$

$$\zeta = \tfrac{1}{2} a\phi^2 + \tfrac{1}{2} w_0\phi^2 - \tfrac{1}{2}\frac{d^2w}{dz_0^2} z^2 - \frac{d^2w}{dz_0 d\phi_0} z\phi$$

$$-\tfrac{1}{2}\frac{d^2w}{d\phi_0^2}\phi^2 + \frac{dv}{dz_0} z\phi + \frac{dv}{d\phi_0}\phi^2 ;$$

or since by (5) $d^2w/dz^2 = 0$, and $dv/d\phi = -w$,

$$\zeta = \tfrac{1}{2} a\phi^2 - \tfrac{1}{2} w_0\phi^2 - \frac{d^2w}{dz_0 d\phi_0} z\phi - \tfrac{1}{2}\frac{d^2w}{d\phi_0^2}\phi^2 + \frac{dv}{dz_0} z\phi.$$

The equation of the deformed surface after transference is thus

$$\zeta = xy\left\{\frac{1}{a}\frac{dv}{dz_0} - \frac{1}{a}\frac{d^2w}{dz_0 d\phi_0}\right\} + y^2\left\{\frac{1}{2a} - \frac{1}{2a^2} w_0 - \frac{1}{2a^2}\frac{d^2w}{d\phi_0^2}\right\} \ldots (6).$$

Comparing with (2) § 235 f we see that

$$\delta\frac{1}{\rho_1} = 0, \qquad \delta\frac{1}{\rho_2} = -\frac{1}{a^2}\left(w + \frac{d^2w}{d\phi^2}\right), \qquad \tau = \frac{1}{a}\left(\frac{dv}{dz} - \frac{d^2w}{dz\,d\phi}\right) \ldots(7);$$

so that by (11) § 235 f

$$U = \frac{4nh^3}{3a^2}\left\{\frac{m}{m+n}\frac{1}{a^2}\left(w + \frac{d^2w}{d\phi^2}\right)^2 + \left(\frac{dv}{dz} - \frac{d^2w}{dz\,d\phi}\right)^2\right\} \ldots\ldots(8).$$

This is the potential energy of bending reckoned per unit of area. It can, if desired, be expressed by (5) entirely in terms of v^1.

We will now apply (8) to calculate the whole potential energy of a complete cylinder, bounded by plane edges $z = \pm l$, and of thickness which, if variable at all, is a function of z only. Since u, v, w are periodic when ϕ increases by 2π, their most general expression in accordance with (5) is [compare (10), &c., § 235 c]

$$v = \Sigma \left[(A_s a + B_s z) \cos s\phi - (A_s' a + B_s' z) \sin s\phi \right] \dots \dots (9),$$

$$w = \Sigma \left[s (A_s a + B_s z) \sin s\phi + s (A_s a + B_s' z) \cos s\phi \right] \dots (10),$$

$$u = \Sigma \left[- s^{-1} B_s a \sin s\phi - s^{-1} B_s' a \cos s\phi \right] \dots \dots \dots \dots (11),$$

in which the summation extends to all integral values of s from 0 to ∞. But the displacements corresponding to $s = 0$, $s = 1$ are such as a rigid body might undergo, and involve no absorption of energy. When the values of u, v, w are substituted in (8) all the terms containing products of sines or cosines with different values of s vanish in the integration with respect to ϕ, as do also those which contain $\cos s\phi \sin s\phi$. Accordingly

$$\int_0^{2\pi} U a \, d\phi = \frac{4\pi n h^3}{3a} \left[\frac{m}{m+n} \frac{1}{a^2} \Sigma (s^3 - s)^2 \right.$$
$$\left. \{ (A_s a + B_s z)^2 + (A_s' a + B_s' z)^2 \} + \Sigma (s^2 - 1)^2 (B_s^2 + B_s'^2) \right] \dots (12).$$

Thus far we might consider h to be a function of z; but we will now treat it as a constant. In the integration with respect to z the odd powers of z will disappear, and we get as the energy of the whole cylinder of radius a, length $2l$, and thickness $2h$,

$$V = \int_{-l}^{+l} \int_0^{2\pi} U a \, d\phi \, dz = \frac{8\pi n h^3 l}{3a} \Sigma (s^2 - 1)^2 \left[\frac{m \cdot s^2}{m+n} \left\{ A_s^2 + A_s'^2 \right. \right.$$
$$\left. \left. + \frac{l^2}{3a^2} (B_s^2 + B_s'^2) \right\} + B_s^2 + B_s'^2 \right] \dots \dots \dots \dots (13),$$

in which $s = 2, 3, 4, \dots$.

The expression (13) for the potential energy suffices for the solution of statical problems. As an example we will suppose that the cylinder is compressed along a diameter by equal forces F, applied at the points $z = z_1$, $\phi = 0$, $\phi = \pi$, although it is true that so highly localised a force hardly comes within the scope of

---

[1] From the general equations of Mr Love's memoir already cited a concordant result may be obtained on introduction of the appropriate conditions.

the investigation, in consequence of the stretchings of the middle surface, which will occur in the immediate neighbourhood of the points of application[1].

The work done upon the cylinder by the forces F during the hypothetical displacement indicated by δA_s, &c., will be by (10)

$$- F \Sigma s \left(a \delta A_s' + z_1 \delta B_s' \right) (1 + \cos s\pi),$$

so that the equations of equilibrium are

$$\frac{dv}{dA_s} = 0, \qquad\qquad \frac{dv}{dB_s} = 0.$$

$$\frac{dv}{dA_s'} = - (1 + \cos s\pi) \, saF, \qquad \frac{dv}{dB_s'} = - (1 + \cos s\pi) \, sz_1 F.$$

Thus for all values of s,

$$A_s = B_s = 0 \, ;$$

and for odd values of s, $\quad A_s' = B_s' = 0.$

But when s is even,

$$\frac{ms^2}{m+n} A_s' = - \frac{3sa^2 F}{8\pi n h^3 l \, (s^2 - 1)^2} \quad\ldots\ldots\ldots(14),$$

$$\left\{ \frac{ms^2}{m+n} \frac{l^2}{3a^2} + 1 \right\} B_s' = - \frac{3saz_1 F}{8\pi n h^3 l \, (s^2 - 1)^2} \quad\ldots\ldots\ldots(15):$$

and the displacement w at any point (z, ϕ) is given by

$$w = 2 \left(A_2' a + B_2' z \right) \cos 2\phi + 4 \left(A_4' a + B_4' z \right) \cos 4\phi + \ldots (16),$$

where A_2', B_2', A_4',... are determined by (14), (15).

A further discussion of this solution will be found in the memoir[2] from which the preceding results have been taken.

We will now proceed with the calculation for the frequencies of vibration of the complete cylindrical shell of length $2l$. If the volume-density[3] be ρ, we have as the expression of the kinetic energy by means of (9), (10), (11),

$$T = \tfrac{1}{2} \cdot 2h\rho \cdot \iint (\dot{u}^2 + \dot{v}^2 + \dot{w}^2) \, a \, d\phi \, dz$$

$$= 2\pi\rho h l a \, \Sigma \, \{ a^2 (1 + s^2) (\dot{A}_s + \dot{A}_s'^2)$$
$$+ [\tfrac{1}{3} l^2 (1 + s^2) + s^{-2} a^2] (\dot{B}_s^2 + \dot{B}_s'^2) \} \ldots\ldots\ldots(17).$$

---

[1] Whatever the curvature of the surface, an area upon it may be taken so small as to behave like a plane, and therefore bend, in violation of Gauss's condition, when subjected to a force which is so nearly discontinuous that it varies sensibly within the area.

[2] *Proc. Roy. Soc.* vol. 45, p. 105, 1888.

[3] This can scarcely be confused with the notation for the curvature in the preceding parts of the investigation.

From the expressions for V and T in (13), (17) the types and frequencies of vibration can be at once deduced. The fact that the squares, and not the products, of A_s, B_s, are involved, shews that these quantities are really the normal co-ordinates of the vibrating system. If A_s, or A_s', vary as $\cos p_s t$, we have

$$p_s^2 = \tfrac{4}{3} \frac{mn}{m+n} \frac{h^2}{\rho a^4} \frac{(s^3-s)^2}{s^2+1} \quad\dots\dots\dots\dots(18).$$

This is the equation for the frequencies of vibration in two dimensions, § 233. For a given material, the frequency is proportional directly to the thickness and inversely to the square on the diameter of the cylinder[1].

In like manner if B_s, or B_s', vary as $\cos p_s' t$, we find

$$p_s'^2 = \tfrac{4}{3} \frac{mn}{m+n} \frac{h^2}{\rho a^4} \frac{(s^3-s)^2}{s^2+1} \frac{1 + \dfrac{3a^2}{s^2 l^2} \dfrac{m+n}{m}}{1 + \dfrac{3a^2}{(s^4+s^2)\, l^2}} \quad\dots\dots\dots(19).$$

If the cylinder be at all long in proportion to its diameter, the difference between p_s' and p_s becomes very small. Approximately in this case

$$p_s'/p_s = 1 + \frac{3a^2}{2s^2 l^2}\left(\frac{m+n}{m} - \frac{1}{s^2+1}\right) \quad\dots\dots\dots(20);$$

or, if we take $m = 2n$, $s = 2$,

$$p_2'/p_2 = 1 + \frac{39a^2}{80 l^2}.$$

235 h. We now pass on to the consideration of spherical shells. The general theory of the extensional vibrations of a complete shell has been given by Lamb[2], but as the subject is of small importance from an acoustical point of view, we shall limit our investigation to the very simple case of symmetrical radial vibrations.

If w be the normal displacement, the lengths of all lines upon the middle surface are altered in the ratio $(a+w):a$. In calculating the potential energy we may take in (10) § 235 d

$$\epsilon_1 = \epsilon_2 = w/a, \quad \varpi = 0;$$

[1] There is nothing in these laws special to the cylinder. In the case of similar shells of any form, vibrating by pure bending, the frequency will be as the thicknesses and inversely as corresponding areas. If the similarity extend also to the thickness, then the frequency is inversely as the linear dimension, in accordance with the general law of Cauchy.

[2] *Proc. Lond. Math. Soc.* xiv. p. 50, 1882.

so that the energy per unit area is

$$4nh\frac{3m-n}{m+n}\frac{w^2}{a^2},$$

or for the whole sphere

$$V = 4\pi a^2 \cdot 4nh\frac{3m-n}{m+n}\frac{w^2}{a^2}\dots\dots\dots\dots(1).$$

Also for the kinetic energy, if ρ denote the volume density,

$$T = \tfrac{1}{2}\cdot 4\pi a^2 \cdot 2h\cdot\rho\cdot\dot{w}^2\dots\dots\dots\dots(2).$$

Accordingly if $w = W\cos pt$, we have

$$p^2 = \frac{4n}{a^2\rho}\frac{3m-n}{m+n}\dots\dots\dots\dots\dots(3),$$

as the equation for the frequency $(p/2\pi)$.

As regards the general theory Prof. Lamb thus summarizes his results. "The fundamental modes of vibration fall into two classes. In the modes of the *First Class*, the motion at every point of the shell is wholly tangential. In the nth species of this class, the lines of motion are the contour lines of a surface harmonic S_n (Ch. XVII.), and the amplitude of vibration at any point is proportional to the value of $dS_n/d\epsilon$, where $d\epsilon$ is the angle suotended at the centre by a linear element drawn on the surface of the shell at right angles to the contour line passing through the point. The frequency $(p/2\pi)$ is determined by the equation

$$\mathrm{k}^2 a^2 = (n-1)(n+2)\dots\dots\dots\dots(i),$$

where a is the radius of the shell, and $\mathrm{k}^2 = p^2\rho/\mathbf{n}$, if ρ denote the density, and $\mathbf{n}$ the rigidity, of the substance."

"In the vibrations of the *Second Class*, the motion is partly radial and partly tangential. In the nth species of this class the amplitude of the radial component is proportional to S_n, a surface harmonic of order n. The tangential component is everywhere at right angles to the contour lines of the harmonic S_n on the surface of the shell, and its amplitude is proportional to $\Lambda dS_n/d\epsilon$, where Λ is a certain constant, and $d\epsilon$ has the same meaning as before." Prof. Lamb finds

$$\Lambda = -\frac{\mathrm{k}^2 a^2 - 4\gamma}{2n(n+1)\gamma}\dots\dots\dots\dots(ii),$$

where k retains its former meaning, and $\gamma = (1+\sigma)/(1-\sigma)$, σ

denoting Poisson's ratio. " Corresponding to each value of n there are *two* values of k^2a^2, given by the equation

$$k^4a^4 - k^2a^2 \{(n^2 + n + 4)\gamma + n^2 + n - 2\} + 4(n^2 + n - 2)\gamma = 0...(iii).$$

Of the two roots of this equation, one is $<$ and the other $> 4\gamma$. It appears, then, from (ii) that the corresponding fundamental modes are of quite different characters. The mode corresponding to the lower root is always the more important."

" When $n = 1$, the values of k^2a^2 are 0 and 6γ. The zero root corresponds to a motion of translation of the shell as a whole parallel to the axis of the harmonic S_1. In the other mode the radial motion is proportional to $\cos \theta$, where θ is the co-latitude measured from the pole of S_1; the tangential motion is along the meridian, and its amplitude (measured in the direction of θ increasing) is proportional to $\frac{1}{2} \sin \theta$.

" When $n = 2$, the values of ka corresponding to various values of σ are given by the following table :—

$\sigma = 0$	$\sigma = \frac{1}{4}$	$\sigma = \frac{3}{10}$	$\sigma = \frac{1}{3}$	$\sigma = \frac{1}{2}$
1·120	1·176	1·185	1·190	1·215
3·570	4·391	4·601	4·752	5·703

The most interesting variety is that of the zonal harmonic. Making $S = \frac{1}{2}(3\cos^2 \theta - 1)$, we see that the polar diameter of the shell alternately elongates and contracts, whilst the equator simultaneously contracts and expands respectively. In the mode corresponding to the lower root, the tangential motion is towards the poles when the polar diameter is lengthening, and *vice versâ*. The reverse is the case in the other mode. We can hence understand the great difference in frequency."

Prof. Lamb calculates that a thin glass globe 20 cm. in diameter should, in its gravest mode, make about 5350 vibrations per second.

As regards inextensional modes, their form has already been determined, (39) &c. § 235 *c*, at least for the case where the bounding curve and the thickness are symmetrical with respect to an axis, and it will further appear in the course of the present investigation. What remains to be effected is the calculation of the potential energy of bending corresponding thereto, as dependent upon the alterations of curvature of the middle surface. The process is similar to that followed in § 235 *g* for the case of the cylinder, and consists in finding the equation of the deformed surface when referred to rectangular axes in and perpendicular to the original surface.

The two systems of co-ordinates to be connected are the usual polar co-ordinates r, θ, ϕ, and rectangular co-ordinates x, y, ζ, measured from the point P, or (a, θ_0, ϕ_0), on the undisturbed sphere. Of these x is measured along the tangent to the meridian, y along the tangent to the circle of latitude, and ζ along the normal inwards.

Since the origin of ϕ is arbitrary, we may take it so that $\phi_0 = 0$. The relation between the two systems is then

$$x = r \left\{ -\sin(\theta - \theta_0) + \sin\theta \cos\theta_0 (1 - \cos\phi) \right\} \dots\dots(4),$$

$$y = r \sin\theta \sin\phi \dots\dots\dots\dots\dots\dots\dots\dots\dots\dots\dots\dots\dots\dots(5),$$

$$\zeta = -r \left\{ \cos(\theta - \theta_0) - \sin\theta_0 \sin\theta (1 - \cos\phi) \right\} + a \dots(6).$$

If we suppose $r = a$, these equations give the rectangular co-ordinates of the point (a, θ, ϕ) on the undisturbed sphere. We have next to imagine this point displaced so that its polar co-ordinates become $a + \delta r$, $\theta + \delta\theta$, $\phi + \delta\phi$, and to substitute these values in (4), (5), (6), retaining only the first power of δr, $\delta\theta$, $\delta\phi$. Thus

$$x = (a + \delta r) \left\{ -\sin(\theta - \theta_0) + \sin\theta \cos\theta_0 (1 - \cos\phi) \right\}$$
$$+ a\delta\theta \left\{ -\cos(\theta - \theta_0) + \cos\theta \cos\theta_0 (1 - \cos\phi) \right\}$$
$$+ a\delta\phi \sin\theta \cos\theta_0 \sin\phi \dots\dots\dots\dots\dots\dots\dots\dots\dots\dots\dots(7),$$

$$y = (a + \delta r) \sin\theta \sin\phi$$
$$+ a\delta\theta \cos\theta \sin\phi + a\delta\phi \sin\theta \cos\phi \dots\dots\dots\dots\dots\dots(8),$$

$$\zeta = a - (a + \delta r) \left\{ \cos(\theta - \theta_0) - \sin\theta_0 \sin\theta (1 - \cos\phi) \right\}$$
$$+ a\delta\theta \left\{ \sin(\theta - \theta_0) + \sin\theta_0 \cos\theta (1 - \cos\phi) \right\}$$
$$+ a\delta\phi \sin\theta_0 \sin\theta \sin\phi \dots\dots\dots\dots\dots\dots\dots\dots\dots\dots\dots(9).$$

These equations give the co-ordinates of any point Q of the sphere after displacement; but we shall only need to apply them in the case where Q is in the neighbourhood of P, or $(a, \theta_0, 0)$, and then the higher powers of $(\theta - \theta_0)$ and ϕ may be neglected.

In pursuance of our plan we have now to imagine the displaced and deformed sphere to be brought back as a rigid body so that the parts about P occupy as nearly as possible their former positions. We are thus in the first place to omit from (7), (8), (9) the terms (involving δ) which are independent of $(\theta - \theta_0)$, ϕ. Further we must add to each equation respectively the terms which represent an arbitrary rotation, viz.

$$y\omega_3 - \zeta\omega_2, \quad \zeta\omega_1 - x\omega_3, \quad x\omega_2 - y\omega_1,$$

determining ω_1, ω_2, ω_3 in such a manner that, so far as the first powers of $(\theta - \theta_0)$, ϕ, there shall be coincidence between the original and displaced positions of the point Q.

If we omit all terms of the second order in $(\theta - \theta_0)$ and ϕ, we get from (7) &c.

$$x = -a(\theta - \theta_0) - \delta r_0(\theta - \theta_0)$$
$$- a\left\{[\delta\theta_0] + \frac{d\delta\theta}{d\theta_0}(\theta - \theta_0) + \frac{d\delta\theta}{d\phi_0}\phi\right\} + a\delta\phi_0 \sin\theta_0 \cos\theta_0 . \phi \ \dots \ (10),$$

$$y = a\sin\theta_0 . \phi + \delta r_0 \sin\theta_0 . \phi + a\delta\theta_0 \cos\theta_0 . \phi$$
$$+ a\sin\theta_0\left\{[\delta\phi_0] + \frac{d\delta\phi}{d\theta_0}(\theta - \theta_0) + \frac{d\delta\phi}{d\phi_0}\phi\right\}$$
$$+ a\delta\phi_0 \cos\theta_0 (\theta - \theta_0) \ \dots\dots\dots\dots\dots\dots\dots\dots\dots\dots\dots(11),$$

$$\zeta = [-\delta r_0] - \frac{d\delta r}{d\theta_0}(\theta - \theta_0) - \frac{d\delta r}{d\phi_0}\phi$$
$$+ a\delta\theta_0 (\theta - \theta_0) + a\delta\phi_0 \sin^2\theta_0 . \phi \dots\dots\dots\dots\dots\dots(12),$$

δr_0 &c. representing the values appropriate to P, where $(\theta - \theta_0)$ and ϕ vanish. The translation of the deformed surface necessary to bring P back to its original position is represented by the omission of the terms included in square brackets. The arbitrary rotation is represented by the additions respectively of

$$a\sin\theta_0 . \phi . \omega_3, \quad a(\theta - \theta_0)\omega_3, \quad -a(\theta - \theta_0)\omega_2 - a\sin\theta_0 . \phi . \omega_1;$$

and thus the destruction of the terms of the first order requires that

$$\delta r/a + d\delta\theta/d\theta = 0 \ \dots\dots\dots\dots\dots\dots(13),$$
$$-d\delta\theta/d\phi + \sin\theta\cos\theta\,\delta\phi + \sin\theta\,\omega_3 = 0 \ \dots\dots\dots(14);$$
$$\sin\theta\,d\delta\phi/d\theta + \cos\theta\,\delta\phi + \omega_3 = 0 \ \dots\dots\dots\dots(15);$$
$$(\delta r/a)\sin\theta + \delta\theta\cos\theta + \sin\theta\,d\delta\phi/d\phi = 0 \dots\dots\dots(16);$$
$$-d\delta(r/a)/d\theta + \delta\theta - \omega_2 = 0 \ \dots\dots\dots\dots\dots(17),$$
$$-d\delta(r/a)/d\phi + \sin^2\theta\,\delta\phi - \sin\theta\,\omega_1 = 0 \ \dots\dots\dots(18);$$

the suffixes being omitted.

These six equations determine ω_1, ω_2, ω_3, giving as the three conditions of inextension

$$\delta r/a + d\delta\theta/d\theta = 0 \quad\dots\dots\dots\dots\dots(19),$$

$$d\delta\theta/d\phi + \sin^2\theta\, d\delta\phi/d\theta = 0 \quad\dots\dots\dots\dots(20),$$

$$\delta r/a + \cot\theta\,\delta\theta + d\delta\phi/d\phi = 0.\dots\dots\dots\dots(21).$$

From (19), (20), (21), by elimination of δr,

$$\frac{d}{d\phi}\left(\frac{\delta\theta}{\sin\theta}\right) + \sin\theta\,\frac{d\delta\phi}{d\theta} = 0\dots\dots\dots\dots(22),$$

$$\frac{d}{d\phi}\delta\phi - \sin\theta\,\frac{d}{d\theta}\left(\frac{\delta\theta}{\sin\theta}\right) = 0.\dots\dots\dots\dots(23);$$

or, since $\sin\theta\, d/d\theta = d/d\log\tan\tfrac12\theta$,

$$\frac{d}{d\phi}\left(\frac{\delta\theta}{\sin\theta}\right) + \frac{d\delta\phi}{d\log\tan\tfrac12\theta} = 0 \quad\dots\dots\dots\dots(24),$$

$$\frac{d\delta\phi}{d\phi} - \frac{d}{d\log\tan\tfrac12\theta}\left(\frac{\delta\theta}{\sin\theta}\right) = 0.\dots\dots\dots\dots(25).$$

From (24), (25) we see that both $\delta\phi$ and $\delta\theta/\sin\theta$ satisfy an equation of the second order of the same form, viz.

$$\frac{d^2u}{d(\log\tan\tfrac12\theta)^2} + \frac{d^2u}{d\phi^2} = 0\dots\dots\dots\dots(26).$$

If the material system be symmetrical about the axis, u is a periodic function of ϕ, and can be expanded by Fourier's theorem in a series of sines and cosines of ϕ and its multiples. Moreover each term of the series must satisfy the equations independently. Thus, if u varies as $\cos s\phi$, (26) becomes

$$\frac{d^2u}{d(\log\tan\tfrac12\theta)^2} - s^2u = 0\dots\dots\dots\dots(27);$$

whence $\qquad u = A'\tan^s\tfrac12\theta + B'\cot^s\tfrac12\theta\dots\dots\dots\dots(28),$

where A' and B' are independent of θ. If we take

$$\delta\phi = \cos s\phi\,[A_s\tan^s\tfrac12\theta + B_s\cot^s\tfrac12\theta]\dots\dots\dots\dots(29),$$

we get for the corresponding value of $\delta\theta$ from (24)

$$\delta\theta/\sin\theta = -\sin s\phi\,[A_s\tan^s\tfrac12\theta - B_s\cot^s\tfrac12\theta]\dots\dots(30);$$

and thence from (21)

$$\delta r/a = \sin s\phi\,[A_s(s+\cos\theta)\tan^s\tfrac12\theta + B_s(s-\cos\theta)\cot^s\tfrac12\theta]\dots(31),$$

as in (39), (40), (41) § 235 c.

The second solution (in B_s) may be derived from the first (in A_s) in two ways which are both worthy of notice. The manner of derivation from (27) shews that it is sufficient to alter the sign of s, $\tan^s \frac{1}{2}\theta$ becoming $\cot^s \frac{1}{2}\theta$, $\sin s\phi$ becoming $-\sin s\phi$, while $\cos s\phi$ remains unchanged. The other method depends upon the consideration that the general solution must be similarly related to the two poles. It is thus legitimate to alter the first solution by writing throughout $(\pi - \theta)$ in place of θ, changing at the same time the sign of $\delta\theta$.

If we suppose $s = 1$, we get

$$\sin\theta \, \delta\phi = \cos\phi \, [A_1 + B_1 - (A_1 - B_1)\cos\theta],$$

$$\delta\theta = -\sin\phi \, [A_1 - B_1 - (A_1 + B_1)\cos\theta],$$

$$\delta r/a = \sin\phi \, [(A_1 + B_1)\sin\theta].$$

The displacement proportional to $(A_1 - B_1)$ is a rotation of the whole surface as a rigid body round the axis $\theta = \frac{1}{2}\pi$, $\phi = 0$; and that proportional to $(A_1 + B_1)$ represents a translation parallel to the axis $\theta = \frac{1}{2}\pi$, $\phi = \frac{1}{2}\pi$. The complementary translation and rotation with respect to these axes is obtained by substituting $\phi + \frac{1}{2}\pi$ for ϕ.

The two other motions possible without bending correspond to a zero value of s, and are readily obtained from the original equations (19), (20), (21). They are a rotation round the axis $\theta = 0$, represented by

$$\delta\theta = 0, \quad \delta\phi = \text{const.}, \quad \delta r = 0,$$

and a displacement parallel to the same axis represented by

$$\delta\phi = 0, \quad \frac{d}{d\theta}\left(\frac{\delta\theta}{\sin\theta}\right) = 0. \quad \frac{\delta r}{a} = -\cot\theta \, \delta\theta,$$

or $\qquad \delta\phi = 0, \quad \delta\theta = \gamma\sin\theta, \quad \delta r = -\gamma a\cos\theta.$

If the sphere be complete, the displacements just considered, and corresponding to $s = 0, 1$, are the only ones possible. For higher values of s we see from (31) that δr is infinite at one or other pole, unless A_s and B_s both vanish. In accordance with Jellet's theorem[1] the complete sphere is incapable of bending.

If neither pole be included in the actual surface, which for example we may suppose bounded by circles of latitude, finite

[1] "On the Properties of Inextensible Surfaces," *Irish Acad. Trans.*, vol. 22, p. 343, 1855.

values of both A and B are admissible, and therefore necessary for a complete solution of the problem. But if, as would more often happen, one of the poles, say $\theta = 0$, is included, the constants B must be considered to vanish. Under these circumstances the solution is

$$\left.\begin{aligned}
\delta\phi &= A_s \tan^s \tfrac{1}{2}\theta \cos s\phi + \tfrac{1}{2}\pi \\
\delta\theta &= - A_s \sin\theta \tan^s \tfrac{1}{2}\theta \sin s\phi + \tfrac{1}{2}\pi \\
\delta r &= A_s a\,(s + \cos\theta)\tan^s \tfrac{1}{2}\theta \sin s\phi + \tfrac{1}{2}\pi
\end{aligned}\right\} \dots\dots\dots(32),$$

to which is to be added that obtained by writing $s\phi + \tfrac{1}{2}\pi$ for $s\phi$, and changing the arbitrary constant.

From (32) we see that, along those meridians for which $\sin s\phi = 0$, the displacement is tangential and in longitude only, while along the intermediate meridians for which $\cos s\phi = 0$, there is no displacement in longitude, but one in latitude, and one normal to the surface of the sphere.

Along the equator $\theta = \tfrac{1}{2}\pi$,

$$\delta\phi = A_s \cos s\phi, \qquad \delta\theta = - A_s \sin s\phi, \qquad \delta r/a = A_s s \sin s\phi,$$

so that the maximum displacements in latitude and longitude are equal.

Reverting now to the expressions for x, y, ζ in (7), (8), (9), with the addition of the translatory and rotatory terms by which the deformed sphere is brought back as nearly as possible to its original position, we know that so far as the terms of the first order in $(\theta - \theta_0)$ and ϕ are concerned, they are reduced to

$$x = - a\,(\theta - \theta_0), \qquad y = a \sin\theta_0 . \phi, \qquad \zeta = 0 \dots\dots\dots(33).$$

These approximations will suffice for the values of x and y; but in the case of ζ we require the expression complete so as to include all terms of the second order. The calculation is straightforward. For any displacement such as δr in (9) we write

$$\delta r_0 + \frac{d\delta r}{d\theta_0}(\theta - \theta_0) + \frac{d\delta r}{d\phi_0}\phi$$

$$+ \tfrac{1}{2}\frac{d^2\delta r}{d\theta_0{}^2}(\theta - \theta_0)^2 + \frac{d^2\delta r}{d\theta_0 d\phi_0}(\theta - \theta_0)\phi + \tfrac{1}{2}\frac{d^2\delta r}{d\phi_0{}^2}\phi^2.$$

The additional rotatory terms are by (17), (18)

$$x\left\{\delta\theta_0 - \frac{1}{a}\frac{d\delta r}{d\theta_0}\right\} + y\left\{\frac{1}{a\sin\theta_0}\frac{d\delta r}{d\phi_0} - \sin\theta_0\,\delta\phi_0\right\}.$$

In these we are to retain only those terms in x, y, which are of the second order and independent of δ, so that we may write

$$x = \tfrac{1}{2}a\phi^2 \sin \theta_0 \cos \theta_0, \quad y = a(\theta - \theta_0)\phi \cos \theta_0.$$

In the complete expression for ζ as a quadratic function of $(\theta - \theta_0)$ and ϕ thus obtained, we substitute x and y from (33). The final equation to the deformed surface, after simplification by the aid of (19), (20), (21), may be written

$$\zeta = \frac{x^2}{2a}\left\{1 - \frac{\delta r}{a} - \frac{1}{a}\frac{d^2\delta r}{d\theta^2}\right\} + \frac{xy}{a\sin\theta}\left\{-\frac{1}{a}\frac{d^2\delta r}{d\theta d\phi} + \frac{\cot\theta}{a}\frac{d\delta r}{d\phi}\right\}$$

$$+ \frac{y^2}{2a}\left\{1 - \frac{\delta r}{a} - \frac{\cot\theta}{a}\frac{d\delta r}{d\theta} - \frac{1}{a\sin^2\theta}\frac{d^2\delta r}{d\phi^2}\right\}\dots\dots(34),$$

the suffixes being now unnecessary.

Taking the value of $\delta r/a$ from (32) we get

$$-\frac{\delta r}{a} - \frac{1}{a}\frac{d^2\delta r}{d\theta^2} = -\frac{s^3 - s}{\sin^2\theta}A_s \tan^s \tfrac{1}{2}\theta \sin s\phi\dots\dots\dots(35),$$

$$-\frac{1}{a\sin\theta}\frac{d^2\delta r}{d\theta d\phi} + \frac{\cos\theta}{a\sin^2\theta}\frac{d\delta r}{d\phi} = -\frac{s^3 - s}{\sin^2\theta}A_s \tan^s \tfrac{1}{2}\theta \cos s\phi\dots(36),$$

$$-\frac{\delta r}{a} - \frac{\cot\theta}{a}\frac{d\delta r}{d\theta} - \frac{1}{a\sin^2\theta}\frac{d^2\delta r}{d\phi^2} = \frac{s^3 - s}{\sin^2\theta}A_s \tan^s \tfrac{1}{2}\theta \sin s\phi\dots(37).$$

To obtain the more complete solution corresponding to (31), we have only to add new terms, multiplied by B_s, and derived from the above by *changing the sign of s*. As was to be expected, the values in (35) and (37) are equal and opposite.

Introducing the values now found into (5) § 235 f, we obtain as the square of the change of principal curvature at any point

$$\left(\delta\frac{1}{\rho}\right)^2 = \frac{(s^3 - s)^2}{a^2 \sin^4\theta}\{A_s^2 \tan^{2s}\tfrac{1}{2}\theta + B_s^2 \cot^{2s}\tfrac{1}{2}\theta - 2A_sB_s \cos 2s\phi\}\dots(38).$$

It should be remarked that, if either A_s or B_s vanish, (38) is independent of ϕ, so that the change of principal curvature is the same for all points on a circle of latitude, and that in any case (38) becomes independent of the product A_sB_s after integration round the circumference. The change of curvature vanishes if $s = 0$, or $s = 1$, the displacement being that of which a rigid body is capable.

Equations (35) &c. shew that along the meridians where $\delta\phi$ vanishes ($\cos s\phi = 0$) the principal planes of curvature are the meridian and its perpendicular, while along the meridians where δr vanishes, the principal planes are inclined to the meridian at angles of 45°.

The value of the square of the change of curvature obtained in (38) corresponds to that assumed for the displacements in (29) &c., and for some purposes needs to be generalised. We may add terms with coefficients A_s' and B_s' corresponding to a change of $s\phi$ to $(s\phi + \frac{1}{2}\pi)$, and there is further to be considered the summation with respect to s. Putting for brevity t in place of $\tan \frac{1}{2}\theta$, we may take as the complete expression for $[\delta(1/\rho)]^2$,

$$\left[\Sigma \frac{s^3 - s}{a\sin^2\theta} \left\{(A_s t^s + B_s t^{-s})\sin s\phi + (A_s' t^s + B_s' t^{-s})\sin(s\phi + \frac{1}{2}\pi)\right\}\right]^2$$

$$+ \left[\Sigma \frac{s^3 - s}{a\sin^2\theta} \left\{(A_s t^s - B_s t^{-s})\cos s\phi + (A_s' t^s - B_s' t^{-s})\cos(s\phi + \frac{1}{2}\pi)\right\}\right]^2$$

When this is integrated with respect to ϕ round the entire circumference, all products of the generalised co-ordinates A_s, B_s, A_s', B_s' disappear, so that (7) § 235 f we have as the expression for the potential energy of the surface included between two parallels of latitude

$$V = 2\pi \Sigma (s^3 - s)^2 \int H \sin^{-3}\theta \left\{(A_s^2 + A_s'^2) t^{2s}\right.$$
$$\left. + (B_s^2 + B_s'^2) t^{-2s}\right\} d\theta \dots\dots(39),$$

where
$$H = \frac{4}{3}nh^3 \dots\dots\dots\dots\dots\dots\dots\dots\dots\dots\dots(40).$$

In the following applications to spherical surfaces where the pole $\theta = 0$ is included, we may omit the terms in B; and, if the thickness be constant, H may be removed from under the integral sign. We have

$$d\theta = \frac{2dt}{1 + t^2}, \qquad \sin\theta = \frac{2t}{1 + t^2},$$

so that

$$\int_0^\theta \sin^{-3}\theta\, t^{2s} d\theta = \frac{1}{8}\int (1 + t^2)^2 t^{2s-4} dt^2 = \frac{1}{8}\left(\frac{t^{2s-2}}{s-1} + \frac{2t^{2s}}{s} + \frac{t^{2s+2}}{s+1}\right)\dots(41).$$

In the case of the hemisphere $t = 1$, and (41) assumes the value

$$\frac{2s^2 - 1}{4(s^3 - s)} \dots\dots\dots\dots\dots\dots\dots\dots\dots(42).$$

Hence for a hemisphere of uniform thickness

$$V = \frac{1}{2}\pi H \Sigma (s^3 - s)(2s^2 - 1)(A_s^2 + A_s'^2)\dots\dots(43).$$

If the extreme value of θ be 60°, instead of 90°, we get in place of (42)

$$\frac{8s^2 + 4s - 3}{4 \cdot 3^{s+1}(s^3 - s)} \dots\dots\dots\dots\dots\dots(44),$$

and $\quad V = \frac{1}{2}\pi H \Sigma\, 3^{-(s+1)}\,(s^3 - s)\,(8s^2 + 4s - 3)\,(A_s^2 + A_s'^2)\dots(45).$

These expressions for V, in conjunction with (32), are sufficien for the solution of statical problems, relative to the deformation of infinitely thin spherical shells under the action of given impressed forces. Suppose, for example, that a string of tension F connects the opposite points on the edge of a hemisphere, represented by $\theta = \frac{1}{2}\pi$, $\phi = \frac{1}{2}\pi$ or $\frac{3}{2}\pi$, and that it is required to find the deformation. It is evident from (32) that all the quantities A_s vanish, and that the work done by the impressed forces, corresponding to the deformation $\delta A_s'$, is

$$- \delta A_s' as \{ \cos \tfrac{1}{2}s\pi + \cos \tfrac{3}{2}s\pi \}\, F.$$

If s be odd this vanishes, and if s be even it is equal to

$$- 2\delta A_s' as \cos \tfrac{1}{2}s\pi . F.$$

Hence if s be odd A_s' vanishes; and by (43), if s be even,

$$dV/dA_s' = \pi H\,(s^3 - s)\,(2s^2 - 1)\,A_s' = - 2as \cos\tfrac{1}{2}s\pi . F;$$

whence $\qquad A_s' = - \dfrac{2aF \cos\frac{1}{2}s\pi}{\pi H\,(s^2 - 1)\,(2s^2 - 1)} \dots\dots\dots\dots (46).$

By (46) and (32) the deformation is completely determined.

If, to take a case in which the force is tangential, we suppose that the hemisphere rests upon its pole with its edge horizontal, and that a rod of weight W is laid symmetrically along the diameter $\theta = \frac{1}{2}\pi$, we find in like manner

$$A_s' = \frac{a W \cos \frac{1}{2}s\pi}{\pi H\,(s^3 - s)\,(2s^2 - 1)} \dots\dots\dots\dots(47)$$

for all even values of s, and $A_s' = 0$ for all odd values of s.

We now proceed to evaluate the kinetic energy as defined by the formula

$$T = \tfrac{1}{2}\sigma a^2 \iint \left\{ \left(\frac{d\delta r}{dt}\right)^2 + \left(\frac{a\,d\delta\theta}{dt}\right)^2 + \left(\frac{a \sin \theta\, d\delta\phi}{dt}\right)^2 \right\} \sin \theta\, d\theta\, d\phi \dots(48),$$

in which σ denotes the surface density, supposed to be uniform.

If we take the complete value of $\delta\phi$ from (29), as supplemented by the terms in A_s', B_s', we have

$$\frac{d\delta\phi}{dt} = \Sigma\left[\cos s\phi\,(\dot{A}_s t^s + \dot{B}_s t^{-s}) + \cos(s\phi + \tfrac{1}{2}\pi)(\dot{A}_s' t^s + \dot{B}_s' t^{-s})\right].$$

When this expression is squared and integrated with respect to ϕ round the entire circumference, all products of letters with a different suffix, and all products of dashed and undashed letters even with the same suffix, will disappear. Hence replacing $\cos^2 s\phi$ &c. by the mean value $\tfrac{1}{2}$, we may take

$$\sin^2\theta\left(\frac{d\delta\phi}{dt}\right)^2 = \tfrac{1}{2}\sin^2\theta\,\Sigma\,(\dot{A}_s^2 + \dot{A}_s'^2)\,t^{2s}$$

$$+\,\tfrac{1}{2}\sin^2\theta\,\Sigma\,(\dot{B}_s^2 + \dot{B}_s'^2)\,t^{-2s} + \sin^2\theta\,\Sigma\,(\dot{A}_s\dot{B}_s + \dot{A}_s'\dot{B}_s').$$

The mean value (30) of $(d\delta\theta/dt)^2$ is the same as that just written with the substitution throughout of $-B$ for B, so that we may take

$$\left(\frac{d\delta\theta}{dt}\right)^2 + \left(\frac{\sin\theta\,d\delta\phi}{dt}\right)^2 = \sin^2\theta\,\Sigma\,(\dot{A}_s^2 + \dot{A}_s'^2)\,t^{2s}$$

$$+\,\sin^2\theta\,\Sigma\,(\dot{B}_s^2 + \dot{B}_s'^2)\,t^{-2s}\ldots\ldots\ldots\ldots(49),$$

as the mean available for our present purpose. In (49) the products of the symbols have disappeared, and if the expression for the kinetic energy were as yet fully formed, the co-ordinates would be shewn to be *normal*. But we have still to include that part of the kinetic energy dependent upon $d\delta r/dt$. As the mean value, applicable for our purpose, we have from (31)

$$\left(\frac{d\delta r}{a\,dt}\right)^2 = \tfrac{1}{2}\,\Sigma\,(\dot{A}_s^2 + \dot{A}_s'^2)\,(s + \cos\theta)^2\,t^{2s}$$

$$+\,\tfrac{1}{2}\,\Sigma\,(\dot{B}_s^2 + \dot{B}_s'^2)\,(s - \cos\theta)^2\,t^{-2s}$$

$$+\,\Sigma\,(\dot{A}_s\dot{B}_s + \dot{A}_s'\dot{B}_s')\,(s^2 - \cos^2\theta)\ldots\ldots\ldots(50).$$

The expressions (49) and (50) have now to be added. If we set for brevity

$$\int\tan^{2s}\tfrac{1}{2}\theta\,\{(s + \cos\theta)^2 + 2\sin^2\theta\}\,\sin\theta\,d\theta = f(s)\ldots\ldots(51),$$

or putting $x = 1 + \cos\theta$,

$$f(s) = \int^2\left(\frac{2 - x}{x}\right)^s\{(s - 1)^2 + 2x(s + 1) - x^2\}\,dx\ldots\ldots(52),$$

we get

$$T = \tfrac{1}{2}\pi\sigma a^4 \{\Sigma f(s)\,(\dot{A}_s{}^2 + \dot{A}_s{}'^2) + \Sigma f(-s)\,(\dot{B}_s{}^2 + \dot{B}_s{}'^2)$$

$$+ 2\Sigma \int (s^2 - \cos^2\theta) \sin\theta\,d\theta\,(\dot{A}_s\dot{B}_s + \dot{A}_s'\dot{B}_s')\} \ \ldots\ldots\ldots(53)$$

It will be seen that, while V in (39) is expressible by the squares only of the co-ordinates, a like assertion cannot in general be made of T. Hence A_s, B_s &c. are *not* in general the normal co-ordinates. Nor could this have been expected. If, for example, we take the case where the spherical surface is bounded by two circles of latitude equidistant from the equator, symmetry shews that the normal co-ordinates are, not A and B, but $(A + B)$ and $(A - B)$. In this case $f(-s) = f(s)$.

A verification of (53) may readily be obtained in the particular case of $s = 1$, the surface under consideration being the entire sphere. Dropping the dashed letters, we get

$$T = \tfrac{1}{2}\pi\sigma a^4 \{\tfrac{20}{3}(\dot{A}_1{}^2 + \dot{B}_1{}^2) + \tfrac{8}{3}\dot{A}_1\dot{B}_1\}$$

$$= \tfrac{1}{2}\pi\sigma a^4 \{4(\dot{A}_1 + \dot{B}_1)^2 + \tfrac{8}{3}(\dot{A}_1 - \dot{B}_1)^2\} \ \ldots\ldots(54).$$

In this case the displacements are of the purely translatory and rotatory types already discussed, and the correctness of (54) may be confirmed.

Whatever may be the position of the circles of latitude by which the surface is bounded, the true types and periods of vibration are determined by the application of Lagrange's method to (39), (53).

When one pole, e.g. $\theta = 0$, is included within the surface, the co-ordinates B vanish, and A_s, A_s' become the normal co-ordinates. If we omit the dashed letters, the expression for T becomes simply

$$T = \tfrac{1}{2}\pi\sigma a^4 \Sigma f(s)\,\dot{A}_s{}^2 \ldots\ldots\ldots\ldots\ldots\ldots(55).$$

From (43), (55) the frequencies of free vibrations for a hemi-sphere are immediately obtainable. The equation for A_s is

$$\sigma a^4 f(s)\,\ddot{A}_s + H(s^3 - s)(2s^2 - 1)\,A_s = 0 \ldots\ldots\ldots(56);$$

so that, if A_s vary as $\cos p_s t$,

$$p_s{}^2 = \frac{H(s^3 - s)(2s^2 - 1)}{\sigma a^4 f(s)} = \frac{2nh^2}{3\rho a^4} \cdot \frac{(s^3 - s)(2s^2 - 1)}{f(s)}$$

if we introduce the value of H from (40), and express σ by means of the volume density ρ.

In like manner for the saucer of 120°, from (44),

$$p_s^2 = \frac{H(s^3 - s)(8s^2 + 4s - 3)}{\sigma a^4 f(s) \cdot 3^{s+1}} \quad \dots\dots\dots\dots(58).$$

The values of $f(s)$ can be calculated without difficulty in the various cases. Thus, for the hemisphere,

$$f(2) = \int_1^2 x^{-2}(4 - 4x + x^2)(1 + 6x - x^2)\,dx$$

$$= 20 \log 2 - 12\tfrac{1}{3} = 1{\cdot}52961,$$

$$f(3) = 57\tfrac{1}{3} - 80 \log 2 = 1{\cdot}88156,$$

$$f(4) = 200 \log 2 - 136\tfrac{1}{3} = 2{\cdot}29609, \ \&\text{c. ;}$$

so that

$$p_2 = \frac{\sqrt{H}}{a^2\sqrt{\sigma}} \times 5{\cdot}2400, \quad p_3 = \frac{\sqrt{H}}{a^2\sqrt{\sigma}} \times 14{\cdot}726, \quad p_4 = \frac{\sqrt{H}}{a^2\sqrt{\sigma}} \times 28{\cdot}462.$$

In experiment, it is the *intervals* between the various tones with which we are most concerned. We find

$$p_3/p_2 = 2{\cdot}8102, \quad p_4/p_2 = 5{\cdot}4316\dots\dots\dots\dots(59).$$

In the case of glass bells, such as are used with air-pumps, the interval between the two gravest tones is usually somewhat smaller; the representative fraction being nearer to 2·5 than 2·8.

For the saucer of 120°, the lower limit of the integral in (52) is $\tfrac{3}{2}$, and we get on calculation

$$f(2) = {\cdot}12864, \quad f(3) = {\cdot}054884,$$

giving

$$p_2 = \frac{\sqrt{H}}{a^2\sqrt{\sigma}} \times 7{\cdot}9947, \quad p_3 = \frac{\sqrt{H}}{a^2\sqrt{\sigma}} \times 20{\cdot}911,$$

$$p_3 : p_2 = 2{\cdot}6157.$$

The pitch of the two gravest tones is thus decidedly higher than for the hemisphere, and the interval between them is less.

With reference to the theory of tuning bells, it may be worth while to consider the effect of a small change in the angle, for the case of a nearly hemispherical bell. In general

$$p_s^2 = \frac{4H(s^3 - s)^2 \int_0^\theta \sin^{-3}\theta \, \tan^{2s}\tfrac{1}{2}\theta \, d\theta}{a^4\sigma \int_0^\theta \tan^{2s}\tfrac{1}{2}\theta \, \{(s + \cos\theta)^2 + 2\sin^2\theta\}\sin\theta \, d\theta} \ \dots (60).$$

If $\theta = \frac{1}{2}\pi + \delta\theta$, and P_s denote the value of p_s for the exact hemisphere, we get from previous results

$$p_s^2 = P_s^2 \left[1 + \delta\theta \left\{ \frac{4(s^3 - s)}{2s^2 - 1} - \frac{s^2 + 2}{f(s)} \right\} \right] \dots\dots\dots(61).$$

Thus

$$p_2^2 = P_2^2 \left[1 + \delta\theta \left\{ \frac{24}{7} - \frac{6}{1\cdot52961} \right\} \right] = P_2^2 (1 - \cdot49\,\delta\theta)$$

$$p_3^2 = P_3^2 \left[1 + \delta\theta \left\{ \frac{96}{17} - \frac{11}{1\cdot88156} \right\} \right] = P_3^2 (1 - \cdot20\,\delta\theta),$$

shewing that an increase in the angle depresses the pitch. As to the interval between the two gravest tones, we get

$$\left(\frac{p_3}{p_2} \right)^2 = \left(\frac{P_3}{P_2} \right)^2 \times (1 + \cdot29\,\delta\theta),$$

shewing that it increases with θ. This agrees with the results given above for $\theta = 60°$.

The fact that the form of the normal functions is independent of the distribution of density and thickness, provided that they vary only with latitude, allows us to calculate a great variety of cases, the difficulties being merely those of simple integration. If we suppose that only a narrow belt in co-latitude θ has sufficient thickness to contribute sensibly to the potential and kinetic energies, we have simply, instead of (60),

$$p_s^2 = \frac{4H(s^3 - s)^2 \sin^{-4}\theta}{a^4\sigma \{(s + \cos\theta)^2 + 2\sin^2\theta\}} \dots\dots\dots (62),$$

whence

$$\frac{p_3}{p_2} = 4 \sqrt{\left\{ \frac{6 + 4\cos\theta - \cos^2\theta}{11 + 6\cos\theta - \cos^2\theta} \right\}} \dots\dots\dots(63).$$

The ratio varies very slowly from 3, when $\theta = 0$, to 2·954, when $\theta = \frac{1}{2}\pi$.

If $2h$ denote the thickness at any co-latitude θ, $H \propto h^3$, $\sigma \propto h$. I have calculated the ratio of frequencies of the two gravest tones of a hemisphere on the suppositions (1) that $h \propto \cos\theta$, and (2) that $h \propto (1 + \cos\theta)$. The formula used is that marked (60) with H and σ under the integral signs. In the first case, $p_3 : p_2 = 1\cdot7942$, differing greatly from the value for a uniform thickness. On the second more moderate supposition as to the law of thickness,

$$p_3 : p_2 = 2\cdot4591, \qquad p_4 : p_2 = 4\cdot4837.$$

It would appear that the smallness of the interval between the gravest tones of common glass bells is due in great measure to the thickness diminishing with increasing θ.

It is worthy of notice that the curvature of deformation $\delta(\rho^{-1})$, which by (38) varies as $\sin^{-2}\theta \tan^s \frac{1}{2}\theta$, vanishes at the pole for $s = 3$ and higher values, but is finite for $s = 2$.

The present chapter has been derived very largely from various published memoirs by the author[1]. The methods have not escaped criticism, some of which, however, is obviated by the remark that the theory does not profess to be strictly applicable to shells of finite thickness, but only to the limiting case when the thickness is infinitely small. When the thickness increases, it may become necessary to take into account certain "local perturbations" which occur in the immediate neighbourhood of a boundary, and which are of such a nature as to involve extensions of the middle surface. The reader who wishes to pursue this rather difficult question may refer to memoirs by Love[2], Lamb[3], and Basset[4]. From the point of view of the present chapter the matter is perhaps not of great importance. For it seems clear that any extension that may occur must be limited to a region of infinitely small area, and affects neither the types nor the frequencies of vibration. The question of what precisely happens close to a free edge may require further elucidation, but this can hardly be expected from a theory of *thin* shells. At points whose distance from the edge is of the same order as the thickness, the characteristic properties of thin shells are likely to disappear.

[1] *Proc. Lond. Math. Soc.*, XIII. p. 4, 1881 ; xx. p. 372, 1889 ; *Proc. Roy. Soc.*, vol. 45, p. 105, 1888; vol. 45, p. 443, 1888.

[2] *Phil. Trans.*, 179 (A), p. 491, 1888; *Proc. Roy. Soc.*, vol. 49, p. 100, 1891 ; *Theory of Elasticity*, ch. XXI.

[3] *Proc. Lond. Math. Soc.*, vol. XXI. p. 119, 1890.

[4] *Phil. Trans.* 181 (A), p. 433, 1890 ; *Am. Math. Journ.*, vol. XVI. p. 254, 1894.

CHAPTER X B.

ELECTRICAL VIBRATIONS.

235 *i.* The introduction of the telephone into practical use, and the numerous applications to scientific experiment of which it admits, bring the subject of alternating electric currents within the scope of Acoustics, and impose upon us the obligation of shewing how the general principles expounded in this work may best be brought to bear upon the problems presenting themselves. Indeed Electricity affords such excellent illustrations that the temptation to use some of them has already (§§ 78, 92 *a*, 111 *b*) proved irresistible. It will be necessary, however, to take for granted a knowledge of elementary electrical theory, and to abstain for the most part from pursuing the subject in its application to vibrations of enormously high frequency, such as have in recent years acquired so much importance from the researches initiated by Lodge and by Hertz. In the writings of those physicists and in the works of Prof. J. J. Thomson[1] and of Mr O. Heaviside[2] the reader will find the necessary information on that branch of the subject.

The general idea of including electrical phenomena under those of ordinary mechanics is exemplified in the early writings of Lord Kelvin; and in his "Dynamical Theory of the Electro-magnetic Field[3]" Maxwell gave a systematic exposition of the subject from this point of view.

[1] *Recent Researches in Electricity and Magnetism*, 1893.
[2] *Electrical Papers*, 1892.
[3] *Phil. Trans.* vol. 155, p. 459, 1865; *Collected Works*, vol. 1, p. 526.

235 *j*. We commence with the consideration of a simple electrical circuit, consisting of an electro-magnet whose terminals are connected with the poles of a condenser, or *leyden*[1], of capacity *C*. The electro-magnet may be a simple coil of insulated wire, of resistance *R*, and of self-induction or *inductance L*. If there be an iron core, it is necessary to suppose that the metal is divided so as to avoid the interference of internal induced currents, and further that the whole change of magnetism is small[2]. Otherwise the behaviour of the iron is complicated with *hysteresis*, and its effect cannot be represented as a simple augmentation of *L*. Also for the present we will ignore the hysteresis exhibited by many kinds of leydens.

If *x* denote the charge of the leyden at time *t*, $\dot{x}$ is the current, and if $E_1 \cos pt$ be the imposed electro-motive force, the equation is

$$L\ddot{x} + R\dot{x} + x/C = E_1 \cos pt \dots\dots\dots\dots (1).$$

The solution of (1) gives the theory of *forced* electrical vibrations; but we may commence with the consideration of the *free* vibrations corresponding to $E_1 = 0$. This problem has already been treated in § 45, from which it appears that the currents are *oscillatory*, if

$$R < 2\sqrt{(L/C)}\dots\dots\dots\dots\dots\dots(2).$$

The fact that the discharges of leydens are often oscillatory was suspected by Henry and by v. Helmholtz, but the mathematical theory is due to Kelvin[3].

When *R* is much smaller than the critical value in (2), a large number of vibrations occur without much loss of amplitude, and the period τ is given by

$$\tau = 2\pi \sqrt{(CL)} \dots\dots\dots\dots\dots\dots (3).$$

In (2), (3) the data may be supposed to be expressed in C. G. S. electro-magnetic measure. If we introduce practical units, so that L', R', C' represent the inductance, resistance and capacity reckoned respectively in earth-quadrants or henrys, ohms, and microfarads[4], we have in place of (2)

$$R' < 2000 \sqrt{(L'/C')} \dots\dots\dots\dots\dots\dots (2'),$$

---

[1] This term has been approved by Lord Kelvin ("On a New Form of Air Leyden &c." *Proc. Roy. Soc.*, vol. 52, p. 6, 1892).

[2] *Phil. Mag.*, vol. 23, p. 225, 1887.

[3] "On Transient Electric Currents," *Phil. Mag.*, June, 1853.

[4] Ohm = 10^9, henry = 10^9, microfarad = 10^{-15}.

and in place of (3)

$$\tau = 2\pi . 10^{-3} \sqrt{(C'L')} \ldots\ldots\ldots\ldots\ldots\ldots (3').$$

With ordinary appliances the value of τ is very small; but by including a considerable coil of insulated wire in the discharging circuit of a leyden composed of numerous glass plates Lodge[1] has succeeded in exhibiting oscillatory sparks of periods as long as $\frac{1}{500}$ second.

If the leyden be of infinite capacity or, what comes to the same thing, if it be short-circuited, the equation of free motion reduces to

$$L\ddot{x} + R\dot{x} = 0 \ldots\ldots\ldots\ldots\ldots\ldots (4);$$

whence

$$\dot{x} = \dot{x}_0 e^{-(R/L)t} \ldots\ldots\ldots\ldots\ldots\ldots (5)[2],$$

$\dot{x}_0$ representing the value of $\dot{x}$ when $t = 0$. The quantity L/R is sometimes called the time-constant of the circuit, being the time during which free currents fall off in the ratio of $e : 1$.

Returning to equation (1), we see that the problem falls under the general head of vibrations of one degree of freedom, discussed in § 46. In the notation there adopted, $n^2 = (CL)^{-1}$, $\kappa = R/L$, $E = E_1/L$; and the solution is expressed by equations (4) and (5). It is unnecessary to repeat at length the discussion already given, but it may be well to call attention to the case of resonance, where the natural pitch of the electrical vibrator coincides with that of the imposed force ($p^2 LC = 1$). The first and third terms then (§ 46) compensate one another, and the equation reduces to

$$R\dot{x} = E_1 \cos pt \ldots\ldots\ldots\ldots\ldots\ldots (6).$$

In general, if the leyden be short-circuited ($C = \infty$),

$$\dot{x} = \frac{E_1}{L^2 p^2 + R^2} \{R \cos pt + pL \sin pt\} \ldots\ldots\ldots\ldots (7);$$

so that, if p much exceed R/L, the current is greatly reduced by self-induction. In such a case the introduction of a leyden of suitable capacity, by which the self-induction is compensated, results in a large augmentation of current[3]. The imposed electromotive force may be obtained from a coil forming part of the circuit and revolving in a magnetic field.

---

[1] *Proc. Roy. Inst.*, March, 1889.

[2] Helmholtz, *Pogg. Ann.*, LXXXIII., p. 505, 1851.

[3] Maxwell, "Experiment in Magneto-Electric Induction," *Phil. Mag.*, May, 1868.

In any circuit, where vibrations, whether forced or free, proportional to $\cos pt$ are in progress, we have $\ddot{x} = -p^2 x$, and thus the terms due to self-induction and to the leyden enter into the equation in the same manner. The law is more readily expressed if we use the *stiffness* μ, equal to $1/C$, rather than the capacity itself. We may say that a stiffness μ compensates an inductance L, if $\mu = p^2 L$, and that an additional inductance ΔL is compensated by an additional stiffness $\Delta\mu$, provided the above proportionality hold good. This remark allows us to simplify our equations by omitting in the first instance the stiffness of leydens. When the solution has been obtained, we may at any time generalise it by the introduction, in place of L, of $L - \mu p^{-2}$, or $L - (p^2 C)^{-1}$. In following this course we must be prepared to admit negative values of L.

235 *k*. We will next suppose that there are two independent circuits with coefficients of self-induction L, N, and of mutual induction M, and examine what will be the effect in the second circuit of the instantaneous establishment and subsequent maintenance of a current $\dot{x}$ in the first circuit. At the first moment the question is one of the function T only, where

$$T = \tfrac{1}{2} L \dot{x}^2 + M \dot{x} \dot{y} + \tfrac{1}{2} N \dot{y}^2 \dots\dots\dots\dots (1);$$

and by Kelvin's rule (§ 79) the solution is to be obtained by making (1) a minimum under the condition that $\dot{x}$ has the given value. Thus initially

$$\dot{y}_0 = -\frac{M}{N} \dot{x} \dots\dots\dots\dots\dots\dots (2);$$

and accordingly (§ 235 *j*) after time t

$$\dot{y} = -\frac{M}{N} \dot{x} \, e^{-(S/N)t} \dots\dots\dots\dots\dots (3),$$

if S be the resistance of the circuit. The whole induced current, as measured by a ballistic galvanometer, is given by

$$\int_0^\infty \dot{y} \, dt = -\frac{M \dot{x}}{S} \dots\dots\dots\dots\dots (4),$$

in which N does not appear. The current in the secondary circuit due to the cessation of a previously established steady current $\dot{x}$ in the primary circuit is the opposite of the above.

A curious property of the initial induced current is at once evident from Kelvin's theorem, or from equation (2). It appears that, if M be given, the initial current is greatest when N is least. Further, if the secondary circuit consist mainly of a coil of n turns, the initial current increases with diminishing n. For, although $M \propto n$, $N \propto n^2$; and thus $\dot{y}_0 \propto 1/n$. In fact the small current flowing through the more numerous convolutions has the same effect as regards the energy of the field as the larger current in the fewer convolutions. This peculiar dependence upon n cannot be investigated by the galvanometer, at least without commutators capable of separating one part of the induced current from the rest; for, as we see from (4), the galvanometer reading is affected in the reverse direction. It is possible however to render evident the increased initial current due to a diminished n by observing the magnetizing effect upon steel needles. The magnetization depends mainly upon the initial maximum value of the current, and in a less degree, or scarcely at all, upon its subsequent *duration*. [1]

The general equations for two detached circuits, influencing one another only by induction, may be obtained in the usual manner from (1) and

$$F = \tfrac{1}{2} R\dot{x}^2 + \tfrac{1}{2} S\dot{y}^2 \dots\dots\dots \dots\dots\dots\dots (5).$$

Thus
$$\left. \begin{aligned} L\ddot{x} + M\ddot{y} + R\dot{x} &= X \\ M\ddot{x} + N\ddot{y} + S\dot{y} &= Y \end{aligned} \right\} \dots\dots\dots\dots\dots (6).$$

These equations, in a more general form, are considered in § 116. If a harmonic force $X = e^{ipt}$ act in the first circuit, and the second circuit be free from imposed force $(Y = 0)$, we have on elimination of $\dot{y}$

$$\dot{x}\left\{ ip \left(L - p^2 \frac{M^2 N}{p^2 N^2 + S^2} \right) + R + p^2 \frac{M^2 S}{p^2 N^2 + S^2} \right\} = e^{ipt} \dots (7),$$

shewing that the reaction of the secondary circuit upon the first is to *reduce* the inductance by

$$p^2 \frac{M^2 N}{p^2 N^2 + S^2} \dots\dots\dots\dots\dots\dots (8)[2],$$

and to *increase* the resistance by

$$p^2 \frac{M^2 S}{p^2 N^2 + S^2} \dots\dots\dots\dots\dots\dots (9)[2].$$

---

[1] *Phil. Mag.*, vol. 38, p. 1, 1869; vol. 39, p. 428, 1870.

[2] Maxwell, *Phil. Trans.*, vol. 155, p. 459, 1865, where, however. M^2 is misprinted M.

The formulæ (8) and (9) may be applied to deal with a more general problem of considerable interest, which arises when (as in some of Henry's experiments) the secondary circuit acts upon a third, this upon a fourth, and so on, the only condition being that there must be no mutual induction except between immediate neighbours in the series. For the sake of distinctness we will limit ourselves to four circuits.

In the fourth circuit the current is due *ex hypothesi* only to induction from the third. Its reaction upon the third, for the rate of vibration under contemplation, is given at once by (8) and (9); and if we use the complete values applicable to the third circuit under these conditions, we may thenceforth ignore the fourth circuit. In like manner we can now deduce the reaction upon the secondary, giving the effective resistance and inductance of that circuit under the influence of the third and fourth circuits ; and then, by another step of the same kind, we may arrive at the values applicable to the primary circuit, under the influence of all the others. The process is evidently general; and we know by the theorem of § 111 *b* that, however extended the train of circuits, the influence of the others upon the first must be to increase its effective resistance and diminish its effective inertia, in greater and greater degree as the frequency of vibration increases.

In the limit, when the frequency increases indefinitely, the distribution of currents is determined by the induction-coefficients, irrespective of resistance, and, as we shall see presently, it is of such a character that the currents are alternately opposite in sign as we pass along the series.

235 l. Whatever may be the number of independent currents, or degrees of freedom, the general equations are always of the kind already discussed §§ 82, 103, 104, viz.

$$\frac{d}{dt}\frac{dT}{d\dot{x}} + \frac{dF}{d\dot{x}} + \frac{dV}{dx} = X \quad \dots\dots\dots\dots\dots (1),$$

where T, F, V are (§ 82) homogeneous quadratic functions. In (1) the co-ordinates x_1, x_2, ... denote the whole quantity of electricity which has passed at time t, the *currents* being $\dot{x}_1$, $\dot{x}_2$, &c. When $V = 0$, it is simpler to express the phenomena by means of the currents. Thus, in the problem of steady electric flow where all the quantities X, representing electro-motive forces, are constant, the currents are determined directly by the linear equations

$$dF/d\dot{x}_1 = X_1, \quad dF/d\dot{x}_2 = X_2, \text{ &c. } \dots\dots\dots (2).$$

On the other hand when the question under consideration is one of initial impulsive effects, or of forced vibrations of exceedingly high frequency, everything depends upon T, and the equations reduce to

$$\frac{d}{dt}\frac{dT}{d\dot{x}_1} = X_1, \quad \frac{d}{dt}\frac{dT}{d\dot{x}_2} = X_2, \text{ &c. } \dots\dots\dots (3).$$

As an example we may consider the problem, touched upon at the close of § 235 *k*, of a train of circuits where the mutual induction is confined to immediate neighbours, so that

$$T = \tfrac{1}{2}a_{11}x_1^2 + \tfrac{1}{2}a_{22}x_2^2 + \tfrac{1}{2}a_{33}x_3^2 + \dots$$
$$+ a_{12}x_1x_2 + a_{23}x_2x_3 + a_{34}x_3x_4 + \dots \dots\dots\dots (4)^1,$$

coefficients such as a_{13}, a_{14}, a_{24} not appearing. If x_1 be given, either as a current suddenly developed and afterwards maintained constant, or as a harmonic time function of high frequency, while no external forces operate in the other circuits, the problem is to determine x_2, x_3, &c. so as to make T as small as possible, § 79. The equations are easily written down, but the conclusion aimed at is perhaps arrived at more instructively by consideration of the function T itself. For, T being homogeneous in x_1, x_2, &c., we have identically

$$2T = x_1\frac{dT}{dx_1} + x_2\frac{dT}{dx_2} + \dots\dots\dots\dots\dots(5).$$

And, since when T is a minimum, dT/dx_2, dT/dx_3, &c., all vanish,

$$2T_{\min.} = x_1\frac{dT}{dx_1} = a_{11}x_1^2 + a_{12}x_1x_2.$$

But if x_2, x_3, &c., had all been zero, $2T$ would have been equal to $a_{11}x_1^2$. It is clear therefore that $a_{12}x_1x_2$ is negative; or, as a_{12} is taken positive, the sign of x_2 is the *opposite* to that of x_1.

Again supposing x_1, x_2 both given, we must, when T is a minimum, have dT/dx_3, dT/dx_4, &c., equal to zero, and thus

$$2T_{\min.} = a_{11}x_1^2 + 2a_{12}x_1x_2 + a_{22}x_2^2 + 2a_{23}x_2x_3.$$

As before, $2T$ might have been

$$a_{11}x_1^2 + 2a_{12}x_1x_2 + a_{22}x_2^2,$$

---

[1] The dots are omitted as unnecessary.

simply. The minimum value is necessarily less than this, and accordingly the signs of x_2 and x_3 are opposite. This argument may be continued, and it shews that, however long the series may be, the induced currents are alternately opposite in sign[1], a result in harmony with the magnetizations observed by Henry.

In certain cases the minimum value of T may be very nearly zero. This happens when the coils which exercise a mutual inductive influence are so close throughout their entire lengths that they can produce approximately opposite magnetic forces at all points of space. Suppose, for example, that there are two similar coils A and B, each wound with a double wire (A_1, A_2), (B_1, B_2), and combined so that the primary circuit consists of A_1, the secondary of A_2 and B_1 joined by inductionless leads, and the tertiary of B_2 simply closed upon itself. It is evident that T is made approximately zero by taking $x_2 = -x_1$ and $x_3 = -x_2 = x_1$. The argument may be extended to a train of such coils, however long, and also to cases where the number of convolutions in mutually reacting coils is not the same.

In a large class of problems, where leyden effects do not occur sensibly, the course of events is determined by T and F simply. These functions may then be reduced to sums of squares; and the typical equation takes the form

$$a\ddot{x} + b\dot{x} = X \dots\dots\dots\dots\dots (6).$$

If $X = 0$, that is if there be no imposed electro-motive forces, the solution is

$$\dot{x} = \dot{x}_0 e^{-bt/a} \dots\dots\dots\dots\dots (7).$$

Thus any system of initial currents flowing whether in detached or connected linear conductors, or in solid conducting masses, may be resolved into "normal" components, each of which dies down exponentially at its own proper rate.

A general property of the "persistences," equal to a/b, is proved in § 92 *a*. For example, any increase in permeability, due to the introduction of iron (regarded as non-conducting), or any diminution of resistance, however local, will in general bring about a rise in the values of all the persistences[2].

In view of the discussions of Chapter v. it is not necessary to dwell upon the solution of equations (1) when X is retained. The

---

[1] *Phil. Mag.*, vol. 38, p. 13, 1869.

[2] *Brit. Assoc. Report*, 1885, p. 911.

reciprocal theorem of § 109 has many interesting electrical applications; but, after what has there been said, their deduction will present no difficulty.

235 *m*. In § 111 *b* one application of the general formulæ to an electrical system has already been given. As another example, also relating to the case of two degrees of freedom, we may take the problem of two conductors *in parallel*. It is not necessary to include the influence of the leads outside the points of bifurcation; for provided that there be no mutual induction between these parts and the remainder, their inductance and resistance enter into the result by simple addition.

Under the sole operation of resistance, the total current x_1 would divide itself between the two conductors (of resistances R and S) in the parts

$$\frac{S}{R+S}x_1 \text{ and } \frac{R}{R+S}x_1;$$

and we may conveniently so choose the second co-ordinate that the currents in the two conductors are in general

$$\frac{S}{R+S}x_1+x_2 \text{ and } \frac{R}{R+S}x_1-x_2,$$

x_1 still representing the total current in the leads. The dissipation-function, found by multiplying the squares of the above currents by $\frac{1}{2}R$, $\frac{1}{2}S$ respectively, is

$$F = \tfrac{1}{2}\frac{RS}{R+S}x_1^2 + \tfrac{1}{2}(R+S)x_2^2 \dots\dots\dots\dots (1).$$

Also, L, M, N being the induction coefficients of the two branches,

$$T = \tfrac{1}{2}\frac{LS^2+2MRS+NR^2}{(R+S)^2}x_1^2$$

$$+ \frac{(L-M)S+(M-N)R}{R+S}x_1x_2 + \tfrac{1}{2}(L-2M+N)x_2^2 \dots(2).$$

Thus, in the notation of § 111 *b*,

$$a_{11} = \frac{LS^2+2MRS+NR^2}{(R+S)^2}, \quad a_{12} = \frac{(L-M)S+(M-N)R}{R+S},$$

$$a_{22} = L-2M+N;$$

$$b_{11} = \frac{RS}{R+S}, \quad b_{12} = 0, \quad b_{22} = R+S.$$

Accordingly by (5), (8) § 111 *b*,

$$R' = \frac{RS}{R+S} + \frac{p^2}{R+S} \frac{\{(L-M)S+(M-N)R\}^2}{(R+S)^2+p^2(L-2M+N)^2} \dots (3).$$

$$L' = \frac{LS^2+2MRS+NR^2}{(R+S)^2} - \frac{\{(L-M)S+(M-N)R\}^2}{(R+S)^2(L-2M+N)}$$
$$+ \frac{\{(L-M)S+(M-N)R\}^2}{(L-2M+N)\{(R+S)^2+p^2(L-2M+N)^2\}} \dots (4).$$

These are respectively the effective resistance and the effective inductance of the combination[1]. It is to be remarked that $(L-2M+N)$ is necessarily positive, representing twice the kinetic energy of the system when the currents in the two conductors are $+1$ and -1.

The expressions for R' and L' may be put into a form[2] which for many purposes is more convenient, by combining the component fractional terms. Thus

$$R' = \frac{RS(R+S)+p^2\{R(M-N)^2+S(L-M)^2\}}{(R+S)^2+p^2(L-2M+N)^2} \dots\dots\dots(3'),$$

$$L' = \frac{LS^2+2MRS+NR^2+p^2(LN-M^2)(L-2M+N)}{(R+S)^2+p^2(L-2M+N)^2} \dots (4'),$$

in which $(LN-M^2)$ is positive by virtue of the nature of T.

As p increases from zero, we know by the general theorem § 111 *b*, or from the particular expressions (3), (4), that R' continually increases and that L' continually decreases.

When p is very small,

$$R' = \frac{RS}{R+S}, \qquad L' = \frac{LS^2+2MRS+NR^2}{(R+S)^2} \dots\dots (5).$$

In this case the distribution of the main current between the conductors is determined by the resistances, and (§ 111 *b*) the values of R' and L' coincide respectively with $2F/x_1^2$, $2T/x_1^2$. The resistance is manifestly the same as if the currents were steady.

On the other hand, when p is very great,

$$R' = \frac{R(M-N)^2+S(L-M)^2}{(L-2M+N)^2}, \qquad L' = \frac{LN-M^2}{L-2M+N} \dots (6).$$

In this case the distribution of currents is independent of the resistances, being determined in accordance with Kelvin's theorem

[1] *Phil. Mag.*, vol. 21, p. 377, 1886.
[2] J. J. Thomson, *loc. cit.* § 421.

in such a manner that the ratio of the currents in the two conductors is $(N - M) : (L - M)$. As when p is small, the values in (6) coincide with $2F/x_1^2$, $2T/x_1^2$.

When the two wires composing the conductors in parallel are wound closely together, the energy of the field under high frequency may be very small. There is an interesting distinction to be noted here dependent upon the manner in which the connections are made. Consider, for example, the case of a bundle of five contiguous wires wound into a coil, of which three wires, connected in series so as to have maximum inductance, constitute one of the branches in parallel, and the other two, connected similarly in series, constitute the other branch. There is still an alternative as to the manner of connection of the two branches. If steady currents would circulate opposite ways (M negative), the total current is divided into two parts in the ratio 3 : 2, in such a manner that the more powerful current in the double wire nearly neutralises at external points the magnetic effects of the less powerful current in the triple wire, and the total energy of the system is very small. But now suppose that the connections are such that steady currents would circulate the same way in both branches (M positive). It is evident that the condition of minimum energy cannot be satisfied when the currents are in the same direction, but requires that the smaller current in the triple wire should be in the opposite direction to that of the larger current in the double wire. In fact the currents must be as 3 to -2; so that (since on the same scale the total current is unity) the component currents in the branches are both numerically greater than the total current which is algebraically divided between them. And this peculiar feature becomes more and more strongly marked the nearer L and N approach to equality[1].

The unusual development of currents in the branches is, of course, attended by an augmented effective resistance. In the limiting case when the m convolutions of one branch are supposed to coincide geometrically with one another and with the n convolutions of the second branch, we have

$$L : M : N = m^2 : mn : n^2,$$

and from (6)

$$R' = \frac{n^2 R + m^2 S}{(m - n)^2} \quad\dotfill (7),$$

[1] *Phil. Mag.*, vol. 21, p. 376, 1886.

an expression which increases without limit, as m and n approach to equality.

The fact that under certain conditions the currents in both branches of a divided circuit may exceed the current in the mains has been verified by direct experiment[1]. Each of the three currents to be compared traversed short lengths of similar German-silver wire, and the test consisted in finding what lengths of this wire it was necessary in the various cases to include between the terminals of a high resistance telephone in order to obtain sounds of equal intensity. The variable currents were derived from a battery and scraping contact apparatus (§ 235 r), directly included in the main circuit.

The general formulæ (3'), (4') undergo simplification when the conductors in parallel exercise no mutual induction. Thus, when $M = 0$,

$$R' = \frac{RS(R+S) + p^2(RN^2 + SL^2)}{(R+S)^2 + p^2(L+N)^2} \quad \dots\dots\dots\dots (8),$$

$$L' = \frac{LS^2 + NR^2 + p^2 LN(L+N)}{(R+S)^2 + p^2(L+N)^2} \quad \dots\dots\dots\dots (9).$$

If further $N = 0$, (8) and (9) reduce to

$$R' = \frac{S\{R(R+S) + p^2L^2\}}{(R+S)^2 + p^2L^2}, \qquad L' = \frac{LS^2}{(R+S)^2 + p^2L^2} \dots (10).$$

The peculiar features of the combination are brought out most strongly when S, the resistance of the inductionless component, is great in comparison with R. In that case if the current be steady or slowly vibrating, it flows mainly through R, while the resistance and inductance of the combination approximate to R and L respectively; but if on the other hand the current be a rapidly vibrating one, it flows mainly through S, so that the resistance of the combination approximates to S, and the inductance to zero. These conclusions are in agreement with (10).

If the branches in parallel be simple electro-magnets, L and N are necessarily positive, and the numerator in (9) is incapable of vanishing. But, as we have seen, when leydens are admitted, this restriction may be removed. An interesting case arises when the second branch is inductionless, and is interrupted by a leyden of

[1] Phil. Mag., vol. 22, p. 495, 1886.

capacity C, so that $N = -(Cp^2)^{-1}$, while at the same time $R = S$. The latter condition reduces the numerator in (9) to

$$(L + N)\{R^2 + p^2 LN\}.$$

Thus L' vanishes, (i) when $LCp^2 = 1$, and (ii) when $CR^2 = L$. The first alternative is the condition that the loop circuit, considered by itself, should be isochronous with the imposed vibrations. The second expresses the equality of the time-constants of the two branches. If they be equal, the combination behaves like a simple resistance, whatever be the character of the imposed electro-motive force[1].

235 n.   When there are more than two conductors in parallel, the general expressions for the equivalent resistance and induc-tance of the combination would be very complicated; but a few particular cases are worthy of notice.

The first of these occurs when there is no mutual induction between the members. If the quantities relating to the various branches be distinguished by the suffixes 1, 2, 3, ..., and if E be the difference of potentials at the common terminals, we have

$$E = (ipL_1 + R_1)x_1 = (ipL_2 + R_2)x_2 = \dots \quad \dots \dots \dots (1);$$

so that

$$\frac{1}{\Sigma(ipL + R)^{-1}} = \frac{E}{x_1 + x_2 + \dots} = R' + ipL' \dots \dots (2),$$

by which R' and L' are determined. Thus, if we write

$$\Sigma \frac{R}{R^2 + p^2 L^2} = A, \qquad \Sigma \frac{L}{R^2 + p^2 L^2} = B \dots \dots (3),$$

we have from (2)

$$R' = \frac{A}{A^2 + p^2 B^2}, \qquad L' = \frac{B}{A^2 + p^2 B^2} \dots \dots \dots (4).$$

Equations (3) and (4) contain the solution of the problem[2].

When $p = 0$,

$$R' = \frac{1}{\Sigma(R^{-1})}, \qquad L' = \frac{\Sigma(LR^{-2})}{\{\Sigma(R^{-1})\}^2} \dots \dots \dots (5).$$

When on the other hand p is very great,

$$R' = \frac{\Sigma(RL^{-2})}{\{\Sigma(L^{-1})\}^2}, \qquad L' = \frac{1}{\Sigma(L^{-1})} \dots \dots \dots (6).$$

---

[1] Chrystal, "On the Differential Telephone," *Edin. Trans.*, vol. 29, p. 615, 1880.

[2] *Phil. Mag.*, vol. 21, p. 379, 1886.

Even when the mutual induction between various members cannot be neglected, tolerably simple expressions can be found for the equivalent resistance and inductance in the extreme cases of p infinitely small or infinitely large. As has already been proved, (§ 111 b), the above-mentioned quantities then coincide in value with $2F/(x_1 + x_2 + ...)^2$, and $2T/(x_1 + x_2 + ...)^2$, and the calculation of these values is easy, inasmuch as the distribution of currents among the branches is determined in the first case entirely by F and in the second case entirely by T. Thus, when p is infinitely small, F is a minimum, and the currents are in proportion to the conductances of the several branches. Accordingly, if the induction coefficients of the branches be denoted, as in § 111 b, by $a_{11}, a_{22}, ... a_{12}, a_{13}, ...,$ and the resistances by $R_1, R_2,$ &c., we have

$$R' = \frac{R_1(1/R_1)^2 + R_2(1/R_2)^2 + ...}{(1/R_1 + 1/R_2 + ...)^2} = \frac{1}{1/R_1 + 1/R_2 + ...} \quad ......(7),$$

$$L' = \frac{a_{11}/R_1^2 + a_{22}/R_2^2 + ... + 2a_{12}/R_1R_2 + 2a_{13}/R_1R_3 + ...}{(1/R_1 + 1/R_2 + 1/R_3 + ...)^2} \quad ... (8).$$

A similar method applies when $p = \infty$, but the result is less simple on account of the complication in the ratios of currents due to mutual induction[1].

235 o. The induction-balance, originally contrived by Dove for use with the galvanometer, has in recent years been adapted to the telephone by Hughes[2], who has described experiments illustrating the marvellous sensibility of the arrangement. The essential features are a primary, or battery, circuit, in which circulates a current rendered intermittent by a make and break interrupter, or by a simple scraping contact, and a secondary circuit containing a telephone. By suitable adjustments the two circuits are rendered conjugate, that is to say the coefficient of mutual induction is caused to vanish, so as to reduce the telephone to silence. The introduction into the neighbourhood of a third circuit, whether composed of a coil of wire, or of a simple conducting mass, such as a coin, will then in general cause a revival of sound.

The destruction of the mutual induction in the case of two flat coils can be arrived at by placing them at a short distance apart,

---

[1] J. J. Thomson, *loc. cit.* § 422.

[2] *Phil. Mag.* vol. viii., p. 50, 1879.

in parallel planes, and with accurately adjusted overlapping. But in Hughes' apparatus the balance is obtained more symmetrically by the method of duplication. Four similar coils are employed. Of these two A_1, A_2, mounted at some distance apart with their planes horizontal, and connected in series, constitute the primary induction coil. The secondary induction coil consists in like manner of B_1, B_2, placed symmetrically at short distances from A_1, A_2, and also connected in series, but in such a manner that the induction between A_1 and B_1 tends to balance the induction between A_2 and B_2. If the four coils were perfectly similar, balance would be obtained when the distances were equal. This of course is not to be depended upon, but by a screw motion the distance between one pair, e.g. A_1 and B_1, is rendered adjustable, and in this way a balance between the two inductions is obtained. Wooden cups, fitting into the coils, are provided in such situations that a coin resting in one of them is situated symmetrically between the corresponding primary and secondary coils. The balance, previously adjusted, is of course upset by the introduction of a coin upon one side, but if a perfectly similar coin be introduced upon the other side also, balance may be restored. Hughes found that very minute differences between coins could be rendered evident by outstanding sound in the telephone.

The theory of this apparatus, when the primary currents are harmonic, is simple[1], especially if we suppose that the primary current x_1 is given. If x_1, x_2, ... be the currents; b_1, b_2, ... the resistances; a_{11}, a_{22}, a_{12}, ... the inductances, the equations for the case of three circuits are

$$\left. \begin{array}{l} a_{22}\dot{x}_2 + a_{23}\dot{x}_3 + b_2 x_2 = -a_{12}\dot{x}_1 \\ a_{23}\dot{x}_2 + a_{33}\dot{x}_3 + b_3 x_3 = -a_{13}\dot{x}_1 \end{array} \right\} \dots\dots\dots\dots (1)$$

We now assume that x_1, x_2, &c. are proportional to e^{ipt}, where $p/2\pi$ is the frequency of vibration. Thus,

$$ip\,(a_{22}x_2 + a_{23}x_3) + b_2 x_2 = -ipa_{12}x_1,$$

$$ip\,(a_{23}x_2 + a_{33}x_3) + b_3 x_3 = -ipa_{13}x_1\,;$$

whence by elimination of x_3

$$x_2\left\{ipa_{22} + b_2 + \frac{p^2 a_{23}{}^2}{ipa_{33} + b_3}\right\} = -ipa_{12}x_1 - \frac{p^2 a_{13}a_{23}x_1}{ipa_{33} + b_3}\dots\dots(2).$$

[1] Brit. Assoc. Rep. 1880, p. 472.

From this it appears that a want of balance depending on a_{12} cannot compensate for the action of the third circuit, so as to produce silence in the secondary circuit, unless b_3 be negligible in comparison with pa_{33}, that is unless the time-constant of the third circuit be very great in comparison with the period of the vibration. Otherwise the effects are in different phases, and therefore incapable of balancing.

We will now introduce a fourth circuit, and suppose that the primary and secondary circuits are accurately conjugate, so that $a_{12} = 0$, and also that the mutual induction a_{34} between the third and fourth circuits may be neglected. Then

$$ip \left(a_{22}x_2 + a_{23}x_3 + a_{24}x_4\right) + b_2 x_2 = 0,$$

$$ip \left(a_{32}x_2 + a_{33}x_3\right) + b_3 x_3 = - ipa_{13}x_1,$$

$$ip \left(a_{42}x_2 + a_{44}x_4\right) + b_4 x_4 = - ipa_{14}x_1;$$

whence

$$x_2 \left\{ ipa_{22} + b_2 + \frac{p^2 a_{23}^2}{ipa_{33} + b_3} + \frac{p^2 a_{24}^2}{ipa_{44} + b_4} \right\}$$

$$= - p^2 x_1 \left\{ \frac{a_{13} a_{23}}{ipa_{33} + b_3} + \frac{a_{14} a_{24}}{ipa_{44} + b_4} \right\} \dots \dots (3).$$

It appears that *two* conditions must be satisfied in order to secure a balance, since both the phases and the intensities of the separate effects must be the same. The first condition requires that the time-constants of the third and fourth circuits be equal, unless indeed both be very great, or both be very small, in comparison with the period. If this condition be satisfied, balance ensues when

$$\frac{a_{13} a_{23}}{a_{33}} + \frac{a_{14} a_{24}}{a_{44}} = 0 \dots \dots \dots (4);$$

and it is especially to be noted that the adjustment is independent of pitch, so that (by Fourier's theorem) it suffices whatever be the nature of the variable currents operative in the primary.

As regards the position of the third and fourth circuits, usually represented by coins in illustrative experiments, it will be seen from the symmetry of the right-hand member of (3) that the middle position between the primary and secondary coils is suitable, inasmuch as the product $a_{13}a_{23}$ is stationary in value when the coin is moved slightly so as to be nearer say to the primary and further from the secondary[1]. Approximate independence of other displacements is secured by the geometrical symmetry of the coils round the axis.

235 *p*. For the accurate comparison of electrical quantities the "bridge" arrangement of Wheatstone is usually the most convenient, and is equally available with the galvanometer in the case of steady or transitory currents, or with the telephone in the case of periodic currents. Similar effects may be obtained in most cases without a bridge by the employment of the differential galvanometer or the differential telephone[2].

In the ordinary use of the bridge the four members *a*, *b*, *c*, *d* combined in a quadrilateral Fig. (53 *a*) are simple resistances. The battery branch *f* joins one pair of opposite corners, and the indicating instrument is in the "bridge" *e* joining the other pair. "Balance" is obtained, when $ad = bc$. But for our purpose we have to suppose that any member, e.g. *a*, is not merely a resistance, or even a combination of resistances. It may include an electromagnet, and it may be interrupted by a leyden. But in any case, so long as the current *x* is strictly harmonic, proportional to e^{ipt}, the general relation between it and the difference of potentials V at the extremities is given by

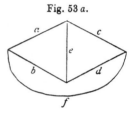

Fig. 53 *a*.

$$V = (a_1 + ia_2) x \ \dots\dots\dots\dots\dots\dots (1),$$

where a_1 and ia_2 are the real and imaginary parts of a complex coefficient a, and are functions of the frequency $p/2\pi$. In the particular case of a simple conductor, endowed with inductance L, a_1 represents the resistance, and a_2 is equal to pL. In general, a_1 is positive; but a_2 may be either positive, as in the above example, or negative. The latter case arises when a resistance R is interrupted by a leyden of capacity C. Here $a_1 = R$, $a_2 = -1/pC$. If there be also inductance L,

$$a_1 = R, \qquad a_2 = pL - 1/pC \ \dots\dots\dots\dots (2).$$

As we have already seen, § 235 *j*, a_2 may vanish for a particular frequency, and the combination is then equivalent to a simple

[1] See Lodge, *Phil. Mag.*, vol. 9, p. 123, 1880.

[2] Chrystal, *Edin. Trans.*, loc. cit.

resistance. But a variation of frequency gives rise to a positive or negative a_2.

In all electrical problems, where there is no mutual induction, the generalized quantities, a, b, &c., combine, just as they do when they represent simple resistances[1]. Thus, if a, a' be two complex quantities representing two conductors in series, the corresponding quantity for the combination is $(a + a')$. Again, if a, a' represent two conductors in parallel, the reciprocal of the resultant is given by addition of the reciprocals of a, a'. For, if the currents be x, x', corresponding to a difference of potentials V at the common terminals,

$$V = ax = a'x',$$

so that
$$x + x' = V(1/a + 1/a').$$

In the application to Wheatstone's combination of the general theory of forced vibrations, we will limit the impressed forces to the battery and the telephone branches. If x, y be the currents in these branches, X, Y the corresponding electro-motive forces, we have, § 107, linear relations between x, y, and X, Y, which may be written

$$\left. \begin{array}{l} X = Ax + By \\ Y = Bx + Cy \end{array} \right\} \quad \dots\dots\dots\dots\dots\dots (3),$$

the coefficient of y in the first equation being identical with that of x in the second equation, by the reciprocal property. The three constants A, B, C are in general complex quantities, functions of p.

The reciprocal relation may be interpreted as follows. If $Y = 0$, $Bx + Cy = 0$, and

$$y = \frac{BX}{B^2 - AC} \quad \dots\dots\dots\dots\dots\dots (4).$$

In like manner, if we had supposed $X = 0$, we should have found

$$x = \frac{BY}{B^2 - AC} \quad \dots\dots\dots\dots\dots\dots (5),$$

shewing that the ratio of the current in one member to the electromotive force operative in the other is independent of the way in which the parts are assigned to the two members.

---

[1] For a more complete discussion of this subject see Heaviside " On Resistance and Conductance Operators," *Phil. Mag.*, vol. 24, p. 479, 1887; *Electrical Papers*, vol. II., p. 355.

We have now to determine the constants A, B, C in terms of the electrical properties of the system. If y be maintained zero by a suitable force Y, the relation between x and X is $X = Ax$. A therefore denotes the (generalized) resistance to any electromotive force in the battery member, *when the telephone member is open*. This resistance is made up of f, the resistance in the battery member, and of that of the conductors $a + c$, $b + d$, combined in parallel. Thus

$$A = f + \frac{(a+c)(b+d)}{a+b+c+d} \quad\dots\dots\dots\dots (6).$$

In like manner

$$C = e + \frac{(a+b)(c+d)}{a+b+c+d} \quad\dots\dots\dots\dots (7).$$

To determine B let us consider the force Y which must act in e in order that the current through it may be zero, in spite of the operation of X. We have $Y = Bx$. The total current x flows partly along the branch $(a+c)$, and partly along $(b+d)$. The current through $(a+c)$ is

$$\frac{x/(a+c)}{1/(a+c) + 1/(b+d)} = \frac{(b+d)x}{a+b+c+d} \quad\dots\dots\dots (8),$$

and that through $(b+d)$ is

$$\frac{(a+c)x}{a+b+c+d} \quad\dots\dots\dots\dots\dots (9).$$

The difference of potentials at the terminals of e, supposed to be interrupted, is thus

$$\frac{c(b+d)x - d(a+c)x}{a+b+c+d};$$

and accordingly

$$B = \frac{bc - ad}{a+b+c+d} \quad\dots\dots\dots\dots (10).$$

By (6), (7), (10) the relationship of X, Y to x, y is completely determined.

The problem of the bridge requires the determination of the current y as proportional to X, when $Y = 0$, that is when no electro-motive force acts in the bridge itself; and the solution is given at once by the introduction into (4) of the values of A, B, C from (6), (7), (10).

If there be an approximate " balance," the expression simplifies. For $(bc - ad)$ is then small, and B^2 may be neglected relatively to

AC in the denominator of (4). Thus, as a sufficient approximation in this case, we may write

$$\frac{y}{X} = -\frac{B}{AC} = \frac{(10)}{(6) \times (7)} \dots\dots\dots\dots\dots (11).$$

The following interpretation of the process leads very simply to the approximate form (11), and is independent of the general theory. Let us first inquire what electro-motive force is necessary in the telephone member to stop any current through it. If such a force act, the conditions are, externally, the same as if the member were open; and the current x in the battery member due to a force equal to X in that member is X/A, where A is written for brevity as representing the right-hand member of (6). The difference of potentials at the terminals of e, still supposed to be open, is found at once when x is known. It is given by

$$c \times (8) - d \times (9) = Bx,$$

where B is defined by (10). In terms of X the difference of potentials is thus BX/A. If e be now closed, the same fraction expresses the force necessary in e in order to prevent the generation of a current in that member.

The case with which we have to deal is when X acts in f and there is no force in e. We are at liberty, however, to suppose that two opposite forces, each of magnitude BX/A, act in e. One of these, as we have seen, acting in conjunction with X in f, gives no current in e; so that, since electro-motive forces act independently of one another, the actual current in e, closed without internal electro-motive force, is simply that due to the other component. The question is thus reduced to the determination of the current in e due to a given force in that member.

So far the argument is rigorous; but we will now suppose that we have to deal with an approximate balance. In this case a force in e gives rise to very little current in f, and in calculating the current in e, we may suppose f to be broken. The total resistance to the force in e is then given simply by C of equation (7), and the approximate value for y is derived by dividing $-BX/A$ by C, as we found in (11).

A continued application of the foregoing process gives y/X in the form of an infinite geometric series:—

$$\frac{y}{X} = -\frac{B}{AC}\left\{1 + \frac{B^2}{AC} + \frac{B^4}{A^2C^2} + \dots\right\} = \frac{B}{B^2 - AC}.$$

This is the rigorous solution already found in (4); but the first term of the series suffices for practical purposes.

The form of (11) enables us at once to compare the effects of increments of resistance and of inductance in disturbing a balance. For let $ad = bc$, and then change d to $d + d'$, where $d' = d_1' + id_2'$. The value of y/X is proportional to d', and the amplitude of the vibratory current in the bridge is proportional to mod. d', that is, to $\sqrt{(d_1'^2 + d_2'^2)}$. Thus d_1', d_2' are equally efficacious when numerically equal[1]. In most cases where a telephone is employed, the balance is more sensitive to changes of inductance than to changes of resistance.

In the use of the Wheatstone balance for purposes of measurement, it is best to make a equal to c. The equality of b and d can then be tested by interchange of a and c, independently of the exactitude of the equality of these quantities. Another advantage lies in the fact that balance is independent of mutual induction between a and c or between b and d.

235 *q*. In the formulæ of § 235 *p* it has been assumed that there is no mutual induction between the various members of the combination. The more general theory has been considered very fully by Heaviside[2], but to enter upon it would lead us too far. It may be well, however, to sketch the theory of the arrangement adopted by Hughes, which possesses certain advantages in dealing with the electrical properties of wires in short lengths[3].

The apparatus consists of a Wheatstone's quadrilateral, Fig. 53 *b*, with a telephone in the bridge, one of the sides of the quadrilateral being the wire or coil under examination (P), and the other three being the parts into which a single German-silver wire is divided by two sliding contacts. If the battery-branch (B) be closed, and a suitable interrupter be introduced into the telephone-branch (T), balance may be obtained by shifting the contacts. *Provided that the interrupter introduces no electro-motive*

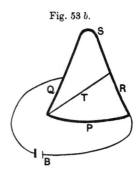

Fig. 53 *b*.

[1] "On the Bridge Method in its Application to Periodic Electric Currents." *Proc. Roy. Soc.*, vol. 49, p. 203, 1891.

[2] "On the Self-Induction of Wires," Part vi.; *Phil. Mag.*, Feb. 1887; *Electrical Papers*, 1892, vol. ii., p. 281.

[3] *Journ. Tel. Eng.*, vol. xv. (1886) p. 1; *Proc. Roy. Soc.*, vol. xl. (1886) p. 451.

force of its own[1], the balance indicates the proportionality of the four resistances. If P be the unknown resistance of the conductor under test, Q, R the resistances of the adjacent parts of the divided wire, S that of the opposite part (between the sliding contacts), then, by the ordinary rule, $PS = QR$; while Q, R, S are subject to the relation

$$Q + R + S = W,$$

W being a constant. If now the interrupter be transferred from the telephone to the battery-branch, the balance is usually disturbed on account of induction, and cannot be restored by any mere shifting of the contacts. In order to compensate the induction, another influence of the same kind must be introduced. It is here that the peculiarity of the apparatus lies. A coil (not shewn in the figure) is inserted in the battery and another in the telephone-branch which act inductively upon one another, and are so mounted that the effect may be readily varied. The two coils may be concentric and relatively movable about the common diameter. In this case the action vanishes when the planes are perpendicular. If one coil be very much smaller than the other, the coefficient of mutual induction M is proportional to the cosine of the angle between the planes. By means of the *two* adjustments, the sliding of the contacts and the rotation of the coil, it is usually possible to obtain a fair silence.

Hughes interpreted his observations on the basis of an assumption that the inductance of P was represented by M, irrespective of resistance, and that the resistance to variable currents could (as in the case of steady currents) be equated to QR/S. But the matter is not quite so simple. The true formulæ are, however, readily obtained for the case where the only sensible induction among the sides of the quadrilateral is the inductance L of the conductor P.

Since there is no current through the bridge, there must be the same current (x) in P and in one of the adjacent sides (say) R, and for a like reason the same current y in Q and S. The difference of potentials at time t between the junction of P and R and the junction of Q and S may be expressed by each of the three following equated quantities :—

$$Qy - Px - L\frac{dx}{dt} = -M\frac{d(x + y)}{dt} = Rx - Sy.$$

---

[1] A condition not always satisfied in practice.

Introducing the assumption that all the quantities vary harmonically with frequency $p/2\pi$, and eliminating the ratio $y : x$, we find as the conditions required for silence in the telephone

$$QR - SP = p^2 ML \ldots\ldots\ldots\ldots\ldots (1),$$

$$M(P + Q + R + S) = SL \ldots\ldots\ldots\ldots (2).$$

It will be seen that the ordinary resistance balance $(SP = QR)$ is departed from. The change here considered is peculiar to the apparatus and, so far as its influence is concerned, it does not indicate a real alteration of resistance in the wire. Moreover, since p is involved, the disturbance depends upon the rapidity of vibration, so that in the case of ordinary mixed sounds silence can be attained only approximately. Again, from the second equation we see that M is not in general a correct measure of the value of L[1].

If however, P be known, the application of (2) presents no difficulty. In many cases we may be sure beforehand that P, viz. the effective resistance of the conductor, or combination of conductors, to the variable currents, is the same as if the currents were steady, and then P may be regarded as known. But there are other cases,—some of them will be alluded to below—in which this assumption cannot be made; and it is impossible to determine the unknown quantities L and P from (2) alone. We may then fall back upon (1). By means of the two equations P and L can always be found in terms of the other quantities. But among these is included the frequency of vibration; so that the method is practically applicable only when the interrupter is such as to give an absolute periodicity. A scraping contact, otherwise very convenient, is thus excluded; and this is undoubtedly an objection to the method.

If the member P be without inductance, but be interrupted by a leyden of capacity C, the same formulæ may be employed, with substitution of $-1/p^2C$ for L. Equation (1) then gives a measure of C which is independent of the frequency.

235 *r*. The success of experiments with this kind of apparatus depends very largely upon the action of the interrupter by which the currents are rendered variable. When periodicity is not

---

[1] "Discussion on Prof. Hughes' Address." *Journ. Tel. Eng.*, vol. xv., p. 54. Feb., 1886.

necessary, a scraping contact, actuated by a clock or by a small motor, answers very well; but it is advisable, following Lodge and Hughes, so to arrange matters that the current is suspended altogether at short intervals. The faint scraping sound heard in the neighbourhood of a balance, is more certainly identified when thus rendered intermittent.

But for many of the most interesting experiments a scraping contact is unsuitable. When the inductance and resistance under observation are rapidly varying functions of the frequency, it is evident that no sharp results are possible without an interrupter giving a perfectly regular electrical vibration. With proper appliances an absolute silence, or at least one disturbed only by a slight sensation of the octave of the principal tone, can be arrived at under circumstances where a scraping contact would admit of no approach to a balance at all.

Tuning-forks, driven electromagnetically with liquid or solid contacts (§ 64), answer well so long as the frequency required does not exceed (say) 300 per second; but for experiments with the telephone we desire frequencies of from 500 to 2000 per second. Good results may be obtained with harmonium reed interrupters, the vibrating tongue making contact once during each period with a stop, which can be adjusted exactly to the required position by means of a screw[1].

But perhaps the best interrupter for use with the telephone is obtained by taking advantage of the instability of a jet of fluid. If the diameter and the speed be chosen suitably, the jet may be caused to resolve itself into drops under the action of a tuning-fork in a perfectly regular manner, one drop corresponding to each complete vibration of the fork. Each drop, as it passes, may be made to complete an electric circuit by squeezing itself between the extremities of two fine platinum wires. If the electro-motive force of the battery be pretty high, and if the jet be salted to improve its conductivity, sufficient current passes, especially if the aid of a small step-down transformer be invoked. Finally the apparatus is made self-acting by bringing the fork under the influence of an electro-magnet, itself traversed by the same intermittent current. Such an apparatus may be made to work with frequencies up to 2000 per second, and it possesses many advantages, among which may be mentioned almost absolute

[1] *Phil. Mag.*, vol. 22, p. 472, 1886

constancy of pitch, and the avoidance of loud aerial disturbance. The principles upon which the action of this interrupter depends will be further considered in a subsequent chapter.

235 *s*. Scarcely less important than the interrupter are the arrangements for measuring induction, whether mutual induction, as required in § 235 *q*, or self-induction. Inductometers, as Heaviside calls them, may be conveniently constructed upon the pattern of Hughes. A small coil is mounted so that one diameter coincides with a diameter of a larger coil, and is movable about that diameter. The mutual induction M between the two circuits depends upon the position given to the smaller coil, which is read by a pointer attached to it, and moving over a graduated circle. If the smaller coil were supposed to be infinitely small, the value of M, as has already been stated, would be proportional to the sine of the displacement from the zero position ($M = 0$). But an approximation to this state of things is not desirable. If the mean radius of the small coil be increased until it amounts to ·55 of that of the larger, not only is the efficiency much enhanced, but the scale of M is brought to approximate coincidence, over almost the whole practical range, with the scale of degrees[1]. The absolute value of each degree may be arrived at in various ways, perhaps most simply by adjusting the mutual induction of the instrument to balance a standard of mutual induction.

For experiments upon the plan of § 235 *q* the one coil is included in the telephone and the other in the battery branch, but when the object is to secure a variable and measurable inductance, the two coils are connected in series. The inductance of the combination is then $L + 2M + N$, of which the first and third terms are independent of the relative position of the coils.

235 *t*. Good results by the method of § 235 *q* have been obtained by Weber[2], and by the author[3] using a reed interrupter of frequency 1050 per second; but the fact that inductance and resistance are mixed up in the measurements is a decided drawback, if it be only because the readings require for their interpretation calculations not readily made upon the spot.

[1] *Phil. Mag.*, vol. 22, p. 498, 1886.

[2] *Electrical Review*, April 9, July 9, 1886.

[3] *Phil. Mag.*, loc. cit.

The more obvious arrangement is one in which both the induction and the resistance of the branch containing the subject under examination are in every case brought up to the given totals necessary for a balance.   To carry this out conveniently we require to be able to add inductance without altering resistance, and resistance without altering inductance, and both in a measurable degree.   The first demand is easily met.   If we include in the circuit the *two* coils of an inductometer, connected in series, the inductance of the whole can be varied in a known manner by rotating the smaller coil.   On the other hand the introduction, or removal, of resistance without alteration of inductance cannot well be carried out with rigour.   But in most cases the object can be sufficiently attained with the aid of a resistance-slide of thin German-silver wire which may be in the form of a nearly close loop.

In the Wheatstone's quadrilateral, as arranged for these experiments, the adjacent sides R, S may be made of similar wires of German silver of equal resistance ($\frac{1}{2}$ ohm).   If doubled they give rise to little induction, but the accuracy of the method is independent of this circumstance.   The side P includes the conductor, or combination of conductors, under examination, an inductometer, and the resistance-slide.   The other side, Q, must possess resistance and inductance greater than any of the conductors to be compared, but need not be susceptible of ready and measurable variations.   In order to avoid mutual induction between the branches, P and Q should be placed at some distance away, being connected with the rest of the apparatus by leads of doubled wire.

It will be evident that when the interrupter acts in the battery branch, balance can be obtained at the telephone in the bridge only under the conditions that both the inductance and the resistance in P are equal in the aggregate to the corresponding quantities in Q.   Hence when one conductor is substituted for another in P, the alterations demanded at the inductometer and in the slide give respectively the changes of inductance and of resistance.   In this arrangement inductance and resistance are well separated, so that the results can be interpreted without calculation; but the movable contacts of the slide appear to introduce uncertainty into the determination of resistance.

In order to get rid of the objectionable movable contacts some sacrifice of theoretical simplicity seems unavoidable.   We can no longer keep the total resistances P and Q constant; but by reverting to the arrangement adopted in a well-known form of Wheatstone's bridge, we cause the resistances taken from P to be added to Q, and *vice versâ*. The transferable resistance is that of a straight wire of German-silver, with which one telephone terminal makes contact at a point whose position is read off on a divided scale. Any uncertainty in the resistance of *this* contact does not influence the measurements.

Fig. 53 c.

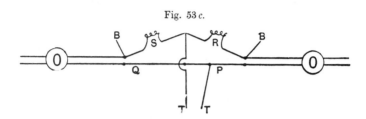

The diagram Fig. (53 c) shows the connection of the parts. One of the telephone terminals T goes to the junction of the ($\frac{1}{2}$ ohm) resistances R and S, the other to a point upon the divided wire. The branch P includes one inductometer (with coils connected in series), the subject of examination, and part of the divided wire. The branch Q includes a second inductometer (replaceable by a simple coil possessing suitable inductance), a rheostat, or any resistance roughly adjustable from time to time, and the remainder of the divided wire. The battery branch B, in which may also be included the interrupter, has its terminals connected, one to the junction of P and R, the other to the junction of Q and S. When it is desired to use steady currents, the telephone can of course be replaced by a galvanometer.

In this arrangement, as in the other, balance requires that the branches P and Q be similar in respect both of inductance and of resistance. The changes in inductance due to a shift in the movable contact may usually be disregarded, and thus any alteration in the subject (included in P) is measured by the rotation necessitated at the inductometer. As for the resistance, it is evident that (R and S being equal) the value for any additional conductor interposed in P is measured by twice the displacement of the sliding contact necessary to regain the balance.

Experimental details of the application of this method to the measurement of various combinations will be found in the paper[1] from which the above sketch is derived. Among these may be mentioned the verification of Maxwell's formulæ, (8), (9) § 235 k, as to the influence of a neighbouring circuit, especially in the extreme case where the equivalent inductance is almost destroyed, and of the formula (10) § 235 m relating to the behaviour of an electro-magnet shunted by a relatively high simple resistance. But the most interesting in many respects is the application to the phenomena presently to be considered, where the conductors in question are no longer approximately linear but must be regarded as solid masses in which the currents are distributed in a manner that needs to be determined by general electrical theory.

As has already been remarked more than once, a leyden may always be supposed to be included in the circuit, the stiffness thereof having the effect of a negative inductance. If there be no hysteresis in the action of the leyden, the whole effect is thus represented; but when the dielectric employed is solid, it appears that dissipative loss cannot be avoided. The latter effect manifests itself as an augmentation of apparent resistance, indistinguishable, unless the frequency be varied, from the ordinary resistance of the leads. A similar treatment may be applied to an electrolytic cell, the stiffness and resistance being presumably both functions of the frequency.

235 u. It was proved by Maxwell[2] that a perfectly conducting sheet, forming a closed or an infinite surface, acts as a magnetic screen, no magnetic actions which may take place on one side of the sheet producing any magnetic effect on the other side. "In practice we cannot use a sheet of perfect conductivity : but the above described state of things may be approximated to in the case of periodic magnetic changes, if the time-constants of the sheet circuits be large in comparison with the periods of the changes."

"The experiment is made by connecting up into a primary circuit a battery, a microphone-clock, and a coil of insulated wire. The secondary circuit includes a parallel coil and a telephone. Under these circumstances the hissing sound is heard almost as well as if the telephone were inserted in the primary circuit

---

[1] *Phil. Mag., loc. cit.*
[2] *Electricity and Magnetism*, 1873, § 655.

itself. But if a large and stout plate of copper be interposed between the two coils, the sound is greatly enfeebled. By a proper choice of battery and of the distance between the coils, it is not difficult so to adjust the strength that the sound is conspicuous in the one case and inaudible in the other "[1].

One of the simplest applications of Maxwell's principle is to the case of a long cylindrical shell placed within a coaxal magnetizing helix. The condition of minimum energy requires that such currents be developed in the shell as shall neutralize at internal points the action of the coil. Thus, if the conductivity of the shell be sufficiently high, the interior space is screened from the magnetizing force of periodic currents flowing in the outer helix, and conducting circuits situated within the shell must be devoid of induced currents. An obvious deduction is that the currents induced in a solid conducting core will be more and more confined to the neighbourhood of the surface as the frequency of electrical vibration is increased.

The point at which the concentration of current towards the surface becomes important depends upon the relative values of the imposed vibration-period and the principal time-constant of the core circuit. If ρ be the specific resistance of the material, μ its magnetic permeability, a the radius of the cylinder, the expression for the induction (c) parallel to the axis, during the progress of the subsidence of free currents in a normal mode, is

$$c = e^{\lambda t} J_0(kr) \dots\dots\dots\dots\dots\dots(1),$$

where
$$k^2 = -\frac{4\pi\lambda\mu}{\rho} \dots\dots\dots\dots\dots\dots(2),$$

and ka is determined by the condition that

$$J_0(ka) = 0 \dots\dots\dots\dots\dots\dots(3).$$

The roots of (3) are, § 206,

$$2\cdot404, \quad 5\cdot520, \quad 8\cdot654, \quad 11\cdot792, \text{ &c.,}$$

so that for the principal mode of greatest persistence

$$c = e^{\lambda t} J_0(2\cdot404 \, r/a) \dots\dots\dots\dots(4),$$

where
$$\lambda = -\frac{(2\cdot404)^2 \rho}{4\pi\mu a^2} \dots\dots\dots\dots(5).$$

Acoustical Observations, *Phil. Mag.*, vol. 13, p. 344, 1882.

For copper in C.G.S. measure $\rho = 1642$, $\mu = 1$, and thus

$$\tau = (-\lambda)^{-1} = \frac{a^2}{800} \text{ nearly}^1.$$

In the case of iron we may take as approximate values, $\mu = 100$, $\rho = 10^4$. Thus for an iron wire of diameter $(2a)$ equal to ·33 cm., the value of τ is about $\frac{1}{2000}$ of a second, and is therefore comparable with the periods concerned in telephonic experiments.

Regarded from an analytical point of view the theory of forced vibrations in a conducting core is equally simple, and was worked out almost simultaneously by Lamb[2], Oberbeck[3] and Heaviside[4]. In this case we are to regard λ as given, equal (say) to ip, where $p/2\pi$ is the frequency. If Ie^{ipt} be the imposed magnetizing force, the solution is

$$c = \frac{J_0(kr)}{J_0(ka)} \mu I e^{ipt} \dots\dots\dots\dots\dots(6),$$

the value of k being given by (2).

"When the period in the field is long in comparison with the time of decay of free currents, we have $J_0(kr) = 1$, nearly, so that c is approximately constant and $= \mu I$ throughout the section of the cylinder. But, in the opposite extreme, when the oscillations in the intensity of the field are rapid in comparison with the decay of free currents, the induced currents extend only to a small depth beneath the surface of the cylinder, the inner strata (so to speak) being almost completely sheltered from electromotive force by the outer ones. Writing $k^2 = (1 - i)^2 q^2$, where

$$q^2 = \frac{2\pi\mu p}{\rho},$$

we have, when qr is large,

$$J_0(kr) = \text{const.} \times \frac{e^{qr} \cdot e^{i(qr - \frac{1}{4}\pi)}}{\sqrt{r}},$$

approximately, and thence

$$c = \mu I \cdot \sqrt{(a/r)} \cdot e^{q(r-a) + iq(r-a)} \left[e^{ipt} \right].$$

This indicates that the electrical disturbance in the cylinder

[1] "On the Duration of Free Electric Currents in an Infinite Conducting Cylinder," *Brit. Assoc. Report* for 1882, p. 446.

[2] *Proc. Math. Soc.*, vol. xv., p. 139, Jan. 1884.

[3] *Wied. Ann.*, vol. xxi., p. 672, Ap. 1884.

[4] *Electrician*, May, 1884. *Electrical Papers*, vol. ii., p. 353.

consists in a series of waves propagated inwards with rapidly diminishing amplitude[1]."

For experimental purposes what we most require to know is the reaction of the core currents upon the helix, in which alone we can directly measure electrical effects. This problem is fully treated by Heaviside[2], but we must confine ourselves here to a mere statement of results. These are most conveniently expressed by the changes of effective inductance L and resistance R due to the core. If m be the number of turns per unit length in the magnetizing helix, and if δL, δR be the apparent alterations of L and R due to the introduction of the core, also reckoned per unit length, we have

$$\left. \begin{array}{l} \delta L = 4m^2\pi^2a^2(\mu P - 1) \\ \delta R = 4m^2\pi^2a^2\mu . \, pQ \end{array} \right\} \quad \dots\dots\dots\dots\dots (7),$$

where P and Q are defined by

$$P - iQ = \phi'/\phi \dots\dots\dots\dots\dots\dots (8),$$

the function ϕ being of the form

$$\phi(x) = J_0(2i\sqrt{x}) = 1 + x + \frac{x^2}{1^2 . 2^2} + \dots + \frac{x^n}{1^2 . 2^2 \dots n^2} + \dots\dots (9),$$

and the argument x being

$$ip\mu . \pi a^2/\rho \dots\dots\dots\dots\dots\dots (10).$$

If the material composing the core be non-conducting, $x = 0$, and therefore

$$P = 1, \qquad Q = 0.$$

Accordingly $\qquad \delta L = 4m^2\pi^2a^2(\mu - 1), \qquad \delta R = 0 \dots\dots\dots (11).$

These values apply also, whatever be the conductivity of the core, if the frequency be sufficiently low.

At the other extreme, when $p = \infty$, we require the ultimate form of ϕ'/ϕ. From the value of J_0 given in (10) § 200, or otherwise, it may be shewn that in the limit

$$\phi'/\phi = x^{-\frac{1}{2}} \dots\dots\dots\dots\dots\dots (12),$$

so that $\qquad\qquad P = Q = \dfrac{1}{\sqrt{\{2p\mu . \pi a^2/\rho\}}} \dots\dots\dots\dots (13).$

The introduction of these values into (7) shews that in the limit, when the frequency is exceedingly high,

$$\delta L = -4m^2\pi^2a^2, \qquad \delta R = 0 \dots\dots\dots\dots (14),$$

---

[1] Lamb, *loc. cit.*, where is also discussed the problem of the currents induced by the sudden cessation of a previously constant field.

[2] *loc. cit.*

as might also have been inferred from the consideration that the induced currents are then confined to the surface of the core.

An example of the application of these formulæ to an intermediate case and a comparison with experiment will be found in the paper already referred to[1].

235 v. The application of Maxwell's principle to the case of a wire, in which a longitudinal electric current is induced, is less obvious; and Heaviside[2] appears to have been the first to state distinctly that the current is to be regarded as propagated inwards from the exterior. The relation between the electromotive force E and the total current C had, however, been given many years earlier by Maxwell[3] in the form of a series. His result is equivalent to

$$\frac{E}{RC} = ipl/R \cdot A + \frac{\phi(ipl\mu/R)}{\phi'(ipl\mu/R)} \quad\text{...............} (1),$$

in which R denotes the whole resistance of the length l to steady currents, μ the permeability, and $p/2\pi$ the frequency. The function ϕ is that defined by (9) § 235 u, and A is a constant dependent upon the situation of the return current[4].

The most convenient form of the results is that which we have already several times employed. If we write

$$E = R'C + ipL'C \quad\text{.....................}(2),$$

in which R' and L' are real, these quantities will represent the effective resistance and inductance of the wire. When the argument in (1) is small, that is when the frequency is relatively low, we thus obtain

$$R' = R\left\{1 + \tfrac{1}{12}\frac{p^2l^2\mu^2}{R^2} - \tfrac{1}{180}\frac{p^4l^4\mu^4}{R^4} + \ldots\right\} \quad\text{.........} (3),$$

$$L'/l = A + \mu\left\{\tfrac{1}{2} - \tfrac{1}{48}\frac{p^2l^2\mu^2}{R^2} + \tfrac{13}{8640}\frac{p^4l^4\mu^4}{R^4} + \ldots\right\} \quad\text{......}(4)^5.$$

[1] *Phil. Mag.*, vol. 22, p. 493, 1886.
[2] *Electrician*, Jan., 1885; *Electrical Papers*, vol. i., p. 440.
[3] *Phil. Trans.*, 1865; *Electricity and Magnetism*, vol. ii., § 690.
[4] The simplest case arises when the dielectric, which bounds the cylindrical wire of radius a, is enclosed within a second conducting mass extending outwards to infinity and bounded internally at a cylindrical surface $r=b$. We then have $A = 2\log(b/a)$. See J. J. Thomson, *loc. cit.*, § 272.
[5] *Phil. Mag.*, vol. 21, p. 387, 1886. It is singular that Maxwell (*loc. cit.*) seems to have regarded his solution as conveying a correction to the self-induction only of the wire.

When p is very small, these equations give, as was to be expected,

$$R' = R, \qquad L' = l\,(A + \tfrac{1}{2}\mu)\,\ldots\ldots\ldots\ldots\ldots(5).$$

If we include the next terms, we recognise that, in accordance with the general rule, L' begins to diminish and R' to increase.

When p is very great, we have to make use of the limiting form of ϕ'/ϕ. As in § 235 u,

$$\phi/\phi' = (1 + i)\,\sqrt{(\tfrac{1}{2}pl\mu/R)}\,\ldots\ldots\ldots\ldots\ldots(6);$$

and thus ultimately

$$R' = \sqrt{(\tfrac{1}{2}pl\mu R)}\,\ldots\ldots\ldots\ldots\ldots\ldots(7),$$

$$L'/l = A + \sqrt{(\mu R/2pl)}\,\ldots\ldots\ldots\ldots\ldots(8),$$

the first of which increases without limit with p, while the second tends to the finite limit A, corresponding to the total exclusion of current from the interior of the wire.

Experiments[1] upon an iron wire about 18 metres long and 3·3 millimetres in diameter led to the conclusion that the resistance to variable currents of frequency 1050 was such that $R'/R = 1·9$. A calculation based upon (1) shewed that this result is in harmony with theory, if $\mu = 99·5$. Such is about the value indicated by other telephonic experiments.

235 w. The theory of electric currents in such wires as are commonly employed in laboratory experiments is simple, mainly in consequence of the subordination of electrostatic capacity. When this element can be neglected, the current is necessarily the same at all points along the length of the wire, so that whatever enters a wire at the sending end leaves it unimpaired at the receiving end. In this case the whole electrical character of the wire can be expressed by two quantities, its resistance R and inductance L, and these may usually be treated as constants, independent of the frequency. The relation of the current to the electromotive force under such circumstances has already been discussed (7) § 235 j. When we have occasion to consider only the amplitude of the current, irrespective of phase, we may regard it as determined by $\sqrt{[R^2 + p^2 L^2]}$, a quantity which is called by Heaviside the *impedance*. Thus in circuits devoid of capacity the impedance is always increased by the existence of L.

[1] *Phil. Mag.*, vol. 22, p. 488, 1886.

Circuits employed for practical telephony may often be re-garded as coming under the above description, especially when the wires are suspended and are of but moderate length. But there are other cases in which electrostatic capacity is the domi-nating feature. The theory of electric cables was established many years ago by Lord Kelvin[1] for telegraphic purposes. If S be the capacity and R the resistance of the cable, reckoned per unit length, V and C the potential and the current at the point z, we have

$$S\, dV/dt = -\, dC/dz, \qquad RC = -\, dV/dz \, \ldots\ldots\ldots (1),$$

whence

$$RS\, dC/dt = d^2C/dz^2 \ldots\ldots\ldots\ldots\ldots\ldots\ldots(2),$$

the well known equation for the conduction of heat discussed by Fourier. On the assumption that C is proportional to e^{ipt}, it reduces to

$$d^2C/dz^2 = \{\surd(\tfrac{1}{2}pRS).(1+i)\}^2 C \ldots\ldots\ldots\ldots (3);$$

so that the solution for waves propagated in the positive direc-tion is

$$C = C_0 e^{-\surd(\frac{1}{2}pRS)\cdot z}\cos\{pt - \surd(\tfrac{1}{2}pRS).z\} \, \ldots\ldots\ldots (4).$$

The distance in traversing which the current is attenuated in the ratio of e to 1 is thus

$$z = \surd(2/pRS) \, \ldots\ldots\ldots\ldots\ldots\ldots\ldots (5).$$

A very slight consideration of the magnitudes involved is sufficient to give an idea of the difficulty of telephoning through a long cable. If, for example, the frequency $(p/2\pi)$ be that of a note rather more than an octave above middle c, and the cable be such as are used to cross the Atlantic, we have in C.G.S. measure

$$\surd p = 60, \qquad (RS)^{-1} = 2 \times 10^{16},$$

and accordingly from (5)

$$z = 3 \times 10^6\, \text{cm.} = 20 \text{ miles approximately.}$$

A distance of 20 miles would thus reduce the intensity of sound, measured by the square of the amplitude, to about a tenth, an operation which could not be repeated often without rendering it inaudible. With such a cable the practical limit would not be likely to exceed fifty miles, more especially as the easy intelligibility of speech requires the presence of tones still higher than is supposed in the above numerical example[2].

[1] *Proc. Roy. Soc.*, 1855; *Mathematical and Physical Papers*, vol. II. p. 61.
[2] "On Telephoning through a Cable." *Brit. Ass. Report* for 1884, p. 632.

235 x. In the above theory the insulation is supposed to be perfect and the inductance to be negligible. It is probable that these conditions are sufficiently satisfied in the case of a cable, but in other telephonic lines the inductance is a feature of great importance. The problem has been treated with full generality by Heaviside, but a slight sketch of his investigation is all that our limits permit.

If R, S, L, K be the resistance, capacity or permittance, inductance, and leakage-conductance respectively per unit of length, V and C the potential-difference and current at distance z, the equations, analogous to (1) § 235 w, are

$$KV + S\frac{dV}{dt} = -\frac{dC}{dz}, \qquad RC + L\frac{dC}{dt} = -\frac{dV}{dz} \dots\dots (1).$$

Thus, if the currents are harmonic, proportional to e^{ipt},

$$\frac{d^2C}{dz^2} = (R + ipL)(K + ipS)\,C \dots\dots\dots\dots (2),$$

with a similar equation for V.

It might perhaps have been expected that a finite leakage K would always act as a complication; but Heaviside[1] has shewn that it may be so adjusted as to simplify the matter. This case, which is remarkable in itself and also serves to throw light upon the general question, arises when $R/L = K/S$. We will write

$$LSv^2 = 1, \qquad R/L = K/S = q \dots\dots\dots\dots (2),$$

where v is a velocity of the order of the velocity of light. The equation for V is then by (1)

$$v^2\,d^2V/dz^2 = (d/dt + q)^2 V \dots\dots\dots\dots (3);$$

or if we take U so that

$$V = e^{-qt}\,U \dots\dots\dots\dots\dots (4),$$

$$v^2\,d^2U/dz^2 = d^2U/dt^2 \dots\dots\dots\dots (5),$$

the well-known equation of undisturbed wave propagation § 144. "Thus, if the wave be positive, or travel in the direction of increasing z, we shall have, if $f_1(z)$ be the state of V initially,

$$V_1 = e^{-qt}f_1(z - vt), \qquad C_1 = V_1/Lv \dots\dots\dots(6).$$

If V_2, C_2 be a negative wave, travelling the other way,

$$V_2 = e^{-qt}f_2(z + vt), \qquad C_2 = -V_2/Lv \dots\dots\dots(7).$$

[1] *Electrician*, June 17, 1887. *Electrical Papers*, vol. II. pp. 125, 309.

Thus, any initial state being the sum of V_1 and V_2 to make V, and of C_1 and C_2 to make C, the decomposition of an arbitrarily given initial state of V and C into the waves is effected by

$$V_1 = \tfrac{1}{2}(V + vLC), \qquad V_2 = \tfrac{1}{2}(V - vLC) \dots\dots\dots (8).$$

We have now merely to move V_1 bodily to the right at speed v, and V_2 bodily to the left at speed v, and attenuate them to the extent e^{-qt}, to obtain the state at time t later, provided no changes of condition have occurred. The solution is therefore true for all future time in an infinitely long circuit. But when the end of a circuit is reached, a reflected wave usually results, which must be added on to obtain the real result."

As in § 144, the precise character of the reflection depends upon the terminal conditions. "One case is uniquely simple. Let there be a resistance inserted of amount vL. It introduces the condition $V = vLC$ if at say B, the positive end of the circuit, and $V = -vLC$ if at the negative end, or beginning. These are the characteristics of a positive and of a negative wave respectively; it follows that any disturbance arriving at the resistance is at once absorbed. Thus, if the circuit be given in any state whatever, without impressed force, it is wholly cleared of electrification and current in the time l/v at the most, if l be the length of the circuit, by the complete absorption of the two waves into which the initial state may be decomposed."

"But let the resistance be of amount R_1 at say B; and let V_1 and V_2 be corresponding elements in the incident and reflected waves. Since we have

$$V_1 = vLC_1, \qquad V_2 = -vLC_2, \qquad V_1 + V_2 = R_1(C_1 + C_2)\dots(9),$$

we have the reflected wave given by

$$\frac{V_2}{V_1} = \frac{R_1 - vL}{R_1 + vL} \dots\dots\dots\dots\dots\dots (10).$$

If R_1 be greater than the critical resistance of complete absorption, the current is negatived by reflection, whilst the electrification does not change sign. If it be less, the electrification is negatived, whilst the current does not reverse."

"Two cases are specially notable. They are those in which there is no absorption of energy. If $R_1 = 0$, meaning a short circuit, the reflected wave of V is a perverted and inverted copy of the incident. But if $R = \infty$, representing insulation, it is C that is inverted and perverted[1]."

The cases last mentioned are evidently analogous to the reflection of a sonorous aerial wave travelling in a pipe. If the end of the pipe be closed, the reflection is of one character, and if it be open of another character. In both cases the whole energy is reflected, § 257. The waves reflected at the two ends of an electric circuit complicate the general solution, especially when the simplifying condition (2) does not hold. But in many cases of practical interest they may be omitted without much loss of accuracy. One passage over a long line usually introduces considerable attenuation, and then the effect of the reflected wave, which must traverse the line three times in all, becomes insignificant.

In proceeding to the general solution of (2) for a positive wave, we will introduce, after Heaviside, the following abbreviations,

$$v^2 LS = 1, \qquad R/Lp = f, \qquad K/Sp = g \dots\dots\dots(11).$$

In terms of these quantities (2) may be written

$$d^2 C/dz^2 = (P + iQ)^2 C \dots\dots\dots (12),$$

where

$$P^2 \text{ or } Q^2 = \tfrac{1}{2}(p/v)^2 \{(1 + f^2)^{\frac{1}{2}}(1 + g^2)^{\frac{1}{2}} \pm (fg - 1)\} \dots (13).$$

Thus, if P and Q be taken positively, the solution for a wave travelling in the positive direction is

$$C = C_0 e^{-Pz} \cos(pt - Qz) \dots\dots\dots(14),$$

the current at the origin being $C_0 \cos pt$.

The cable formula, § 235 w, is the particular case arrived at by supposing in (13) $f = \infty$, $g = 0$, which then reduces to

$$P^2 = Q^2 = \tfrac{1}{2} pRS \dots\dots\dots (15).$$

Again, the special case of equation (3) is derivable by putting $f = g = q/p$. The result is

$$P = q/v, \qquad Q = p/v \dots\dots\dots(16).$$

If the insulation be perfect, $g = 0$, and (13) becomes

$$P^2 \text{ or } Q^2 = \tfrac{1}{2}(p/v)^2 \{(1 + f^2)^{\frac{1}{2}} \mp 1\} \dots\dots\dots (17).$$

---

[1] Heaviside, *Collected Works*, vol. II. p. 312.

In certain examples of long copper lines of high conductivity, f may be regarded as small so far as telephonic frequencies are concerned. Equation (17) then gives

$$P = pf/2v = R/2vL, \qquad Q = p/v \dots\dots\dots (18).$$

For a further discussion of the various cases that may arise the reader must be referred to the writings of Heaviside already cited. The object is to secure, as far as may be, the propagation of waves without alteration of type. And here it is desirable to distinguish between simple attenuation and distortion. If, as in (16) and (18), P is independent of p, the amplitudes of all components are reduced in the same ratio, and thus a complex wave travels without *distortion*. The cable formula (15) is an example of the opposite state of things, where waves of high frequency are attenuated out of proportion to waves of low frequency. It appears from Heaviside's calculations that the distortion is lessened by even a moderate inductance.

The effectiveness of the line requires that neither the attenuation nor the distortion exceed certain limits, which however it is hard to lay down precisely. A considerable amount of distortion is consistent with the intelligibility of speech, much that is imperfectly rendered being supplied by the imagination of the hearer.

235 y. It remains to consider the transmitting and receiving appliances. In the early days of telephony, as rendered practical by Graham Bell, similar instruments were employed for both purposes. Bell's telephone consists of a bar magnet, or battery of bar magnets, provided at one end with a short pole-piece which serves as the core of a coil of fine insulated wire. In close proximity to the outer end of the pole-piece is placed a circular disc of thin iron, held at the circumference. Under the influence of the permanent magnet the disc is magnetized radially, the polarity at the centre being of course opposite to that of the neighbouring end of the steel magnet.

The operation of the instrument as a transmitter is readily traced. When sonorous waves impinge upon the disc, it responds with a symmetrical transverse vibration by which its distance from the pole-piece is alternately increased and diminished. When the interval is diminished, more induction passes through the pole-piece, and a corresponding electro-motive force acts in the enveloping coil. The periodic movement of the disc thus gives rise to a periodic current in any circuit connected with the telephone coil.

The electro-motive force is in the first instance proportional to the permanent magnetism to which it is due; and this law would continue to hold, were the behaviour of the pole-piece and of the disc conformable to that of the " soft iron " of approximate theory. But as the magnetism rises, and the state of saturation is more nearly approached, the response to periodic changes of force becomes feebler, and thus the efficiency falls below that indicated by the law of proportionality. If we could imagine the state of saturation in the pole-piece to be actually attained, the induction through the coil would become almost incapable of variation, being reduced to such as might occur were the iron removed. There is thus a point, dependent upon the properties of magnetic matter, beyond which it is pernicious to raise the amount of the permanent magnetism; and this point marks the maximum efficiency of the transmitter. It is probable that the most favourable condition is not fully reached in instruments provided with steel magnets; but the considerations above advanced may serve to explain why an electro-magnet is not substituted.

The action of the receiving instrument may be explained on the same principles. The periodic current in the coil alternately opposes and cooperates with the permanent magnet, and thus the iron disc is subjected to a periodic force acting at its centre. The vibrations are thence communicated to the air, and so reach the ear of the observer. As in the case of the transmitter, the efficiency attains a maximum when the magnetism of the pole-piece is still far short of saturation.

The explanation of the receiver in terms of magnetic forces pulling at the disc is sometimes regarded as inadequate or even as altogether wide of the mark, the sound being attributed to " molecular disturbances " in the pole-piece and disc. There is indeed every reason to suppose that molecular movements accompany the change of magnetic state, but the question is how do these movements influence the ear. It would appear that they can do so only by causing a transverse motion of the surface of the disc, a motion from which nodal subdivisions are not excluded.

In support of the "push and pull theory" it may be useful to cite an experiment tried upon a bipolar telephone. In this instrument each end of a horse-shoe magnet is provided with a pole-piece and coil, and the two pole-pieces are brought into proximity with the disc at places symmetrically situated with regard to the centre. In the normal use of the instrument the two coils are permanently connected as in an ordinary horse-shoe electro-magnet, but for the purposes of the experiment provision was made whereby one of the coils could be reversed at pleasure by means of a reversing key. The sensitiveness of the telephone in the two conditions was tested by including it in the circuit of a Daniell cell and a scraping contact apparatus, resistance from a box being added until the sound was but just easily audible. The resistances employed were such as to dominate the self-induction of the circuit, and the comparison shewed that the reversal of the coil from its normal connection lowered the sensitiveness to current in the ratio of 11 : 1. That the reduction was not still greater is readily explained by outstanding failures of symmetry; but on the "molecular disturbance" theory it is not evident why there should be any reduction at all.

Dissatisfaction with the ordinary theory of the action of a receiving telephone may have arisen from the difficulty of understanding how such very minute motions of the plate could be audible. This is, however, a question of the sensitiveness of the ear, which has been proved capable of appreciating an amplitude of less than 8×10^{-8} cm.[1]. The subject of the audible minimum will be further considered in the second volume of this work.

The calculation *a priori* of the minimum current that should be audible in the telephone is a matter of considerable difficulty; and even the determination by direct experiment has led to widely discrepant numbers. In some recent experiments by the author a unipolar Bell telephone of 70 ohms resistance was employed. The circuit included also a resistance box and an induction coil of known construction, in which acted an electro-motive force capable of calculation. Up to a frequency of 307 this could be obtained from a revolving magnet of known moment and situated at a measured distance from the induction coil. For the higher frequencies magnetized tuning-forks, vibrating with measured amplitudes, were substituted. In either case the

[1] *Proc. Roy. Soc.* vol. xxvi. p. 248, 1877.

resistance of the circuit was increased until the residual sound was but just easily audible. Care having been taken so to arrange matters that the self-induction of the circuit was negligible, the current could then be deduced from the resistance and the calculated electro-motive force operating in the induction coil. The following are the results, in which it is to be understood that the currents recorded might have been halved without the sounds being altogether lost:

Pitch	Source	Current in 10^{-8} amperes
128	Fork	2800
192	Revolving Magnet	250
256	Fork	83
307	Revolving Magnet	49
320	Fork	32
384	....	15
512	....	7
640	....	4·4
768	....	10

The effect of a given current depends, of course, upon the manner in which the telephone is wound. If the same space be occupied by the copper in the various cases, the current capable of producing a particular effect is inversely as the square root of the resistance.

The numbers in the above table giving the results of the author's experiments are of the same order of magnitude as those found by Ferraris[1], whose observations, however, related to sounds that were not pure tones. But much lower estimates have been put forward. Thus Tait[2] gives 2×10^{-12} amperes, and Preece a still lower figure, 6×10^{-13}. These discrepancies, enormous as they stand, would be still further increased were the comparison made to refer to the amounts of energy absorbed.

According to the calculations of the author the above tabulated sensitiveness to a periodic current of frequency 256 is about what might reasonably be expected on the push and pull theory[3]. At

---

[1] *Atti della Accad. d. Sci. Di Torino*, vol. XIII. p. 1024, 1877.
[2] *Edin. Proc.* vol. IX. p. 551, 1878.
[3] I propose shortly to publish these calculations.

this frequency, which is below those proper to the telephone plate (§ 221 a), the motion of the plate is governed by elasticity rather than by inertia, and an equilibrium theory (§ 100) is applicable as a rough approximation. The greater sensitiveness of the telephone at frequencies in the neighbourhood of 512 would appear to depend upon resonance (§ 46). It is doubtful whether the much higher sensitiveness claimed by Tait and Preece could be reconciled with theory.

It appears to be established that the iron plate of a telephone may be replaced by one of copper, or even of non-conducting material, without absolute loss of sound; but these effects are probably of a different order of magnitude. In the case of copper induced currents may confer the necessary magnetic properties. For a description of the ingenious receiver invented by Edison and for other information upon telephonic appliances the reader may consult Preece and Stubbs' *Manual of Telephony*.

In existing practice the transmitting instrument depends upon a variable contact. The first carbon transmitter was constructed by Edison in 1877, but the instruments now in use are modifications of Hughes' microphone[1]. A battery current is led into the line through pieces of metal or of carbon in loose juxtaposition, carbon being almost universally employed in practice. Under the influence of sonorous vibration the electrical resistance of the contacts varies, and thus the current in the line is rendered representative of the sound to be reproduced at the receiving end.

That the resistance of the contact should vary with the pressure is not surprising. If two clean convex pieces of metal are forced together, the conductivity between them is represented by the diameter of the circle of contact (§ 306). The relation between the circle of contact and the pressure with which the masses are forced together has been investigated in detail by Hertz[2]. His conclusion for the case of two equal spheres is that the cube of the radius of the circle of contact is proportional to the pressure and to the radii of the spheres. But it has not yet been shewn that the action of the microphone can be adequately explained upon this principle.

---

[1] *Proc. Roy. Soc.*, vol. XXVII. p. 362, 1878.

[2] Crelle, *Journ. Math.* XCII. p. 156, 1882.

APPENDIX.

ON PROGRESSIVE WAVES.

From the Proceedings of the London Mathematical Society,
Vol. IX., p. 21, 1877.

It has often been remarked that, when a group of waves advances
into still water, the velocity of the group is less than that of the individual waves of which it is composed; the waves appear to advance
through the group, dying away as they approach its anterior limit.
This phenomenon was, I believe, first explained by Stokes, who regarded the group as formed by the superposition of two infinite trains
of waves, of equal amplitudes and of nearly equal wave-lengths, advancing in the same direction. My attention was called to the subject
about two years since by Mr Froude, and the same explanation then
occurred to me independently[*]. In my book on the "Theory of
Sound" (§ 191), I have considered the question more generally, and
have shewn that, if V be the velocity of propagation of any kind of
waves whose wave-length is λ, and $k = 2\pi/\lambda$, then U, the velocity of
a group composed of a great number of waves, and moving into an undisturbed part of the medium, is expressed by

$$U = \frac{d\,(kV)}{dk} \dots\dots\dots\dots\dots\dots\dots\dots\dots\dots\dots(1),$$

[*] Another phenomenon, also mentioned to me by Mr Froude, admits of a similar
explanation. A steam-launch moving quickly through the water is accompanied by
a peculiar system of diverging waves, of which the most striking feature is the
obliquity of the line containing the greatest elevations of successive waves to the
wave-fronts. This wave pattern may be explained by the superposition of two (or
more) infinite trains of waves, of slightly differing wave-lengths, whose directions
and velocities of propagation are so related in each case that there is no change of
position relatively to the boat. The mode of composition will be best understood by
drawing on paper two sets of parallel and equidistant lines, subject to the above
condition, to represent the crests of the component trains. In the case of two trains
of slightly different wave-lengths, it may be proved that the tangent of the angle
between the line of maxima and the wave-fronts is half the tangent of the angle
between the wave-fronts and the boat's course.

or, as we may also write it,

$$U \; : \; V = 1 + \frac{d \log V}{d \log k} \quad\dots\dots\dots\dots\dots\dots(2).$$

Thus, if $V \propto \lambda^n$, $\qquad U = (1 - n) \; V \dots\dots\dots\dots\dots\dots(3).$

In fact, if the two infinite trains be represented by $\cos k \, (Vt - x)$ and $\cos k' \, (V't - x)$, their resultant is represented by

$$\cos k \, (Vt - x) + \cos k' \, (V't - x),$$

which is equal to

$$2 \cos \left\{ \frac{k' V' - k V}{2} \, t - \frac{k' - k}{2} \, x \right\} \; . \; \cos \left\{ \frac{k' V' + k V}{2} \, t - \frac{k' + k}{2} \, x \right\}.$$

If $k' - k$, $V' - V$ be small, we have a train of waves whose amplitude varies slowly from one point to another between the limits 0 and 2, forming a series of groups separated from one another by regions comparatively free from disturbance. The position at time t of the middle of that group, which was initially at the origin, is given by

$$(k' V' - k V) \, t - (k' - k) \, x = 0,$$

which shews that the velocity of the group is $(k' V' - k V) \div (k' - k)$. In the limit, when the number of waves in each group is indefinitely great, this result coincides with (1).

The following particular cases are worth notice, and are here tabu lated for convenience of comparison :—

$V \propto \lambda$,	$U = 0$,	Reynolds' disconnected pendulums.
$V \propto \lambda^{\frac{1}{2}}$,	$U = \frac{1}{2} V$,	Deep-water gravity waves.
$V \propto \lambda^0$,	$U = V$,	Aërial waves, &c.
$V \propto \lambda^{-\frac{1}{2}}$,	$U = \frac{3}{2} V$,	Capillary water waves.
$V \propto \lambda^{-1}$,	$U = 2V$,	Flexural waves.

The capillary water waves are those whose wave-length is so small that the force of restitution due to capillarity largely exceeds that due to gravity. Their theory has been given by Thomson (Phil. Mag., Nov. 1871). The flexural waves, for which $U = 2V$, are those corresponding to the bending of an elastic rod or plate ("Theory of Sound," § 191).

In a paper read at the Plymouth meeting of the British Association (afterwards printed in "Nature," Aug. 23, 1877), Prof. Osborne Reynolds gave a dynamical explanation of the fact that a group of deep-water waves advances with only half the rapidity of the individual waves. It appears that the energy propagated across any point, when a train of waves is passing, is only one-half of the energy necessary to supply the waves which pass in the same time, so that, if the train of waves be limited, it is impossible that its front can be propagated with the full velocity of the waves, because this would imply the acquisition of more energy than can in fact be supplied. Prof. Reynolds did not contemplate the cases where *more* energy is propagated than corresponds to the waves passing in the same time; but his argument, applied conversely to the results already given, shews that such cases must exist. The ratio of the energy propagated to that of the passing waves is $U : V$; thus the energy propagated in the unit time is $U : V$ of that existing in a length V, or U times that existing in the unit length. Accordingly

Energy propagated in unit time : Energy contained (on an average) in unit length $\qquad = d\,(kV) \;:\; dk,$ by (1).

As an example, I will take the case of small irrotational waves in water of finite depth l*. If z be measured downwards from the surface, and the elevation (h) of the wave be denoted by

$$h = H \cos (nt - kx) \quad\dots\dots\dots\dots\dots\dots(4),$$

in which $n = kV$, the corresponding velocity-potential (ϕ) is

$$\phi = - \, VH \frac{e^{k(z-l)} + e^{-k(z-l)}}{e^{kl} - e^{-kl}} \sin (nt - kx) \quad\dots\dots\dots\dots (5).$$

This value of ϕ satisfies the general differential equation for irrotational motion ($\nabla^2 \phi = 0$), makes the vertical velocity $d\phi/dz$ zero when $z = l$, and $- \, dh/dt$ when $z = 0$. The velocity of propagation is given by

$$V^2 = \frac{g}{k} \frac{e^{kl} - e^{-kl}}{e^{kl} + e^{-kl}} \quad\dots\dots\dots\dots\dots\dots\dots (6).$$

We may now calculate the energy contained in a length x, which is supposed to include so great a number of waves that fractional parts may be left out of account.

For the potential energy we have

$$V_1 = g\rho \iint_0^h z \, dz \, dx = \tfrac{1}{2} g\rho \int h^2 \, dx = \tfrac{1}{4} g\rho H^2 \,.\, x \quad\dots\dots\dots\dots(7).$$

For the kinetic energy,

$$T = \tfrac{1}{2} \rho \iint \left\{ \left(\frac{d\phi}{dx}\right)^2 + \left(\frac{d\phi}{dz}\right)^2 \right\} dx \, dz$$

$$= \tfrac{1}{2}\rho - \int \left(\phi \, \frac{d\phi}{dz} \right)_{z=0} dx = \tfrac{1}{4} g\rho H^2 \,.\, x \quad\dots\dots\dots\dots\dots(8),$$

by (5) and (6). If, in accordance with the argument advanced at the

* Prof. Reynolds considers the trochoidal wave of Rankine and Froude, which involves molecular rotation.

end of this paper, the equality of V_1 and T be assumed, the value of the velocity of propagation follows from the present expressions. The whole energy in the waves occupying a length x is therefore (for each

unit of breadth) $\qquad V_1 + T = \frac{1}{2}g\rho H^2 \cdot x \dots\dots\dots\dots\dots(9)$,

H denoting the maximum elevation.

We have next to calculate the energy propagated in time t across a plane for which x is constant, or, in other words, the work (W) that must be done in order to sustain the motion of the plane (considered as a flexible lamina) in the face of the fluid pressures acting upon the front of it. The variable part of the pressure (δp), at depth z, is given by

$$\delta p = -\rho \frac{d\phi}{dt} = n\rho VH \frac{e^{k(z-l)} + e^{-k(z-l)}}{e^{kl} - e^{-kl}} \cos (nt - kx),$$

while for the horizontal velocity

$$\frac{d\phi}{dx} = kVH \frac{e^{k(z-l)} + e^{-k(z-l)}}{e^{kl} - e^{-kl}} \cos (nt - kx);$$

so that $\qquad W = \iint \delta p \frac{d\phi}{dx} dz \, dt = \frac{1}{4}g\rho H^2 \cdot Vt \cdot \left[1 + \frac{4kl}{e^{2kl} - e^{-2kl}} \right] \dots\dots (10)$,

on integration. From the value of V in (6) it may be próved that

$$\frac{d(kV)}{dk} = \frac{1}{2}V\left\{ 1 + \frac{1}{V^2} \frac{d(kV^2)}{dk} \right\} = \frac{1}{2}V\left\{ 1 + \frac{4kl}{e^{2kl} - e^{-2kl}} \right\};$$

and it is thus verified that the value of W for a unit time

$$= \frac{d(kV)}{dk} \times \text{energy in unit length.}$$

As an example of the direct calculation of U, we may take the case of waves moving under the joint influence of gravity and cohesion.

It is proved by Thomson that

$$V^2 = \frac{g}{k} + T''k \dots\dots\dots\dots\dots\dots (11),$$

where T'' is the cohesive tension. Hence

$$U = \frac{1}{2}V\left\{ 1 + \frac{1}{V^2} \frac{d(kV^2)}{dk} \right\} = \frac{1}{2}V \frac{g + 3k^2T}{g + k^2T} \dots\dots\dots\dots(12).$$

When k is small, the surface tension is negligible, and then $U = \frac{1}{2}V$; but when, on the contrary, k is large, $U = \frac{3}{2}V$, as has already been stated. When $Tk^2 = g$, $U = V$. This corresponds to the minimum velocity of propagation investigated by Thomson.

Although the argument from interference groups seems satisfactory, an independent investigation is desirable of the relation between energy existing and energy propagated. For some time I was at a loss for a method applicable to all kinds of waves, not seeing in particular why the comparison of energies should introduce the consideration of a variation of wave-length. The following investigation, in which the increment of wave-length is *imaginary*, may perhaps be considered to meet the want :—

Let us suppose that the motion of every part of the medium is resisted by a force of very small magnitude proportional to the mass and to the velocity of the part, the effect of which will be that waves generated at the origin gradually die away as x increases. The motion, which in the absence of friction would be represented by $\cos(nt - kx)$, under the influence of friction is represented by $e^{-\mu x} \cos(nt - kx)$, where μ is a small positive coefficient. In strictness the value of k is also altered by the friction; but the alteration is of the second order as regards the frictional forces and may be omitted under the circumstances here supposed. The energy of the waves per unit length at any stage of degradation is proportional to the square of the amplitude, and thus the whole energy on the positive side of the origin is to the energy of so much of the waves at their greatest value, $i.e.$, at the origin, as would be contained in the unit of length, as $\int_0^\infty e^{-2\mu x} dx : 1$, or as $(2\mu)^{-1} : 1$. The energy transmitted through the origin in the unit time is the same as the energy dissipated ; and, if the frictional force acting on the element of mass m be hmv, where v is the velocity of the element and h is constant, the energy dissipated in unit time is $h\Sigma mv^2$ or $2hT$, T being the kinetic energy. Thus, on the assumption that the kinetic energy is half the whole energy, we find that the energy transmitted in the unit time is to the greatest energy existing in the unit length as $h : 2\mu$. It remains to find the connection between h and μ.

For this purpose it will be convenient to regard $\cos(nt - kx)$ as the real part of $e^{int} e^{ikx}$, and to inquire how k is affected, when n is given, by the introduction of friction. Now the effect of friction is represented in the differential equations of motion by the substitution of $d^2/dt^2 + h\,dk/dt$ in place of d^2/dt^2, or, since the whole motion is proportional to e^{int}, by substituting $-n^2 + ihn$ for $-n^2$. Hence the introduction of friction corresponds to an alteration of n from n to $n - \tfrac{1}{2}ih$ (the square of h being neglected); and accordingly k is altered from k to $k - \tfrac{1}{2}ih\,dk/dn$. The solution thus becomes $e^{-\frac{1}{2}hx\,dk/dn} e^{i(nt-kx)}$, or, when the imaginary part is rejected, $e^{-\frac{1}{2}hx\,dk/dn} \cos(nt - kx)$; so that $\mu = \tfrac{1}{2}h\,dk/dn$, and $h : 2\mu = dn/dk$. The ratio of the energy transmitted in the unit time to the energy existing in the unit length is therefore expressed by dn/dk or $d\,(kV)/dk$, as was to be proved.

It has often been noticed, in particular cases of progressive waves, that the potential and kinetic energies are equal; but I do not call to mind any general treatment of the question. The theorem is not usually true for the individual parts of the medium*, but must be understood to refer either to an integral number of wave-lengths, or to a space so considerable that the outstanding fractional parts of waves may be left out of account. As an example well adapted to give insight into the question, I will take the case of a uniform stretched circular membrane ("Theory of Sound," § 200) vibrating with a given number of nodal circles and diameters. The fundamental modes are not quite determinate in consequence of the symmetry, for any diameter may be made nodal. In order to get rid of this indeterminateness, we may suppose the membrane to carry a small load attached to it anywhere except on a nodal circle. There are then two definite fundamental modes, in one of which the load lies on a nodal diameter, thus producing no effect, and in the other midway between nodal diameters, where it produces a maximum effect ("Theory of Sound," § 208). If vibrations of both modes are going on simultaneously, the potential and kinetic energies of the whole motion may be calculated by *simple addition* of those of the components. Let us now, supposing the load to diminish without limit, imagine that the vibrations are of equal amplitude and differ in phase by a quarter of a period. The result is a *progressive* wave, whose potential and kinetic energies are the sums of those of the stationary waves of which it is composed. For the first component we have $V_1 = E \cos^2 nt$, $T_1 = E \sin^2 nt$; and for the second component, $V_2 = E \sin^2 nt$, $T_2 = E \cos^2 nt$; so that $V_1 + V_2 = T_1 + T_2 = E$, or the potential and kinetic energies of the progressive wave are equal, being the same as the whole energy of either of the components. The method of proof here employed appears to be sufficiently general, though it is rather difficult to express it in language which is appropriate to all kinds of waves.

* Aërial waves are an important exception.

END OF VOL. I.

A CATALOG OF SELECTED
DOVER BOOKS
IN SCIENCE AND MATHEMATICS

A CATALOG OF SELECTED
DOVER BOOKS
IN SCIENCE AND MATHEMATICS

QUALITATIVE THEORY OF DIFFERENTIAL EQUATIONS, V.V. Nemytskii and V.V. Stepanov. Classic graduate-level text by two prominent Soviet mathematicians covers classical differential equations as well as topological dynamics and ergodic theory. Bibliographies. 523pp. 5⅜ × 8½. 65954-2 Pa. $14.95

MATRICES AND LINEAR ALGEBRA, Hans Schneider and George Phillip Barker. Basic textbook covers theory of matrices and its applications to systems of linear equations and related topics such as determinants, eigenvalues and differential equations. Numerous exercises. 432pp. 5⅜ × 8½. 66014-1 Pa. $10.95

QUANTUM THEORY, David Bohm. This advanced undergraduate-level text presents the quantum theory in terms of qualitative and imaginative concepts, followed by specific applications worked out in mathematical detail. Preface. Index. 655pp. 5⅜ × 8½. 65969-0 Pa. $14.95

ATOMIC PHYSICS (8th edition), Max Born. Nobel laureate's lucid treatment of kinetic theory of gases, elementary particles, nuclear atom, wave-corpuscles, atomic structure and spectral lines, much more. Over 40 appendices, bibliography. 495pp. 5⅜ × 8½. 65984-4 Pa. $12.95

ELECTRONIC STRUCTURE AND THE PROPERTIES OF SOLIDS: The Physics of the Chemical Bond, Walter A. Harrison. Innovative text offers basic understanding of the electronic structure of covalent and ionic solids, simple metals, transition metals and their compounds. Problems. 1980 edition. 582pp. 6⅛ × 9¼. 66021-4 Pa. $16.95

BOUNDARY VALUE PROBLEMS OF HEAT CONDUCTION, M. Necati Özisik. Systematic, comprehensive treatment of modern mathematical methods of solving problems in heat conduction and diffusion. Numerous examples and problems. Selected references. Appendices. 505pp. 5⅜ × 8½. 65990-9 Pa. $12.95

A SHORT HISTORY OF CHEMISTRY (3rd edition), J.R. Partington. Classic exposition explores origins of chemistry, alchemy, early medical chemistry, nature of atmosphere, theory of valency, laws and structure of atomic theory, much more. 428pp. 5⅜ × 8½. (Available in U.S. only) 65977-1 Pa. $11.95

A HISTORY OF ASTRONOMY, A. Pannekoek. Well-balanced, carefully reasoned study covers such topics as Ptolemaic theory, work of Copernicus, Kepler, Newton, Eddington's work on stars, much more. Illustrated. References. 521pp. 5⅜ × 8½. 65994-1 Pa. $12.95

PRINCIPLES OF METEOROLOGICAL ANALYSIS, Walter J. Saucier. Highly respected, abundantly illustrated classic reviews atmospheric variables, hydrostatics, static stability, various analyses (scalar, cross-section, isobaric, isentropic, more). For intermediate meteorology students. 454pp. 6⅛ × 9¼. 65979-8 Pa. $14.95

RELATIVITY, THERMODYNAMICS AND COSMOLOGY, Richard C. Tolman. Landmark study extends thermodynamics to special, general relativity; also applications of relativistic mechanics, thermodynamics to cosmological models. 501pp. 5⅜ × 8½. 65383-8 Pa. $13.95

APPLIED ANALYSIS, Cornelius Lanczos. Classic work on analysis and design of finite processes for approximating solution of analytical problems. Algebraic equations, matrices, harmonic analysis, quadrature methods, much more. 559pp. 5⅜ × 8½. 65656-X Pa. $13.95

INTRODUCTION TO ANALYSIS, Maxwell Rosenlicht. Unusually clear, accessible coverage of set theory, real number system, metric spaces, continuous functions, Riemann integration, multiple integrals, more. Wide range of problems. Undergraduate level. Bibliography. 254pp. 5⅜ × 8½. 65038-3 Pa. $8.95

INTRODUCTION TO QUANTUM MECHANICS With Applications to Chemistry, Linus Pauling & E. Bright Wilson, Jr. Classic undergraduate text by Nobel Prize winner applies quantum mechanics to chemical and physical problems. Numerous tables and figures enhance the text. Chapter bibliographies. Appendices. Index. 468pp. 5⅜ × 8½. 64871-0 Pa. $12.95

ASYMPTOTIC EXPANSIONS OF INTEGRALS, Norman Bleistein & Richard A. Handelsman. Best introduction to important field with applications in a variety of scientific disciplines. New preface. Problems. Diagrams. Tables. Bibliography. Index. 448pp. 5⅜ × 8½. 65082-0 Pa. $12.95

MATHEMATICS APPLIED TO CONTINUUM MECHANICS, Lee A. Segel. Analyzes models of fluid flow and solid deformation. For upper-level math, science and engineering students. 608pp. 5⅜ × 8½. 65369-2 Pa. $14.95

ELEMENTS OF REAL ANALYSIS, David A. Sprecher. Classic text covers fundamental concepts, real number system, point sets, functions of a real variable, Fourier series, much more. Over 500 exercises. 352pp. 5⅜ × 8½. 65385-4 Pa. $11.95

PHYSICAL PRINCIPLES OF THE QUANTUM THEORY, Werner Heisenberg. Nobel Laureate discusses quantum theory, uncertainty, wave mechanics, work of Dirac, Schroedinger, Compton, Wilson, Einstein, etc. 184pp. 5⅜ × 8½. 60113-7 Pa. $6.95

INTRODUCTORY REAL ANALYSIS, A.N. Kolmogorov, S.V. Fomin. Translated by Richard A. Silverman. Self-contained, evenly paced introduction to real and functional analysis. Some 350 problems. 403pp. 5⅜ × 8½. 61226-0 Pa. $10.95

PROBLEMS AND SOLUTIONS IN QUANTUM CHEMISTRY AND PHYSICS, Charles S. Johnson, Jr. and Lee G. Pedersen. Unusually varied problems, detailed solutions in coverage of quantum mechanics, wave mechanics, angular momentum, molecular spectroscopy, scattering theory, more. 280 problems plus 139 supplementary exercises. 430pp. 6½ × 9¼. 65236-X Pa. $13.95

CATALOG OF DOVER BOOKS

ASYMPTOTIC METHODS IN ANALYSIS, N.G. de Bruijn. An inexpensive, comprehensive guide to asymptotic methods—the pioneering work that teaches by explaining worked examples in detail. Index. 224pp. 5⅜ × 8½. 64221-6 Pa. $7.95

OPTICAL RESONANCE AND TWO-LEVEL ATOMS, L. Allen and J.H. Eberly. Clear, comprehensive introduction to basic principles behind all quantum optical resonance phenomena. 53 illustrations. Preface. Index. 256pp. 5⅜ × 8½.
65533-4 Pa. $8.95

COMPLEX VARIABLES, Francis J. Flanigan. Unusual approach, delaying complex algebra till harmonic functions have been analyzed from real variable viewpoint. Includes problems with answers. 364pp. 5⅜ × 8½. 61388-7 Pa. $9.95

ATOMIC SPECTRA AND ATOMIC STRUCTURE, Gerhard Herzberg. One of best introductions; especially for specialist in other fields. Treatment is physical rather than mathematical. 80 illustrations. 257pp. 5⅜ × 8½. 60115-3 Pa. $6.95

APPLIED COMPLEX VARIABLES, John W. Dettman. Step-by-step coverage of fundamentals of analytic function theory—plus lucid exposition of five important applications: Potential Theory; Ordinary Differential Equations; Fourier Transforms; Laplace Transforms; Asymptotic Expansions. 66 figures. Exercises at chapter ends. 512pp. 5⅜ × 8½. 64670-X Pa. $12.95

ULTRASONIC ABSORPTION: An Introduction to the Theory of Sound Absorption and Dispersion in Gases, Liquids and Solids, A.B. Bhatia. Standard reference in the field provides a clear, systematically organized introductory review of fundamental concepts for advanced graduate students, research workers. Numerous diagrams. Bibliography. 440pp. 5⅜ × 8½. 64917-2 Pa. $11.95

UNBOUNDED LINEAR OPERATORS: Theory and Applications, Seymour Goldberg. Classic presents systematic treatment of the theory of unbounded linear operators in normed linear spaces with applications to differential equations. Bibliography. 199pp. 5⅜ × 8½. 64830-3 Pa. $7.95

LIGHT SCATTERING BY SMALL PARTICLES, H.C. van de Hulst. Comprehensive treatment including full range of useful approximation methods for researchers in chemistry, meteorology and astronomy. 44 illustrations. 470pp. 5⅜ × 8½. 64228-3 Pa. $11.95

CONFORMAL MAPPING ON RIEMANN SURFACES, Harvey Cohn. Lucid, insightful book presents ideal coverage of subject. 334 exercises make book perfect for self-study. 55 figures. 352pp. 5⅜ × 8¼. 64025-6 Pa. $11.95

OPTICKS, Sir Isaac Newton. Newton's own experiments with spectroscopy, colors, lenses, reflection, refraction, etc., in language the layman can follow. Foreword by Albert Einstein. 532pp. 5⅜ × 8½. 60205-2 Pa. $11.95

GENERALIZED INTEGRAL TRANSFORMATIONS, A.H. Zemanian. Graduate-level study of recent generalizations of the Laplace, Mellin, Hankel, K. Weierstrass, convolution and other simple transformations. Bibliography. 320pp. 5⅜ × 8½. 65375-7 Pa. $8.95

THE ELECTROMAGNETIC FIELD, Albert Shadowitz. Comprehensive undergraduate text covers basics of electric and magnetic fields, builds up to electromagnetic theory. Also related topics, including relativity. Over 900 problems. 768pp. 5⅜ × 8¼. 65660-8 Pa. $18.95

FOURIER SERIES, Georgi P. Tolstov. Translated by Richard A. Silverman. A valuable addition to the literature on the subject, moving clearly from subject to subject and theorem to theorem. 107 problems, answers. 336pp. 5⅜ × 8½. 63317-9 Pa. $9.95

THEORY OF ELECTROMAGNETIC WAVE PROPAGATION, Charles Herach Papas. Graduate-level study discusses the Maxwell field equations, radiation from wire antennas, the Doppler effect and more. xiii + 244pp. 5⅜ × 8½. 65678-0 Pa. $6.95

DISTRIBUTION THEORY AND TRANSFORM ANALYSIS: An Introduction to Generalized Functions, with Applications, A.H. Zemanian. Provides basics of distribution theory, describes generalized Fourier and Laplace transformations. Numerous problems. 384pp. 5⅜ × 8½. 65479-6 Pa. $11.95

THE PHYSICS OF WAVES, William C. Elmore and Mark A. Heald. Unique overview of classical wave theory. Acoustics, optics, electromagnetic radiation, more. Ideal as classroom text or for self-study. Problems. 477pp. 5⅜ × 8½. 64926-1 Pa. $12.95

CALCULUS OF VARIATIONS WITH APPLICATIONS, George M. Ewing. Applications-oriented introduction to variational theory develops insight and promotes understanding of specialized books, research papers. Suitable for advanced undergraduate/graduate students as primary, supplementary text. 352pp. 5⅜ × 8½. 64856-7 Pa. $9.95

A TREATISE ON ELECTRICITY AND MAGNETISM, James Clerk Maxwell. Important foundation work of modern physics. Brings to final form Maxwell's theory of electromagnetism and rigorously derives his general equations of field theory. 1,084pp. 5⅜ × 8½. 60636-8, 60637-6 Pa., Two-vol. set $23.90

AN INTRODUCTION TO THE CALCULUS OF VARIATIONS, Charles Fox. Graduate-level text covers variations of an integral, isoperimetrical problems, least action, special relativity, approximations, more. References. 279pp. 5⅜ × 8½. 65499-0 Pa. $8.95

HYDRODYNAMIC AND HYDROMAGNETIC STABILITY, S. Chandrasekhar. Lucid examination of the Rayleigh-Benard problem; clear coverage of the theory of instabilities causing convection. 704pp. 5⅜ × 8¼. 64071-X Pa. $14.95

CALCULUS OF VARIATIONS, Robert Weinstock. Basic introduction covering isoperimetric problems, theory of elasticity, quantum mechanics, electrostatics, etc. Exercises throughout. 326pp. 5⅜ × 8½. 63069-2 Pa. $8.95

DYNAMICS OF FLUIDS IN POROUS MEDIA, Jacob Bear. For advanced students of ground water hydrology, soil mechanics and physics, drainage and irrigation engineering and more. 335 illustrations. Exercises, with answers. 784pp. 6⅛ × 9¼. 65675-6 Pa. $19.95

CATALOG OF DOVER BOOKS

HANDBOOK OF MATHEMATICAL FUNCTIONS WITH FORMULAS, GRAPHS, AND MATHEMATICAL TABLES, edited by Milton Abramowitz and Irene A. Stegun. Vast compendium: 29 sets of tables, some to as high as 20 places. 1,046pp. 8 × 10½. 61272-4 Pa. $24.95

MATHEMATICAL METHODS IN PHYSICS AND ENGINEERING, John W. Dettman. Algebraically based approach to vectors, mapping, diffraction, other topics in applied math. Also generalized functions, analytic function theory, more. Exercises. 448pp. 5⅜ × 8¼. 65649-7 Pa. $10.95

A SURVEY OF NUMERICAL MATHEMATICS, David M. Young and Robert Todd Gregory. Broad self-contained coverage of computer-oriented numerical algorithms for solving various types of mathematical problems in linear algebra, ordinary and partial, differential equations, much more. Exercises. Total of 1,248pp. 5⅜ × 8½. Two volumes. Vol. I 65691-8 Pa. $14.95
Vol. II 65692-6 Pa. $14.95

TENSOR ANALYSIS FOR PHYSICISTS, J.A. Schouten. Concise exposition of the mathematical basis of tensor analysis, integrated with well-chosen physical examples of the theory. Exercises. Index. Bibliography. 289pp. 5⅜ × 8½. 65582-2 Pa. $8.95

INTRODUCTION TO NUMERICAL ANALYSIS (2nd Edition), F.B. Hildebrand. Classic, fundamental treatment covers computation, approximation, interpolation, numerical differentiation and integration, other topics. 150 new problems. 669pp. 5⅜ × 8½. 65363-3 Pa. $15.95

INVESTIGATIONS ON THE THEORY OF THE BROWNIAN MOVEMENT, Albert Einstein. Five papers (1905-8) investigating dynamics of Brownian motion and evolving elementary theory. Notes by R. Fürth. 122pp. 5⅜ × 8½. 60304-0 Pa. $4.95

CATASTROPHE THEORY FOR SCIENTISTS AND ENGINEERS, Robert Gilmore. Advanced-level treatment describes mathematics of theory grounded in the work of Poincaré, R. Thom, other mathematicians. Also important applications to problems in mathematics, physics, chemistry and engineering. 1981 edition. References. 28 tables. 397 black-and-white illustrations. xvii + 666pp. 6⅛ × 9¼. 67539-4 Pa. $17.95

AN INTRODUCTION TO STATISTICAL THERMODYNAMICS, Terrell L. Hill. Excellent basic text offers wide-ranging coverage of quantum statistical mechanics, systems of interacting molecules, quantum statistics, more. 523pp. 5⅜ × 8½. 65242-4 Pa. $12.95

STATISTICAL PHYSICS, Gregory H. Wannier. Classic text combines thermodynamics, statistical mechanics and kinetic theory in one unified presentation of thermal physics. Problems with solutions. Bibliography. 532pp. 5⅜ × 8½. 65401-X Pa. $12.95

CATALOG OF DOVER BOOKS

NUMERICAL METHODS FOR SCIENTISTS AND ENGINEERS, Richard Hamming. Classic text stresses frequency approach in coverage of algorithms, polynomial approximation, Fourier approximation, exponential approximation, other topics. Revised and enlarged 2nd edition. 721pp. 5⅜ × 8½.
65241-6 Pa. $15.95

THEORETICAL SOLID STATE PHYSICS, Vol. I: Perfect Lattices in Equilibrium; Vol. II: Non-Equilibrium and Disorder, William Jones and Norman H. March. Monumental reference work covers fundamental theory of equilibrium properties of perfect crystalline solids, non-equilibrium properties, defects and disordered systems. Appendices. Problems. Preface. Diagrams. Index. Bibliography. Total of 1,301pp. 5⅜ × 8½. Two volumes. Vol. I 65015-4 Pa. $16.95
Vol. II 65016-2 Pa. $14.95

OPTIMIZATION THEORY WITH APPLICATIONS, Donald A. Pierre. Broadspectrum approach to important topic. Classical theory of minima and maxima, calculus of variations, simplex technique and linear programming, more. Many problems, examples. 640pp. 5⅜ × 8½. 65205-X Pa. $14.95

THE CONTINUUM: A Critical Examination of the Foundation of Analysis, Hermann Weyl. Classic of 20th-century foundational research deals with the conceptual problem posed by the continuum. 156pp. 5⅜ × 8½. 67982-9 Pa. $6.95

ESSAYS ON THE THEORY OF NUMBERS, Richard Dedekind. Two classic essays by great German mathematician: on the theory of irrational numbers; and on transfinite numbers and properties of natural numbers. 115pp. 5⅜ × 8½.
21010-3 Pa. $5.95

THE FUNCTIONS OF MATHEMATICAL PHYSICS, Harry Hochstadt. Comprehensive treatment of orthogonal polynomials, hypergeometric functions, Hill's equation, much more. Bibliography. Index. 322pp. 5⅜ × 8½. 65214-9 Pa. $9.95

NUMBER THEORY AND ITS HISTORY, Oystein Ore. Unusually clear, accessible introduction covers counting, properties of numbers, prime numbers, much more. Bibliography. 380pp. 5⅜ × 8½. 65620-9 Pa. $9.95

THE VARIATIONAL PRINCIPLES OF MECHANICS, Cornelius Lanczos. Graduate level coverage of calculus of variations, equations of motion, relativistic mechanics, more. First inexpensive paperbound edition of classic treatise. Index. Bibliography. 418pp. 5⅜ × 8½. 65067-7 Pa. $12.95

MATHEMATICAL TABLES AND FORMULAS, Robert D. Carmichael and Edwin R. Smith. Logarithms, sines, tangents, trig functions, powers, roots, reciprocals, exponential and hyperbolic functions, formulas and theorems. 269pp. 5⅜ × 8½. 60111-0 Pa. $6.95

THEORETICAL PHYSICS, Georg Joos, with Ira M. Freeman. Classic overview covers essential math, mechanics, electromagnetic theory, thermodynamics, quantum mechanics, nuclear physics, other topics. First paperback edition. xxiii + 885pp. 5⅜ × 8½. 65227-0 Pa. $21.95

ORDINARY DIFFERENTIAL EQUATIONS, Morris Tenenbaum and Harry Pollard. Exhaustive survey of ordinary differential equations for undergraduates in mathematics, engineering, science. Thorough analysis of theorems. Diagrams. Bibliography. Index. 818pp. 5⅜ × 8½. 64940-7 Pa. $18.95

STATISTICAL MECHANICS: Principles and Applications, Terrell L. Hill. Standard text covers fundamentals of statistical mechanics, applications to fluctuation theory, imperfect gases, distribution functions, more. 448pp. 5⅜ × 8½. 65390-0 Pa. $11.95

ORDINARY DIFFERENTIAL EQUATIONS AND STABILITY THEORY: An Introduction, David A. Sánchez. Brief, modern treatment. Linear equation, stability theory for autonomous and nonautonomous systems, etc. 164pp. 5⅜ × 8¼. 63828-6 Pa. $6.95

THIRTY YEARS THAT SHOOK PHYSICS: The Story of Quantum Theory, George Gamow. Lucid, accessible introduction to influential theory of energy and matter. Careful explanations of Dirac's anti-particles, Bohr's model of the atom, much more. 12 plates. Numerous drawings. 240pp. 5⅜ × 8½. 24895-X Pa. $6.95

THEORY OF MATRICES, Sam Perlis. Outstanding text covering rank, non-singularity and inverses in connection with the development of canonical matrices under the relation of equivalence, and without the intervention of determinants. Includes exercises. 237pp. 5⅜ × 8½. 66810-X Pa. $8.95

GREAT EXPERIMENTS IN PHYSICS: Firsthand Accounts from Galileo to Einstein, edited by Morris H. Shamos. 25 crucial discoveries: Newton's laws of motion, Chadwick's study of the neutron, Hertz on electromagnetic waves, more. Original accounts clearly annotated. 370pp. 5⅜ × 8½. 25346-5 Pa. $10.95

INTRODUCTION TO PARTIAL DIFFERENTIAL EQUATIONS WITH AP-PLICATIONS, E.C. Zachmanoglou and Dale W. Thoe. Essentials of partial differential equations applied to common problems in engineering and the physical sciences. Problems and answers. 416pp. 5⅜ × 8½. 65251-3 Pa. $11.95

BURNHAM'S CELESTIAL HANDBOOK, Robert Burnham, Jr. Thorough guide to the stars beyond our solar system. Exhaustive treatment. Alphabetical by constellation: Andromeda to Cetus in Vol. 1; Chamaeleon to Orion in Vol. 2; and Pavo to Vulpecula in Vol. 3. Hundreds of illustrations. Index in Vol. 3. 2,000pp. 6⅛ × 9¼. 23567-X, 23568-8, 23673-0 Pa., Three-vol. set $44.85

CHEMICAL MAGIC, Leonard A. Ford. Second Edition, Revised by E. Winston Grundmeier. Over 100 unusual stunts demonstrating cold fire, dust explosions, much more. Text explains scientific principles and stresses safety precautions. 128pp. 5⅜ × 8½. 67628-5 Pa. $5.95

AMATEUR ASTRONOMER'S HANDBOOK, J.B. Sidgwick. Timeless, compre-hensive coverage of telescopes, mirrors, lenses, mountings, telescope drives, micrometers, spectroscopes, more. 189 illustrations. 576pp. 5⅜ × 8¼. (Available in U.S. only) 24034-7 Pa. $11.95

SPECIAL FUNCTIONS, N.N. Lebedev. Translated by Richard Silverman. Famous Russian work treating more important special functions, with applications to specific problems of physics and engineering. 38 figures. 308pp. 5⅜ × 8½.
60624-4 Pa. $9.95

OBSERVATIONAL ASTRONOMY FOR AMATEURS, J.B. Sidgwick. Mine of useful data for observation of sun, moon, planets, asteroids, aurorae, meteors, comets, variables, binaries, etc. 39 illustrations. 384pp. 5⅜ × 8¼. (Available in U.S. only)
24033-9 Pa. $8.95

INTEGRAL EQUATIONS, F.G. Tricomi. Authoritative, well-written treatment of extremely useful mathematical tool with wide applications. Volterra Equations, Fredholm Equations, much more. Advanced undergraduate to graduate level. Exercises. Bibliography. 238pp. 5⅜ × 8½.
64828-1 Pa. $8.95

POPULAR LECTURES ON MATHEMATICAL LOGIC, Hao Wang. Noted logician's lucid treatment of historical developments, set theory, model theory, recursion theory and constructivism, proof theory, more. 3 appendixes. Bibliography. 1981 edition. ix + 283pp. 5⅜ × 8½.
67632-3 Pa. $8.95

MODERN NONLINEAR EQUATIONS, Thomas L. Saaty. Emphasizes practical solution of problems; covers seven types of equations. ". . . a welcome contribution to the existing literature. . . ."—*Math Reviews.* 490pp. 5⅜ × 8½. 64232-1 Pa. $11.95

FUNDAMENTALS OF ASTRODYNAMICS, Roger Bate et al. Modern approach developed by U.S. Air Force Academy. Designed as a first course. Problems, exercises. Numerous illustrations. 455pp. 5⅜ × 8½.
60061-0 Pa. $9.95

INTRODUCTION TO LINEAR ALGEBRA AND DIFFERENTIAL EQUATIONS, John W. Dettman. Excellent text covers complex numbers, determinants, orthonormal bases, Laplace transforms, much more. Exercises with solutions. Undergraduate level. 416pp. 5⅜ × 8½.
65191-6 Pa. $10.95

INCOMPRESSIBLE AERODYNAMICS, edited by Bryan Thwaites. Covers theoretical and experimental treatment of the uniform flow of air and viscous fluids past two-dimensional aerofoils and three-dimensional wings; many other topics. 654pp. 5⅜ × 8½.
65465-6 Pa. $16.95

INTRODUCTION TO DIFFERENCE EQUATIONS, Samuel Goldberg. Exceptionally clear exposition of important discipline with applications to sociology, psychology, economics. Many illustrative examples; over 250 problems. 260pp. 5⅜ × 8½.
65084-7 Pa. $8.95

LAMINAR BOUNDARY LAYERS, edited by L. Rosenhead. Engineering classic covers steady boundary layers in two- and three-dimensional flow, unsteady boundary layers, stability, observational techniques, much more. 708pp. 5⅜ × 8½.
65646-2 Pa. $18.95

LECTURES ON CLASSICAL DIFFERENTIAL GEOMETRY, Second Edition, Dirk J. Struik. Excellent brief introduction covers curves, theory of surfaces, fundamental equations, geometry on a surface, conformal mapping, other topics. Problems. 240pp. 5⅜ × 8½.
65609-8 Pa. $8.95

ROTARY-WING AERODYNAMICS, W.Z. Stepniewski. Clear, concise text covers aerodynamic phenomena of the rotor and offers guidelines for helicopter performance evaluation. Originally prepared for NASA. 537 figures. 640pp. 6⅛ × 9¼.
64647-5 Pa. $15.95

DIFFERENTIAL GEOMETRY, Heinrich W. Guggenheimer. Local differential geometry as an application of advanced calculus and linear algebra. Curvature, transformation groups, surfaces, more. Exercises. 62 figures. 378pp. 5⅜ × 8½.
63433-7 Pa. $9.95

INTRODUCTION TO SPACE DYNAMICS, William Tyrrell Thomson. Comprehensive, classic introduction to space-flight engineering for advanced undergraduate and graduate students. Includes vector algebra, kinematics, transformation of coordinates. Bibliography. Index. 352pp. 5⅜ × 8½.   65113-4 Pa. $9.95

A SURVEY OF MINIMAL SURFACES, Robert Osserman. Up-to-date, in-depth discussion of the field for advanced students. Corrected and enlarged edition covers new developments. Includes numerous problems. 192pp. 5⅜ × 8½.
64998-9 Pa. $8.95

ANALYTICAL MECHANICS OF GEARS, Earle Buckingham. Indispensable reference for modern gear manufacture covers conjugate gear-tooth action, gear-tooth profiles of various gears, many other topics. 263 figures. 102 tables. 546pp. 5⅜ × 8½.   65712-4 Pa. $14.95

SET THEORY AND LOGIC, Robert R. Stoll. Lucid introduction to unified theory of mathematical concepts. Set theory and logic seen as tools for conceptual understanding of real number system. 496pp. 5⅜ × 8¼.   63829-4 Pa. $12.95

A HISTORY OF MECHANICS, René Dugas. Monumental study of mechanical principles from antiquity to quantum mechanics. Contributions of ancient Greeks, Galileo, Leonardo, Kepler, Lagrange, many others. 671pp. 5⅜ × 8½.
65632-2 Pa. $14.95

FAMOUS PROBLEMS OF GEOMETRY AND HOW TO SOLVE THEM, Benjamin Bold. Squaring the circle, trisecting the angle, duplicating the cube: learn their history, why they are impossible to solve, then solve them yourself. 128pp. 5⅜ × 8½.   24297-8 Pa. $4.95

MECHANICAL VIBRATIONS, J.P. Den Hartog. Classic textbook offers lucid explanations and illustrative models, applying theories of vibrations to a variety of practical industrial engineering problems. Numerous figures. 233 problems, solutions. Appendix. Index. Preface. 436pp. 5⅜ × 8½.   64785-4 Pa. $11.95

CURVATURE AND HOMOLOGY, Samuel I. Goldberg. Thorough treatment of specialized branch of differential geometry. Covers Riemannian manifolds, topology of differentiable manifolds, compact Lie groups, other topics. Exercises. 315pp. 5⅜ × 8½.   64314-X Pa. $9.95

HISTORY OF STRENGTH OF MATERIALS, Stephen P. Timoshenko. Excellent historical survey of the strength of materials with many references to the theories of elasticity and structure. 245 figures. 452pp. 5⅜ × 8½. 61187-6 Pa. $12.95

GEOMETRY OF COMPLEX NUMBERS, Hans Schwerdtfeger. Illuminating, widely praised book on analytic geometry of circles, the Moebius transformation, and two-dimensional non-Euclidean geometries. 200pp. 5⅜ × 8¼.
63830-8 Pa. $8.95

MECHANICS, J.P. Den Hartog. A classic introductory text or refresher. Hundreds of applications and design problems illuminate fundamentals of trusses, loaded beams and cables, etc. 334 answered problems. 462pp. 5⅜ × 8½. 60754-2 Pa. $10.95

TOPOLOGY, John G. Hocking and Gail S. Young. Superb one-year course in classical topology. Topological spaces and functions, point-set topology, much more. Examples and problems. Bibliography. Index. 384pp. 5⅜ × 8¼.
65676-4 Pa. $10.95

STRENGTH OF MATERIALS, J.P. Den Hartog. Full, clear treatment of basic material (tension, torsion, bending, etc.) plus advanced material on engineering methods, applications. 350 answered problems. 323pp. 5⅜ × 8½. 60755-0 Pa. $9.95

ELEMENTARY CONCEPTS OF TOPOLOGY, Paul Alexandroff. Elegant, intuitive approach to topology from set-theoretic topology to Betti groups; how concepts of topology are useful in math and physics. 25 figures. 57pp. 5⅜ × 8¼.
60747-X Pa. $3.95

ADVANCED STRENGTH OF MATERIALS, J.P. Den Hartog. Superbly written advanced text covers torsion, rotating disks, membrane stresses in shells, much more. Many problems and answers. 388pp. 5⅜ × 8½. 65407-9 Pa. $10.95

COMPUTABILITY AND UNSOLVABILITY, Martin Davis. Classic graduate-level introduction to theory of computability, usually referred to as theory of recurrent functions. New preface and appendix. 288pp. 5⅜ × 8½. 61471-9 Pa. $8.95

GENERAL CHEMISTRY, Linus Pauling. Revised 3rd edition of classic first-year text by Nobel laureate. Atomic and molecular structure, quantum mechanics, statistical mechanics, thermodynamics correlated with descriptive chemistry. Problems. 992pp. 5⅜ × 8½. 65622-5 Pa. $19.95

AN INTRODUCTION TO MATRICES, SETS AND GROUPS FOR SCIENCE STUDENTS, G. Stephenson. Concise, readable text introduces sets, groups, and most importantly, matrices to undergraduate students of physics, chemistry, and engineering. Problems. 164pp. 5⅜ × 8½. 65077-4 Pa. $7.95

THE HISTORICAL BACKGROUND OF CHEMISTRY, Henry M. Leicester. Evolution of ideas, not individual biography. Concentrates on formulation of a coherent set of chemical laws. 260pp. 5⅜ × 8½. 61053-5 Pa. $7.95

THE PHILOSOPHY OF MATHEMATICS: An Introductory Essay, Stephan Körner. Surveys the views of Plato, Aristotle, Leibniz & Kant concerning propositions and theories of applied and pure mathematics. Introduction. Two appendices. Index. 198pp. 5⅜ × 8½. 25048-2 Pa. $8.95

THE DEVELOPMENT OF MODERN CHEMISTRY, Aaron J. Ihde. Authoritative history of chemistry from ancient Greek theory to 20th-century innovation. Covers major chemists and their discoveries. 209 illustrations. 14 tables. Bibliographies. Indices. Appendices. 851pp. 5⅜ × 8½. 64235-6 Pa. $18.95

DE RE METALLICA, Georgius Agricola. The famous Hoover translation of greatest treatise on technological chemistry, engineering, geology, mining of early modern times (1556). All 289 original woodcuts. 638pp. 6¾ × 11.
60006-8 Pa. $18.95

SOME THEORY OF SAMPLING, William Edwards Deming. Analysis of the problems, theory and design of sampling techniques for social scientists, industrial managers and others who find statistics increasingly important in their work. 61 tables. 90 figures. xvii + 602pp. 5⅜ × 8½.
64684-X Pa. $15.95

THE VARIOUS AND INGENIOUS MACHINES OF AGOSTINO RAMELLI: A Classic Sixteenth-Century Illustrated Treatise on Technology, Agostino Ramelli. One of the most widely known and copied works on machinery in the 16th century. 194 detailed plates of water pumps, grain mills, cranes, more. 608pp. 9 × 12.
28180-9 Pa. $24.95

LINEAR PROGRAMMING AND ECONOMIC ANALYSIS, Robert Dorfman, Paul A. Samuelson and Robert M. Solow. First comprehensive treatment of linear programming in standard economic analysis. Game theory, modern welfare economics, Leontief input-output, more. 525pp. 5⅜ × 8½.
65491-5 Pa. $14.95

ELEMENTARY DECISION THEORY, Herman Chernoff and Lincoln E. Moses. Clear introduction to statistics and statistical theory covers data processing, probability and random variables, testing hypotheses, much more. Exercises. 364pp. 5⅜ × 8½.
65218-1 Pa. $10.95

THE COMPLEAT STRATEGYST: Being a Primer on the Theory of Games of Strategy, J.D. Williams. Highly entertaining classic describes, with many illustrated examples, how to select best strategies in conflict situations. Prefaces. Appendices. 268pp. 5⅜ × 8½.
25101-2 Pa. $7.95

CONSTRUCTIONS AND COMBINATORIAL PROBLEMS IN DESIGN OF EXPERIMENTS, Damaraju Raghavarao. In-depth reference work examines orthogonal Latin squares, incomplete block designs, tactical configuration, partial geometry, much more. Abundant explanations, examples. 416pp. 5⅜ × 8¼.
65685-3 Pa. $10.95

THE ABSOLUTE DIFFERENTIAL CALCULUS (CALCULUS OF TENSORS), Tullio Levi-Civita. Great 20th-century mathematician's classic work on material necessary for mathematical grasp of theory of relativity. 452pp. 5⅜ × 8½.
63401-9 Pa. $11.95

VECTOR AND TENSOR ANALYSIS WITH APPLICATIONS, A.I. Borisenko and I.E. Tarapov. Concise introduction. Worked-out problems, solutions, exercises. 257pp. 5⅜ × 8¼.
63833-2 Pa. $8.95

THE FOUR-COLOR PROBLEM: Assaults and Conquest, Thomas L. Saaty and Paul G. Kainen. Engrossing, comprehensive account of the century-old combinatorial topological problem, its history and solution. Bibliographies. Index. 110 figures. 228pp. 5⅜ × 8½. 65092-8 Pa. $6.95

CATALYSIS IN CHEMISTRY AND ENZYMOLOGY, William P. Jencks. Exceptionally clear coverage of mechanisms for catalysis, forces in aqueous solution, carbonyl- and acyl-group reactions, practical kinetics, more. 864pp. 5⅜ × 8½. 65460-5 Pa. $19.95

PROBABILITY: An Introduction, Samuel Goldberg. Excellent basic text covers set theory, probability theory for finite sample spaces, binomial theorem, much more. 360 problems. Bibliographies. 322pp. 5⅜ × 8½. 65252-1 Pa. $9.95

LIGHTNING, Martin A. Uman. Revised, updated edition of classic work on the physics of lightning. Phenomena, terminology, measurement, photography, spectroscopy, thunder, more. Reviews recent research. Bibliography. Indices. 320pp. 5⅜ × 8¼. 64575-4 Pa. $8.95

PROBABILITY THEORY: A Concise Course, Y.A. Rozanov. Highly readable, self-contained introduction covers combination of events, dependent events, Bernoulli trials, etc. Translation by Richard Silverman. 148pp. 5⅜ × 8¼. 63544-9 Pa. $6.95

AN INTRODUCTION TO HAMILTONIAN OPTICS, H. A. Buchdahl. Detailed account of the Hamiltonian treatment of aberration theory in geometrical optics. Many classes of optical systems defined in terms of the symmetries they possess. Problems with detailed solutions. 1970 edition. xv + 360pp. 5⅜ × 8½. 67597-1 Pa. $10.95

STATISTICS MANUAL., Edwin L. Crow, et al. Comprehensive, practical collection of classical and modern methods prepared by U.S. Naval Ordnance Test Station. Stress on use. Basics of statistics assumed. 288pp. 5⅜ × 8½. 60599-X Pa. $7.95

DICTIONARY/OUTLINE OF BASIC STATISTICS, John E. Freund and Frank J. Williams. A clear concise dictionary of over 1,000 statistical terms and an outline of statistical formulas covering probability, nonparametric tests, much more. 208pp. 5⅜ × 8½. 66796-0 Pa. $7.95

STATISTICAL METHOD FROM THE VIEWPOINT OF QUALITY CONTROL, Walter A. Shewhart. Important text explains regulation of variables, uses of statistical control to achieve quality control in industry, agriculture, other areas. 192pp. 5⅜ × 8½. 65232-7 Pa. $7.95

THE INTERPRETATION OF GEOLOGICAL PHASE DIAGRAMS, Ernest G. Ehlers. Clear, concise text emphasizes diagrams of systems under fluid or containing pressure; also coverage of complex binary systems, hydrothermal melting, more. 288pp. 6½ × 9¼. 65389-7 Pa. $10.95

STATISTICAL ADJUSTMENT OF DATA, W. Edwards Deming. Introduction to basic concepts of statistics, curve fitting, least squares solution, conditions without parameter, conditions containing parameters. 26 exercises worked out. 271pp. 5⅜ × 8½. 64685-8 Pa. $9.95

TENSOR CALCULUS, J.L. Synge and A. Schild. Widely used introductory text covers spaces and tensors, basic operations in Riemannian space, non-Riemannian spaces, etc. 324pp. 5⅜ × 8¼. 63612-7 Pa. $9.95

A CONCISE HISTORY OF MATHEMATICS, Dirk J. Struik. The best brief history of mathematics. Stresses origins and covers every major figure from ancient Near East to 19th century. 41 illustrations. 195pp. 5⅜ × 8½. 60255-9 Pa. $7.95

A SHORT ACCOUNT OF THE HISTORY OF MATHEMATICS, W.W. Rouse Ball. One of clearest, most authoritative surveys from the Egyptians and Phoenicians through 19th-century figures such as Grassman, Galois, Riemann. Fourth edition. 522pp. 5⅜ × 8½. 20630-0 Pa. $11.95

HISTORY OF MATHEMATICS, David E. Smith. Nontechnical survey from ancient Greece and Orient to late 19th century; evolution of arithmetic, geometry, trigonometry, calculating devices, algebra, the calculus. 362 illustrations. 1,355pp. 5⅜ × 8½. 20429-4, 20430-8 Pa., Two-vol. set $26.90

THE GEOMETRY OF RENÉ DESCARTES, René Descartes. The great work founded analytical geometry. Original French text, Descartes' own diagrams, together with definitive Smith-Latham translation. 244pp. 5⅜ × 8½.
60068-8 Pa. $7.95

THE ORIGINS OF THE INFINITESIMAL CALCULUS, Margaret E. Baron. Only fully detailed and documented account of crucial discipline: origins; development by Galileo, Kepler, Cavalieri; contributions of Newton, Leibniz, more. 304pp. 5⅜ × 8½. (Available in U.S. and Canada only) 65371-4 Pa. $9.95

THE HISTORY OF THE CALCULUS AND ITS CONCEPTUAL DEVELOP-MENT, Carl B. Boyer. Origins in antiquity, medieval contributions, work of Newton, Leibniz, rigorous formulation. Treatment is verbal. 346pp. 5⅜ × 8½.
60509-4 Pa. $9.95

THE THIRTEEN BOOKS OF EUCLID'S ELEMENTS, translated with introduction and commentary by Sir Thomas L. Heath. Definitive edition. Textual and linguistic notes, mathematical analysis. 2,500 years of critical commentary. Not abridged. 1,414pp. 5⅜ × 8½. 60088-2, 60089-0, 60090-4 Pa., Three-vol. set $31.85

GAMES AND DECISIONS: Introduction and Critical Survey, R. Duncan Luce and Howard Raiffa. Superb nontechnical introduction to game theory, primarily applied to social sciences. Utility theory, zero-sum games, n-person games, decision-making, much more. Bibliography. 509pp. 5⅜ × 8½. 65943-7 Pa. $12.95

THE HISTORICAL ROOTS OF ELEMENTARY MATHEMATICS, Lucas N.H. Bunt, Phillip S. Jones, and Jack D. Bedient. Fundamental underpinnings of modern arithmetic, algebra, geometry and number systems derived from ancient civilizations. 320pp. 5⅜ × 8½. 25563-8 Pa. $8.95

CALCULUS REFRESHER FOR TECHNICAL PEOPLE, A. Albert Klaf. Covers important aspects of integral and differential calculus via 756 questions. 566 problems, most answered. 431pp. 5⅜ × 8½. 20370-0 Pa. $8.95

CATALOG OF DOVER BOOKS

CHALLENGING MATHEMATICAL PROBLEMS WITH ELEMENTARY SOLUTIONS, A.M. Yaglom and I.M. Yaglom. Over 170 challenging problems on probability theory, combinatorial analysis, points and lines, topology, convex polygons, many other topics. Solutions. Total of 445pp. 5⅜ × 8½. Two-vol. set.

Vol. I 65536-9 Pa. $7.95
Vol. II 65537-7 Pa. $7.95

FIFTY CHALLENGING PROBLEMS IN PROBABILITY WITH SOLUTIONS, Frederick Mosteller. Remarkable puzzlers, graded in difficulty, illustrate elementary and advanced aspects of probability. Detailed solutions. 88pp. 5⅜ × 8½.
65355-2 Pa. $4.95

EXPERIMENTS IN TOPOLOGY, Stephen Barr. Classic, lively explanation of one of the byways of mathematics. Klein bottles, Moebius strips, projective planes, map coloring, problem of the Koenigsberg bridges, much more, described with clarity and wit. 43 figures. 210pp. 5⅜ × 8½. 25933-1 Pa. $6.95

RELATIVITY IN ILLUSTRATIONS, Jacob T. Schwartz. Clear nontechnical treatment makes relativity more accessible than ever before. Over 60 drawings illustrate concepts more clearly than text alone. Only high school geometry needed. Bibliography. 128pp. 6⅛ × 9¼. 25965-X Pa. $7.95

AN INTRODUCTION TO ORDINARY DIFFERENTIAL EQUATIONS, Earl A. Coddington. A thorough and systematic first course in elementary differential equations for undergraduates in mathematics and science, with many exercises and problems (with answers). Index. 304pp. 5⅜ × 8½. 65942-9 Pa. $8.95

FOURIER SERIES AND ORTHOGONAL FUNCTIONS, Harry F. Davis. An incisive text combining theory and practical example to introduce Fourier series, orthogonal functions and applications of the Fourier method to boundary-value problems. 570 exercises. Answers and notes. 416pp. 5⅜ × 8½. 65973-9 Pa. $11.95

AN INTRODUCTION TO ALGEBRAIC STRUCTURES, Joseph Landin. Superb self-contained text covers "abstract algebra": sets and numbers, theory of groups, theory of rings, much more. Numerous well-chosen examples, exercises. 247pp. 5⅜ × 8½. 65940-2 Pa. $8.95

Prices subject to change without notice.
Available at your book dealer or write for free Mathematics and Science Catalog to Dept. GI, Dover Publications, Inc., 31 East 2nd St., Mineola, N.Y. 11501. Dover publishes more than 175 books each year on·science, elementary and advanced mathematics, biology, music, art, literature, history, social sciences and other areas.